INTRODUCTORY
BIOLOGY

D1540602

INTRODUCTORY BIOLOGY

Paul R. Ehrlich
Professor of Biology
Stanford University

Richard W. Holm
Professor of Biology
Stanford University

Michael E. Soulé
Assistant Professor of Biology
University of California, San Diego

**McGraw-Hill
Book Company**

New York
St. Louis
San Francisco
Düsseldorf
Johannesburg
Kuala Lumpur
London
Mexico
Montreal
New Delhi
Panama
Rio de Janeiro
Singapore
Sydney
Toronto

**INTRODUCTORY
BIOLOGY**

Copyright © 1973 by McGraw-Hill,
Inc. All rights reserved.
Printed in the United States of America.
No part of this publication may be reproduced,
stored in a retrieval system,
or transmitted,
in any form or by any means,
electronic, mechanical, photocopying, recording, or otherwise,
without the prior written permission of the publisher.

1234567890MURM798765432

This book was set in Helvetica Light Roman Linofilm by Black Dot, Inc.
The editors were James R. Young, Jr. and Diane Drobnis;
the designer was J. Paul Kirouac;
and the production supervisors were Matt Martino and Joe Campanella.
The drawings were done by F. W. Taylor Company and Danmark & Michaels, Inc.
The printer was The Murray Printing Company;
the binder, Rand McNally & Company.

LIBRARY OF CONGRESS CATALOGING IN PUBLICATION DATA

Ehrlich, Paul R
INTRODUCTORY BIOLOGY

1. Biology. I. Holm, Richard W., joint author.
II. Soulé, Michael E., joint author. III. Title.
QH308.2.E42   574   72-3484
ISBN 0-07-019128-X
ISBN 0-07-019127-1 (pbk.)

To **Berenice E. Bluestone**
Ruth R. Ehrlich
Beryle J. Holm

CONTENTS

PREFACE

Biology is the most exciting of the sciences today. In recent years spectacular advances have been made in the understanding of the basic mechanisms of life. Working with tools and techniques adapted from physics and chemistry, biologists have begun to unravel the secrets of heredity. Ecology, the branch of biology that deals with the interrelationships of all living things with each other and with their environments, has suddenly become the most widely discussed of all scientific disciplines.

People come naturally by their interest in biology. Children ask their parents myriad basic biological questions: "Why do birds sing?" "Where did I come from?" "Why do I breathe?" "Why do I have blue eyes when Sammy has brown eyes?" "Why is milk good for me?" The interest of most professional biologists grows out of such childhood curiosity about biological phenomena. Man is a curious animal, and his scientific endeavors provide a route to the satisfaction of curiosity.

From some courses in biology, students might easily get the impression that the intimate details of the hereditary mechanism and how they operate in intestinal bacteria and viruses are the most important biological phenomena. Indeed, they are important. It is somewhat unfortunate, however, that the recent successes of molecular biology, especially the elucidation of chemical processes in biological systems, have led to a distorted view of the science of biology as a whole. The story of DNA has occupied center stage for more than a decade. Here, for the first time, the biological sciences were making the kind of hard progress that characterizes physics and chemistry. The thrill of biological discovery for the first time spilled from the pages of scientific journals to the public press and the daily newspapers. Diagrams of the double helix of DNA became familiar to casual readers of weekly news magazines.

The excitement of research on the genetic code had its impact on the teaching of biology in the colleges and even high schools. Publishers and authors strove to modernize their textbooks to emphasize the discoveries made in ultracentrifuges and amino acid analyzers. The overall result was a serious imbalance in the way in which the biological sciences were presented to both students and the public. We are now beginning to reap the rewards of that imbalance. Many educated people know more about DNA than they know about the functioning life-support systems of our planet.

In this text we have attempted partially to reduce this imbalance. As scientists, we share with our colleagues in molecular biology the excitement over the spectacular advances made in that field. We feel it to be very important, however, to transmit to students some of the excitement of other areas of biology. These areas may be much more complex and difficult than molecular biology. The scientific problems involved in them are still much larger than the progress that has been made in solving them. Nevertheless, these areas are of the most crucial importance to man today.

We have not yet, in disciplines such as ecology and neurophysiology, unraveled the basic secrets. We may still be years, if not decades, from writing the chemical and mathematical descriptions of some of the most important biological processes. To a large degree both the functioning of the brain and the functioning of ecosystems remain mysterious. Unfortunately, it has become all too clear to us that human health and happiness in the next few decades depend upon a better appreciation for, and understanding of, ecology and the human nervous system.

As striking as advances in molecular biology have been, they hold little promise for improvement of the human lot in that time. If we should survive the next two or three decades—and there are many who say we probably shall not—then we may enter into a peaceful age. As the life sciences flourish, we can expect tremendous benefits for man to derive from the progress already made in molecular biology. The problem for all human beings, biologists and nonbiologists alike, is clear and immediate: How can we possibly accomplish this transition to that golden age? We are going to have to learn more about ecology so that we may know how men should behave in order to ensure the continued functioning of the life-support systems of spaceship *Earth*. And we must learn much more about human behavior in order to change the attitudes of men toward the world in which they live and toward their fellow human beings.

In writing this book, we have had a number of aims. The book is not designed solely for those who intend to have a career as professional biologists. It is written for people who want to be well-informed citizens of a troubled world. We have tried to cover the full sweep of biology, giving emphasis to those things which require understanding by all who are involved in important decision making.

The pertinence of many areas will be obvious. One cannot determine the amount of food that can be grown in a given area without some understanding of the process of photosynthesis and the role that water plays in the economy of plants. Similarly, if one is to discuss the genetics of intelligence, one must understand something about hereditary mechanisms. The question we have had to answer for ourselves, however, is: How much detail is it necessary for the average well-educated person to know about the chemical mechanisms of photosynthesis or gene action? We have adopted, as a guideline, the rule that we should not expect a student to know more about any area of biology than does the average competent biologist who works in a different area. In other words, if professionals do not carry certain facts around in their heads, there would seem to be little reason to expect a layman to do so. This applies with equal force to the details of biochemical pathways and to the nearly infinite variety of technical names for organisms. We have tried to indicate in our summary paragraphs (set in capital italic type) the basic ideas which we hope each student will retain.

In general, we have tried to outline biochemical processes without going into great detail about the actual molecules involved and how they interact. Biochemists may complain that this will leave the student with only a superficial understanding of the process, but this is a risk which must be accepted. Most human beings will go through their entire lives with superficial understandings of virtually all areas of our culture. There is just too much information at the disposal of mankind for each person to have in-depth knowledge about every subject.

Our goal is not to produce a generation of students familiar in detail with all the accumulated knowledge of the biological sciences. Nor is it to indoctrinate a generation with the goal of producing ecological activists. It is, rather, to produce citizens with enough knowledge of the life sciences to permit them to make rational and responsible decisions about the many biological questions that are facing us and will be facing us in the years ahead.

We wish to express our deeply sincere gratitude to Ward B. Watt of Stanford University and Burton S. Guttman of the University of Kentucky for their excellent and detailed criticism of the manuscript in its preliminary stages of preparation.

PAUL R. EHRLICH
RICHARD W. HOLM
MICHAEL E. SOULÉ

NOTE TO THE STUDENT **BOXES**

From time to time you will find references to "Boxes." Boxes are used for review material or for in-depth consideration of selected topics. Boxes are placed at the end of chapters and are given numbers corresponding to those chapters.

1

WAYS
OF LOOKING
AT LIVING THINGS

INTRODUCTORY BIOLOGY is divided into four parts: Introduction: Ways of Looking at Living Things; The Component Systems of Organisms; Strategies for Survival; and The Biology of Populations.

This first part will introduce you to the ways in which biologists approach the diversity and complexity of living systems. It deals with fundamental questions about the living world as a whole and outlines the genetic and evolutionary background that is necessary for the comprehension of the remainder of the text.

In Chap 1, taxonomic and ecological relationships are discussed. A brief survey of the kinds of plants, animals, and microorganisms is also presented. Chapter 2 discusses the laws of heredity and the mechanisms of cell division to provide a basis for understanding the nature of the genetic material and its functioning in individuals and populations.

*(Illustration on proceeding pages
Courtesy of J. Paul Kirouac.)*

CHAPTER 1

DIVERSITY AND ITS ORIGIN

All of us know that there is more to a forest than a collection of trees. We know intuitively, for instance, that an orchard is not quite the same as a forest. And we will admit that a petrified forest is no longer a forest in the usual sense. The exact nature of the differences among these three groupings (Fig. 1–1) may prove elusive, however. The reason for this elusiveness lies in the many properties of life which we do not easily perceive.

A group of trees forms a prominent landmark in our environment. Our eyes take in the size of the trees, their shape, and the greenness of their leaves. Our hands feel their rough bark; our ears detect the rustling of their leaves in the wind. If the trees are in flower, the fragrance of their blossoms may attract our attention. We are aware of a pleasant coolness if we move into their shade on a hot day, or we may seek shelter under them if it rains. These properties of trees are obvious; they impinge directly upon our senses. And they are shared by the living forest and the apple orchard. Clearly, if living forests and orchards are different, we must search elsewhere for the difference. Of course, we feel the petrified forest to be most distinct; after all, the trees are not alive, the leaves are gone, and the wood has changed dramatically. Most of us, however, cannot specify clearly all the differences between the living and the once-living.

THE DIFFERENCES AMONG AN ORCHARD, A LIVING FOREST, AND A PETRIFIED FOREST RESULT FROM PROPERTIES OF LIFE WHICH ARE DIFFICULT TO PERCEIVE DIRECTLY.

WEBS OF RELATIONSHIP

A very important difference between a living forest and an orchard lies in the complex web of relationships which binds all living things into an integrated, coherent system that interacts with the nonliving world. It lies in the subtle bonds that interconnect plants, animals, and microorganisms

Fig. 1–1 Three groupings of trees.
(*A*) A forest in the Eastern United
States. *(United States Department
of the Interior, National Park Service.)*

A

(*B*) An olive orchard in Spain.
(Courtesy of Diane Drobnis.)

B

*(C) A petrified forest in Arizona.
(United States Department of the
Interior, National Park Services.)*

C

with each other and with their environment. These bonds elude us because, in general, they are not immediately detectable by our senses.

Identifying the interconnections of organisms and their environment is often very difficult. Discrete, particulate elements, whether these are trees or animals or stones, attract our attention. Only after painstaking study and careful experimentation do we detect the bonds and unifying principles that combine the disparate elements into a meaningful whole. In the same way that a blind man who gains his sight must learn to see, we must learn to see all the forest instead of only its individual trees.

A forest differs from an apple orchard in many ways. Some of these need only to be pointed out to become obvious: for instance, the even spacing of the trees and the lack of undergrowth in the orchard. Perceiving other differences requires a more specialized knowledge of kinds of organisms—of **taxonomy.** Taxonomy is the science of classifying organisms on the basis of their similarities and differences. Understanding how these similarities and differences came about requires a knowledge of the process of gradual change in organisms. This process is called **evolution** and will be discussed in detail later. With such knowledge, a difference in the diversity of plants, animals, and microorganisms would become apparent. For example, the orchard has fewer kinds of organisms than does the forest.

Chemical analyses and observation would permit us to chart roughly the flow of energy and materials through the forest and the orchard. We would see energy moving from the sun into green plants, from the plants into animals and microorganisms, and from animals into other animals and microorganisms. We would see materials follow pathways from air, water, and soil through plants, animals, and microorganisms and back to the non-living world. These patterns, too, would be different in orchard and forest. In a biological sense, then, the expression "you can't see the forest for the trees" is accurate; the trees *do* keep us from seeing the forest. They deflect our attention from the less apparent but more important factors which are the essence of the forest.

PLANTS, ANIMALS, AND MICROORGANISMS ARE INTERCONNECTED WITH EACH OTHER AND WITH THEIR ENVIRONMENT BY AN INTRICATE WEB OF BONDS AND RELATIONSHIPS WHICH USUALLY CANNOT BE DETECTED IMMEDIATELY BY OUR SENSES.

Let us suppose that our awareness of the world included information not normally available in the human repertoire of sense organs. What if not only such things as form, color, sound, and odor were apparent, but also biological relationships, physical and chemical relationships, and relationships in time? Suppose you could perceive not only the sleek form and purr of a cat but also the energy it derives from the mouse it has just eaten. What if, in addition to the redness of a rose and its pleasant perfume, you were aware of the biological significance of its color and odor, which attract the insects necessary for pollination and, ultimately, seed formation? What if, when you look at your own arm, you could also see that it was derived, over hundreds of millions of years, from the fleshy, lobed fin of now-extinct fishes? With such extended perception, your study of living systems, of biology, would be much more simple. The difference between a forest and an orchard would be as striking as the difference you now perceive between forest and meadow. The difference between a living forest and a petrified forest would become even more striking.

Imagine that you can look at the world of life with your perception extended to include some of these vital relationships—for example, the ones involving the flow of matter and energy. Standing in a sunlit forest clearing, you immediately would be aware that plants, animals, and microorganisms show basic similarities in the way they handle energy. For instance, you would see that, like batteries, they all maintain an energy differential between themselves and their environment. Organisms are clearly more highly ordered than their nonliving surroundings, and this order is maintained by a constant expenditure of energy. You would see also that life processes involve a highly efficient transfer of energy— an efficiency that is not often attained, for example, in man-made machines.

Energetic Relationships

Green plants would be all around you, and you would undoubtedly be impressed by the differences between them and the myriad animals in their environment. Certainly they are stationary, seem relatively unresponsive, and perhaps appear more variable in form than animals. But details of their energetic, chemical, and biological relationships also prove to be very different. Eating the plants are green caterpillars, which were hard to spot when your perception was restricted. They now stand out prominently. You can see that the green pigments are carrying out chemically related functions in all the plants. But the green pigment of the caterpillars has a very different functon.

You can now see that energy is flowing from the plants into the caterpillars feeding upon them. The green pigments of the plants are intimately involved in reactions that permit them to trap energy from the sun and store it in usable form. The plants use some of this energy for their own growth and reproduction. The caterpillars are not able to carry out these reactions of trapping light energy. Therefore, they are dependent upon the plants for energy. In fact, all animals ultimately depend upon green plants for energy. Even animals that eat other animals for food obtain energy that originally came from the sun via green plants. More oxygen enters the animals than leaves them, whereas the reverse is true for plants. Plants give off more oxygen than they take in. There are many other chemical differences between plants and animals which would be apparent to your heightened senses. In short, the larger, more complex organisms of the forest clearing fall into clear-cut groups—plants and animals—on many bases other than the most obvious ones related to form and movement. The tiny microorganisms are less readily categorized. Some of these relationships are shown diagrammatically in Fig. 1–2.

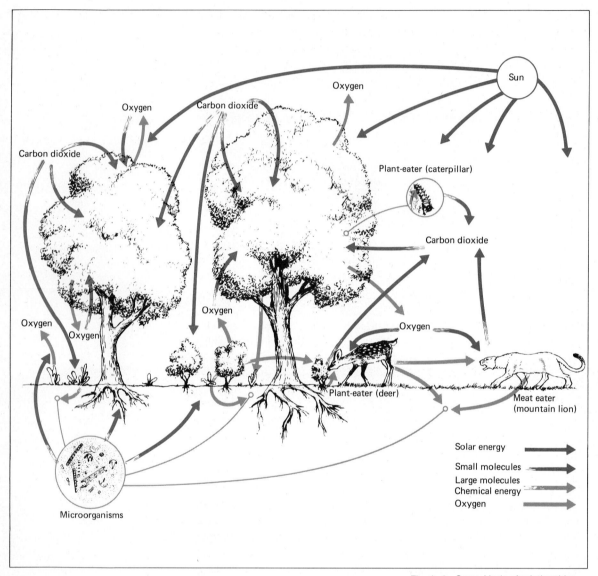

Fig. 1–2 Some kinds of relationships in a forest.

GREEN PLANTS CHANGE LIGHT ENERGY FROM THE SUN INTO CHEMICAL ENERGY, WHICH THEY USE IN GROWTH AND REPRODUCTION. ANIMALS CANNOT DO THIS, AND THEREFORE ALL ANIMALS, WHETHER THEY EAT PLANTS OR OTHER ANIMALS, ARE DEPENDENT UPON PLANTS FOR ENERGY.

ENERGY AND COMMUNITIES   Let us suppose you now leave the forest clearing and begin to survey the distribution of the different kinds of living things over the surface of the planet. The plants and animals in a particular place may be called a **community.** The place in which organisms occur is their **habitat.** Animals and microorganisms seem to be found in a greater variety of situations than do plants. Plants, at least the green ones, do not occur beneath large rocks or deep in caves, although animals and microorganisms are found in these places. Both plants and animals are relatively uncommon in land areas where there is little available moisture. In fact, you notice that a large number of physical features that vary over the face of the earth, such as rainfall, temperature, wind patterns, and chemical composition of the soil, are intimately related to the distribution of the organisms. They make up the differences among habitats and determine, in part, the kinds of organisms in a community.

When you turn your attention from the land to the portions of the planetary surface that are covered with water, many differences between the terrestrial and aquatic environments are immediately apparent. In the sea, there are fewer anchored green plants and many more anchored passive animals. You are not deceived by the superficial resemblance between these stationary plants and animals. Although some of the passive animals may be green, their pigment is not the same as that in plants. The energy-trapping reactions which you now know are characteristic of green land plants, but not of animals, are going on at a great rate in the sea. These reactions are not, however, being carried out primarily by the underwater equivalent of grasses and trees, and not at all by the anchored animals. They occur principally in the very numerous microorganisms floating in the water near the surface.

This does not surprise you since, obviously, radiation from the sun is necessary for these reactions and you can see that this radiant energy does not penetrate very far into the water. The lack of solar energy helps to explain the absence of green plants under rocks and in caves; indeed, there are many resemblances between deep-sea and cave environments. The presence of animals in the darkness of marine depths confirms your earlier observations about their energy sources. You see that chemical energy is constantly raining down from the surface in a highly concentrated form—the living and dead bodies of surface organisms, most of which are microscopic. The lighted zone supports the unlighted deeper reaches in the sea as on the land.

ALTHOUGH THERE ARE MANY DIFFERENCES BETWEEN COMMUNITIES ON LAND AND THOSE IN THE SEA, THE ENERGETIC RELATIONSHIP BETWEEN GREEN PLANTS AND ANIMALS IS FOUND IN BOTH HABITATS.

In general, if your perceptions of energy relations really could be extended virtually without limit, you would probably reach the point at which you would "not see the trees for the forest." You would be able to see too much. Every individual organism would appear not as a totally discrete package but rather as a sort of whirlpool, with a continuous input and output of energy and materials. All the individual whirlpools would be bound to one another and to their physical surroundings. The entire planet might appear as a worldwide whirlpool with millions of local eddies. Indeed, it might be possible for you to detect the form and functioning of the individual organisms only by mentally suppressing both your perception of the blanketing network of energy relationships and the realization that this vast network was very different yesterday and will be different tomorrow. Above all, you probably would be most impressed with the impermanence of all that you see. And if you permitted yourself simultaneously to "see" all the other kinds of relationships that exist among organisms—evolutionary, taxonomic, social, competitive, and so forth, your confusion would be complete. An extremely complex and ever-changing series of webs would be superimposed upon the whirlpools. Obviously, in order to study the living world, a biologist must treat it as if it were made up of discrete organisms, more or less separable from their physical environment.

A biologist is not personally equipped with extended perception, except for that available through various instruments such as microscopes and Geiger counters. However, as a result of his training, he does in a sense view the world with alien eyes. His picture of the world is basically that described above; obviously, it is a big jump from the layman's view of nature to that of the trained biologist. Since you cannot directly perceive many features of the biological world, you must learn of their existence indirectly. Some of the functional systems and relationships you will see readily; others will require special knowledge or equipment. We cannot provide new sense organs which will permit you directly to detect the bonds of communication among organisms or their relationship by descent or the flow of energy through them. We can describe these things to you. We can give you some feeling for how scientists, using the same array of sense organs as you possess, have been able to detect or infer a multidimensional network in which all organisms are components in one complex and ever-changing system. Whether they are strange or familiar to us, all organisms are functioning units in a worldwide system that biologists call the **ecosphere.** The biologist wishes to determine and understand the interrelationships of all the units in the ecosphere. We are beginning to realize that such an understanding is important for everyone.

Let us look at one small segment of the ecosphere, a troop of baboons on the plains of Africa (Fig. 1–3). A troop of baboons is a good starting place for our study of biology. Biology is the science of life, and baboons are clearly "alive" in every sense of the word. In addition, baboons are rather close relatives of ours, and their society has been thoroughly studied by anthropologists (S. Washburn and I. DeVore, among others). This work has provided us with considerable insight into the lives of these large monkeys and perhaps has told us something of ourselves as well.

Baboons travel about in troops that number between ten and several hundred members. These troops may wander over 3 to 6 sq miles of countryside. The social life of baboons is based on a framework of bonds that interconnect the individuals. In large part, the organization and functioning of the troop are the sum of these bonds.

In the life of a baboon, the first bond to develop is that between the mother and the child. The newborn infant can cling to the fur of the mother's chest upon birth, but soon thereafter it learns to ride piggyback. All the infant's nourishment comes from its mother's milk, and the mother, in turn, obviously derives satisfaction from the infant. Infants, in fact, are a focus of attention, and a mother-offspring pair attracts other baboons, including both adults and juveniles, all of whom attempt to groom the mother and the infant. As shown in Fig. 1–4, the mother-infant bond can be thought of as generating a series of other bonds to the older members of the troop. Soon the young baboon begins to take solid food and to leave the mother. The youngster enters into play groups with other juveniles—the first peer group in the life of a baboon. Strong "friendship" bonds develop in conjunction with the chasing, tail-pulling, and other

**Social Relationships:
A Baboon Troop**

Fig. 1–3 A troop of baboons in Nairobi, Africa. Males are seen along the edges of the road. Between them are females and juveniles. (*Courtesy of I. De Vore.*)

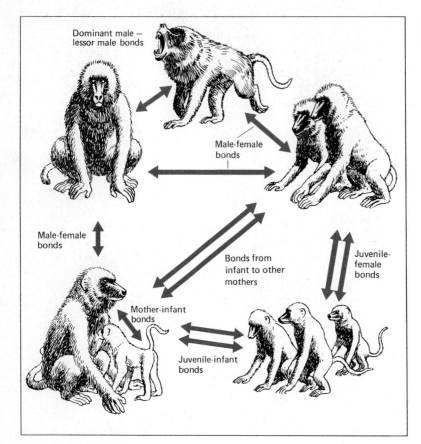

horseplay of these groups. If a juvenile is hurt and cries out during the turmoil of such roughhousing, adults hurriedly terminate the play. Gradually, the mother-child bond attenuates.

The bonds among interacting juveniles in the play group generate another set of bonds to the adults, who, to some degree, regulate juvenile activities. Furthermore, communication between juveniles and adults is undoubtedly important in the socialization (education) of the young in that juvenile baboons tend to mimic the behavior of adults. Peer groups of "friendship" bonds are not restricted to juveniles. Adult females form "preference pairs," and in large troops, some males stay together much of the time and defend any of their group that may be threatened by other males of the troop (Fig. 1–5).

Another set of relationships that is extremely important in baboon society is the status system, called a ***dominance hierarchy.*** The dominant males in this hierarchy are very "attractive" (the word is used here in a purely descriptive sense) to less dominant baboons; they are groomed often, especially by females, and are the unquestioned leaders. The order of dominance is more or less dependent on strength and fighting ability, but a relatively weak male may rank higher than a stronger male if the former belongs to a friendship group. Once the rankings in a group are stabilized, there is very little fighting. Every baboon knows to whom he must give way and who must give way to him. If a squabble should develop, a dominant male will rush in and stop it. Such a system, established through force but maintained by convention, leads to stability and security.

Finally, bonds between males and females form for sexual reasons, but these bonds are usually short-lived. Male-female pairs may last for only an hour, or the couple may remain together for up to several days. The sexual bond is the vehicle for hereditary bonds—the bonds of family relationship, such as those between progeny (offspring) and parents, between siblings (brothers and sisters), and between more distant relatives. These bonds are relatively unimportant in the lives of baboons, and there is nothing like a family unit as it exists in humans. We can theoretically trace these hereditary bonds back over thousands or millions of generations to the remote ancestors of baboons and even to the origin of life itself. Such long-range hereditary continuity is called *evolutionary* relationship.

Fig. 1-5 Bonds among individuals are strengthened by individuals grooming one another. The male on the right-hand side has two females to his left, one of which is grooming him. *(Courtesy of I. De Vore.)*

INDIVIDUALS IN A BABOON TROOP ARE HELD TOGETHER BY MANY KINDS OF BONDS. AMONG THESE ARE SOCIAL BONDS BETWEEN MOTHER AND CHILD, BETWEEN JUVENILE PEERS, AND BETWEEN ADULTS OF THE GROUP. SHORT-LIVED SEXUAL BONDS LEAD TO HEREDITARY BONDS, WHICH TRANSCEND THE TROOP AND MAY BE EXTENDED TO EVOLUTIONARY BONDS. SOCIAL BONDS GIVE STA-BILITY AND ORGANIZATION AND PROVIDE A FRAMEWORK FOR THE EDUCATION (SOCIALIZATION) OF THE YOUNG.

The baboon troop is not an independent entity. The troop acts on its environment, and its environment, in turn, affects the troop. Baboons are preyed upon by lions, leopards, and various kinds of parasites. Similarly, the baboons occasionally devour smaller animals such as small monkeys, newborn gazelles, birds' eggs, and insects, although most of

Food Webs

Fig. 1–6 Energy-matter bonds in a baboon troop. Energy flows from the sun to the green plants. Matter flows from the plants to the baboons and also from green plants to insects to baboons.

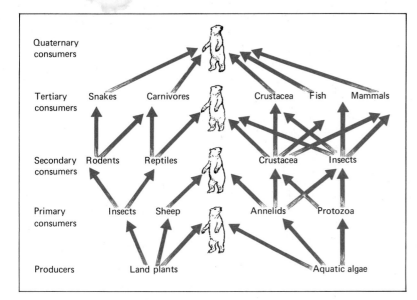

Fig. 1–7 A generalized diagram of a food web. The bear acts as a consumer at four levels and is thus an omnivore (eating both plants and animals). *(Adapted from Relis B. Brown, GENERAL BIOLOGY. Copyright McGraw-Hill Book Company, 1970. Used by permission.)*

their diet consists of plants. One way or another, though, the energy utilized by a baboon and the material that forms its body are derived from plants. The animals on which a baboon feeds either eat plants themselves or consume other animals that eat plants (Fig. 1–6). Of course, the ultimate source of energy or nourishment is the energy radiated by the sun and trapped by green plants. Energy "flows" from the sun to plants to baboons, although occasionally other animals are interposed.

Energy-matter bonds, collectively called the **food web,** are shown in Fig. 1–7. Note that organisms can be ranked according to levels in the food or energy hierarchy. Plants are called **producers** since they produce the complex substances characteristic of life from the sun's energy and simple chemical substances in the environment. Or, as chemists would say, they *synthesize organic compounds from inorganic compounds.* Animals also can do this, but only to a very limited extent. Herbivores (plant eaters) are called **primary consumers,** carnivores (flesh eaters) who eat herbivores are called **secondary consumers,** and carnivores that eat other carnivores are called **tertiary consumers.** The consumers as a group lack the chemical virtuosity of the plants, having only the ability to synthesize organic compounds from other organic compounds, which they eat and transform into their own. Baboons are best considered primary consumers, although they also function as secondary and tertiary consumers at times, as you can see by considering their diet mentioned above.

EVERY FOOD WEB CONSISTS OF A SERIES OF LEVELS. THESE ARE THE LEVELS OF PRODUCERS, PRIMARY CONSUMERS, SECONDARY CONSUMERS, AND SO ON. PRODUCERS ARE GREEN PLANTS, PRIMARY CONSUMERS ARE HERBIVORES, SECONDARY CONSUMERS ARE CARNIVORES.

In summary, a baboon troop can be described by considering three distinct sets of bonds: social bonds, hereditary-evolutionary bonds, and matter-energy bonds. The latter two kinds of bonds, of course, would be found in an examination of any group of organisms. None of these bonds can be perceived *directly* by our senses, although we are able to infer their existence by what we see or hear. If we *could* perceive them, we would be able instantly to understand the fundamental features of baboon biology. The baboons would not appear as isolated individuals nor as an isolated group of individuals. Instead, we would realize that a troop is but one element in a vast, earthwide complex of life. We would realize that many changes in an individual baboon affect the troop, that changes in the troop affect other segments of the ecosphere, and that any other element in the entire global network *could* be affected.

Ecological and Taxonomic Relationships

We have thus far stressed relationships among organisms which fall into the broad class of **ecological relationships.** These are the energetic and behavioral relationships involved in the everyday functioning of organisms: the relationships of predators and their prey, the exchange of gases, parent-offspring bonding, and so forth.

Taxonomic relationships make up another broad class of relationships which biologists recognize. These are the relationships among kinds of organisms which are based on their similarity. You are dealing with ecological relationships when you discuss forests, meadows, and grazing animals. You express taxonomic relationships when you speak of plants and animals or birds, bees, and butterflies. Taxonomic relationships seem more intuitively "real" than most ecological relationships. We deal with classes based on similarity all the time; chairs and tables, trucks and cars, knives and forks are pairs in similarity classes, just as are dogs and cats, apples and oranges, and butterflies and moths.

But neither class of relationship is more real or more important than the other. As a human being, you have some very important points of similarity with the whale. In spite of superficial differences due to the streamlining necessary for swimming, the basic body plan of whales is very much like that of a human being. And similarities in body function between men and whales are very strong as well. In virtually every respect, you are more similar to a whale than to any plant—you and the whale have a much closer taxonomic relationship than you and, say, the grass plants in your front lawn. As you will learn later, the basic reason that you are more similar to the whale than to the grass is evolutionary. About 100 million years ago, you and the whale had a common ancestor, one that was a warm-blooded animal with four limbs, a highly developed nervous system, and all the other appurtenances of warm-blooded animals. Since that time, the ancestors of whales and men went separate evolutionary ways. But in order to find a common ancestor of man and a grass plant, one would have to journey backward in time perhaps 1,000 million years—and the common ancestor, when found, would be a tiny, single-celled organism showing no obvious similarity to either you or the grass plant (Fig. 1–8).

So you and the whale are taxonomically close, but what about ecologically? Most people have only the slightest ecological relationship with whales—although men are the major killers of whales. But all humanity

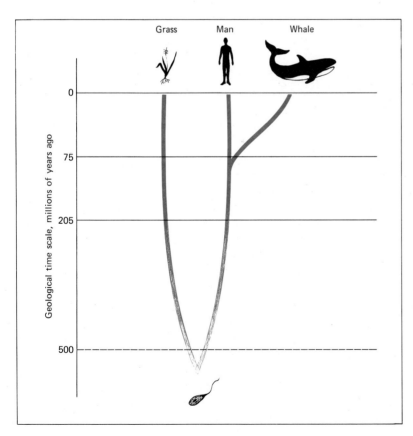

Fig. 1–8 Taxonomic and evolutionary relationships among a grass, a man, and a whale. Man and the whale have a common ancestor some 100 million years ago. The common ancestor of plants and animals existed much farther back in time.

has an intimate and critical ecological relationship with plants, depending on them for, among other things, all the food we eat and all the oxygen we breathe.

In this book we usually will refer to organisms using a system of similarity relationships. The normal taxonomic system is based on these relationships, and the Latin or Greek names assigned to organisms often reflect them. But you should always keep in mind that this is just one kind of relationship, neither more nor less important than the ecological. In turn, although our approach to life will not be primarily taxonomic, you should also remember that underlying the names of organisms and groups of similar organisms is a vast and important tree of evolutionary relationships (Fig. 1–9).

Fig. 1–9 A very much simplified view of the supposed evolutionary relationships among the major groups of plants and animals. This view is largely hypothetical, the more so the farther back in time one goes. Many kinds of plants and animals that existed in the past but are extinct today are not shown in this diagram.

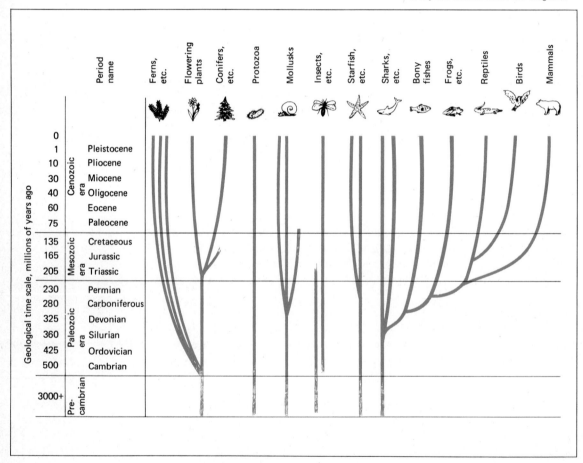

*TWO BASIC KINDS OF RELATIONSHIPS RECOGNIZED BY BIOLOGISTS
ARE ECOLOGICAL RELATIONSHIPS AND TAXONOMIC RELATIONSHIPS.
ECOLOGICAL RELATIONSHIPS INCLUDE THE WAYS IN WHICH AN OR-
GANISM FUNCTIONS IN ITS ENVIRONMENT AND HOW IT AFFECTS AND
IS AFFECTED BY OTHER ORGANISMS. TAXONOMIC RELATIONSHIPS
DEAL WITH THE SIMILARITIES AMONG ORGANISMS AND ARE USUALLY
BASED UPON EVOLUTIONARY DESCENT.*

The "Game of Life"

Our approach to biology will focus on how organisms maintain and change their positions, individually and collectively, in the ever-dynamic ecosphere. We may think of organisms as being in a game with nature. This is merely a way of talking about how organisms function, but it may lead to some useful insights. The goal of the game of life is, of course, survival. Those plants and animals which survive "win"; those which do not, "lose." We shall view different kinds of organisms as employing different strategies in the game of life, and we will see that this game is played not only against the physical environment but also against other organisms. For instance, the breathing of air, sexual reproduction, and speech will emerge as important strategies arising in the course of ev-olution. They and many other strategies contribute to the present success of one important kind of living organism, the one we call *Homo sapiens*— man. We will end our text with a consideration of possible future strategies for this organism. Before going on, however, let us consider, on a rather broad scale, the diversity of living things.

THE DIVERSITY OF LIFE

In very general terms, plants, animals, and microorganisms are distributed more or less as a thin shell on the surface of this planet. They are rarely found at a soil depth of more than a few feet and, except for some very small organisms floating or flying in the air, are rarely more than a few hundred feet above land or water. Organisms are found more than 30,000 ft below the surface of the sea; but with the exception of the ocean floor, they are much more abundant within 100 ft of the surface. This shell of organisms, and the space they occupy, is the ecosphere—a term that refers to the totality of living things and their environment.

The Discreteness of Organisms

Our first look at living things led us to think of them as units or packages of living material, packages that we call organisms. If we had to choose between describing the living material as consisting of *discrete units* and describing it as a *continuous film,* the former would seem more apt.

The living material of the ecosphere obviously is not a gigantic sheet of living substance blanketing the earth. It consists largely of separate organisms, ranging in size from bacteria to whales and redwood trees. As we have seen, these organisms, while remaining physically discrete, are connected by many kinds of bonds.

The discreteness of living systems usually is taken for granted because there are few if any exceptions to it. The business of science, however, is not simply to explain exceptions; it is to explain the rules as well. Why should living material be organized into packages instead of a continuous film? Any part of a continuous living film, by definition, would be physically connected to all other parts, and this connection would inevitably lead to certain consequences. A continuous film having the properties characteristic of living systems would have to have a source of energy and materials and a way for them to enter the living film, but there also would have to be a distribution system and some method for disposing of wastes.

One reason that life is organized into packages is a consequence of the mathematical relationship between the surface area of a solid and its volume. As shown in Fig. 1–10, if the living film were 1 meter (m) thick, a cubic segment of the film 1 m long and 1 m wide would have 1 sq m of the surface exposed to the atmosphere and 1 sq m exposed to the surface (substrate) on which the film rested. A 1-m cube of living material separate from a film would have 5 sq m of surface exposed to the atmosphere and 1 sq m to the substrate. By successive division of the cube, the area exposed to the atmosphere could be greatly increased. The area exposed by dividing the original cube into others only 1 cm on each side would be 100 times the original surface. (The surface of a 1-m cube is 60,000 sq cm; that of the 1 million 1-cm cubes into which the meter cube could be divided is 6 million sq cm.)

When the first living systems appeared on earth 3 billion or so years ago, they presumably faced problems of exchanging materials with their environment. Living systems carry out such exchanges across their surfaces. Most students of the origin of life believe that these remote ancestors of contemporary organisms obtained their energy from simple organic molecules. Later, some mastered the complex chemical trick of photosynthesis. You can now appreciate some of the advantages of systems which did not take the form of thick continuous films. One of these advantages obviously is related to exchange with the environment. In proportion to their volume, fragments of a living sheet would have more surface for exchange than would the intact film.

SINCE LIVING SYSTEMS MUST INTERACT WITH THE ENVIRONMENT

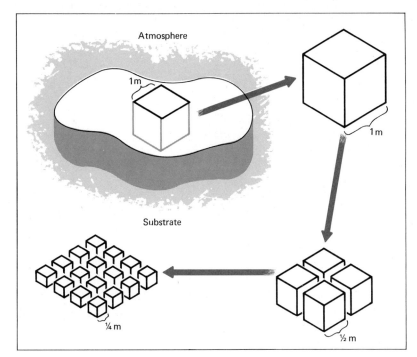

ACROSS THEIR SURFACE, THE RELATIONSHIP BETWEEN THE AMOUNT OF SURFACE AND THE VOLUME IT MUST SUPPORT IS VERY IMPORTANT. THIS IS ONE OF THE REASONS LIFE DOES NOT OCCUR IN A CONTINUOUS LIVING SHEET. DISCRETE ORGANISMS HAVE MUCH GREATER SURFACE AREA.

There would be other advantages to fragmentation as opposed to continuity. A life form in a continuous sheet would have "all its eggs in one basket." Environmental change might easily destroy the entire sheet. Fortuitous changes in one part of the film—permitting, say, more efficient utilization of energy—might be swamped out by movements of material from adjacent regions. Continuity might hinder the development of more efficient ways of maintaining and perpetuating living systems. Progressive evolution might be difficult. Furthermore, a continuous film starting (as it must) in one place presumably would have more difficulty extending itself over barriers than would a living system which could disperse in small pieces. Biologists today believe that even the very earliest living systems were not filmlike but were discrete, microscopic units, probably roughly spherical in shape.

That organisms are physically discrete entities but with very significant interconnections, hereditary and nonhereditary, has now been established. We have already noted that there are many *kinds* of organisms. Simple observation, for instance, led to the formulation of the fundamental plant-animal distinction. This diversity of kinds raises a number of additional questions.

The Many Kinds of Organisms

One important question is why there are so many kinds of organisms instead of just one or a few. Each of us is accustomed to viewing a fair sample of the 2 million or so kinds of organisms known to biologists. Consequently, the reason for diversity, like the reason for water running downhill, is rarely questioned. Yet if one were to encounter, on some distant planet, only *one* kind of organism, he would be struck by this difference from the earth in diversity of kinds of organisms. Although it is relatively easy to see why life is "packaged" rather than continuous in space, the complexity of the pattern of packaging is not so easily understood. Why are there not just one or two or twenty different kinds of packages, instead of millions? At present, there are no definite answers to this question. Some educated guesses can be made, though. Once given the discreteness of organisms plus some kind of hereditary machinery which, in general, ensures that descendant organisms will resemble their progenitors, diversity seems to be an inevitable result of the evolutionary process—the process by which organisms change through time.

One factor weighing against "uniform packages" or a film kind of organization is inefficiency. All the packages or parts of a film would be more or less alike. As a result, all parts would by necessity have to be jacks-of all-trades; there could be very little specialization. Such an organization would be analogous to that of a primitive human society of hunters and gatherers. In such a society, each man must be adept at performing all the tasks necessary for survival. He must be able to construct shelters, to hunt, to find the widely scattered berries, seeds, and roots that make up the vegetable portion of his diet, to preserve his catch, to make whatever utensils, tools, and weapons he needs, and to fight if necessary. Individuals in this kind of society have relatively little genuine leisure time. They must spend most of their time at activities directed toward the gathering and preservation of food, for example, mending nets, tools, etc. In more complex societies, labor and crafts are portioned out among specialists and artisans, allowing greater productivity of the society and more leisure time for all the individuals as a result of this more efficient use of resources.

Similarly, it seems that more of the solar energy impinging on an area can flow through biological systems if the area is occupied by a collection of many different kinds of organisms than if it is occupied by just one or a few kinds. Green plants, for instance, are good at trapping solar energy but not at extracting it from other green plants. In a community consisting only of cabbages, the energy and materials present in a living or a dead cabbage will be forever trapped, never to be returned to the soil or used by another biological system. In a community of cabbages and rabbits, the energy and materials can be passed through two systems (although some will be lost in the transfer). Add hawks and bacteria, and the complexity of energy pathways rapidly increases. At the same time the quantity of living material supportable by the same input of solar energy increases. Although diversity seems to be both the inevitable result of evolution and also energetically efficient from the point of view of use of available solar energy, this does not mean that diversity has been in any sense a "goal" of the evolutionary process. Evolution has no goal, ultimate or otherwise—it simply happens as the consequence of the processes of reproduction we will discuss in later chapters.

The Size Range of Organisms

Diversity is part of the picture of life we are slowly developing. Does this diversity have bounds, or are there no limits to the forms organisms may assume and no limits to their size? There seem to be both upper and lower limits to size. From what we know of present and past life, the viruses and the blue whale mark these extreme positions. Though bounded, this size range is still impressive. It is easier to conceive of this range if it is magnified to correspond with our visual abilities. For instance, if the period at the end of this sentence represented the size of a virus, a German shepherd dog magnified by the same factor (5,000X) would be about 4 miles long and 2 miles tall.

This range in size of organisms is obviously much too large to represent by scale lines printed in a book of this size; it would require a rather long foldout. To overcome this and similar difficulties, scientists often use a device called a logarithmic transformation. A logarithm is an exponent, such as the 3 in 10^3. The base, the number 10 in this example, is chosen to fit the situation. The number 10 is commonly used to express the idea of **order of magnitude.** For example, 100, or 10^2, is one order of magnitude greater than 10, or 10^1, and 1,000,000 (10^6) is five orders of magnitude greater than 10. With this in mind, it is possible to make a graph using orders of magnitude as a scale instead of a scale based on the common arithmetic numbers. For instance, if the order of magnitude 1 were represented by 10¢, as in Fig. 1–11, $1 million would be the eighth order of magnitude. Fig. 1–12 shows the impracticality of using a simple arithmetic plot.

It is important to remember that a logarithmic scale *is not* some kind of a distortion of the more conveniently used arithmetic scale. Both are

Cents	Exponential form	Order of magnitude
10	10^1	1
100	10^2	2
1,000	10^3	3
10,000	10^4	4
100,000	10^5	5
1,000,000	10^6	6
10,000,000	10^7	7
100,000,000 ¢	10^8	8
(= $ 1 million)		

Fig. 1-11 Orders of magnitude and logarithmic transformation. If the order of magnitude 1 is 10¢, then the eighth order of magnitude would be $1 million.

equally legitimate metrics (standards of measurements), neither one more real or more natural than the other. It is convenient to measure some things on an arithmetic scale and some on a logarithmic scale and to use entirely different metrics for still other phenomena. We can utilize the con-

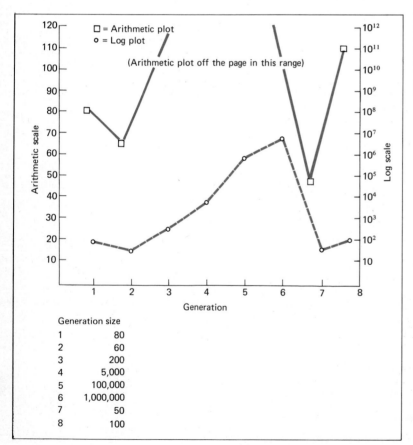

Fig. 1-12 Comparison of arithmetic and logarithmic scales. The number of individuals of a hypothetical fly population is plotted for eight generations. On the arithmetic scale (left axis) the plot quickly goes off the graph. The logarithmic scale makes it possible to show all eight generations.

Generation size

Generation	Size
1	80
2	60
3	200
4	5,000
5	100,000
6	1,000,000
7	50
8	100

cept of orders of magnitude to represent the size scale of organisms, as in Fig. 1–13.

Apparently then, size limits do exist for organisms. What factor or factors might determine these limits? Why, for instance, have there been no land animals larger than the giant brontosaurus-type dinosaurs? These may have reached weights of 50 tons and lengths (with a long tail) of 90 ft. Why are there few trees taller than the 340-ft California redwoods? On the other hand, why are there no warm-blooded animals smaller than shrews and hummingbirds and no viruses smaller than 0.017 micron (μm)?

The main reason for many maximum size limits was clearly described by Galileo more than 300 years ago. He set forth the principle that the force due to gravity acting on a body is proportional to the mass of the body. Mass—that is, the amount of material present—of course is usually proportional to volume, and volume increases as the cube of linear dimension. This means that an apple tree with a trunk diameter of 1 ft and

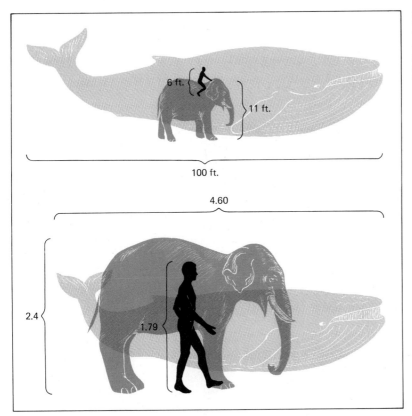

6 ft.

11 ft.

100 ft.

4.60

2.4

1.79

Fig. 1–13 Effect of logarithmic transformation on the sizes of animals. The upper drawing shows a whale, an elephant, and a man in arithmetic scale. Below, the same animals are shown in logarithmic scale. Assuming the whale is 100 ft long, the elephant is 11 ft tall, and the man is 6 ft tall, logarithms of these are 4.60, 2.40 and 1.79, respectively.

bearing fifty 3-in. apples, each weighing $\frac{1}{3}$ lb, would have about 17 lb of fruit. If the diameter of the apples were doubled, the fruit would weigh eight times as much, or about 130 lb. The strength of a vertical support, such as a tree trunk, is proportional to its cross-sectional area, and cross-sectional area increases only as the *square* of linear dimensions. In order to support eight times the weight of fruit, the trunk of an apple tree would have to increase its area by a factor of 8. This means a diameter of almost 3 ft. This is why pumpkins and watermelons do not grow on trees.

Using this same reasoning, it can be seen why elephants are not twice as large as we find them. To support their great bulk, elephants must walk stiff-legged on thick, pillarlike legs ending in broad padded feet. An animal essentially the same shape as an elephant but with double the elephant's linear dimensions would weigh about eight times as much, or about 40 tons! It is thought that many millions of years ago, giant dinosaurs in this weight range lived partially submerged in swamps and lakes, so that water supported much of their weight. Therefore, it seems that a 40-ton elephant could not possibly have the same shape and habits as its normal-sized model. The legs of such a terrestrial giant would have to be about three times the typical 15-in. diameter of the legs of the African elephant just to support the added bulk.

Returning from the realm of science fiction, the same principle can be observed in the skeletons of familiar animals: the bones of a mouse make up about 8 percent of its body weight, those of a dog about 14 percent, and those of a man about 18 percent. The clumsiness that accompanies increasing bulk obviously limits overall size, and bulk determines the size of legs and the length of necks. A law of diminishing returns applies to hugeness. The giant inhabitants of Jonathan Swift's Brobdingnag, in "Gulliver's Travels," would have weighed 90 tons. To make them move, Swift had to repeal the laws of physics.

Smallness, too, has its limits, but for different reasons. Surface area relative to volume increases as size decreases. Consequently, small animals (birds and mammals) that expend energy to maintain their body temperatures above that of their surroundings have a relatively greater surface from which heat radiates. This is why shrews, just to stay alive, must eat one or two times their weight of food each day; a cat requires 0.06 times its weight in food, and a horse 0.02 times. It is interesting that hummingbirds have evolved a strategy that allows the luxury of small size without this major handicap. Hummingbirds, unlike shrews, sleep all night. A shrew would starve to death if it behaved in this way. Hummingbirds manage to do so because their body temperature drops at night to that of the air and they become torpid. In this state they use very little of their energy reserves.

*UPPER AND LOWER SIZE LIMITS EXIST FOR THE PACKAGES IN WHICH
LIFE OCCURS. THE SURFACE-VOLUME RATIO AFFECTS THE LIMIT AT
BOTH EXTREMES. GRAVITY AFFECTS THE UPPER LIMIT.*

Plants are not limited in size by the heat problem, nor are many animals.
The minimum size of these organisms is determined by problems of
miniaturization. The absolute minimal requirements for the simplest
organisms we know—the viruses—are two structures: a molecule containing the information necessary for the reproduction of the organism and
a capsular coat around this molecule. Viruses, which meet these minimal
requirements, cannot reproduce without the help of a larger host that has
more biological machinery. For this reason, many biologists do not consider them to be "alive."

The structural unit of all living things other than viruses is the cell. We
will be discussing the structure and functioning of cells in succeeding
chapters. Let us here merely describe their general structure. Cells are
separated from the environment or from other cells by a **cell membrane.**
Plant cells usually have a stiff cell wall outside this membrane. The vast
majority of cells include a number of smaller structures of characteristic
shape and size. These are called **organelles,** and each type has a specialized function within the cell. The most prominent organelle is the
nucleus, which regulates the activities of the entire cell. A cell containing
a nucleus and other organelles is called a **eukaryotic** cell. The cells of
bacteria and blue-green algae lack a nucleus and organelles of the same
kind as those of eukaryotes. They are said to be **prokaryotic** cells.
(See Fig. 1–14.)

The Cell: Basic Structural Unit

Fig. 1–14 Diagrams to show basic
differences among eukaryotic plant
and animal cells and prokaryotic
bacterial cells.

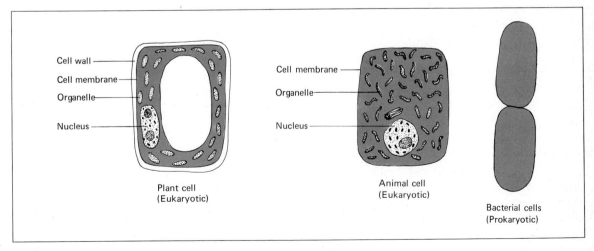

Cell wall

Cell membrane

Organelle

Nucleus

Plant cell
(Eukaryotic)

Cell membrane

Organelle

Nucleus

Animal cell
(Eukaryotic)

Bacterial cells
(Prokaryotic)

Many plants and animals exist as single cells. In some forms of plants and animals, single cells are put together in more or less loosely organized colonies. All other plants and animals are multicellular; their bodies consist of from a relatively few to an immense number of highly integrated cells.

The basic package of living systems, then, is the cell. Increase in size is not achieved by increasing the size of cells. This would upset the by now familiar surface-volume ratio. Increase in size is accomplished by putting many cells together to form a multicellular plant or animal body. The advantages of multicellularity are several. Large size is useful in obtaining more food (energy). It may also permit an organism to escape predation. With a multicellular body, there is a basis for specialization of individual cells. A division of labor may take place, which results in increased efficiency of the organism as a whole.

THE BASIC STRUCTURAL UNIT OF LIFE IS THE CELL. THE MINIMUM SIZE OF A CELL IS THAT WHICH CAN CONTAIN THE NECESSARY MOLECULES FOR FUNCTIONING, GROWTH, AND REPRODUCTION. THE MAXIMUM SIZE OF A CELL IS THAT AT WHICH THE NECESSARY EXCHANGES CAN TAKE PLACE ACROSS THE CELL MEMBRANE. TO EXCEED THIS SIZE, ORGANISMS MUST BE MADE UP OF MANY CELLS.

THE CLASSIFICATION OF LIVING THINGS

Having briefly considered some aspects of the diversity of organisms, it now seems appropriate to introduce some of the major kinds. We have already mentioned the most basic of the generally recognized taxonomic groups of organisms: plants, animals, and microorganisms. Before going any further with the introductions, it is necessary to explain the hierarchical nature of the taxonomic system. A **hierarchy** is a system in which elements are categorized into progressively larger, mutually exclusive classes. Military units are normally arranged hierarchically (Fig. 1–15). Each soldier belongs to a squad, squads are arranged into platoons, platoons into companies, companies into regiments, regiments into divisions, and so forth. No soldier, however, belongs to more than one squad, platoon, company, regiment, or division. These progressively larger classes are mutually exclusive.

Classification of Animals

Let us first look at the classification of animals. Each human individual belongs to one species, and that species and others (now extinct) make up a genus (pl. genera). Genera are grouped into families, families into orders, orders into classes, classes into phyla (sing. phylum), and phyla into kingdoms. Each human belongs to one species *(Homo sapiens),* one genus *(Homo),* one family (Hominidae), one order (Primates), one class

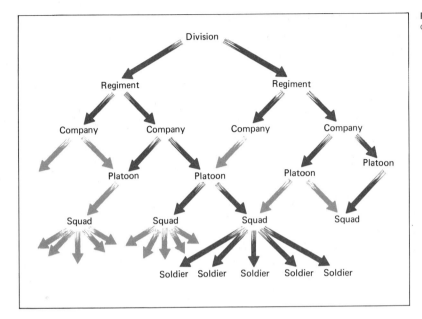

Fig. 1-15 The hierarchical
organization of military units.

(Mammalia), one phylum (Chordata), and one kingdom (Animalia). Man's taxonomic position, together with that of some other animals, is shown in Fig. 1-16.

Perhaps you noticed some rather confusing conventions in the names at the lower end of the hierarchy. Modern men belong to the species *Homo sapiens* but to the genus *Homo.* By convention, the name of the basic kind (species) is made up of two parts, the generic name and the trivial name. The name of the genus comes first and is always capitalized. Together, the two names make up the species name ("species" is both singular and plural). It is as if the order of your given name and surname were reversed—so that, for example, instead of George Jones, you would write *Jones george* (another convention is that specific names are always set in italic type). The family name always contains the root of the name of one of its genera, followed by the ending "-idae." Hominidae contains the genus *Homo.* Another family of the order Primates is the Cercopithecidae, which contains many genera of Old World monkeys, including *Cercopithecus* and another genus you have already met, *Papio,* the baboons. The classification of some common animals is shown in Table 1-1 to help you familiarize yourself with the system.

Classification of Plants

The classification of plants proceeds along exactly the same major lines, but there are minor differences in detail. The term "phylum" is not used in botanical classification. Instead, the word "division" is

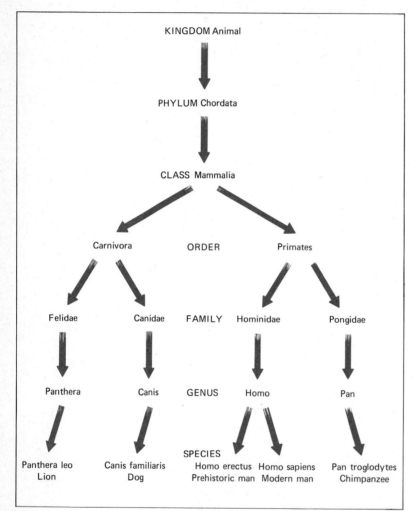

Fig. 1-16 The hierarchical organization of animal classification. The taxonomic relationship of modern and prehistoric man, a chimpanzee, a dog, and a lion are shown.

TABLE 1-1.
Animal Classification

KINGDOM	Animal	Animal	Animal	Animal	Animal	Animal
PHYLUM	Chordata	Chordata	Chordata	Arthropoda	Arthropoda	Coelenterata
CLASS	Mammalia	Mammalia	Mammalia	Insecta	Crustacea	Hydrozoa
ORDER	Primates	Carnivora	Carnivora	Lepidoptera	Decapoda	Hydroida
FAMILY	Hominidae	Canidae	Felidae	Nymphalidae	Homaridae	Hydridae
GENUS	*Homo*	*Canis*	*Panthera*	*Danais*	*Homaris*	*Chlorohydra*
SPECIES	*Homo sapiens*	*Canis familiaris*	*Panthera leo*	*Danais plexippus*	*Homaris americanus*	*Chlorohydra viridissima*
COMMON NAME	Man	Dog	Lion	Monarch butterfly	American lobster	Hydra

KINGDOM	Plant	Plant	Plant	Plant
DIVISION	Eumycota	Bryophyta	Coniferophyta	Anthophyta
CLASS	Basidiomycetes	Musci		Monocotyledonae
ORDER	Agaricales	Eubrya	Coniferales	Graminales
FAMILY	Agaricaceae	Funariaceae	Pinaceae	Gramineae
GENUS	*Agaricus*	*Funaria*	*Pinus*	*Zea*
SPECIES	*Agaricus campestris*	*Funaria hygrometrica*	*Pinus strobus*	*Zea mays*
COMMON NAME	Mushroom	Moss	White pine	Corn

TABLE 1–2.
Plant Classification

employed for this category. As with animals, the species name of a plant is always binomial, consisting of the generic name followed by the specific epithet (that is, the trivial name). As you can see in Table 1–2, the endings of family and ordinal names are somewhat different in botany.

INTRODUCTION TO THE DIVERSITY OF ORGANISMS

There are three main divisions of life: animals, plants, and microorganisms. Some workers use the taxonomic system to express their views about the evolutionary and genetic relationships among organisms. Others construct a system based solely on similarities and differences and without regard for hypothetical evolutionary relationships. Here we shall give only a very broad and general taxonomic survey. Taxonomists may not always agree on the purposes of classifying organisms, but the results are amazingly similar, regardless of the philosophy or the method. We will begin our brief survey of diversity with the animals, which are the most familiar of the three groups, and we will not treat microorganisms separately from animals and plants.

Animals

Sponges (phylum Porifera) and a related group of animals are so different from the rest of the animal kingdom that taxonomists often set them off in a group by themselves. Sponges are rather like hollow sacs and are made up of three layers of cells: an outer layer of flat cells, which make up a sort of skin, an inner layer of cells with movable projections (flagella), and a middle group of cells, which move around in the space between the other two layers. Food enters the sponge through holes (pores) in its sides, and wastes move out through large holes (excurrent siphon) at the top. Sponges show little arrangement of cells into the specialized tissues, organs, or organ systems that are so characteristic of most animals. For instance, they do not have a complex digestive system of tubes and glands such as men have; they break their food down inside individual cells in the same manner as single-celled organisms. They also lack the well-defined nervous system which permits the highly integrated activities of most animals. But then, a sponge's activities for most of its life consist simply of sitting on the ocean floor sucking in tiny plants and animals as food and spewing out waste. The sponge's main claim to fame, as a source of mopping devices for man, is rapidly disappearing before the assault of artificial sponges.

Somewhat more complex than sponges are animals of the phyla Coelenterata (Fig. 1–17) and Ctenophora: hydra, jellyfishes, Portuguese men-

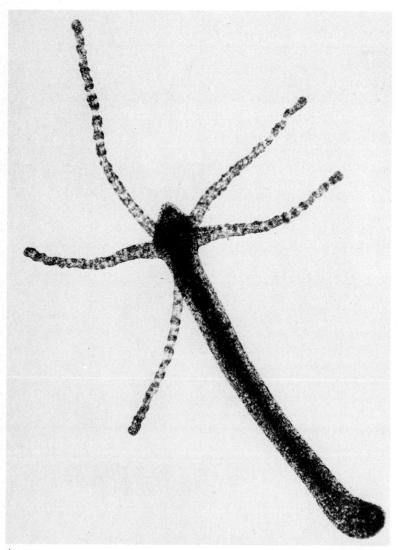

A

Fig. 1–17 Animals of the phylum Coelenterata. (A) Hydra. *(Courtesy of Philip Feinberg.)*

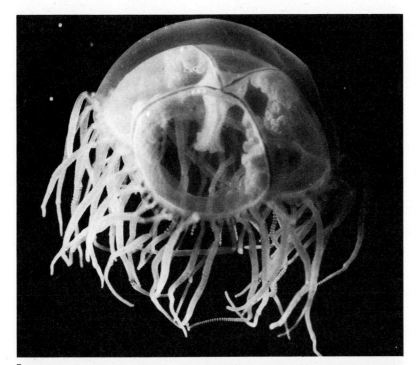

(*B*) A jellyfish. *(Carolina Biological Supply House.)*

(*C*) A sea anemone. *(Jeraboam, N. Y.)*

B

C

of-war, sea anemones, corals, and comb jellies. Most of these organisms all have radial symmetry. Hydras, which are among the few freshwater coelenterates, display many of the main features of the phylum: a two-cell-layer body plan featuring a central digestive cavity with a combined input-output opening surrounded by tentacles. Sea anemones may be familiar to you if you have ever wandered around tide pools at the ocean's edge. They look rather like squat hydras. Many marine coelenterates drift or swim around in a position that is upside down relative to that of the sedentary (attached) forms. They are more or less umbrella-shaped, with dangling tentacles. The tentacles of coelenterates bear specialized stinging cells that inject poison into their prey. Accidental contact with the tentacles of some jellyfish may prove extremely uncomfortable, or even fatal, for man.

The flatworms (phylum Platyhelminthes), like most of the other animals we will discuss, are bilaterally symmetrical. Free-living forms, such as *Planaria* (Fig. 1–18), have a digestive tract with a single opening. Many of the flatworms are parasitic, and some of these have lost their digestive tracts. Tapeworms, which may grow to be 50 to 75 ft long, live in the intestinal tracts of other animals and absorb nutrients directly through their skin. Most of the interior of their body segments is occupied by reproductive organs.

Nematodes (roundworms) and rotifers are sometimes grouped together in the phylum Aschelminthes. Nematodes (Fig. 1–19) are virtually

Fig. 1–18 Free-living flatworms called *Planaria* which belong to the phylum Platyhelminthes. (*Carolina Biological Supply House.*)

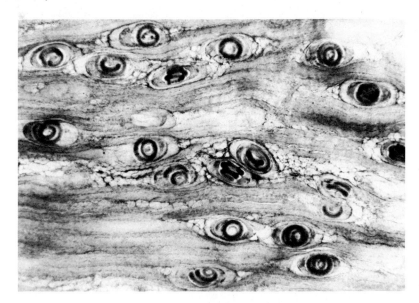

Fig. 1–19 An example of the phylum Aschelminthes is the parasitic worm *Trichinella,* here shown in muscle tissue. (*Jeroboam, N. Y.*)

ubiquitous, free-living in soil and water and occurring as parasites in most animals and plants. They are exceedingly abundant; it would not be astounding to find 100,000 tiny roundworms in a handful of soil. Aschelminthes have a digestive tract with separate mouth and anus and distinct excretory systems. Rotifers (Fig. 1–20) look quite different from the

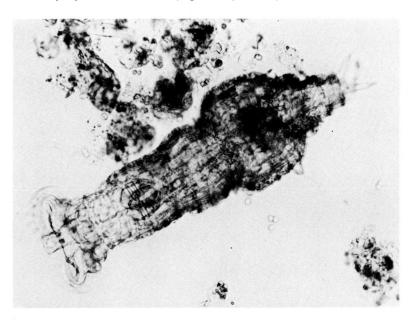

Fig. 1–20 A rotifer (phylum Aschelminthes) is shown live. The rapidly moving cilia are visible as a blur at the forward (left) end of the animal. (*Jeroboam, N. Y.*)

doubly tapered roundworms and are often put in a different phylum. They live in fresh water and draw food into their mouths by means of a ring of cilia (movable projections of cells) around their forward (anterior) end. They normally attach to a leaf or other solid substrate with their posterior "foot."

One of the most diverse phyla is the Mollusca, or molluscs. This group includes oysters, clams, snails, slugs, octopuses, squid, and a variety of lesser known creatures (Fig. 1–21). They have well-developed digestive, nervous, excretory, and circulatory systems. Molluscs are most varied in the oceans, although they are also found on land and in fresh

A

Fig. 1–21 Three members of the phylum Mollusca. *(A)* Oysters *(Crassostrea)* attach themselves in groups to underwater surfaces; here they are exposed at low tide; in oysters, one shell is generally convex, the other concave. *(Courtesy of Jack Dermid.)*

(B) The garden snail *(Helix)* crawls with a flattened muscular "foot;" sensory organs are located in the antennae. *(Courtesy of Ross Hutchins.)*

B

(C) The common American squid *(Loligo)* in which the foot is modified into 10 grasping arms provided with suction cups, the eyes of squid are highly developed and the organism is a highly mobile carnivore. *(Marineland of Florida.)*

C

water. Most of them have special glands which are used to produce a characteristic limestonelike shell.

Closely related to the molluscs are the segmented worms of the phylum Annelida (Fig. 1–22), which includes leeches and earthworms, among

Fig. 1–22 One of the many types of Annelida, the sandworm (*Nereis*) has spiny appendages. (*Courtesy of Lynwood M. Chase*)

others. They also have a rather complete repertoire of organ systems and are found in the sea, on land, and in fresh water.

The largest of all the phyla is the Arthropoda, or "jointed legs." Nearly a million species have been named in this group alone, and some people claim that as many as 9 million more await discovery. The arthropods include crustaceans (crabs, lobsters, and their numerous tiny relatives), millipedes, centipedes, spiders, mites, and insects. They are all characterized by a horny external skeleton and jointed legs (Fig. 1–23). When one thinks of the diversity of life, insects come immediately to mind. There are more named species of insects, some 750,000, than of all other organisms put together, and there may be many times that number as yet undiscovered and unnamed. Many families of insects contain more species than most classes of other organisms. There are, by coincidence, almost exactly the same number of species of butterflies (about one-tenth of the insect order Lepidoptera) as there are species of birds (the entire class Aves). With the exception of the sea, where they are rare, insects occur in virtually every habitat. They range in size from moths with wingspreads greater than the height of this book to beetles the size of a comma on this page. They live on snowbanks, in leaves, in people's hair, in flowers, inside dead wood, in kitchens. They feed on everything from library books to small green plants growing in the hair of sloths. Nematodes rival them in abundance of individuals but not in number of known kinds. Increasing knowledge may show that the mites equal or exceed the insects in number of species. Mapping the diversity of these tiny relatives of spiders has barely begun, however.

The phylum Echinodermata, including starfishes, sea urchins, and other marine organisms (Fig. 1–24), has radially symmetrical adults but bilaterally symmetrical immature forms. The skeletons of echinoderms consist of internal plates. Although these animals have a well-developed digestive system, their nervous system and circulatory system are poorly devel-

Fig. 1–23 (*A*) The diversity of Arthropoda includes scorpions, bees, flies, butterflies, lobsters, and spiders; (*B*) a representation of the diversity of one group of insects, the beetles. There are more described species of beetles than of any other group. (*Adapted from Relis B. Brown, GENERAL BIOLOGY. Copyright McGraw-Hill Book Company, 1970. Used by permission*)

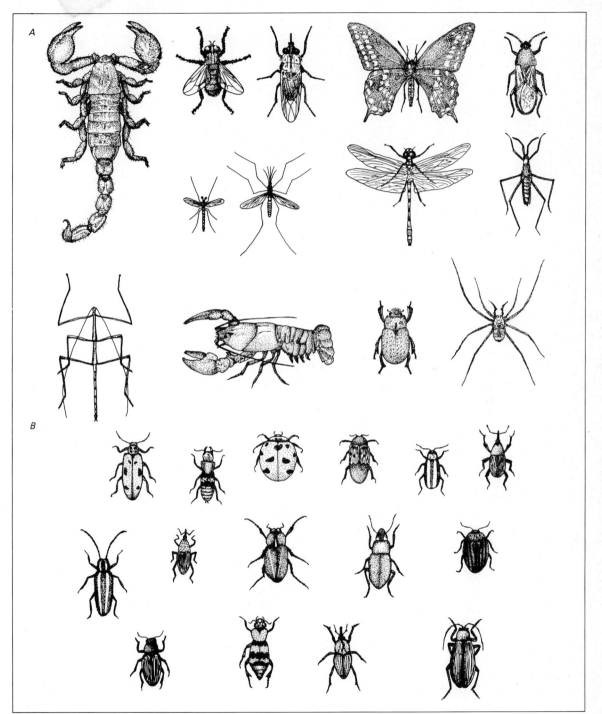

A

B

A   B   C

D   E

Fig. 1–24 The diversity of phylum Echinodermata. (*A*) A starfish (*Asterias*); (*B*) the underside of the green starfish shows the very numerous "tube feet" used in locomotion; the mouth (not seen) is centrally located; (*C*) other starfish do not have long arms (*Oreaster*); (*D*) some starfish have much divided arms and are called basket starfish; (*E*) a sea urchin with long spines (*Diadema*). (*A, courtesy of Daniel A. Gotshall; B, courtesy of Jim Annan; C, D, and E, courtesy of Patrick Cohn.*)

oped and they have no excretory system. Starfishes have a remarkable capacity to regenerate lost arms.

Finally, we come to man's "home phylum," the phylum Chordata. The Chordata include fishes, amphibians, reptiles, birds, and mammals, as well as some less familiar organisms. All these possess, at least early in life, a notochord. The notochord is a rodlike structure running along the back, and around it, in most chordates, the vertebrae of the backbone form. Several groups of chordates, however, never develop a backbone made of bone or cartilage. One of these groups includes the sea squirts and their relatives (Fig. 1–25), and they, along with all the other phyla of such animals, are **invertebrates.** The remainder of the phylum Chordata comprises the **vertebrates** (Fig. 1–26). The vertebrates include three living classes of fishes, one extinct class of fishes, and the classes Amphibia, Reptilia, Aves, and Mammalia.

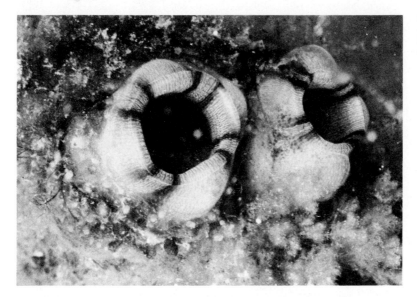

Fig. 1–25   A sea squirt (a Tunicate);
a chordate which does not have a
backbone of bone or cartilage in the
adult form. (*Courtesy of Jim Annan.*)

One important group of microorganisms has not yet been mentioned. This
is the group of single-celled organisms called Protozoa (Fig. 1–27).
Protozoans are single-celled, although some of them form colonies. You
have probably heard of the *Amoeba*, a protozoan which moves by extending its cell into a protrusion into which the rest of the body flows. Other
protozoans have minute movable extensions of the cell (cilia) covering
the entire cell. By coordinated beating, these propel the cell through
water. *Paramecium* is an example of such a ciliate protozoan. There
are other kinds of protozoans, some of which are parasitic—for example,
the protozoan which causes malaria.

The plant kingdom may be subdivided into two main groups: the pro-   **Plants**
karyotic organisms and the eukaryotic organisms. Prokaryotic organisms
have relatively simple cells that lack nuclei and the other organelles
usually present in eukaryotic cells. There are two important groups of
prokaryotes: the bacteria (Schizomycophyta) and the blue-green algae
(Cyanophyta). You are familiar with the bacteria (Fig. 1–28) as agents of
many diseases in man and other organisms. They also play a number of
indispensable roles in the economy of nature. The blue-green algae
(Fig. 1–29) are aquatic or terrestrial prokaryotes which are also important
in the functioning of ecosystems. Not all biologists classify the prokaryotes as plants. They are sometimes placed in a kingdom of their
own called Monera.

Aa   Ab

Ac

Fig. 1–26 The vertibrates are so numerous and so diverse that it is difficult to portray their variety; this photo essay shows a range of types. (*A*) The fishes: (*a*) a primitive jawless fish; the lamprey (*Petromyzon*), clinging to a rock with its suckerlike mouth (*Carolina Biological Supply House*); (*b*) a shark (*Courtesy of Hank Meyer Associates Inc.*); (*c*) a typical bony fish (*Courtesy of Ross Hutchins.*)

Ba

(*B*) The amphibia: (*a*) a red-spotted newt (*Diemictylus*) (*Courtesy of Jack Dermid*); (*b*) a male green tree frog (*Hyla*) calling (*Courtesy of Ross Hutchins*).

Bb

Ca

Cb

(*C*) The reptiles: (*a*) the giant land tortoise (*Testudo*) of the Galapagos Islands (*Courtesy of Grant Haist*); (*b*) the American alligator (*Alligator*). (*Courtesy of San Diego Zoo*).

Cc

Cd

Ce

(c) A lizard, the five-lined skink
(*Eumeces*) with eggs (*Courtesy of
Jack Dermid*); (d) the horned "toad"
which is actually a lizard (*Phrynosoma*)
(*Courtesy of Jack Dermid*); (e) a king
snake (*Lampropeltis*) with eggs
(*Courtesy of Leonard Lee Rue III*).

(*D*) The birds (*class Aves*); (*a*) the heron (*Casmerodius*) (*Courtesy of Grant Haist*); (*b*) the wood thrush (*Hylocichla*) (*Courtesy of Jack Dermid*).

Da

Db

Ea

Eb

(*E*) The mammals: (*a*) the oppossum (*Didelphis*) is a marsupial (pouched mammal) (*Jeroboam*); (*b*) the nose of the star-nosed mole (*Condylura*) is modified into senstivie appendages (*Courtesy of Lynwood M. Chase*); (*c*) the silver-haired bat (*Lasionycteris*) emerging from its daytime resting place (*Courtesy of Jack Dermid*).

Ec

(*d*) Mother and offspring of the bottlenose porpoise (*Tursiops*), which has a blow hole visible on the head of the adult (*Marineland of Florida*); (*e*) a mother zebra (*Equus*) nurses her foal (*Jeroboam, N.Y.*).

Ed

Ee

(*f*) An African bull elephant
(*Loxodonta*) (*Courtesy of Leonard
Lee Rue III*); (*g*) a pair of white-faced
monkeys; (*Jeroboam, N. Y.*).

Ef

Eg

There is a great diversity of eukaryotic plant groups, and no general agreement has been reached about how they should be classified at higher taxonomic levels. The most conspicuous land plants are the vascular plants. These have specialized cells and tissues that conduct water and food materials throughout the plant. The Anthophyta, or flowering plants (Fig. 1–30), are probably most familiar. There are some 300,000 species of flowering plants, and they occur in virtually every hab-

A

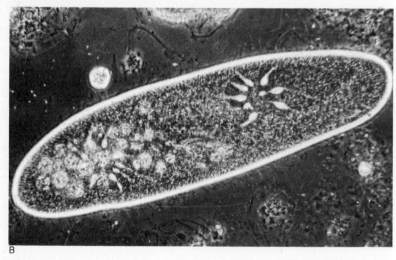

B

Fig. 1–27 Several types of Protozoa. (A) *Amoeba* showing amoeboid movement; (B) *Paramecium,* with cilia barely visible along the margins of the cell. The conspicuous star-shaped structure is a contractile vacuole; (*A & B courtesy of Eric V. Gravé*).

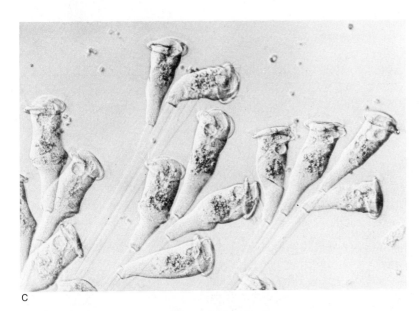

C

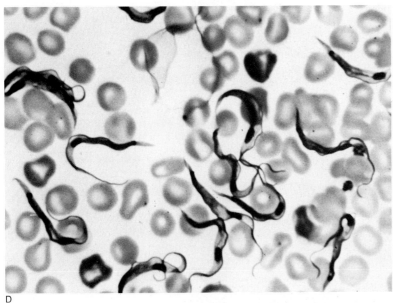

D

(C) Vorticella feeds by producing a current of water using a ring of cilia visible as blurs at one end of the cell *(Carolina Biological Supply House)*; *(D) Trypanosoma,* a parasitic protozoan seen here among human red blood cells *(Jeroboam, N. Y.).*

itat except the deep sea. The other groups of vascular plants are older, geologically, and many have become extinct. The Coniferophyta, or conifers (Fig. 1–31), Pterophyta, or ferns (Fig. 1–32), Arthrophyta, or horsetails (Fig. 1–33), and Lycophyta, or club mosses (Fig. 1–34), were

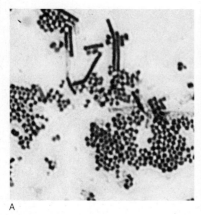

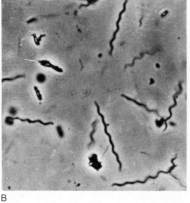

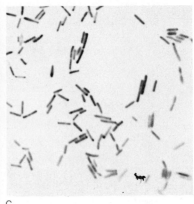

A   B   C

Fig. 1-28 Three major kinds of bacteria: (*A*) round cells called cocci (sing. coccus) with a few rod-shaped forms; (*B*) spiral forms; (*C*) rod-shaped forms called bacilli (sing. bacillus). (*A, B, courtesy of Eric V. Grave; C, Carolina Biological Supply House.*)

all more abundant in past times than they are today. The mosses and liverworts, Bryophyta (Fig. 1-35), are mainly terrestrial plants lacking specialized tissues for conducting water and nutrients. Their evolutionary relationship to the vascular plants and to other kinds of plants is obscure.

The remaining plants fall into two groups: the fungi and the algae. The algae are largely aquatic, marine or freshwater, although a few occur on land. Each of the major groups of algae represents a very distinct and very old evolutionary line. They are unicellular, colonial, or multicellular

Fig. 1-29 A filamentous blue-green alga, *Oscillatoria*. (*Courtesy of Eric V. Grave.*)

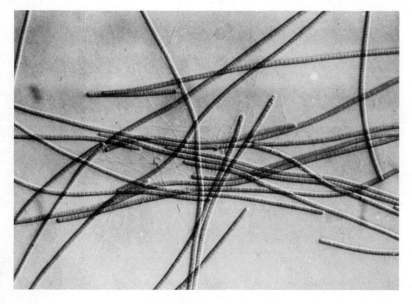

A   B   C

Fig. 1-30 A sampling of flowering plants: *(A) Magnolia,* a common tree in forests of southeastern United States; *(B) Opuntia,* a cactus from southwestern United States; *(C)* the Indian cup *(Silphium)* is a member of the daisy family; *(D) Lilium; (E)* a grass *(Bouteloua).* (*A, courtesy of Ross Hutchins; B, C, and D, Roche; E, United States Department of Agriculture.*)

D   E

plants that differ greatly in their chemistry, specifically in the nature of the pigments involved in photosynthesis and in their food-storage products. The Chlorophyta, or green algae (Fig. 1-36), are chemically similar to the vascular plants and probably are ancestral to them. The Rhodophyta, or red algae, are largely marine. Their constitution of pigments usually gives them a deep red color. They are found on rocky shores around the world, although they are especially rich in the tropics. The Phaeophyta, or brown algae (Fig. 1-37), are also almost exclusively marine. Popularly known as rockweeds and kelps, brown algae are found along virtually all shores, and one of them, *Sargassum,* forms large rafts in warm seas—for example, in the Sargasso Sea of the Gulf Stream—thousands of miles from any shore.

Many biologists believe that the fungi are so distinct from plants and

Fig. 1-31 (*A*) A mixed forest of conifers; the central tree is *Pinus ponderosa;* note the two men standing at the base for scale. (*Courtesy of United States Forest Service.*)

(*B*) Seed-bearing cones of the Austrian pine (*Pinus*). (*Roche.*)

animals that they should be considered a separate group. It is only for convenience that we group them with the plants here. The simplest fungi are unicellular or filamentous. Some lack cell walls across their filaments, or these may be perforated so that the contents can flow from cell to cell. Many of these simpler fungi are pests, especially of plants. Some are of economic importance, such as those that produce antibiotics (e.g., *Penicillium*) and those that flavor cheese (e.g., *Aspergillus*).

The two major groups of fungi are the sac fungi (Ascomycetes) and the club fungi (Basidiomycetes). In these fungi, the basic unit of structure is a multicellular filament called a **hypha.** This filament begins with the

Fig. 1-32 (*A*) *Adiantum*, the maidenhair fern; (*B*) *Polypodium*, which grows on the branches of trees. (*A, B, courtesy of Ross Hutchins.*)

Fig. 1-33 A horsetail (*Equisetum*). The branched stems carry on photosynthesis while the club-shaped branches are reproductive. (*Roche.*)

germination of a **spore.** Its growth may be diffuse, branching and spreading throughout the substrate. The total mass of hyphae make up the **mycelium** of the fungus (Fig. 1-38). At the time of reproduction, specialized spore-producing structures are formed. These are variously formed, often saclike structures in the Ascomycetes (Fig. 1-39). The spore-producing parts of Basidiomycetes include the familiar mushrooms, as well as other forms (Fig. 1-40). We will discuss later some of the rather strange aspects of the life cycles of some of the fungi.

Fig. 1–34 The foxtail clubmoss (*Lycopodium*). (*Courtesy of Jack Dermid.*)

Change is a phenomenon of the universe. The universe is dynamic, not static. The universe itself expands and contracts; its stars are born out of dust clouds, mature, grow old, and die. The planets of a solar system, apparently formed at the same time and out of the same dust as the stars, about which they revolve, cool. Life emerges on the surface of one or more of them, mountains are built, continents may move like floating islands, and eventually, the planets, too, die along with their star. The sun and the earth have life cycles not altogether unlike those of organisms.

CHANGE: THE SOURCE OF DIVERSITY

Fig. 1–35 Two representative bryophytes: (*A*) a liverwort (*Conocephalum*); (*B*) a moss (*Polytrichum*) with club-shaped reproductive structures. (*A, courtesy of Ross Hutchins; B, courtesy of Grant Haist.*)

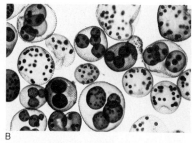

A

B

Fig. 1–36 Two kinds of *Chlorophyta* (green algae): (*A*) the filamentous *Spirogyra*; (*B*) *Volvox*, which forms spherical colonies of cells. (*A, B, Jeroboam, N. Y.*)

Fig. 1–37 A brown alga (*Sargassum*) stranded on a beach; it normally is found in great masses held up by the balloon-like gas-filled floats. (*Courtesy of Jack Dermid.*)

Fig. 1–38 A portion of the mycelium of a fungus growing on the underside of a rotting pine log; it is made up of numerous fine filaments called hyphae. (*Courtesy of Jack Dermid.*)

Fig. 1-39 An edible and delicious member of the Ascomycetes (the morel fungus); spore-producing structures are scattered over the honey-combed portion. (*Courtesy of Grant Haist.*)

Every spot on the surface of the earth is also changing. The rocks are eroded by wind and water, soil is built up, and organisms grow and die contributing their tissues to the soil. In the oceans and lakes, currents constantly mix the water, and winds ruffle the surface. If long periods of time are considered, many more extensive changes could be observed. Glaciers move toward the equator from the poles at widely spaced intervals. The transgression of seas across submerging continents alternates with the building of mountain ranges.

A

B

Fig. 1–40 Spore-producing structures of three typical fungi of the Basidiomycetes; most of their mycelium is not visible: (*A*) the fly agaric (*Amanita*), a very poisonous fungus (*Courtesy of Lynwood M. Chase*); (*B*) a bracket fungus, growing on and in a tree (*Courtesy of Grant Haist*).

(*C*) A stinkhorn fungus (*Phallus*) which produces an odor attracting flies (*Courtesy of Grant Haist*).

C

ALL ASPECTS OF THE ENVIRONMENT ARE CONTINUALLY CHANGING, AND IT IS IN THESE CHANGING HABITATS THAT COMMUNITIES OF ORGANISMS OCCUR.

Organisms also are subject to change in this inconstant environment. These changes can, for convenience, be divided into four categories: (1) ephemeral or reversible changes, (2) developmental changes, (3)

Kinds of Change

cyclic changes, and (4) evolutionary changes. Let us briefly describe each of these kinds of changes.

1 Relatively slight, reversible fluctuations are characteristic of many processes and systems in organisms. Human body temperature, for instance, averages 37°C (98.6°F) in most humans, but it is rarely exactly at this level. Fluctuations of a degree or so are not uncommon and depend on the time of day, on the amount of physical exertion an individual is undergoing, and even on a person's emotional state (Fig. 1–41). In females, there is usually a temperature change associated with egg production. Blood pressure fluctuates in rhythm with the heartbeat and changes with emotions such as excitement and fear. In many plants, tiny pores in the leaves close when the plant gets too dry, reducing the loss of water.

2 Developmental changes are alterations of form and function that follow in sequence, usually on schedule. Ripening of fruit, the metamorphosis of a tadpole into a frog (Fig. 1–42), and the greying of human hair are changes of this kind.

3 We have learned to expect to see moths and owls at night, but not during the day, and to expect butterflies and hawks on the opposite schedule. In fact, many of the activities of organisms are keyed to cyclic changes of the environment (such as the daily alternation of light and dark), most of which are astronomical in origin. The phases of the moon, the ebb and flow of the tides, and the sequence of sea-

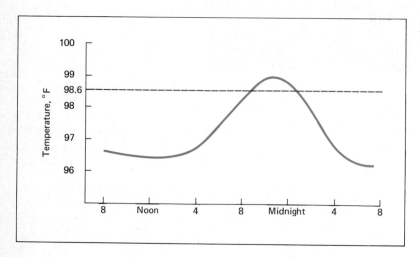

Fig. 1–41 Fluctuation of body temperature in man with time.

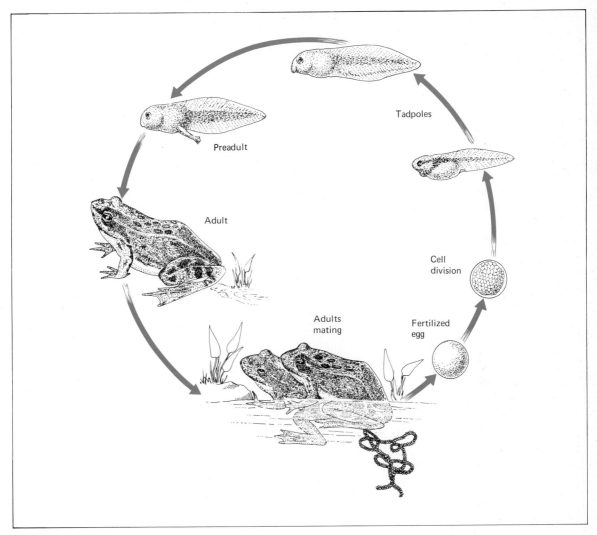

Preadult

Adult

Tadpoles

Cell
division

Adults
mating

Fertilized
egg

Fig. 1-42 Stages of metamorphosis
in frogs.

sons in the temperate and arctic regions are all examples of astro-
nomic cycles with profound biological impact. The breeding activities of many organisms are precisely regulated by solar and lunar
cycles. Twice each month during the spring, at the time of the highest
tides of the year, certain stretches of beach in southern California
swarm with a small, smeltlike fish called grunion. Swimming and
slithering high up on the beach, the fish breed as the females deposit
their eggs in holes made by squirming movements of their bodies.
The particular night on which these fish breed is connected with a

lunar or tidal cycle, but the months during which breeding occurs depend on the annual or solar cycle.

Other cyclic biological changes are the flowering of cherry trees, the migration of waterfowl, and color changes in subarctic and arctic animals (Fig. 1–43). During the fall, the fur of the snowshoe rabbit and the plumage of the ptarmigan turn from brown or grey to white. When the transformation is complete, the animals are almost invisible against the snow-covered landscape. In the spring the coats of these animals again turn dark.

The fourth kind of change, evolutionary change, is neither cyclic nor predictable and requires much longer periods of time. The idea that organisms change gradually over long periods of time is not new. The cosmologies of many ancient civilizations assumed, for instance, that men arose from some "lower" form of life such as fish. But the concept of evolution was largely a metaphysical one until Charles Darwin (1809–1882) amassed voluminous data to support the hypothesis. At the same time, and more importantly, he provided a reasonable causal explanation of the process. Nowadays it is common to witness evolution in progress, both in the laboratory and in nature.

FOUR BASIC KINDS OF CHANGE CAN BE OBSERVED IN ORGANISMS. THESE ARE REVERSIBLE CHANGES IN INDIVIDUALS, DEVELOPMENTAL CHANGES IN INDIVIDUALS, CYCLIC CHANGES IN INDIVIDUALS, AND EVOLUTIONARY CHANGES IN POPULATIONS.

Evolutionary Change

One of the most elegant analyses of an evolutionary change is the work of a group of British biologists studying a gradual increase in the frequency of a dark form of a moth common around the industrial areas of England. The typical, or "peppered," form of the peppered moth, *Biston betularia,* made up more than 99 percent of the population in the area of Manchester in 1848, whereas the dark, sooty form was so rare as to be a collector's item. Today in many areas of England, it is the peppered form that is absent or nearly so. This evolutionary change parallels in time the industrial revolution in Europe and the consequent pollution of the atmosphere of urban areas.

Evolutionary change differs in an important way from the first three classes of change considered. Ephemeral, cyclic, and developmental changes are changes in and of an *individual.* An evolutionary change is a change in a *population* of individuals and occurs over a number of generations. One way to appreciate this distinction is to consider a familiar phenomenon, the darkening of human hair that often occurs with time. The hair of many blond children darkens as they grow older; this is a developmen-

A

B

Fig. 1-43 Seasonal change in color in the snowshoe rabbit or varying hare (*Lepus*): (*A*) summer phase with brown pelt; (*B*) winter phase with white fur. (*A, B, courtesy of Leonard Lee Rue III.*)

tal change. A decrease in the **frequency** of blonds has been observed in the United States. This results from the marriage of blonds to brunettes at a rate higher than was prevalent in Europe. (It has been predicted that natural blonds will be very rare in the United States in a few generations. Alas.) This change, occurring over many generations, is an evolutionary change, a change in a population through time.

To understand how evolutionary change occurs, it is necessary to know something about reproduction—*the process that links one generation with the next.* Returning to the peppered moth, it, like many organisms, has an annual **life cycle.** The adults die as winter approaches, and survival of the population is left to a stage of the life cycle (caterpillars) which can withstand the rigors of the cold season. The eggs are the products of the reproductive process. An evolutionary change means a change in the *kinds* of fertilized eggs that are laid from one generation to the next. Appreciation of how this comes about depends on an understanding of the two simple rules of reproduction.

EVOLUTIONARY CHANGE OCCURS IN POPULATIONS, NOT IN INDIVID-UALS. IT IS THE RESULT OF CHANGES IN THE KINDS OF ORGANISMS THAT ARE PRODUCED IN A POPULATION FROM GENERATION TO GENERATION.

THE FIRST RULE OF REPRODUCTION

The first rule of reproduction is that *like begets like.* This rule is so much a part of our experience that it is not usually thought of as a great scientific principle. It is a simple, commonplace observation that cats give birth to cats, not to cows, cockatoos, or kangaroos. But everyday observations have a way of being transformed into profound principles in the minds of great thinkers. Humanity had observed apples falling from trees and the planets traveling in their orbits long before the principle of gravity occurred to Newton.

But, unlike the law of gravity, the rule of like begets like is not a rigorous mathematical principle. Rather, it is a statement that is true within limits. There are upper and lower limits to the likeness of offspring to their parents. Children are never exact replicas of their parents; neither are they so different from their parents that they could be mistaken for another kind of animal or even for another, very different kind of person. The children of Pygmy people never grow up to look like Watusi, and the children of Watusi never develop into Peruvian Indians.

The rule that like begets like, then, is not absolute; rather it is an example of a **probabilistic** statement (Box 1–1). A probabilistic statement is one that assumes that a phenomenon occurs over a range of possibilities and that, therefore, experiments may have more than one possible outcome. Probabilistic thinking is extremely important to the biologist. Probability theory and its offspring, statistical inference, give the biologist the tools necessary for dealing rigorously with vague concepts such as "like begets like, but not absolutely." However, the probabilistic nature of the concept should be obvious to even a casual observer. Brown-eyed parents do not always have brown-eyed children. Sons do not usually grow to exactly the same height as their fathers. Indeed, exact resemblance, such as may occur between identical twins, is unknown between parent and offspring in human populations. Children ordinarily do look much more like their parents than they look like unrelated people, and they always look more like children than calves or chicks. Like begets like, but not exactly.

THE FIRST RULE OF REPRODUCTION IS THAT LIKE BEGETS LIKE. BUT THIS IS A PROBABILISTIC STATEMENT; OFFSPRING ARE RARELY EXACTLY LIKE THEIR PARENTS.

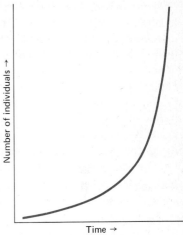

Fig. 1–44 Geometric increase in population size.

THE SECOND RULE OF REPRODUCTION

The second rule of reproduction is the *rule of surplus,* which means that the number of progeny produced by a parent generation is virtually always greater than the number of parents.

Left unchecked by social and other factors that inhibit reproduction, the population size of any organism in time would be doubled, then quadrupled, and so on. Robert Thomas Malthus (1766–1834) saw this rule very clearly and wrote, in his essay on population (1798), that population, when unchecked, increases in a geometric ratio. Graphically (Fig. 1–44), this means that the size of a population when plotted against time would produce an ever-ascending curve. If population growth were a straight-line function with time, it would mean that a constant number of individuals are added each year, rather than a constant proportion of the number already existing. This is the difference between simple interest (a constant increment) and compound interest (a constant proportion).

THE SECOND RULE OF REPRODUCTION IS THAT MORE OFFSPRING ARE PRODUCED BY PARENTS THAN WOULD BE NECESSARY TO REPLACE THE PARENT GENERATION.

CONSEQUENCES OF THE LAWS OF REPRODUCTION

What are the inevitable consequences of the two rules of reproduction, assuming for the moment that no intrinsic or extrinsic factors are affecting the expression of these rules? The answer can be found by again turning to the peppered moths.

To begin with, a population of imaginary moths could be started with 100 founders—50 typical moths and 50 dark moths. Assuming that the average reproductive potential of the individuals, regardless of color, is four (that is, each moth, on the average, contributes four progeny to the next generation), the population will then increase by a factor of 4 each generation. After 10 generations, there would be more than 100 million moths. During this period of fantastic population growth, the frequencies of dark moths and typical moths would remain constant. That is, the population would have changed in size but not in kind. There would have been no evolution. So far, there has been nothing to account for the evolutionary replacement of typical moths with black moths in England.

But Malthus went further, in his essay on population, than merely pointing out the natural fecundity of men and other organisms:

> Through the animal and vegetable kingdoms, nature has scattered the seeds of life abroad, with the most profuse and liberal hand. She has been comparatively sparing in the room and the nourishment necessary to rear them. . . . Necessity, that imperious all pervading law of nature, restrains them within the prescribed bounds.

To paraphrase, the surplus of population leads irrevocably to a "struggle for existence" among individuals. Here is the crucial point. Everyone knows that not every individual that is born survives to adulthood. In fact, populations usually do not increase at Malthusian (geometric) rates. In the rare cases in which the growth rate of a population is Malthusian, this phase endures for a relatively short time. Soon the "struggle" begins to take its toll, and the population levels out or decreases in size.

Although a surplus of offspring are produced by each adult (each year, in species that reproduce annually), many of those produced do not survive to reproduce. On the average, the survivors of this natural thinning process, called **mortality,** number about the same, year in and year out.

This still does not explain how evolution occurs. Some light can be shed on this dilemma by examining Malthus' phrase, "struggle for existence." It helps to imagine such a struggle as a kind of obstacle course that each organism has to run. The end of this obstacle course, its goal, is reproduction. If an individual reaches the goal, it passes on (begets) its characteristics (likeness) to the next generation. This is essentially a game with nature. An individual loses the game if it does not live long enough to reproduce or is sterile. If it survives long enough and successfully reproduces, it "wins" the game.

At this point one might ask why some organisms live so long if reproduction is the goal of life. In fact, many organisms do not have long life-spans. It is not uncommon for organisms to reproduce only once in their life and then to die immediately thereafter. This kind of life history is especially common in such plants as annual wildflowers, such insects as the peppered moth, and such fish as salmon.

But other species employ a different strategy in the game of life, spreading their reproductive activity over long periods. Even for these organisms, however, death and the termination of reproduction generally coincide. (Humans are exceptional in this regard, as will be discussed later.) What this boils down to is that there are degrees of winning. The relative success of an individual in the reproductive derby will depend in part on the characteristics of that individual. Individuals are not all the same, as we have noted. The more a given kind of individual reproduces, the more of its "likeness" is contributed to the next generation. The more successful an individual's strategy in survival and reproduction, the greater its relative reproductive contribution to the next generation. Different strategies in the game of life, such as having a few offspring and caring for them well (humans) or having multitudes of offspring and ignoring them (e.g., many marine animals), can be considered analogous to gambling strategies at the racetrack. For instance, a $10 bet on a favorite and five $2 bets on long shots are two strategies for achieving a common goal— winning money. Winning in the game of life, at least in the evolutionary sense, is achieving the maximum possible perpetuation of one's kind.

VARIABILITY EXISTS IN NATURAL POPULATIONS, AND EACH INDIVIDUAL TENDS TO HAVE OFFSPRING THAT RESEMBLE IT MORE THAN THEY RESEMBLE OTHERS. THE OBSTACLE-COURSE MODEL SAYS THAT THESE DIFFERENT KINDS OF INDIVIDUALS WILL NOT ALL HAVE THE SAME CHANCE FOR SURVIVAL OR THE SAME PROBABILITY OF REPRODUCTIVE SUCCESS IF THEY SURVIVE TO REPRODUCTIVE AGE. THE DIFFERENTIAL LIKELIHOOD OF REPRODUCTIVE SUCCESS IS THE "STRUGGLE FOR EXISTENCE."

NATURAL SELECTION AND DIVERSITY

The final step in our argument is the crucial one, and it is this step that is the creative contribution of Charles Darwin, Alfred Russell Wallace, and a few others. Darwin was reading Malthus' discussion about overpopulation and the struggle for existence, "when it at once struck me (Darwin) that under these circumstances favourable variations would tend to be preserved, and unfavourable ones to be destroyed. The result of this would be the formation of new species. Here then I had at last got a theory by which to work. . . ."

In retrospect it seems so simple. But creativity often consists of seeing the underlying simplicity in superficially complex situations. Darwin's and Wallace's great discovery was simply that *differential reproduction will result in change from one generation to the next and this accounts for evolution.* In terms of the moth example, the change from a high frequency of peppered moths in 1848 to a high frequency of dark moths a century later in industrial regions must mean that the dark moths have tended to outreproduce the peppered moths in these areas. In other words, dark moths have been more successful in the "obstacle course."

What is there about the obstacle course, that is, the environment, that permits more dark moths to make it than typical moths? Are the typical moths more susceptible to disease, famine, or possibly predation?

The complete answer may involve all these factors, but the outstanding one appears to be predation. As shown in Fig. 1–45, the typical moths are cryptically patterned (camouflaged) on the lichen-covered trees in unpolluted regions. The dark individuals are cryptically colored on the soot-covered trunks. Any predators, such as birds, relying on vision to find their prey will find more of the dark moths in the unpolluted regions, where they stand out conspicuously. This hypothesis has been tested experimentally in nature by H. B. D. Kettlewell, and the results are compelling. In an unpolluted wood in Dorset, England, equal numbers of the two forms were released. Most of the moths settled on the trunks of trees, where they were kept under observation from blinds nearby. Birds were seen to capture 164 dark moths and 26 peppered ones. In a polluted wood near Birmingham, the opposite results were recorded. Here birds were observed to capture 15 dark moths and 43 peppered moths. There can be little doubt that the predators were **selecting** out the conspicuous moths and that the protectively colored moths have a higher probability of survival and of contributing more moths of their type to the next generation. Again, probability enters the picture. Not all peppered individuals die, and not all dark ones survive in a polluted wood. We are concerned with the *average* survival value of each kind, the average reproductive value. Differential reproductive success of individuals is a populational process and must be studied probabilistically. This process is called **natural selection,** and natural selection is the primary process of evolutionary change. There are two central prerequisites for natural selection: hereditary variability (genetically different kinds of individuals) and differential reproduction of the kinds.

DIFFERENTIAL REPRODUCTION OF VARYING INDIVIDUALS IN A POPULATION LEADS TO EVOLUTIONARY CHANGE, THAT IS, A CHANGE IN THE KINDS OF ORGANISMS PRESENT FROM GENERATION TO GENERATION.

A

B

Fig. 1–45 Light and dark forms of
the peppered moth (*Biston*): (*A*) two
moths on lichen-covered bark; (*B*) two
moths on pollution-darkened bark of
industrial areas. (*Courtesy of the
American Museum of Natural History.*)

The dark moths and the peppered moths can be thought of as being in a
game with nature. The dark moths use one strategy (being dark), and
the typical moths use an alternative strategy (being peppered). For the
moth, success of the strategy depends on the fit of the strategy to the
obstacle course (the environment). For racetrack gamblers, success also
depends on the fit of the strategy to the environment—the environment in
this case being an array of horses, jockeys, and racetrack conditions.

Let us look at two other examples of evolutionary strategies. Some snakes
lay eggs, and other snakes bear live young, retaining the eggs in their

bodies until they hatch. The live bearers extend much farther north, both in Europe and in North America. The explanation for this seems to be that the live-bearing snakes bask in the sun and thereby maintain the eggs in their bodies at a relatively high temperature (20 to 30°C) during the development of the young, even when the temperature of the air is near freezing. Eggs laid in the soil or under rocks or logs would not develop fast enough or hatch soon enough to survive in areas which have short summers. Of the two strategies, egg laying and egg retaining, only the latter leads to survival in arctic and subarctic regions. The strategy of retaining the eggs resulted from natural selection. Could it have resulted from natural selection in a population of snakes in which the reproductive system of every female snake was absolutely identical? Egg laying also is a strategy with advantages. The female is not handicapped by getting fat, and all her eggs are not "in one basket." If she dies, her eggs may still hatch. But in cooler climates, the advantages of egg retaining outweigh these.

Many species of plants grow in a variety of habitats, ranging from the seashore to high elevations in mountains. Often the form and physiology of individuals vary from habitat to habitat (Fig. 1–46). By use of techniques we shall describe later, it can be determined whether such differences are the result of individual plasticity or are inherited. Those that are inherited are examples of evolutionary strategy.

Natural selection is the single principle that makes the rest of biology intelligible. It accounts for the diversity of organisms (via the gradual change of populations), and at the same time it provides an explanation for the myriad different characteristics of organisms, the strategies they employ in their game with nature.

Much of the rest of this book will be a discussion of these characteristics or strategies—how and why they evolved, how and why they work.

SUMMARY

All organisms are literally part of a "web of life." Biologists study the intricate bonds which interconnect plants, animals, and microorganisms with each other and with their environment. For example, green plants change light energy from the sun into chemical energy, which they use in growth and reproduction. Animals, which cannot do this, are all dependent upon green plants, whether they eat plants or other animals. The green plants are called producers, herbivores are primary consumers, and carnivores are secondary or higher-order consumers.

Other kinds of bonds between organisms include social bonds between mother and child, juvenile peers, and adults; sexual bonds; hereditary bonds; and evolutionary bonds.

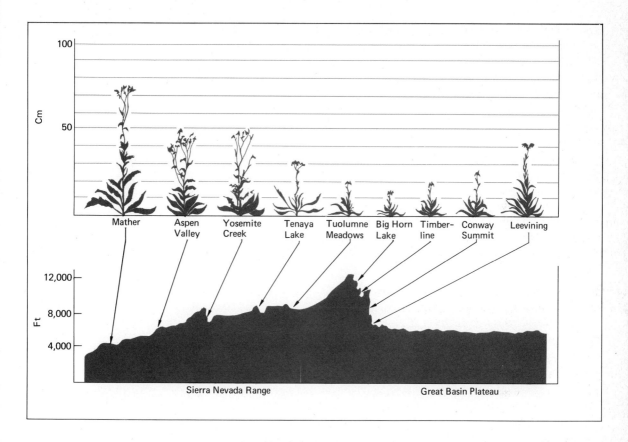

Fig. 1–46 Variation in size of the plant milfoil (*Achillea*) along a transect across the Sierra Nevada Range of California. (*Courtesy of Carnegie Institution.*)

Biologists recognize two very broad kinds of relationship among organisms: taxonomic relationships and ecological relationships. Ecological relationships include the ways in which an organism functions in its environment and how it affects, and is affected by, other organisms. Taxonomic relationships are those which deal with the structural and functional similarities among organisms, usually based upon evolutionary descent.

Living systems must interact with the environment across their surfaces. Therefore, the relationship between the amount of surface and the volume it must support is very important. This is one of the reasons life does not occur in a continuous living sheet. Discrete organisms have much greater surface area. In addition to existing as separate packages, life occurs as many different kinds of packages: plant and animal, aquatic and terrestrial, large and small. One of the reasons for this variety is that it makes possible specialization in ways of using the resources of the environment. Life in the form of many different kinds of organisms is more efficient.

Upper and lower size limits are imposed by the surface-volume effect at both extremes. Of course, the force of gravity also affects the upper limit. In addition, size is affected by the nature of the structural units of life—the cells. The minimum size of cells is that which can contain the necessary molecules for functioning, growth, and reproduction. The maximum size of cells is that at which the necessary exchanges can take place across the cell membrane. To exceed this size, organisms must be made up of many cells.

Organisms undergo reversible changes in response to minor environmental changes, and in the course of time, they also experience developmental changes. As their environment changes cyclically, many organisms also show cyclic changes. Finally, the great diversity of organisms is the result of evolutionary change. While the first three types of change occur in individuals, evolutionary change occurs over many generations in populations of reproducing individuals.

There are two basic rules of reproduction. The first is that like begets like. Offspring, however, are never exactly like their parents. The second rule is that the number of progeny produced by the parents of one generation is nearly always greater than the number of parents themselves. Offspring vary somewhat from their parents and from each other, and it is the most fit that survive. This results in differential reproduction of types, and it can cause changes in the population from generation to generation. Variation in fertility or reproductive success which does not involve death also results in differential reproduction. Differential reproduction is called natural selection. It is the dominant factor leading to evolutionary change. Natural selection is also the single principle that makes biology intelligible.

BOX 1-1. PROBABILITY

In a probabilistic approach to a problem, interest ordinarily centers on making statements about certain kinds of averages of repeated events, measurements, or results of experiments. Often it is assumed that there is a series of discrete possible outcomes or events, and some probability is associated with each one of the possibilities. As an example, consider flipping "fair" coins—ones in which the probability of flipping a head (H) is $\frac{1}{2}$ and of flipping a tail (T) is also $\frac{1}{2}$. One rule of probability theory states that the probability of two independent events occurring together is the product of the probability of each occurring alone. Since honest coins have no way of watching or acting on each other's behavior, the flipping of two such coins constitutes two independent events. Therefore, if two "fair" coins are flipped simultaneously, the probability of the event that both come up heads (HH) is the product of the individual probabilities of coming up heads, or $\frac{1}{2} \times \frac{1}{2} = \frac{1}{4}$. The probability of both turning up tails (TT) is the same, $\frac{1}{4}$. The probability of a third kind of event, one of the coins being tails and one being heads, is $\frac{1}{2}$ (the probability of HT plus the probability of TH). In this example we have used the classical, or a priori, definition of probability. In essence, we counted the possible number of outcomes (four in this case: HH, HT, TH, TT) and calculated the probabilities by taking the ratio of "favorable" outcomes to the possible outcomes. For instance, for the outcome "one head and one tail," two of the four outcomes are favorable to the result.

Biologists today do not ordinarily use the classical definition of probability because it has a number of limitations. A ***relative-frequency*** definition is employed instead. In essence, this definition states that the probability of an event A, P(A), is determined by dividing the number of times that the event is observed (N_o) by the number of times that the event could have been observed (N), provided that N is very large. Thus the definition is based on experience and is therefore an a posteriori definition. More technically, the probability is the limit approached by the fraction as N approaches infinity. In mathematical notation, this simple idea is abbreviated

$$P(A) = \lim_{N \to \infty} \frac{N_0}{N}$$

This definition actually is the basis of the a priori definition; the a priori probabilities are determined on the basis of long human experience with relative frequencies. The $P(H) = P(T) = \frac{1}{2}$ probability can be viewed as the result of man's accumulated relative-frequency data on a tremendous number of coin tosses (assuming, of course, that crooks use two-tailed coins as frequently as two-headed coins).

Darwin, C.: "Charles Darwin's Autobiography." Many editions available.
Darwin, C.: "The Voyage of the Beagle." Many editions available.
DeVore, I. (ed.): "Primate Behavior," Holt, Rinehart and Winston, New York, 1965.
Ehrlich, P. R., R. Holm, and D. Parnell: "The Process of Evolution," 2 ed., McGraw-Hill Book Co., New York, 1973.
Gates, D. M.: "Energy Exchange in the Biosphere," Harper & Row, New York, 1962.
Malthus, T. R.: "An Essay on the Principle of Population," Antony Flew (ed.), Penguin Books, Baltimore, 1970.

SUPPLEMENTARY READINGS
Books

Thompson, D'Arcy: "On Growth and Form," Cambridge University Press, London, 1961.

Gates, D. M.: Animal Climates (Where Animals Must Live), *Environ. Res.* **3:**132-144 (1970)

Moog, F.: Gulliver Was a Bad Biologist, *Sci. Am.,* **179** (No. 5) (1948).

Washburn, S. L., and I. DeVore: "The Social Life of Baboons," *Sci. Am.,* **204** (No 6), Offprint 614 (1961).

Articles

CHAPTER 2

REPRODUCTION

In the last chapter, we learned that an important characteristic of organisms is their ability to make copies of themselves—to reproduce. Why is reproduction characteristic of all living systems? The answer must be expressed in terms of the survival value of reproduction. In fact, the answer can be derived from our discussion, in the last chapter, of environmental change. If the environment were totally free of destructive agents and if it were frozen in this beneficent state, so that no harmful elements were introduced, immortality might be possible. But such constancy is not the state of nature. (If there is one word that describes nature, that word is "change.") Even if optimal conditions existed today, catastrophe might strike tomorrow in the form of a storm, flood, fire, volcanic eruption, or earthquake. Every spot on the surface of the planet eventually suffers violent changes. Land areas are drowned by the transgression of seas. Seas dry up. Mountains rise and then are leveled. Climatic patterns change. Aside from these short- and long-range nonbiological changes, every area witnesses biological changes. Organisms come and go and evolve.

Reproduction, it is clear, is necessary for replacement of lost individuals if individuals are not immortal. But the reality of environmental change (including the evolution of other organisms living in the same place) suggests that reproduction is necessary for an altogether different reason than mere replacement of lost individuals.

The Necessity of Reproduction

Using the analogy of life in a game with nature, it is evident that what is a satisfactory strategy for survival under one set of conditions may be unsatisfactory under a different set. To adjust to environmental changes, such as increasing precipitation during a glacial period, requires plasticity often beyond that in the repertoire of change in a single individual. Survival of a species, then, depends on having available a variety of *kinds* of individuals, and long-term survival must depend on the continued generation of variation so that natural selection can result in evolutionary change.

*REPRODUCTION, CHARACTERISTIC OF ALL LIVING SYSTEMS, IS NEC-
ESSARY BOTH FOR THE REPLACEMENT OF INDIVIDUALS AND FOR THE
CONTINUAL PRODUCTION OF NEW KINDS OF INDIVIDUALS WHICH
MAY BE BETTER ABLE TO SURVIVE AND REPRODUCE (BE MORE
FIT) IN A CHANGED ENVIRONMENT.*

On these a priori grounds, reproduction of similar but not identical
individuals would appear to be a good strategy for a species, since
without it, environmental change would be expected to result eventually
in extinction. Only those forms of life that have the capacity to make not
quite faithful copies of themselves have a long-range probability of
survival. This raises the question of how much variation is enough and how
much is too much. Presumably, too little would limit evolutionary plasticity and too much would result in many abnormal, "unfit" individuals
that would die at an early age. A population with too many abnormal
members would die out in a few generations.

Having established to our satisfaction why reproduction is necessarily
a universal property of organisms, we can proceed to the interesting "how"
questions: (1) How do organisms produce offspring that resemble their
parents? (2) How is the variation characteristic of this process generated
and maintained?

**THE BASIC FEATURES
OF REPRODUCTION**

The first stage in the life of most individuals is the fertilized egg, or
zygote. This term is preferable to the ambiguous word "egg" because the
latter can mean either an unfertilized egg, such as the commercial
breakfast variety, or the fertilized kind that, if put in a warm place, hatches
out a chick.

Production of the Zygote

As we have seen, organisms are made of one or more variously specialized structures called cells. The simplest organisms, such as bacteria,
are single prokaryotic cells, and the most complex, such as man, are
multicellular and contain many billions of eukaryotic cells. The life of
many organisms begins by the joining together, or fusion, of two cells, one
originating in one parent and the other in the second parent. Because
the two parents can usually be distinguished as belonging to different
types, or **sexes,** this form of reproduction is called **sexual reproduction.**
The two reproductive cells, or **gametes,** that participate in fertilization
are often very different in size and form. Usually, the gametes of one
sex are smaller than those of the other and commonly are also passively
or actively motile. The small, motile gametes of animals are called

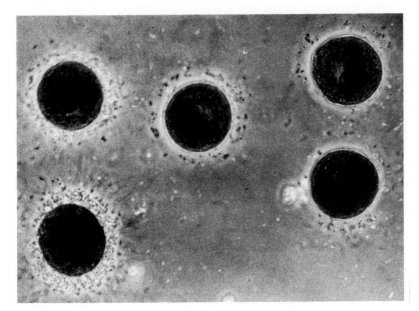

Fig. 2-1 Eggs of the sea urchin (Arbacia) in a suspension of sperm in water. All of the eggs have been fertilized except for the one surrounded by the most sperm (lower left). Approximately × 240. (Jeroboam-Zuzolo, N. Y.)

sperm, and they are the sexual products of males. (Males, by definition, are the sperm-producing sex.) The small gametes of some plants are produced by cells called **pollen.** Pollen grains are not self-propelling, as are sperm, but are specialized for deployment by wind and by animal agents such as insects, birds, and bats. Gametes produced by the female parent are commonly nonmotile and much larger than the sperm. Called eggs, or **ova** (sing. **ovum**), they usually contain large amounts of stored food.

SEXUAL REPRODUCTION IS USUALLY ACCOMPANIED BY THE FUSION OF TWO GAMETES, CALLED EGG AND SPERM. THIS PROCESS OF FERTILIZATION RESULTS IN A ZYGOTE.

Every zygote, whether it is a newly fertilized frog egg, robin egg, or spider egg or the fertilized ovum of a coconut palm or pine tree, contains the adult form in a *potential* state. The word "potential" is used advisedly because it is now understood that the adult organism is *not* contained in miniature in the zygote—a theory popularized long ago by a school of embryologists called preformationists. The perceptions of these scientists were so affected by their preconceptions that they actually convinced themselves they could see little men, or *homunculi,* in the heads of sperm or in eggs. In the seventeenth and eighteenth centuries, this school waged a bitter struggle with an opposing group called the epigeneticists. The latter group believed that the zygote is the *undifferentiated* potential organism and that the egg undergoes a process of *development,* or **morphogenesis** in present-day terminology.

As is often the case in scientific controversies, neither side was altogether incorrect. Of course, no little man (or woman) resides in the sperm as the spermists thought or in the egg as ovists believed. Nevertheless, the organism does exist in the zygote in a symbolic form, just as an airplane exists in a symbolic form in the blueprints and plans that are used to construct it. All the symbols, or coded information, necessary to make and maintain the complete organism in all its stages of development are present. A great deal more will be said about these coded instructions in Chap. 4.

Variation among Generations

To return to the questions posed above: (1) How do organisms produce offspring which resemble themselves? (2) How is the variation characteristic of this process controlled? The twentieth century has seen the answers to these questions gradually assume explicit form. Nevertheless, the ghosts of long-dead theories of inheritance still linger in our language. The most tenacious of these theories is the "paint-pot," or "blood-mixing," theory. The phrases "blood line," "pure blood," and "half blood," to mention a few, stem from the age-old concept that the blood of parents contains the inheritance (the word *"heredity"* was born by analogy from the word "inheritance"). It was thought that the two parental bloods mix together in the formation of offspring. This theory of inheritance holds that offspring will be intermediate between their parents, like a blend of different-colored paints. Such a theory tends to be supported by many commonplace observations. In some plants, such as four-o'clocks, the crossing of a white-flowered individual with a red-flowered individual yields pink offspring. Children are often intermediate in coloring and size between their parents.

The paint-pot theory, however, has a serious flaw. If offspring are always intermediate between their parents, then there should be a gradual diminution of variation, and after a few generations, a state of homogeneity should be reached. In fact, no such diminution of variability occurs. A cross between two pink four-o'clocks yields white, pink, and red-flowered offspring (Fig. 2-2). Rather than behaving as pigments irretrievably lost in a mixture on a painter's palette, the hereditary determinants behave more like a limited number of discrete particles capable of being segregated out in various combinations. The apparent mixing is reversible.

THE INFORMATION NECESSARY FOR THE DEVELOPMENT AND MAIN-TENANCE OF AN ORGANISM IS FOUND IN THE ZYGOTE IN THE FORM

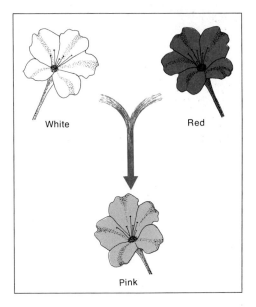

White

Red

Pink

Fig. 2-2 A cross between white and red four-o'clocks produces pink offspring. However, crosses between two pink-flowered plants always yields white, pink, and red offspring.

OF HEREDITARY DETERMINANTS WHICH ACT MORE OR LESS AS INDIVIDUAL UNITS WHOSE BEHAVIOR MAY BE FOLLOWED FROM GENERATION TO GENERATION.

THE LAWS OF HEREDITY

The first significant steps in resolving the question of heredity were made in the garden of an Augustinian monastery near Brünn, Austria (now Brno, Czechoslovakia), in 1856. There a Moravian monk named Gregor Mendel (1822–1884) began experiments which, when they were rediscovered early in this century, were to become the most revolutionary contribution to our understanding of genetics. Mendel, by the way, was never able to achieve public certification as a teacher because, it seems, of a constitutional inability to take examinations.

A combination of luck, foresight, and educational background contributed to the success of Mendel's experiments. Mendel chose, as his experimental material, the garden pea *(Pisum sativum),* a plant that exhibits an especially clear-cut inheritance for many of its characters. The flowers in the strains he used were either white or red (actually violet-red) and were either clustered at the tip of the vine or spread out along the stalk. The seeds were either smooth or wrinkled and either green or yellow. The seed pods were either inflated or constricted and either green or yellow. Finally, the plant was either tall or dwarfed.

Apparently the pea plant was a studied choice. He must have known that **quantitative** characters, that is, characters that vary gradually along a continuum, such as weight or height in man, would be much more difficult to analyze than would **qualitative** characters, that is, characters whose states are clearly distinguishable. (Mendel had studied mathematical physics in Vienna and was therefore in a position to appreciate the importance of such distinctions.)

The Monohybrid Cross

Mendel's first experiments are summarized below. In a cross between plants differing in one character, the offspring showed the character state of only one of the parents. The first generation of offspring in such a genetic experiment is conventionally labeled the F_1 (first filial, from the Latin *filius,* meaning "son") generation. The second step in these experiments was the crossing of the F_1 plants with each other, thereby producing the F_2 generation. The result, as you might have anticipated, was the reappearance of the white-flowered plants that had disappeared in the F_1 generation, plus many red-flowered individuals. In the language of the day, some of these second-generation offspring were "throwbacks" to one of the grandparents. For the sake of brevity, two terms that describe this phenomenon will be introduced. If a character—flower color, for example—in an F_1 generation always resembles the same parent (in a cross between two individuals from different strains that always breed true), the state of the character that is always expressed in the F_1 is called **dominant.** The character state never expressed in the F_1 generation of such a cross is called **recessive.** (These terms will be defined more precisely below.)

Mendel's actual data for his experiment on flower color were as follows:

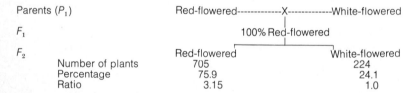

Parents (P_1)	Red-flowered-------------X-------------White-flowered	
F_1	100% Red-flowered	
F_2	Red-flowered	White-flowered
Number of plants	705	224
Percentage	75.9	24.1
Ratio	3.15	1.0

In the F_2 generation, then, about one-fourth of the plants resembled the original parent with the recessive character state (white flowers).

To summarize: (1) the F_1 plants resembled only one parent, but (2) the recessive character was not lost or swamped, as it appeared again in the F_2 plants. From these elementary observations, Mendel was able to solve the riddle of genetics and thereby formulate the principles on which the science still rests. These principles are basically probabilistic— so if you do not remember about elementary coin-flipping probabilities (Box 1–1), you might wish to refer back to it now.

IN THE F_1 GENERATION OF A CROSS BETWEEN TWO INDIVIDUALS DIFFERING IN SOME TRAIT, THE OFFSPRING USUALLY RESEMBLE ONLY ONE OF THE PARENTS. IN THE F_2 GENERATION, ONE-FOURTH OF THE OFFSPRING RESEMBLE THE ORIGINAL PARENT (P_1) WITH THE RECESSIVE CHARACTER AND THREE-FOURTHS RESEMBLE THE ORIGINAL PARENT WITH THE DOMINANT CHARACTER.

If you are puzzled about how coin flipping relates to Mendel's 3:1 ratio, ask yourself the following question: In what percentage of flips of two honest coins is one or more heads expected to appear? What is the ratio of the event of at least one head appearing to the event of no heads appearing? Since two heads will appear one-fourth of the time and one head will appear one-half of the time, at least one head will appear in three-fourths of the cases. The ratio of at least one head appearing to no heads appearing is therefore 3:1 (75:25 percent).

What this has been building up to is that heredity of flower color (as well as of the other characters Mendel studied) in the common pea mathematically fits a model of flipping two coins simultaneously. Mendel assumed that each F_1 plant had *two* kinds of color-determining factors, a red one and a white one. Each parent was, in a sense, flipping a coin to "decide" which kind of factor to give each gamete. Further, *if each F_1 parent contributed one factor, the factors should be distributed among the offspring as shown in Fig. 2–3.*

TWO OF MENDEL'S FIRST CONCLUSIONS WERE THAT:
1 AN INDIVIDUAL PLANT HAS TWO CHARACTER-DETERMINING FACTORS FOR EACH CHARACTER; ONE IS DERIVED FROM EACH PARENT.
2 THE NATURE OF EACH OF THESE FACTORS MAY NOT BE DETECTABLE IN AN INDIVIDUAL WITHOUT BREEDING EXPERIMENTS. FOR EXAMPLE, THE FACTOR THAT IS EXPRESSED IN AN F_1 PLANT IS CALLED DOMINANT. THE OTHER, NOT OBVIOUS IN THE F_1, IS CALLED RECESSIVE.

Testing the Hypothesis: The Backcross

How is it possible to test these hypotheses? (One essential criterion for a good hypothesis is that it leads to testable predictions.) According to hypothesis 1, the red-flowered plants in the F_1 generation are thought to have two flower-color determinants, one from each parent (one red one and one white one). Suppose that one of these F_1 plants is crossed with the white-flowered parent. What is the predicted outcome according to the laws of probability?

If, indeed, the passing on of character-determining factors is analogous to tossing coins, the probability of the F_1 individual contributing a red-

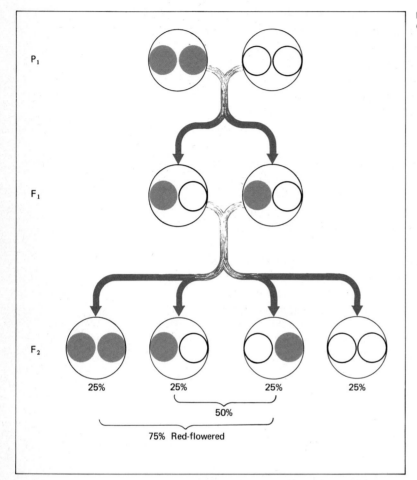

Fig. 2-3 Behavior of color-determining factors in two generations.

flower factor is ½, which is the same as the probability of passing on a white-flower factor. What is the probability of the original white-flowered parent passing on a white-flower factor? It is 1, since both of its hypothetical factors for this character are white. All the flower-color factors in the gametes of the white-flowered parents will be white-flower factors. Like the two coins, these two plants, the F_1 parent and the white-flowered parent, are behaving independently. This means that the contribution of one plant does not affect the contribution of the other.

Knowing these probabilities, one can predict the probability of getting a red-flowered plant from a cross between a red F_1 and the white parent. Since red is dominant, a plant with either both factors red or only one factor red will have red flowers. But since one factor comes from each parent according to hypothesis 1 and since the white-flowered parent has *no* red-flower factors, the only kind of red-flowered offspring in this new cross will be the mixed kind, with a red factor from the F_1 red-flowered plant and a white factor from the white-flowered parent. So the probability of observing a red-flowered offspring in the cross between red offspring and white parent is equal to the product of the probability of the F_1 plant contributing a red factor (½) and the probability of the white-flowered parent contributing a white factor (1), or $½ \times 1 = ½$. All the other offspring, 50 percent, will be white-flowered.

The hypotheses predict, then, that red-flowered and white-flowered plants will occur in equal proportions from this cross of an F_1 with a parent—a **backcross,** as it is called. Mendel performed just this experiment. Out of 166 plants from the backcross, 85 were red-flowered and 81 were white-flowered—pretty close. Statistical procedures exist whereby it is possible to test whether such experimental results are unexpectedly different from the predicted values. And in this case, they are not. Continued experimental verification of the predictions of a hypothesis eventually elevates the hypothesis to the position of a theory or even a law. In fact, those and others of Mendel's hypotheses are now called Mendel's laws.

OBSERVATIONS OF THE PATTERNS OF INHERITANCE OF TRAITS FROM GENERATION TO GENERATION LEAD TO THE HYPOTHESIS THAT EACH PLANT HAS TWO FACTORS DETERMINING THE STATE OF A CHARACTER—ONE FROM EACH PARENT. THIS HYPOTHESIS PREDICTS THAT THE BACKCROSS OF AN F_1 INDIVIDUAL WITH THE RECESSIVE PARENT WILL PRODUCE EQUAL PROPORTIONS OF THE PARENT-GENERATION TYPES. THIS PREDICTION IS FULFILLED BY THE DATA.

Genes and Alleles

At this point, some biological terminology must be introduced. What Mendel called a "factor"—for example, the flower-color factor—is now called a **gene.** The different forms of factors affecting the same trait are called **alleles** of a gene. We say that, in garden peas, there is a *gene* influencing flower color. This gene has two alternative states, or *alleles*. In organisms such as pea plants, two alleles of a gene are found in an individual. If an individual has two identical alleles, as would occur if each of its parents contributed an allele determining red flowers, that individual is said to be **homozygous.** A pea plant which received a white-flower allele from both parents would also be homozygous. Thus an individual can be homozygous for either the red-flower allele or the white-flower allele. If a plant received a red-flower allele from one parent and a white-flower allele from the other, that individual is called **heterozygous.**

*GENES EXIST IN ALTERNATIVE STATES CALLED ALLELES. AN INDI-
VIDUAL CARRIES TWO ALLELES OF EACH GENE. IF THESE ARE IDEN-
TICAL, THE INDIVIDUAL IS HOMOZYGOUS; IF DIFFERENT, THE INDI-
VIDUAL IS HETEROZYGOUS.*

If you remember Mendel's experiments and convert his terminology to
the modern usage, you will see that a heterozygous pea plant has the same
flower color as a plant that is homozygous for the red-flower allele. The
red-flower allele is dominant over the white-flower allele when the alleles
are combined in a heterozygous plant. The white-flower allele is recessive, and white-flowered plants are those which are homozygous for this
allele. When dominance occurs, it clearly is not always possible to predict which alleles are present in an individual merely by looking at it.
A red-flowered plant may have two identical alleles or two different
ones. Since such situations are very common, words are needed to distinguish the genetic constitution of an individual from the appearance of
the individual. The term **genotype** is used to mean the genetic constitution of an organism. The appearance of the organism with respect to
a particular trait is called its **phenotype.** In garden peas, the red-flower
phenotype can be the expression of either of two different genotypes.
There is only one genotype which produces the white-flower phenotype—
the homozygous recessive genotype.

*THE GENOTYPE OF AN ORGANISM IS ITS GENETIC CONSTITUTION.
THE PHENOTYPE OF AN ORGANISM IS ITS APPEARANCE—THE EX-
PRESSION OF THE GENOTYPE IN THE COURSE OF DEVELOPMENT.*

The Dihybrid Cross

In the course of his experiments, Mendel studied a number of different
characteristics of peas. All the traits he studied followed the pattern of
inheritance shown by flower color. Each trait obeyed the predictions
of his hypothesis. Mendel then asked the question: What are the patterns
of inheritance when considering *two different* traits in the same individuals? In peas, the seeds may be either round or wrinkled in shape and
their color may be either yellow or green. Mendel designed experiments
in which these two different seed characteristics could be followed at
the same time. A cross designed to study the inheritance of two characteristics at a time is called a **dihybrid cross.**

To simplify the description of what Mendel found, we will symbolize a
gene affecting a particular trait by a letter referring to the phenotype
produced by the dominant allele. For example, the gene affecting seed
shape will be symbolized by *R*—round being the expression of the dominant allele. Wrinkled is the phenotype of individuals homozygous for

the recessive allele, symbolized by the lowercase *r*. Therefore, wrinkled seeds have the genotype *rr* and round seeds have either the *RR* or the *Rr* genotype.

Similarly, we may symbolize the seed-color gene as *Y*. The genotypes *YY* and *Yy* produce yellow seeds. The genotype *yy* results in a green-seeded phenotype. According to Mendel's hypothesis, every pea plant will have a pair of alleles *for each gene*.

IN GENETIC SYMBOLISM, THE INITIAL LETTER OF A WORD REFERRING TO THE PHENOTYPE IS NORMALLY USED TO REPRESENT THE GENE DETERMINING THE TRAIT. IF UPPERCASE, IT REFERS TO THE DOMINANT ALLELE; IF LOWERCASE, TO THE RECESSIVE ALLELE.

A plant that is homozygous dominant for both these genes is symbolized by *RRYY*. Its seeds will be round and yellow. Plants with the genotype *rryy* will have the phenotype wrinkled and green. Mendel asked what would be the genotype and phenotype of the offspring of a cross between these two kinds of plants. Recall that each parent contributes one allele of each gene to each gamete and therefore to the zygote of each of its offspring. Further, according to the coin-flipping model, the distribution of alleles among the gametes is such that each allele has a 50 percent chance of occurring in any gamete. Since, in this case, both plants have identical alleles for each gene, the gametes of the round-yellow parent will always be *RY* and those of the wrinkled-green parent will always be *ry*. The F_1 offspring, as a result, will all be heterozygous for both genes, *RrYy* (Fig. 2–4). Their phenotypes will all be the same also; they will all have round yellow seeds.

What will be the genotypes and phenotypes of the F_2 generation obtained by crossing two of the F_1 individuals? Again, the first step is to determine the kinds and the probabilities of the gametes. Given that the genotype of both parents is *RrYy*, the kinds of gametes will be *RY, Ry, rY,* and *ry*. (In counting genotypes for the purpose of predicting genotypic and phenotypic ratios, it is conventional to distinguish between the two equally probable ways that a heterozygote can be produced in the following manner: *Aa* signifies that the sperm carried the *A* allele and the egg the *a* allele; *aA* signifies the converse.) What are the probabilities associated with each of these kinds? Since each allele has a probability of $1/2$ of appearing in a given gamete, the probability that two such independent alleles will appear in the same gamete is the product of their individual probabilities, or $1/4$. This probability applies to all the gametes. You need not take this argument for granted; work it out.

A convenient way of portraying the possible kinds of offspring from such

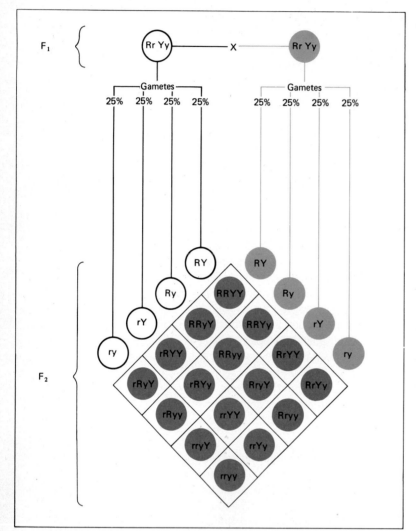

Fig. 2–4 The results of a dihybrid cross between two heterozygous individuals.

a cross is the Punnett square (Fig. 2–4). But the same results can be obtained mathematically. For instance, if we wanted to know the probability of obtaining from this cross an offspring with wrinkled green seeds, we would know that the genotype of such a plant would be doubly recessive for both genes, or *rryy,* and that this genotype results from the union of only one class of gametes—*ry.* Since these *ry* gametes occur among the gametes of each individual with a probability of ¼, the probability of their union is the product of their individual probabilities, or ¹⁄₁₆.

Altogether, there are four phenotypes resulting from the cross *RrYy* and

RrYy. These are round-yellow, round-green, wrinkled-yellow, and wrinkled-green. If you count up the number of genotypes responsible for each phenotype, you will find there are nine for round-yellow, three for round-green, three for wrinkled-yellow, and one for wrinkled-green. This ratio of 9:3:3:1 is characteristic for the F_2 generation of any cross between two individuals which are heterozygous for alleles of genes affecting two different traits.

From results of this kind, Mendel realized that genes for different characters in his experiments behaved independently of one another (Fig. 2–5). That is, the allele for round seeds, for instance, is not found in plants with yellow seeds any more frequently than would be expected by chance. This is why we were able to use the law of independent probabilities to compute the expected frequencies of the various kinds of offspring. Mendel's demonstration of the apparent independence of the different genes when being parceled out among the gametes is called *the law of independent assortment.* Actually, this turns out to be the simplest case of a rather complex phenomenon, and Mendel was quite fortunate in his choice of characters. Genes affecting different characters are not necessarily independent of one another in inheritance.

IN MENDEL'S EXPERIMENTS, THE F_2 PROGENY IN A DIHYBRID CROSS SHOW PHENOTYPES IN A 9:3:3:1 RATIO, DEMONSTRATING THAT TWO GENES MAY BEHAVE INDEPENDENTLY FROM GENERATION TO GENERATION. NOT ALL GENES ARE INDEPENDENT, HOWEVER.

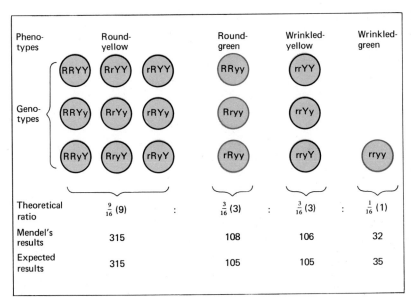

Fig. 2–5 Origin of the 9:3:3:1 ratio in a dihybrid cross.

Phenotypes	Round-yellow			Round-green	Wrinkled-yellow	Wrinkled-green
	RRYY	RrYY	rRYY	RRyy	rrYY	
Genotypes	RRYy	RrYy	rRYy	Rryy	rrYy	
	RRyY	RryY	rRyY	rRyy	rryY	rryy
Theoretical ratio	$\frac{9}{16}$ (9)	:		$\frac{3}{16}$ (3) :	$\frac{3}{16}$ (3) :	$\frac{1}{16}$ (1)
Mendel's results	315			108	106	32
Expected results	315			105	105	35

We have seen that Mendel deliberately chose to study the inheritance of traits which appeared in one of two alternative states. In other words, the gene controlling a trait could exist in one or another allelic state and an individual could be homozygous dominant, heterozygous, or homozygous recessive. In the study of a great many different traits in a wide variety of organisms, it has been found that many different allelic states of a gene may occur. An individual, then, may have any two alleles from this assortment. Flower color in many plants, coat color in cats and dogs, and blood groups in man are examples of characteristics that are controlled by genes with **multiple alleles.**

No individual can have more than a pair of alleles for any gene, except for certain special cases not discussed here, but this does not mean that in a population there can be no more than two alleles. Suppose we were accustomed to wearing shoes that were either different or of the same kind on our right and left feet. No individual could wear more than one pair of shoes (the same or different). At a large party you would expect there to be many different kinds of shoes. In a large population of organisms you might expect there to be many kinds of alleles for a particular gene.

MANY, IF NOT MOST, GENES HAVE MANY ALLELES, AND THE PHENO-TYPE OF AN INDIVIDUAL DEPENDS UPON WHICH TWO ALLELES HE HAS RECEIVED FROM HIS PARENTS. NO INDIVIDUAL ORGANISM CAN HAVE MORE THAN TWO ALLELES OF ANY ONE GENE, BUT IN A POPU-LATION A VARIETY OF ALLELES MAY BE FOUND.

Let us look at the system of the A-B-O blood group. You have probably had your blood type determined for one reason or another, and you know that blood type is important in blood transfusions. If incompatible donor blood is transfused, it may cause red blood cells to form clumps, and this may lead to death. The basis for this reaction lies in certain compounds known as antigens and antibodies that are present in the red blood cells and the blood plasma. We shall discuss them in greater detail in Chap.11. Here it is enough to say that many animals, including man, can detect foreign proteins and a few other substances that may get into their blood. These are **antigens,** and to neutralize or eliminate them, the body makes specific **antibodies,** which combine with the antigen molecules and generally make little clumps. Each antibody is specific for one antigen; it will generally ignore other antigens, even those that are very similar in shape. Red blood cells containing a particular antigen on their surfaces will clump, or **agglutinate,** in plasma containing antibodies directed against that antigen. Your plasma contains antibodies for whatever anti-

Multiple Alleles

gens your own red blood cells *do not contain;* otherwise, your red blood cells would agglutinate.

Each of us has a distinctive set of antigens on the surfaces of our red blood cells; each antigen is really a short chain of various sugar molecules. These are characterized by the antibodies they elicit, and there are many important types, such as Rhesus (Rh) and M-N-S, which we will not consider; let us concentrate on the best-known series: A, B, AB, and O. There are two types of antigen chains on the red blood cells, A and B. The corresponding plasma antibodies directed against them are anti-A and anti-B. Cells containing A antigens will agglutinate in plasma containing anti-A antibodies, and the same is true for B antigens and anti-B antibodies. Cells from a person with type-A blood contain only A antigens; those of type-B blood, B antigens; those with type-AB blood, *both* A and B antigens; and those with type-O blood, *no* A or B antigens. Therefore, you can see that the plasma antibodies of these four types would be, respectively, anti-B, anti-A, no A or B antibodies, and both anti-A and anti-B.

The best medical procedure would be to transfuse blood from a donor who has exactly the same blood type as the recipient. It is possible, however, to use blood from a donor of another type if the plasma antibodies of the recipient are compatible with the red-blood-cell antigens of the donor. Type-O blood has no cellular antigens. Therefore, its red blood cells will not be agglutinated by A, B, or AB plasma. The amount of plasma involved in a typical transfusion, and therefore the amount of anti-A and anti-B antibodies, will not be able to agglutinate the cells of the recipient significantly. A person with type-O blood can donate blood to anyone. He cannot receive blood from just anyone, however. A type-O person has antibodies to both A and B cells, and his antibodies will clump red blood cells from A, B, and AB types. A type-O person can receive blood only from another type O.

Following the same line of reasoning, you can see that AB individuals can receive blood from any type but can donate only to AB persons. Persons with A type can receive from O or A and donate to A and AB. Type-B individuals can receive from B or O and donate to B and AB.

Determination of the A or B antigen is controlled by one gene, for which there are three alleles. These alleles are I^A, I^B, and i. The allele i is recessive to both I^A and I^B; neither of the latter is dominant over the other. An ii individual is type O. The genotype $I^A I^B$ produces both A and B antigens, and the person is therefore type AB. An A individual may have the genotypes $I^A I^A$ or $I^A i$. Try to figure out the possible genotypes for type B.

Let us examine the consequences of mating among the different blood types. A man with AB-type blood produces two kinds of sperm with respect to this gene: I^A and I^B. He could never have a type-O child, who must receive an i from each parent. Could a child of type B be produced by a man and a woman of A and O genotypes, respectively? Although evidence as to blood-group type is sometimes used in paternity suits, it is important to realize that the best it can do is say that a man with a given blood type could *not* have been the father.

THE MAIN BLOOD-GROUP TYPES IN MAN ARE CONTROLLED BY THREE ALLELES OF ONE GENE, WHICH DETERMINE THE NATURE OF THE ANTIGENS IN THE BLOOD PLASMA AND RED BLOOD CELLS.

Quantitative Inheritance

The characteristics that we have discussed up to this point are those which occur in one of several alternative states. A flower may be red or white or possibly pink, but there is no difficulty in deciding which category it belongs in. A person's blood type may be A, B, AB, or O but not something in between one of these four types. Of course, you are aware that not all characteristics of organisms vary qualitatively, that is, in discrete steps. Indeed, most of the traits in which we have a practical interest vary quantitatively, or continuously, from one extreme to another. Height and weight are good examples. Many persons had attempted to study patterns of inheritance before Mendel. Often, however, they attempted to deal with continuously varying traits and were unsuccessful in determining general laws. The celebrated British mathematician Galton was distinguished for his mathematical discoveries. When he turned to studies of heredity, however, he did not get very far because he worked with such things as size and intelligence in man, traits that vary continuously.

Now that we understand how qualitative inheritance occurs, we can understand how quantitatively varying traits must be inherited. Studies of such traits in economically important plants and animals have given a wealth of data. It now seems clear that for most traits, such as size, weight, butterfat content, egg size, length of ear in corn, *many different genes* are involved. Studies of quantitative inheritance are based upon statistical analysis of the parents and their progeny and not upon the counting of individuals with several distinct phenotypes. Usually, it is not possible to specify exactly how many different genes affect a particular characteristic. It is often assumed, for purposes of discussion, that each of many genes contributes a specific increment to the trait—in other words, their effects are additive. There are no known examples, however, of purely additive control of continuously varying traits.

In order to make clear how quantitative inheritance involving many genes with additive effect might work, let's use a hypothetical example: height in some plant. Let us suppose there is a set of 72 different genes affecting height. Any one gene may have an *H* allele or an *h* allele. Each *H* allele contributes 1 in. to the total height. Each *h* contributes ½ in. to height. Therefore, for any one gene, *HH* means 2 in. added, *Hh* means 1½ in. added, and *hh* means a 1-in. increment of height; with 72 genes, the height of an individual plant might theoretically vary from 72 to 144 in. Every plant's height would fall either at an even inch or half-inch mark on the total scale. A plant could be 144 in. tall or 141½ in. tall but not 140¼ in. tall. There are 145 different phenotypes possible: 72, 72½, 73, . . . , 143, 143½, 144. If alleles *H* and *h* were about equally common in a population, a frequency diagram of the heights of a great many plants might look something like Fig. 2–6.

In fact, when heights are measured in the plants, we will find a smooth curve (Fig. 2–7) if enough are sampled. This is because with *many* genes affecting *one* trait, the effects of any one gene will be small. Moreover, height is affected not only by genotype but also by environmental factors, such as nutrition. An individual might have a genotype that would permit it to achieve 100 in. With poor nutrition, it might grow to only 97¾ in. Thus an environmental effect caused a change of 2¼ in. in height, whereas a change from *H* to *h* in one gene caused an effect of ½ in. For simplicity, we have omitted discussion of other factors affecting quantitative inheritance, such as the interaction of different genes.

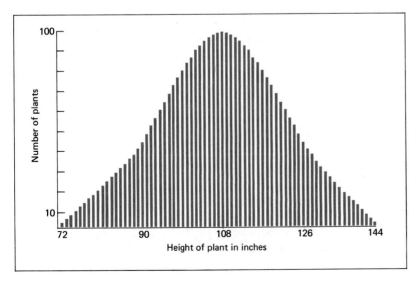

Fig. 2–6 Frequency diagram showing possible distribution of height in a hypothetical plant population.

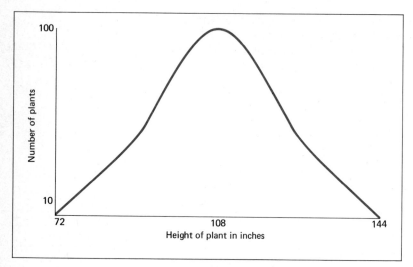

CONTINUOUSLY VARYING TRAITS OF ORGANISMS ARE CONTROLLED BY MANY DIFFERENT GENES, EACH WITH A RELATIVELY SMALL EFFECT. ENVIRONMENTAL EFFECTS SMOOTH OUT THE CURVE.

CYTOGENETIC CORRELATIONS

Mendel's discoveries, as often happens in intellectual history, were ahead of his time. His scientific contemporaries were unable to appreciate the elegance and significance of his mathematical approach to the study of inheritance. It was not until the turn of the century, long after Mendel's death and after biologists had observed the behavior of the rodlike objects called chromosomes, that his work was understood and appreciated. Then, however, the astonishing correlation between the actual behavior of chromosomes in the cells of reproductive tissues and the predicted behavior of hypothetical genes, pointed out by Sutton in 1902, led to a rapid growth of a new field of biology: cytogenetics. Cytogenetics is the combination of genetics, the study of heredity, and cytology, the study of cells.

The Cell

Before going on with our study of reproduction, let's consider briefly some of the properties of cells. Every cell has some sort of boundary separating it from other cells or from the environment. This outer boundary consists of layers of protein, fats, and carbohydrates. In plants the outer layer, which may be composed of additional molecules of various carbohydrates, is called the cell wall. Most of the kinds of plants you are familiar with have a cell wall made up of cellulose. Multicellular plants have intracellular cement (called the middle lamella), which holds the cells together. Plant cells in a multicellular body are commonly interconnected by fine strands called **plasmodesmata.** In both plant and animal cells, the living material inside the cell boundary is called the **protoplast.**

THE PROTOPLAST The protoplast itself is bounded by a thin membrane called a plasma membrane or cell membrane. (The plasma membrane lies inside the cell wall of plant cells.) We will return later to the important properties of this membrane, which, among other things, regulates the flow of materials in and out of the cell. Making up the protoplast are a variety of structures of varying form and function. These are called **organelles,** and we will discuss each of them in greater detail later.

THE NUCLEUS The largest and most conspicuous organelle of all eukaryotic cells is the **nucleus.** It is a roughly spherical organelle which may lie in the center of the protoplast or near the cell boundary in specialized cells. By use of the electron microscope, it can be seen that the nucleus is bounded by a **nuclear envelope,** which consists of two membranes (Fig. 2–8). Each of these membranes is similar to the plasma

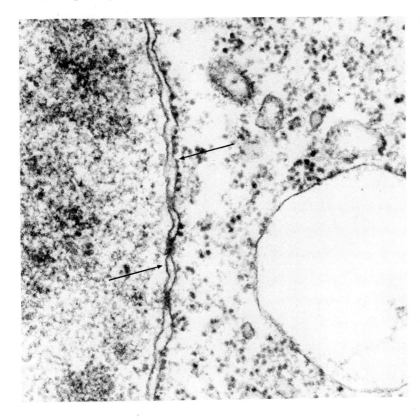

Fig. 2–8 Electron micrograph of a portion of the nuclear envelope in the protozoan *Vorticella.* × 74,000 (*Courtesy of Richard Allen.*)

membrane. The nuclear envelope ordinarily has a regular pattern of minute pores, which connect the inside of the nucleus with the remainder of the protoplast, which is called the **cytoplasm.** These pores can be seen only with the electron microscope.

In sections of plant or animal cells prepared for the light microscope, one to several smaller spherical structures can be seen within the nucleus. These are called **nucleoli,** and they play a role in the synthetic activity of the cell. Aside from the nucleoli, there are often strands of darkly staining material, collectively called **chromatin.** Chromatin may also be granular in appearance.

THE CYTOPLASM   In cells that are not dividing, the region outside the nuclear envelope appears to be much more complex than the nucleus.

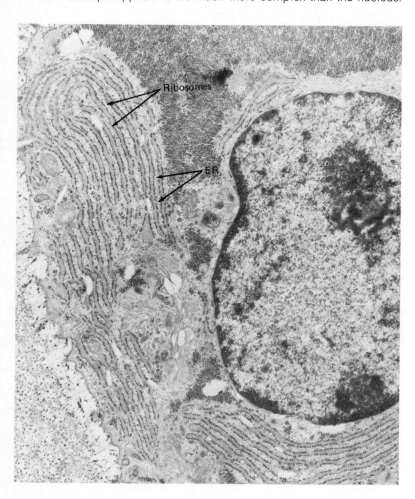

Fig. 2–9 Electron micrograph showing endoplasmic reticulum in *an osteoclast;* the membranes are densely covered with ribosomes. × 13,400. (*Courtesy of the Laboratory for Cell Biology of the Biological Laboratories, Harvard University.*)

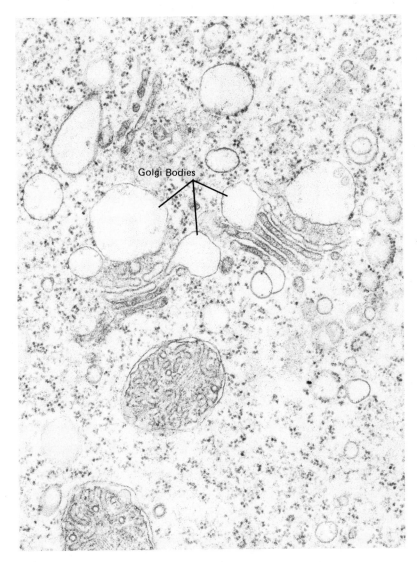

Golgi Bodies

Fig. 2-10 Electron micrograph of two or three Golgi bodies in the protozoan *Vorticella;* the Golgi bodies are associated with large vesicles in the center of the figure. × 46,000. *(Courtesy of Richard Allen.)*

There are a variety of organelles of different form and function. Here we shall mention only those found in most cells of higher organisms. The cytoplasm can be seen with the electron microscope to be permeated by a system of interconnected tubes and vesicles which vary in form with the state of activity of the cell. Collectively, they make up the **endoplasmic reticulum** (Fig. 2-9). Some of the membranes of the endoplasmic reticulum are covered with small granules called **ribosomes.** Ribosomes are the sites of protein synthesis. They may also occur in the cytoplasm between the membranes of the endoplasmic reticulum.

A specialized kind of cellular membrane system is called the **Golgi body** or **dictyosome** (Fig. 2–10). Again from the use of the electron microscope, we know that dictyosomes are made of flattened vesicles that do not have associated ribosomes. Cells may have few or many dictyosomes, depending upon their synthetic activity. Dictyosomes apparently serve as collecting and packaging organelles for the products of protein synthesis. In many cells, the edges of the dictyosome vesicles pinch off enzyme-containing globules, which are subsequently secreted.

Cells of plants and animals commonly contain various organelles enclosing specialized enzymes, which are thus partitioned off from the rest of the cell. Animal cells, for example, often have **lysosomes,** which contain

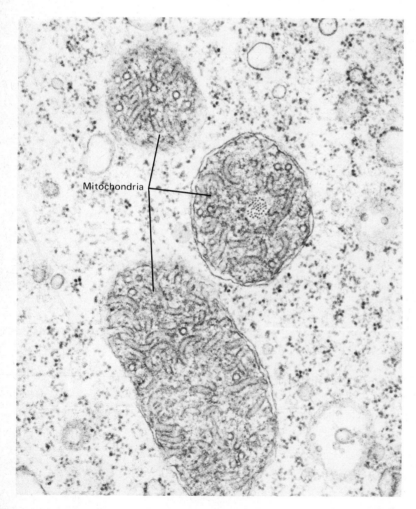

Fig. 2–11 Electron micrograph of mitochondria in the protozoan *Vorticella;* protozoan mitochondria have tubular cristae. × 50,000. (*Courtesy of Richard Allen.*)

Mitochondria

digestive enzymes. Upon the death of a cell, these break open and, by their "self-destruct" effects, facilitate the cleanup and removal of dead tissue. When a carcass is tenderized by allowing it to hang, the meat is being partially digested by its own lysosomes.

All eukaryotic cells have organelles called **mitochondria** scattered throughout the cytoplasm among the tubes of the endoplasmic reticulum. These are the sites of energy mobilization in the cell. Typically, each mitochondrion is a rod-shaped structure with a smooth outer membrane (Fig. 2–11, also Fig. 4–34), within which is a second membrane, of much greater surface area, which is folded into a series of flattened plates, or cristae. These may be likened to the printed circuits of a transistor radio. That is, they are constructed in such a way that electrons move about them along specific pathways. The cells of eukaryotic green plants also contain another, related organelle whose function is energy trapping and mobilization. These are the **chloroplasts** (Fig. 2–12), which have internal vesicles arranged in a complex pattern of layers. These vesicles contain the green pigments called chlorophylls, as well as other pigments.

Specialized cells may contain other kinds of organelles and secreted or stored materials of many different sorts. Crystals, starch grains, oil droplets, and protein particles are examples of such materials and will be discussed in a later chapter.

CELLS OF PLANTS AND ANIMALS ARE NOT MERE BAGS OF COMPOUNDS BUT ARE HIGHLY STRUCTURED, CONTAINING MANY DIFFERENT KINDS OF ORGANELLES SPECIALIZED FOR VARIOUS FUNCTIONS.

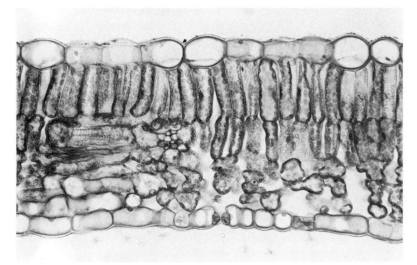

Fig. 2–12 Light micrograph of a cross section of a leaf showing the numerous circular chloroplasts in the cells of the leaf's interior. (*Lester V. Bergman and Assoc., Inc.*)

*REPRODUCTION OF CELLS MUST ENSURE THAT ALL THESE STRUC-
TURES ARE DUPLICATED SO THEY REMAIN FUNCTIONAL.*

A cell that is not dividing is said to be in interphase. It is primarily during this period in its life that synthetic activities are going on and energy is being mobilized and transformed in various ways. Such a cell may become highly specialized for some particular function, or it may divide into daughter cells. The division of the nucleus is a process called **mitosis,** and it is closely followed by division of the cytoplasm.

Division of the Cell

MITOSIS   Cells that are about to divide show characteristic changes in the nucleus. What was originally visible as irregular strands of chromatin now begins to assume definite form. The course of mitosis is continuous, but it is customary to divide it arbitrarily into four stages (Fig. 2–13). The first stage is known as **prophase.** In early prophase, strands of chromatin become thicker and thicker, until the nucleus appears to be filled with threads of darkly staining material. In fact, it can be seen that the apparent thickening of the filaments is the result of their being coiled up like springs (Fig. 2–14). Each of the filaments is known as a **chromosome.** In an organism which has a small number of chromosomes, one can often see by the end of prophase that the chromosomes differ in size and shape. Also it becomes clear that there are two chromosomes of each type in each cell (Fig. 2–15); these pairs of chromosomes, are called **homologous** chromosomes. An organism may, for example, have five pairs of homologous chromosomes, or ten chromosomes in all.

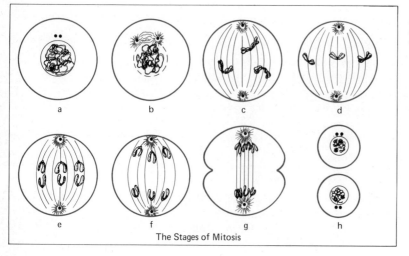

The Stages of Mitosis

Fig. 2–13   Stages of mitosis; *a, b,* and *c,* prophase; *d,* metaphase; *e* and *f,* anaphase; *g,* telophase; *h,* daughter cells.

While the chromosomes are undergoing shortening by coiling, other changes are going on in the cell. The nucleolus and the nuclear envelope disappear. In the cytoplasm, fiberlike microtubules form and become arranged in the shape of a spindle (Fig. 2–13) extending from one end of the cell to the other. These microtubules are called **spindle fibers,** and the ends of the spindle are its **poles.** The middle, or widest part, of the spindle is the equatorial plane. Animal cells characteristically have a pompon of fibers, called **asters,** at each pole of the spindle. Cells of higher plants generally do not have asters.

Prophase ends with the breakdown of the nuclear envelope. At this time it can be seen that each chromosome is actually double (Fig. 2–13), consisting of two longitudinal halves known as **chromatids.** Since they are halves of one chromosome, they are called **sister** chromatids. With the nuclear envelope gone, the chromosomes become attached to spindle fibers, each one at a specific point called the **centromere** of the chromosome, which is single (at least functionally), although the chromosome is double.

Some chromosomes have the centromere at the midpoint of their length. In others, the centromere is toward one end of the chromosome. The lengths of chromosome on either side of the centromere are called the *arms* of the chromosome. When the centromere is at the midpoint, the arms are equal in length and the chromosome may appear V-shaped. Often the arms are of different lengths, and the chromosomes may appear J-shaped or rodlike. Sometimes each chromosome has a shape different from every other except its homologue. In some cells, all the chromosome pairs can be differentiated on the basis of position of the centromere. And, in late prophase, each chromosome of these pairs consists of two sister chromatids. As you can see in Fig. 2–13, the sister chromatids of a chromosome are held together as a pair in the centromere region.

Eventually the chromosomes move, or are pushed about, until they come to lie on the equatorial plane of the spindle. If the arms are very long, only the centromeres may occupy the plane. This stage of mitosis, a short one, is known as **metaphase.**

During the next stage, **anaphase,** the chromatids separate and move away from the equator to opposite poles of the spindle. As soon as they have separated from one another, each chromatid is called a **daughter chromosome.** The exact nature of anaphasic movement is not known, but its results are obvious: a group of daughter chromosomes at each end of the spindle. As the figure shows, the two groups are the same in chromosome number and type and are identical to the chromosome constitution of the parental nucleus.

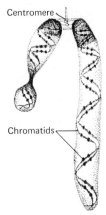

Fig. 2–14 Diagram of a metaphase chromosome showing coiling, chromatids, and centromere. (*After Sharp.*)

Fig. 2–15 Diagram of the set of chromosomes in a cell of the fruit fly *Drosophila melanogaster*. In this insect homologous chromosomes remain close during mitosis. There are four pairs of homologous chromosomes of characteristic size and shape. The chromosomes marked X and Y are discussed later.

The last stage of mitosis, **telophase,** is essentially a reversal of the first. During telophase the spindle fibers disappear. New nuclear envelopes are organized from the endoplasmic reticulum around each group of chromosomes. The chromosomes begin to uncoil and thus lengthen, and the nucleoli reappear. By the end of telophase, there are two daughter nuclei, similar to the parental nucleus in appearance and identical to it chromosomally. Photographs of some of the stages of mitosis in plants and animals are shown in Figs. 2–16 and 2–17.

MITOSIS IS THE DIVISION OF THE NUCLEUS OF A CELL. IT IS ARBI-TRARILY DIVIDED INTO FOUR STAGES: PROPHASE, METAPHASE, ANAPHASE, AND TELOPHASE. IN THE PROCESS, EACH OF THE CHRO-MOSOMES IS DIVIDED INTO EQUAL HALVES, WHICH ARE INCORPO-RATED INTO DAUGHTER NUCLEI, IDENTICAL WITH ONE ANOTHER AND WITH THE PARENTAL NUCLEUS.

DIVISION OF THE CYTOPLASM   Mitosis is a process in which the chromosomes of a parental cell are divided equally between two daughter cells. It is often called **equational** division. Despite the presence of diverse organelles, the division of the cytoplasm is neither as complex or as precise as chromosome division. During telophase in plant cells, a lamella composed of granules produced by the dictyosomes forms in

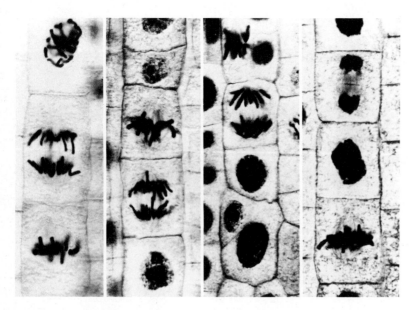

Fig. 2–16 A longitudinal section of an onion (*Allium*) root tip in which several stages of mitosis can be seen. Identify these by comparison with Fig. 2–13. (*Carolina Biological Supply House.*)

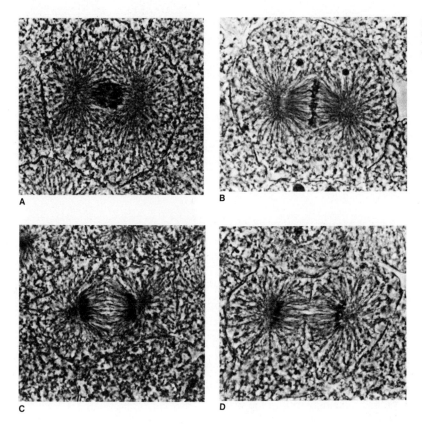

Fig. 2-17 Four stages of mitosis as seen in sections of embryonic cells of the whitefish (*Coregonus*). Identify these by comparison with Fig. 2-13. (*Courtesy of Eric V. Gravé.*)

the equatorial plane (Fig. 2-18). This is called the **cell plate.** Originating in the spindle, it subsequently extends across the cell, dividing it in two. Eventually each of the daughter cells will form a new cell wall against the cell plate, and the latter then becomes the intercellular cement holding them together.

Division of the cytoplasm in animals usually occurs in a very different way. Here, as telophase progresses, the cell begins to pinch in along a circle around the equatorial plane (Fig. 2-19). At first a furrow appears on the surface, and this gradually deepens into a groove. Eventually the connection between the daughter cells is reduced to a slender thread, which soon parts.

In either type of cell division, the various cytoplasmic organelles are apportioned more or less equally between the daughter cells. Some, such as mitochondria and plastids, reproduce themselves. Others, such as ribosomes, are synthesized anew by materials originating in the nucleus.

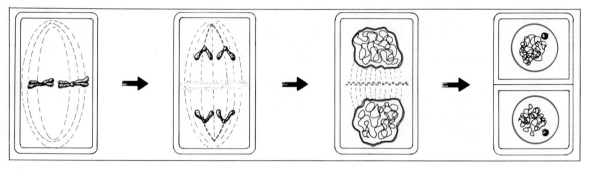

Fig. 2-18 Diagram to show division of the cytoplasm in plants.

Unless the spindle has formed at one end of a large cell, the two daughter cells will be about the same size, have about the same number and kinds of organelles, and have *exactly* the same chromosomal material.

DIVISION OF THE CYTOPLASM USUALLY IS NEITHER AS COMPLEX NOR AS PRECISE AS MITOSIS. MITOSIS RESULTS IN IDENTICAL DAUGHTER NUCLEI. CELL DIVISION USUALLY PRODUCES DAUGHTER CELLS ROUGHLY EQUAL TO THE PARENT CELL IN CYTOPLASMIC CONTENT.

IMPORTANCE OF MITOSIS What is accomplished in mitosis should now be clear. Mitosis provides for the establishment of a group of identical cells. We have said that the nucleus regulates and controls the synthetic activities of the cell. Since mitosis ensures that each daughter cell receives the same chromosomal material as the parental cell contained, you might suspect that the chromosomes play a major role in the control of cellular activities, as indeed they do. Each daughter cell is potentially able to carry out all the activities of the parental cell because it has the same chromosomal material as the parental cell. In the next section, we will consider how the chromosomes function when the cell is in the nondividing condition. First, however, we must discuss another kind of nuclear division, called **meiosis,** which takes place in the formation of gametes or, in some plants, of spores. In other words, meiosis occurs in sexual reproduction of organisms.

We have said that sexual reproduction involves the fusion of gametes to form a zygote and that each cell has a characteristic number of chromosomes present in homologous pairs. Obviously these statements are not compatible. If each gamete contained two sets of chromosomes,

Meiosis

the zygote would contain four sets. If zygotes produced new gametes by mitosis, the fusion of these gametes would produce cells with eight sets. Each generation, the number of chromosomes would double. This problem is resolved by the special kind of nuclear division called meiosis. During the course of meiosis, the chromosomes of the parental cell divide once but the cell itself divides twice. The result is four daughter cells, each with only one set of chromosomes. A cell is **diploid** if there are two chromosomes, a homologous pair, representing each of the chromosomes in the gametic set. Diploid cells thus have two sets of chromosomes. Gametes are **haploid** cells because they contain half the number of chromosomes of a diploid cell. However, this haploid number cannot be made up of any group of chromosomes that add up to half the diploid number. During meiosis, the chromosomes must be divided and apportioned out so that each daughter cell receives a complete set of chromosomes, one member of each homologous pair.

SINCE THE ZYGOTE IS FORMED BY THE FUSION OF TWO GAMETES, IT MUST CONTAIN TWICE THE NUMBER OF CHROMOSOMES AS THE GAMETES. THE TYPE OF NUCLEAR DIVISION WHICH REDUCES THE ZYGOTIC NUMBER TO THE GAMETIC NUMBER IS CALLED MEIOSIS. IT CONSISTS OF TWO DIVISIONS OF THE PARENTAL CELL AND ONLY ONE DIVISION OF ITS CHROMOSOMES. GAMETES HAVE ONE SET OF CHROMOSOMES AND ARE CALLED HAPLOID. THE ZYGOTE AND ITS DERIVATIVES HAVE TWO SETS OF CHROMOSOMES AND ARE CALLED DIPLOID.

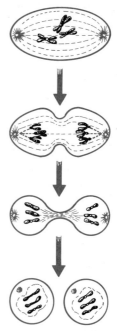

Fig. 2-19 Diagram to show division of the cytoplasm in animals.

If the diploid number of chromosomes is symbolized by $2n$ and if $2n = 10$, then each cell must have five pairs of homologous chromosomes. In meiosis, each gamete will receive one chromosome from each of the pairs of homologues, and this is symbolized by $n = 5$. Obviously, meiosis must be more complex than mitosis. Like the latter, it is customarily divided into stages, but since two divisions of the cell occur, there are eight stages in meiosis. Each of the two divisions of meiosis is divided into the same four stages as mitosis.

The first division of meiosis begins in much the same way as does mitosis. However, before coiling and shortening, homologous chromosomes come together in pairs and become very closely and precisely associated along their length (Fig. 2-20). This **synapsis** results in the formation of what are called **bivalents.** Each bivalent is a pair of homologous chromosomes. During the process of coiling after synapsis, individual chromatids break and rejoin, and it sometimes happens that nonsister chromatids join instead of sister chromatids. Thus the end of a chromatid from one of the pair of homologues may become attached to a chromatid from its

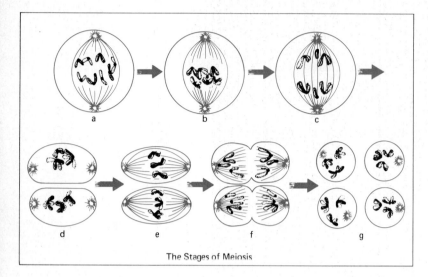

Fig. 2-20 The stages of meiosis: *a,* first prophase; *b,* first metaphase; *c,* first anaphase; *d,* second prophase; *e,* second metaphase; *f,* second telophase; *g,* four daughter cells with haploid number of chromosomes.

The Stages of Meiosis

synaptic partner, and vice versa. This process is called ***crossing-over*** (Fig. 2-20). In crossing-over, homologous chromosomes exchange portions of their chromatids.

As the first meiotic division comes to an end, the paired chromosomes behave as if they were trying to move apart. They are held together at the points of crossing-over. One result of this is the formation of a cross-shaped configuration wherever crossing-over took place. Each cross is called a ***chiasma*** (pl. ***chiasmata***), and in fact, often there are several crossovers, and thus chiasmata, in each bivalent. As the centromeres pull apart, the chiasmata are pulled toward the ends of the chromosome arms.

When the chromosomes are maximally shortened, the nuclear envelope disappears. The bivalents then become arranged on the equatorial plane of the spindle which has formed in the cytoplasm (Fig. 2-20). In a mitotic cell division in an organism with 10 chromosomes, there would be 10 chromosomes on the equator of the spindle. How many bivalents would be found in this organism at the end of a first meiotic prophase? In the stage of meiosis called metaphase, there would, of course, be five bivalents.

In mitosis, at the beginning of anaphase, each chromosome divides into two daughters. A crucial difference in meiosis is that, in the first meiotic anaphase, the homologous chromosomes that synapsed now disengage without dividing. This separation of chromosomes is known as ***disjunction.*** Whole chromosomes now move to opposite poles. When the chromosomes reach the poles, the first meiotic telophase begins and two nuclei are formed, each of which has a haploid set of chromosomes. It is important to note that the chromosomes of each haploid set are *not* the same as those that entered meiosis. This is because crossing-over has taken place earlier. Wherever a crossover occurred, homologous chromosomes have exchanged chromosomal material.

IN THE FIRST DIVISION OF MEIOSIS, THE CHROMOSOMES PAIR TO FORM BIVALENTS CONSISTING OF TWO HOMOLOGUES. HOMOLOGOUS CHROMOSOMES EXPERIENCE ONE OR MORE EXCHANGES OF CHROMATIDS CALLED CROSSING-OVER. AT THE END OF THE FIRST DIVISION, HOMOLOGOUS CHROMOSOMES UNDERGO DISJUNCTION BUT THE CENTROMERES DO NOT SEPARATE.

Let us look more closely at the origin of homologous chromosomes. Since a zygote is formed by the fusion of two haploid gametes, one from each parent, each parent has contributed one set of chromosomes. Each pair of chromosomes, or bivalents, contains one chromosome from the paternal parent and one chromosome from the maternal parent. When crossing-over occurs, parts of chromatids from maternal and paternal chromosomes are exchanged.

When disjunction takes place in anaphase I, the paternal centromere of each bivalent goes to one pole of the spindle and the maternal centromere goes to the opposite pole. Attached to the paternal centromere will be chromatids that are partly paternal and partly maternal, because of crossing-over. The same will be true for the maternal centromere. Thus there has been a recombination of chromosomal material. Furthermore, there is no reason for all the paternal centromeres to go to the same pole and all the maternal centromeres to go to the other. In fact, the probabilities are independent for each centromere of a bivalent. Therefore, the daughter nuclei at the end of telophase I will not only have recombined chromosomes, but each will have a mixture of maternal and paternal centromeres in its haploid set. This is because centromeres *segregate independently* at disjunction.

WHEN DISJUNCTION OF HOMOLOGOUS CHROMOSOMES OCCURS AT THE END OF THE FIRST MEIOTIC DIVISION, EACH BIVALENT BEHAVES INDEPENDENTLY OF THE OTHERS WITH RESPECT TO THE POLE OF THE SPINDLE TO WHICH MATERNAL AND PATERNAL CENTROMERES WILL MOVE. THUS, EVENTUALLY, EACH GAMETE WILL HAVE A FULL SET OF CHROMOSOMES, BUT THIS WILL BE A MIXTURE OF CHROMOSOMES FROM THE TWO PARENTS.

The first division of meiosis is followed in many organisms by division of the cytoplasm. In others it may be delayed until after the second division. The second division usually follows the first rather rapidly, and is similar to a mitotic division. The second division produces four daughter cells, each with a haploid set of chromosomes which are a mixture of maternal and paternal chromosomal material.

IMPORTANCE OF MEIOSIS   Meiosis accomplishes two important things: it reduces the chromosome number from diploid to haploid, and it provides a thorough recombination of paternal and maternal chromosomal material. Because of crossing-over and independent segregation of centromeres, the chances of the four daughter haploid cells having the same chromosomal material are very small, and it is extremely unlikely that zygotes resulting from their fusion will be identical with either parent.

REDUCTION OF THE NUMBER OF CHROMOSOMES FROM THE ZYGOTIC NUMBER TO THE GAMETIC NUMBER WHILE PRESERVING COMPLETE HAPLOID SETS INTACT IS ACCOMPLISHED BY MEIOSIS. EACH HAPLOID SET, HOWEVER, CONTAINS A MIXTURE OF PATERNAL AND MATERNAL CHROMOSOMES, WHICH HAVE EXPERIENCED CROSSING-OVER.

Chromosomal Behavior and the Laws of Heredity

If you have understood the laws of reproduction discussed above and have followed the behavior of chromosomes in meiosis, you should have noticed some striking correlations between the genetics and cytology of organisms. These are diagramed in the following set of figures. Genetic factors, or genes, are present in pairs, as are chromosomes. If we assume that genes are located on chromosomes, then the situation for seed-shape inheritance in Mendel's peas can be diagramed as in Fig. 2–21. Each chromosome is shown as a line, and its centromere is shown as a circle.

Meiosis in each of the homozygous parents produces only one kind of gamete with respect to particular chromosomes. Combination of the two gametes, each carrying a chromosome with a different allele, produces a heterozygous zygote. This zygote is the first cell of the F_1 generation. The F_2 generation is produced by mating two heterozygous F_1 individuals. Meiosis in a heterozygous pea plant results in two kinds of gametes with respect to this chromosome. Half of the gametes that each individual produces have the dominant allele Y and the other half contain the chromosome with the recessive allele y.

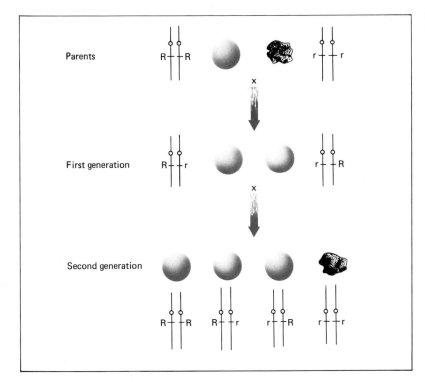

Fig. 2–21 Chromosome behavior in a monohybrid cross.

These gametes can combine in three ways to form the zygotes of the F_2 generation (Fig. 2–21), resulting in the homozygous dominant and recessive types, as well as the heterozygous form. Now it is clear why the recessive phenotype reappears in the F_2 generation after apparently having been lost in the F_1. We are not dealing with Mendel's hypothetical factors, we are dealing with actual structures in the cell—the chromosomes—whose behavior is understood. Try for yourself to work out the behavior of the chromosomes in a backcross of an F_1 individual.

Now let us look at a cross in which two characters are studied. If it is assumed that R and Y are on different chromosomes, then the parental types are as shown in Fig. 2–22. Note that the pairs of homologous chromosomes are distinguished by the position of their centromere. Meiosis in the parental types results in one kind of gamete for each. One parent produces gametes containing chromosomes on which are located Y and R; the other produces y- and r-carrying gametes. Fertilization results in the F_1 heterozygote, which has different alleles on the two pairs of homologous chromosomes.

Meiosis in the F_1 heterozygotes results in four possible gamete types

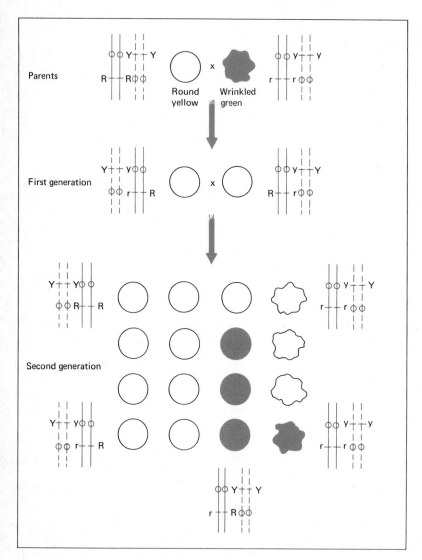

Fig. 2-22 Chromosome behavior in a dihybrid cross.

because the nonhomologous chromosomes behave independently. Upon fertilization, 16 possible genotypes and chromosome combinations can be produced. The phenotypes of the offspring fall into the 9:3:3:1 ratio characteristic of a dihybrid cross. Mendel's law of independent assortment is explained by the independent assortment of nonhomologous chromosomes.

THE BEHAVIOR OF HYPOTHETICAL HEREDITARY DETERMINANTS, OR GENES, AND THAT OF CHROMOSOMES ARE COMPLETELY PARALLEL

IN SEXUAL REPRODUCTION. THIS INDICATES THAT GENES ARE LOCATED ON CHROMOSOMES. PATTERNS OF INHERITANCE CAN BE PREDICTED FROM KNOWLEDGE OF CHROMOSOMAL BEHAVIOR, AND THE MODE OF INHERITANCE OF GENES MAY BE USED TO PREDICT CHROMOSOMAL BEHAVIOR.

Mendel was indeed fortunate that the genes affecting the characters of the phenotype he studied were on different chromosomes. You can see immediately that genes on the same chromosome will be moved about together in meiosis unless a crossover occurs between them. Rather than continue with Mendel's peas, let us consider a hypothetical case of a chromosome with two different genes, *A* and *B*. The location of a gene on a chromosome is called the **locus** (pl. **loci**) of that gene. Therefore, in this discussion we shall consider a chromosome with two loci: one for gene *A* and one for gene *B* (Fig. 2–23) The locus of gene *A* is closer to the centromere. If no crossing-over occurs, *A* and *B* will move together in meiosis; they are *linked*. **Linkage** is the result of two genes being on the same chromosome. Unless the loci are very close together, the chances are that a crossover will periodically occur between them (Fig. 2–24). The frequency of crossing-over between two loci determines the amount of recombination between the two genes. Let us see how this works. Suppose you cross homozygous dominant and homozygous recessive individuals to make a heterozygous F$_1$. You wish to study the inheritance of *A* and *B;* therefore, you must use a kind of cross in which every possible zygote can be identified with respect to its genotype. Such a cross is called a **testcross**. The individual to be studied is crossed with a homozygous recessive individual, which can produce *only one kind of gamete.* An *aabb* organism can produce only gametes containing a chromosome with *a* and *b.*

If *A* and *B* are so tightly linked that no crossing-over occurs between

Linkage and Chromosome Mapping

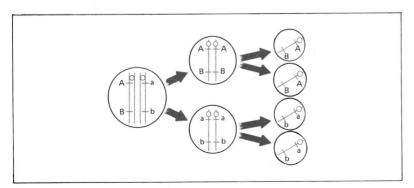

Fig. 2–23 Diagram to show the complete linkage of two genes in meiosis.

them, the testcross should produce the results shown in Fig. 2–25. Off-spring will be identical with the parental types genotypically and pheno-typically. That is, they will be either *AaBb* or *aabb.*

A different result is often found in such a testcross, however. Suppose you make the cross and carefully count the number of offspring that are like their parents in phenotype **(parentals)** and those that show a different combination of traits than is found in either parent **(recombinants)**. The data might look like these:

PARENTS   *AaBb* ×   *aabb*

OFFSPRING   Parental types: *AaBb, aabb* 90%
                          Recombinants: *Aabb, aaBb* 10%

This means that recombination has occurred in 10 percent of the gametes. The only mechanism that can account for this is crossing-over between homologous chromosomes during meiosis. During 90 percent of the meiotic divisions leading to gamete formation, no crossing-over occurred in the region of the chromosome between *A* and *B*. But 10 percent of the time a crossover did occur between these genes.

To see how this works chromosomally, refer to Fig. 2–25. During the prophase of the first division of meiosis, homologous chromosomes synapse and crossing-over takes place. In a certain percentage of bi-valents—in this instance, 10 percent of them—crossovers will occur be-tween *A* and *B*. If a crossover occurs between *A* and *B* in 10 percent of the meiotic divisions that produce gametes, then 10 percent of the ga-metes will have recombinant chromosomes. Conversely, 90 percent of the gametes will contain parental-type chromosomes. As you can see in Fig. 2–25, 45 percent of the gametes will have parental-type *AB* and 45 percent will have type *ab*. Of the recombinant gametes, 5 percent will have *aB* chromosomes and 5 percent will have *Ab* chromosomes. In a testcross with an *aabb* parent that can produce only *ab* chromo-

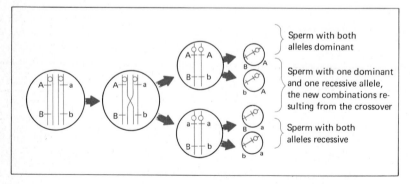

Fig. 2-24 Diagram to show the effects of crossing-over on the linkage of two genes.

Sperm with both
alleles dominant

Sperm with one dominant
and one recessive allele,
the new combinations re-
sulting from the crossover

Sperm with both
alleles recessive

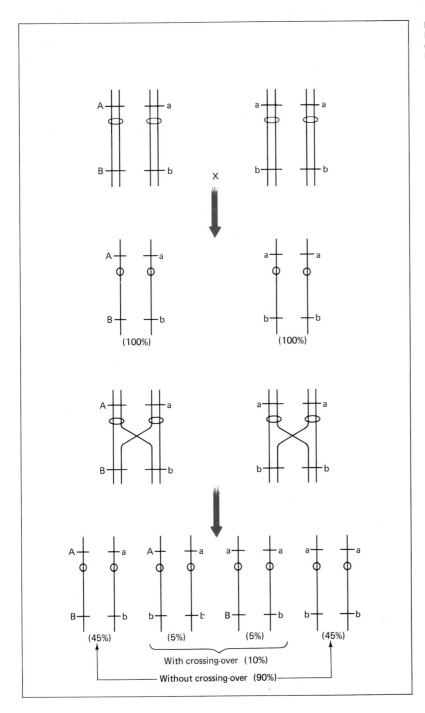

Fig. 2–25 Diagram to show the behavior of chromosomes in a test-cross. Above, without recombination, below, with recombination.

somes, the recombinants will be easily recognized in 10 percent of the offspring.

Ignoring a few complicating factors which need not concern us here, we can say that the frequency of crossing-over between two loci is proportional to their distance apart. Obviously, if *A* is linked with *B* and if *B* is linked with a third locus, *C,* then *A* and *C* are linked (that is, are on the same chromosome). Thus we have a means of making a **genetic map,** which shows the spatial relationships among the loci on a chromosome.

Such a map can be made without even seeing the chromosome. It depends only upon the percentage of recombination between the various loci considered in pairs, since this is proportional to their distance from one another. Suppose you made a series of testcrosses and obtained the following data for genes *A, B,* and *C:*

	Percentage of recombination
A and *B*	10
B and *C*	17

The two possible genetic maps you could construct from these data are shown in Fig. 2–26. To decide between them, you find out the percentage of recombination between *A* and *C.* If you find it to be 27 percent, this would support the first map. What percentage of recombination between *A* and *C* would support the second map?

IF TWO GENES ARE LOCATED ON THE SAME CHROMOSOME, THEY ARE LINKED IN INHERITANCE TO A DEGREE PROPORTIONAL TO THE AMOUNT OF CROSSING-OVER BETWEEN THEM. THE NUMBER OF

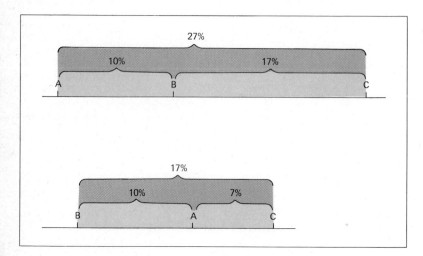

Fig. 2–26 Two possible genetic maps which show the spacing of genes A, B, and C.

CROSSOVERS BETWEEN TWO GENES IS A FUNCTION OF THEIR DISTANCE APART. BY COUNTING THE NUMBER OF RECOMBINANT TYPES IN PLANNED CROSSES, IT IS POSSIBLE TO CONSTRUCT GENETIC MAPS SHOWING THE ORDER OF GENES AND THEIR DISTANCE APART.

Using such techniques as the testcross, as well as other measures, geneticists have prepared genetic maps for many different kinds of plants and animals where large numbers of offspring can be obtained and counted as recombinant or parental types. The genetic map of the fruit fly *(Drosophila)* is shown in Fig. 2–27. In animals such as man, however, it is not possible to make controlled crosses or to obtain large numbers of progeny in a single cross. Genetic maps are therefore very difficult to make. There is one situation, however, in which it is possible at least to localize genes on a particular chromosome of the basic set, and this is discussed in the next section.

Sex-linked Inheritance

The basic chromosome set in man numbers 23 chromosomes. Therefore, a human zygote contains 46 chromosomes, 23 from the egg and 23 from the sperm. In man, as in most other animals and some plants, expression of the sex phenotype is affected in part by special chromosomes called **sex chromosomes.** Modern techniques enable biologists to observe human chromosomes in great detail. In a preparation of chromosomes, you might see something like Fig. 2–28. After the chromosomes have been photographed, they can be cut out and matched in pairs according to their size and shape. As you can see in Fig. 2–28, a human female has 23 pairs of chromosomes in each cell.

A similar preparation made for a male cell, however, gives a different result (Fig. 2–29). All the chromosomes can be paired except for two. By careful matching, it can be seen that there are 22 pairs which are the same as those of the female. The twenty-third couple, however, includes one chromosome of a kind present in the female as a pair and one chromosome which does not occur in the female at all. These are the sex chromosomes. The odd one, occurring only in males, is called the Y chromosome. Its partner, which differs from it in size and shape, is an X chromosome. Therefore, we can refer to the sex-chromosome constitution of a male human as XY and of a female as XX.

Sex chromosomes have been found in most animals and a few plants. In most instances, the exact mode of operation of sex chromosomes is not understood. Clearly they are not the only determinants of sex. Other chromosomes also play an important role in determining the sex phenotype of organisms.

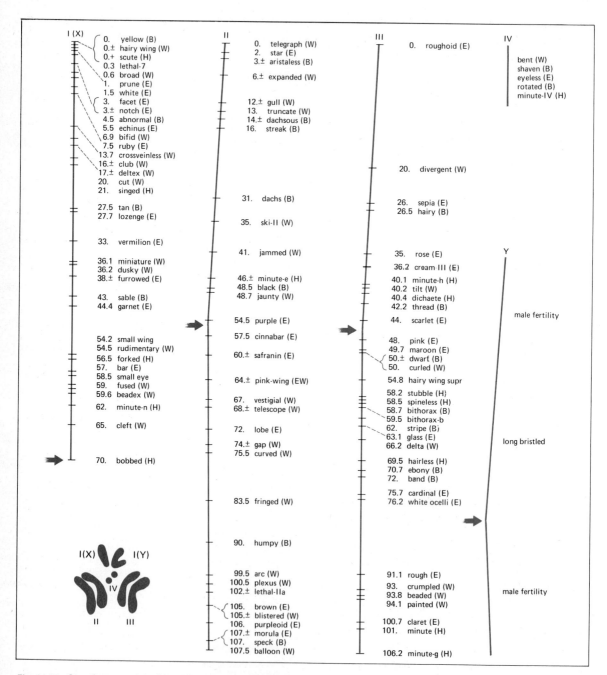

Fig. 2-27 Genetic maps of the four chromosomes of the fruit fly *Drosophila melanogaster* (*X* and *Y* chromosomes are discussed later). The names refer to the phenotypes of individuals homozygous for the recessive alleles of each gene shown. Thus, on chromosome IV, "bent" refers to wing phenotype, "shaven" refers to the hairiness of the body, "eyeless" refers to the eyes, "rotated" refers to the body, and "minute-IV" refers to hairs. The arrows show the position of the centromeres. (*Adopted from Relis R. Brown, GENERAL BIOLOGY. Copyright McGraw-Hill Book Company, 1970. Used by permission.*)

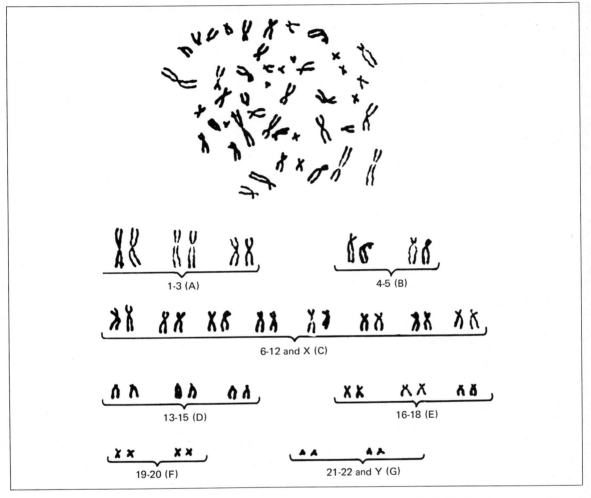

Fig. 2-28 The chromosomes of a normal human female. Above, as seen under the microscope; below, numbered and arranged in groups (indicated by capital letters) according to size and shape. (*Courtesy of Park S. Gerald, M.D.*)

As a result of some cytological accident, rare individuals are produced who have a different sex-chromosome constitution. For example, there are individuals who lack a Y chromosome and yet have only one X (XO). Such persons have many abnormalities of development and usually are mentally retarded. Although they are not able to reproduce, they are phenotypically female. Occasionally persons with an XXY constitution are found. They also show many developmental abnormalities and retarded intelligence, but they are phenotypically male. Thus it appears that the Y chromosome is necessary for maleness. Specific genetic loci have not been identified on the Y chromosome. Because of the behavior of the X chromosome in meiosis and fertilization, it is relatively easy to identify which genes are located on it.

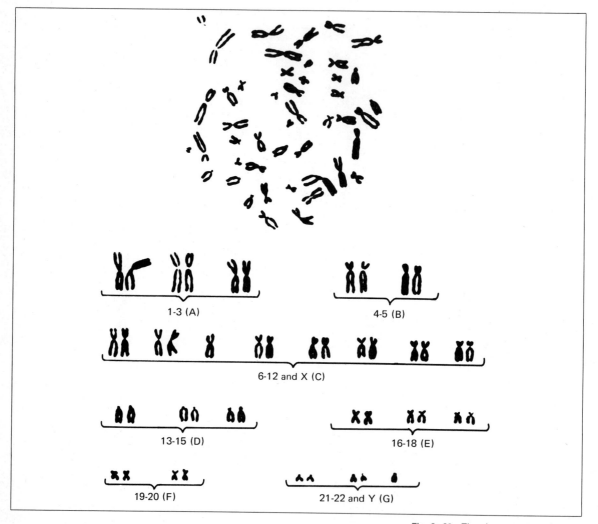

1-3 (A)

4-5 (B)

6-12 and X (C)

13-15 (D)

16-18 (E)

19-20 (F)

21-22 and Y (G)

A female human has two X chromosomes, and therefore after meiosis all her eggs will contain an X. Males, however, have an X and a Y. Meiosis will result in half of the sperm containing X chromosomes and half of the sperm containing Y chromosomes. An X-containing sperm fusing with an egg (which will always contain an X) will produce a zygote that is potentially a female human. A Y-containing sperm will produce a zygote that is potentially male. You can see, therefore, that a male contributes his X chromosome only to his daughters, while his Y goes to his sons. Figure 2–30 shows the same pattern of chromosome behavior in a fruit fly.

Fig. 2–29 The chromosomes of a normal human male. Above, as seen under the microscope; below numbered and arranged in groups (indicated by capital letters) according to size and shape. (*Jeroboam, N.Y.*)

Fig. 2–30 Behavior of the *X* and *Y* chromosomes in the fruit fly *Drosophila melanogaster.* Follow the hypothetical alleles *R* and *r* (red eyes versus white eyes) from generation to generation. (Females indicated by ♀ males, by ♂)

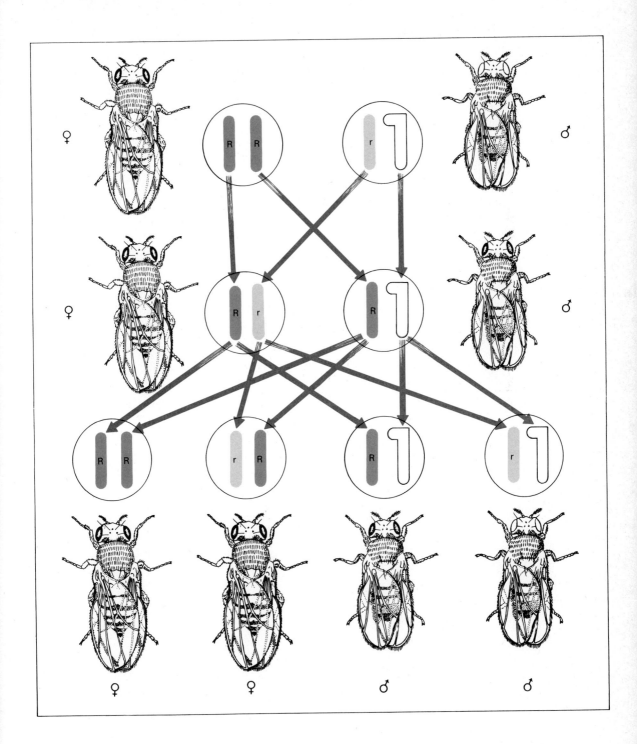

This leads to a peculiar pattern of inheritance of traits controlled by the genes on the X chromosome. This pattern of inheritance is called **sex-linked** inheritance. The inheritance of red-green color blindness is a case in point. Look at the pedigree shown in Fig. 2–31, in which circles represent females and squares males. The pattern of inheritance is the same as the behavior of the X chromosome. The gene for color blindness is symbolized by the letter *C.* *C* is dominant and ensures normal vision. The homozygous recessive, *cc,* has red-green color blindness. Since this gene is on the X chromosome, a female may have the genotype *CC, Cc,* or *cc.* Males, however, can be only *C(Y)* or *c*(Y). The latter genotype, as well as *cc,* results in the individual having difficulty in distinguishing certain shades of red and green.

Now that you understand the basis of inheritance, fill in the genotypes in the pedigree shown in Fig. 2–31. (HINT: Consider the males first.)

THE BEHAVIOR OF THE X CHROMOSOME IN SEXUAL REPRODUCTION IN MAN LEADS TO A PATTERN OF INHERITANCE, FOR GENES ON THIS CHROMOSOME, CALLED SEX-LINKED INHERITANCE. MALES TRANS-MIT THEIR X CHROMOSOME ONLY TO THEIR DAUGHTERS, THEIR Y GOING TO THEIR SONS. SINCE MALES HAVE ONLY ONE X CHROMO-SOME, RECESSIVE ALLELES ON THE X CHROMOSOME WILL ALWAYS BE EXPRESSED IN MEN (SINCE THERE CAN BE NO DOMINANT ALLELE TO AFFECT THEIR EXPRESSION).

There are other traits that show this pattern of sex-linked inheritance and whose genes are therefore located on the X chromosome. An interesting

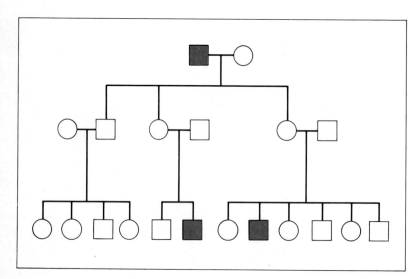

Fig. 2–31 A three-generation pedigree showing the inheritance of color blindness. (Circles represent females and squares represent males.) Assuming that *C* and *c* are the alleles and that *cc* individuals have red-green color blindness, try to deduce the genotype of each individual in the pedigree. [*Adapted by permission of the publisher, from Synder/David: THE PRINCIPLES OF HEREDITY, 5/e (Lexington, Mass.: D.C. Heath and Company, 1957).*]

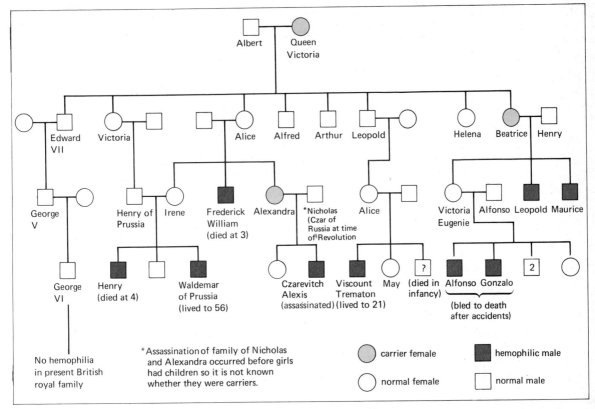

No hemophilia in present British royal family

*Assassination of family of Nicholas and Alexandra occurred before girls had children so it is not known whether they were carriers.

◯ carrier female (shaded) carrier female

◯ normal female

■ hemophilic male

☐ normal male

Fig. 2-32 A pedigree showing the occurrence of hemophilia in the royal families of Europe.

example of such a trait is hemophilia, a disease in which the blood fails to clot properly. A relatively small cut can be a serious matter for a male with the recessive allele. In the royal families of Europe, the spread of this allele after its origin, probably with Queen Victoria, can easily be traced (Fig. 2-32). Hemophilia is certainly one of the most historically important of all diseases. It afflicted the great-grandson of Victoria, Alexis, the son of Tsar Nicholas II of Russia. The horror of the disease attacking their only son strongly affected the behavior of the Tsar and, especially, his wife Alexandra. It was a principal element in the close and fatal relationship between Rasputin and the royal family. A very convincing case has been made for the argument that if Alexis had not had hemophilia, the Russian Revolution would not have occurred.

Importance of Cytogenetic Correlations

Detailed studies of patterns of inheritance and the behavior of chromosomes have established beyond a doubt that the genes controlling the phenotype of an organism are located on the chromosomes. All eukaryotic organisms show basically the same patterns of inheritance. Prokaryotic organisms, as we shall discuss in greater detail later, have different kinds of chromosomes, and their heredity differs in certain details. Nothing that has been found in prokaryotic organisms, however, disagrees in principle with what is known of eukaryotic forms. Indeed, studies of prokaryotic organisms and of the biochemistry of inheritance have given new insights into genetics at the molecular level.

You have now been introduced to the basic laws of genetics and to the cellular mechanisms which make these laws operate. One large area has been left vague, however. We have not discussed exactly what a gene or genetic information is, how it functions, and how new genes arise. We have not discussed the biochemical basis of heredity. The greatest triumph of modern biology has been to find answers to these questions. It is now known that genes are sequences of subunits of molecules of deoxyribonucleic acid (DNA). We know that encoded in the sequence of subunits is the information necessary to control the synthesis of proteins, including those that function as **enzymes** (biochemical catalysts). This process of protein synthesis is now fairly well understood, although the ways in which protein production is regulated are still being elucidated. It is clear, however, that the production of enzymes is the critical step in putting genetic information to work. Enzymes catalyze the myriad reactions that result in development, and they also mediate the replication of the genes themselves.

We now also know the broad outlines of how DNA is reproduced and passed from cell to cell and from generation to generation. New genes are understood to be created by accidental changes in the sequence of DNA subunits during the reproduction of the DNA. This process of *mutation* together with recombination, which was discussed above, is a fundamental source of genetic variability.

Chapter 4 will take up these biochemical processes in the context of the primary function of genes as the ultimate control of biological syntheses.

GENES ARE SUBUNITS OF MOLECULES OF DEOXYRIBONUCLEIC ACID. IN THE ARRANGEMENT OF THESE SUBUNITS IS CODED THE INFORMATION NECESSARY TO SYNTHESIZE PROTEIN MOLECULES. GENES CONTROL PROTEIN SYNTHESIS, INCLUDING THE SYNTHESIS OF ENZYMES, WHICH ARE NECESSARY FOR THE CHEMICAL FUNCTIONING OF CELLS AND ORGANISMS.

Reproduction is a characteristic of all living systems. It is necessary not only for the replacement of individuals lost for one reason or another but also for the production of new kinds of individuals. The environment **SUMMARY** is always in a state of change. If the progeny of organisms are variable, there may be some that are better fit than their parents.

Variability is largely a product of sexual reproduction, the fusion of two gametes, called egg and sperm, in the process of fertilization which results in a zygote. The information necessary for the development and maintenance of an organism is found in the zygote in the form of hereditary determinants which act more or less as individual units whose behavior may be followed from generation to generation.

In the F_1 generation of a cross between two parents differing in some trait, the offspring usually resemble one parent. In the F_2 generation, one-fourth of the offspring resemble the original parent with the recessive character state and three-fourths resemble the original parent with the dominant character.

Alternative states of hereditary factors, or genes, are called alleles. Each individual carries two alleles of each gene. If these are identical, the individual is said to be homozygous; if different, the individual is heterozygous. The term "genotype" refers to the genetic constitution of an organism. The phenotype of an organism is its appearance—the expression of the genotype in the course of development. Many, if not most, genes have many alleles, and the phenotype of an individual depends upon which pair of alleles he has received from his parents. Individual organisms do not ordinarily possess more than two alleles of any one gene, but a variety of alleles may be found in a population of individuals. Many important traits of organisms vary continuously, rather than in discrete steps. These are controlled by many different genes, each with a relatively small additive effect.

The cells of which higher plants and animals are composed are highly structured, containing many different kinds of organelles specialized for various functions. Reproduction of cells must ensure that all these structures are duplicated in such a way as to maintain cell functioning. The nucleus divides by a process called mitosis. In the process, each of the chromosomes is divided into equal longitudinal halves, which are incorporated into daughter nuclei. The daughter nuclei are identical with one another and with the parental nucleus.

Division of the cytoplasm usually is neither as complex nor as precise as mitosis. Cell division usually produces daughter cells roughly equal in cytoplasmic content.

Since the zygote is formed by the fusion of two gametes, it contains twice as many chromosomes as are in each gamete. The type of nuclear division which reduces the zygotic number of chromosomes to the gametic number is called meiosis. Meiosis consists of two divisions of the parental cell and only one division of its chromosomes. Not only is the zygotic number reduced by half, but each gametic set contains a mixture of paternal and maternal chromosomes which have experienced crossing-over. The set of chromosomes found in the gamete is called a haploid set—usually the minimum group of chromosomes, all different, required for the functioning of a cell. The zygote and its derivatives have two sets of chromosomes and are called diploid.

The behavior of hypothetical hereditary determinants, or genes, and that of chromosomes are completely parallel during sexual reproduction. This indicates that genes are located on chromosomes. Patterns of inheritance can be predicted from knowledge of chromosomal behavior, and the mode of inheritance of genes can be used to predict chromosomal behavior. If two genes are located on the same chromosome, they are linked in inheritance to a degree proportional to the amount of crossing-over between them. The number of crossovers between two genes is a function of their distance apart. By counting the number of recombinant types in planned crosses, it is possible to construct genetic maps showing the order of genes and their distance apart.

SUPPLEMENTARY READINGS
Books

Fawcett, D. W.: "An Atlas of Fine Structure," Saunders, Belmont, Ca., 1967.
Kennedy, D.: "The Living Cell," Readings from the *Scientific American,* W. H. Freeman and Company, San Francisco, 1965.
Lerner, I. M.: "Heredity, Evolution and Society," Freeman, San Francisco, 1968.
Loewy, A. G., and P. Siekevitz: Cell Structure and Function, 2d ed. Holt, New York, 1969.
Peters, J. A. (ed.): "Classic Papers in Genetics," Prentice-Hall, Englewood Cliffs, N.J. 1959. *Includes Mendel's paper.*
Srb, A. M., R. D. Owens, and R. S. Edgar: "General Genetics," Freeman, San Francisco, 1965.

2

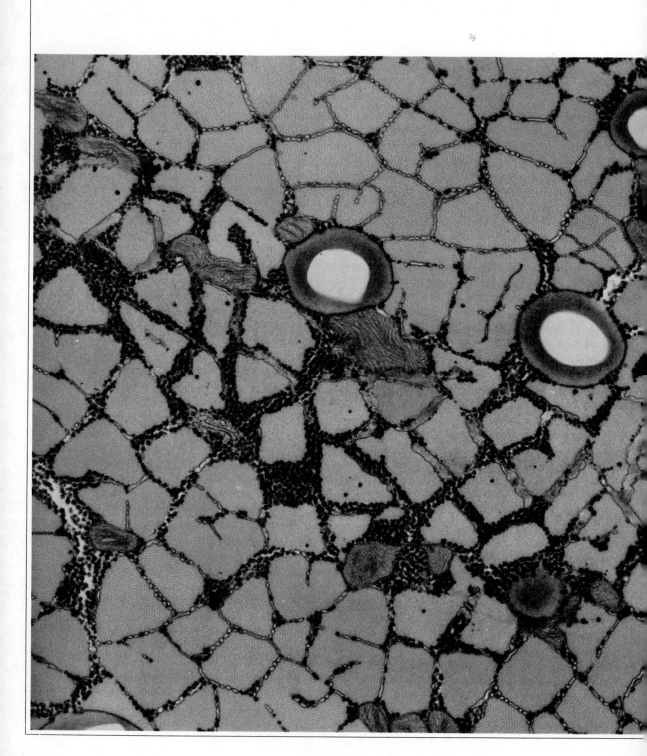

THE
COMPONENT SYSTEMS
OF ORGANISMS

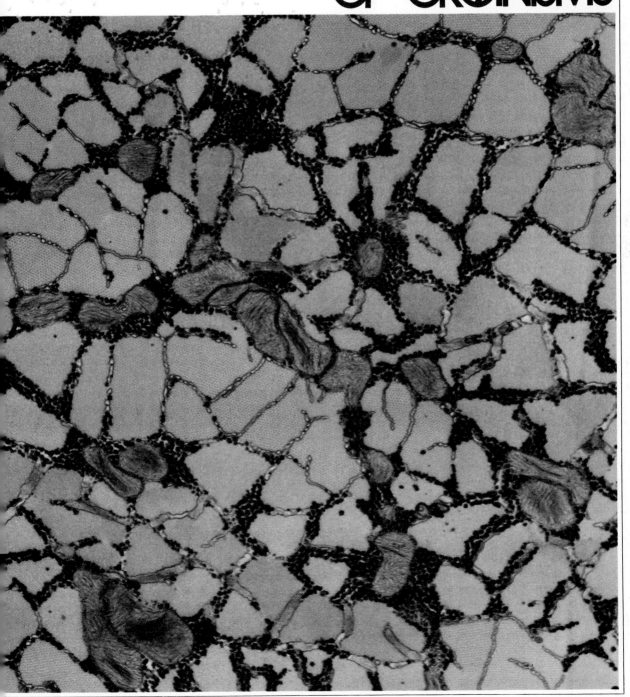

This part dissects the individual organism out of the web of life and examines its functioning over a short time span. How is it organized? How does it use environmental materials to make more of itself? How does it keep from disintegrating? How do its diverse parts manage to function as a harmonious whole?

Chapter 3 is an overall survey of the kinds of systems that make up organisms, and these are compared with man-made machines. In Chap. 4, the sources of energy and matter and their use in synthesis are discussed.

Chapters 5 and 6 discuss the utilization of matter by organisms and how the by-products of this use are dealt with. Homeostasis is a central theme.

Integration and behavior in plants and animals are the subjects of Chap. 7 to 9. Chemical and nervous coordination are considered, together with effectors.

[Illustration on preceeding pages:
Transverse section of Frog muscle.
(*Courtesy of David Soifer, from Omikron.*)]

ORGANISMS: SYSTEMS AND PROCESSES

In the previous chapters, the broad outlines of the distribution, diversity, genetics, and evolution of organisms were introduced. It is now possible to begin asking more penetrating questions about the phenomenon of life. Often, in discussing complex situations, it is convenient to construct hypothetical models of how they work. In this chapter, we will describe organisms as if they were black boxes; that is, organisms will be considered as objects whose internal workings are a mystery. These objects will be compared with complex machines to see if parallels can be drawn. The economy—the input and output of matter and energy—of organisms will be described, and certain obvious aspects of plant and animal behavior and function will be discussed. From these analyses, some conclusions about what goes on inside the black boxes will be drawn, and the operations of certain kinds of systems will be inferred.

Organisms and Machines

We will begin by making an analogy between organisms and machines. This is not because organisms are really machines in disguise but because any similarities between machines and organisms may suggest common underlying mechanisms or principles. On the other hand, differences between the functioning of machines and that of organisms will reveal the unique characteristics of organisms. Such an analysis-by-comparison is called an argument by ***analogy***. Analogies are often very useful because they help in the formulation of meaningful questions and in the detection of distinguishing characteristics.

MAKING ANALOGIES BETWEEN ORGANISMS AND MACHINES MAY LEAD ONE TO THE RECOGNITION OF SIGNIFICANT QUESTIONS ABOUT HOW ORGANISMS FUNCTION.

A fair analogy between organisms and machines should involve a machine of a complex nature. It would hardly be illustrative, for instance, to compare a hammer, a very simple machine, with an object as complex as a horse. A better analogy is that of a bird with an airplane. Both are heavier-than-air craft with a number of "behavior patterns" in common, and the modern airplane has reached a high degree of technological complexity.

At this point, we assume that the reader knows nothing about the internal workings of birds—the bird is still a black box, as it were. Therefore, a logical place to begin the analysis is with something we do know about, the input and the output of birds and airplanes. Almost everyone has seen food disappear into the mouth of a bird and has observed the bird's output of material. Similarly, most people have seen planes being fueled and have seen or heard their exhaust.

ENERGETICS

Fuel, Energy, and the Airplane

The input of a plane is matter in the form of fuel (gasoline or kerosene) and air, and energy in the form of radio signals (electromagnetic energy). The fuel is made up largely of the two elements, carbon and hydrogen— a **hydrocarbon.** (If you feel you need a review of basic chemical terms see Box 3-1. If you have an elementary understanding of this discussion, we suggest you read this chapter through once before reviewing the Box.) The primary component of the air actually consumed (in reality, changed) is oxygen (although other gases, such as nitrogen, may be affected). The balance of the atmospheric gases, mostly nitrogen, are not used as fuel. As shown in Fig. 3-1, the output of planes is composed of matter in the form of gases and various forms of **energy.**

A basic rule of physics, the first law of thermodynamics, states that energy can neither be created nor destroyed, although it may take many different forms (Box 4-1). The total amount of energy on the output side of a process equals the total amount on the input side. Since the input of the airplane includes only matter in the form of hydrocarbon and oxygen (practically speaking), it must be assumed that some of this matter contains energy in a stored, or **potential,** state. It must further be assumed that this potential energy is liberated inside the plane's engine when the oxygen is combined with the hydrocarbon. Actually, it is the arrangement of atoms in the hydrocarbon that accounts for the potential energy of this material. When mixed with oxygen at high temperatures, the hydrocarbon burns, that is, undergoes the process of **oxidation.** The result is that the potential energy is released as heat and as the energy of motion **(kinetic energy)** of the expanding gases. The energy of motion of the expanding gases moves the pistons and, in turn, the propeller of a conventional aircraft. In a jet, these gases are allowed to escape out of the rear of the plane, thus driving the craft forward. According to Newton's second

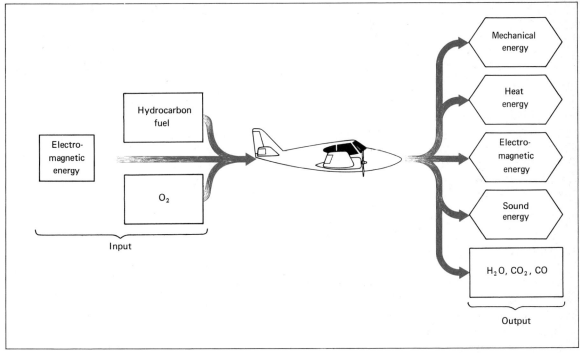

Fig. 3–1 The inputs and outputs of an airplane.

principle, for every action (gases exhausted behind) there is an equal and opposite reaction (plane moves forward).

THE POTENTIAL ENERGY OF THE HYDROCARBON INPUT OF AN AIRPLANE IS RELEASED AS HEAT AND MECHANICAL ENERGY WHEN THE HYDROCARBON IS COMBINED WITH OXYGEN IN THE ENGINE.

But this does not account for all the mechanical energy of the expanding gases or of the moving pistons. A flow-sheet representation (Fig. 3–2) is the most convenient way of expressing the ultimate fate of all the kinetic energy of the engine. The engine of an aircraft is essentially a machine within a machine. It is an energy converter that converts the potential chemical energy of the fuel to kinetic energy. In addition, heat and noise, also kinetic-energy forms, are produced by this process.

The mechanical energy of the moving pistons is transferred to the propeller, where work is done on the air. This leads to the principal output of the airplane: kinetic energy in the form of high-speed movement of the airplane and molecules of the atmosphere. (Large amounts of air are pushed downward and backward by the passage of an airplane.) The energy necessary to do all this work presumably was present in the

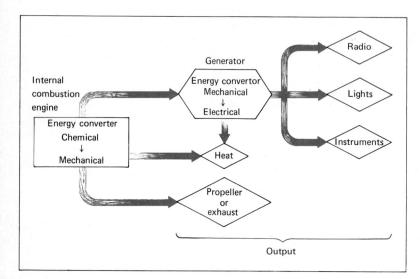

Fig. 3-2 Flow-sheet representation of energy transformations of an airplane engine.

inputs—fuel and oxygen. The energy of the pistons is also transferred to a second-order energy converter called a generator. This is another machine within a machine; it transforms the mechanical energy of the moving parts of the engine to electrical energy (the kinetic energy of moving electrons). In turn, the electrical energy is used directly to power instruments, radios, lights, smaller motors, switches, and other apparatus. All of the latter can be thought of as third-order energy converters, or machines that convert electrical energy into light energy, radio-wave energy, or back to mechanical energy (in the case of switches and the needles on instruments).

MANY DIFFERENT TRANSFORMATIONS OF ENERGY OCCUR IN AN AIRPLANE: POTENTIAL CHEMICAL ENERGY TO HEAT AND MECHANICAL ENERGY, MECHANICAL ENERGY TO ELECTRICAL ENERGY, ELECTRICAL ENERGY TO LIGHT ENERGY, ETC.

The airplane is not typical of many machines in that its output does not include complex material products. A newspaper is the product of a machine called a printing press. A photograph is the product of a machine called a camera—a machine that uses light to trigger a chemical reaction.

Turning our attention to organisms, it can be seen in Fig. 3-3 that the matter-energy flow sheet of an animal is indeed analogous to that of a plane. On the input side of an animal's economy are various forms of

Fuel, Energy, and the Organism

energy. For instance, at this moment you, an animal, have an energy input. Light energy is being received by your eyes as you read this, and sound energy is being detected by your ears, even if your attention is not immediately directed to it. Other items of your input today will include other organisms, water, and oxygen. Are the organisms that are part of an animal's intake its fuel in the same sense that hydrocarbons are a plane's fuel? This is a commonsense question that requires no scientific sophistication to answer.

Recall that many animals live in caves, in the great depths of oceans, and in other dark places. It would seem from this that light energy is not a necessary item of input. And certainly there is no reason to suspect that animals use sound, radio waves, or other forms of electromagnetic energy as a source of energy to drive their life processes. Therefore, for such animals, the bodies of plants and animals, by the above process of elimination, are obviously the fuel. Of course, the bodies of plants and animals are not nearly as simple in structure as are the hydrocarbon fuels of internal-combustion engines. Nevertheless, carbon does form the backbone of most of the molecules in the plant and animal bodies. Furthermore, oxygen is definitely a requirement, and oxygenated carbon (carbon dioxide: CO_2) is a large part of the output and accounts for most of the carbon that enters on the input side of the economy. This would

Fig. 3-3 Inputs and outputs of an animal shown as a black box.

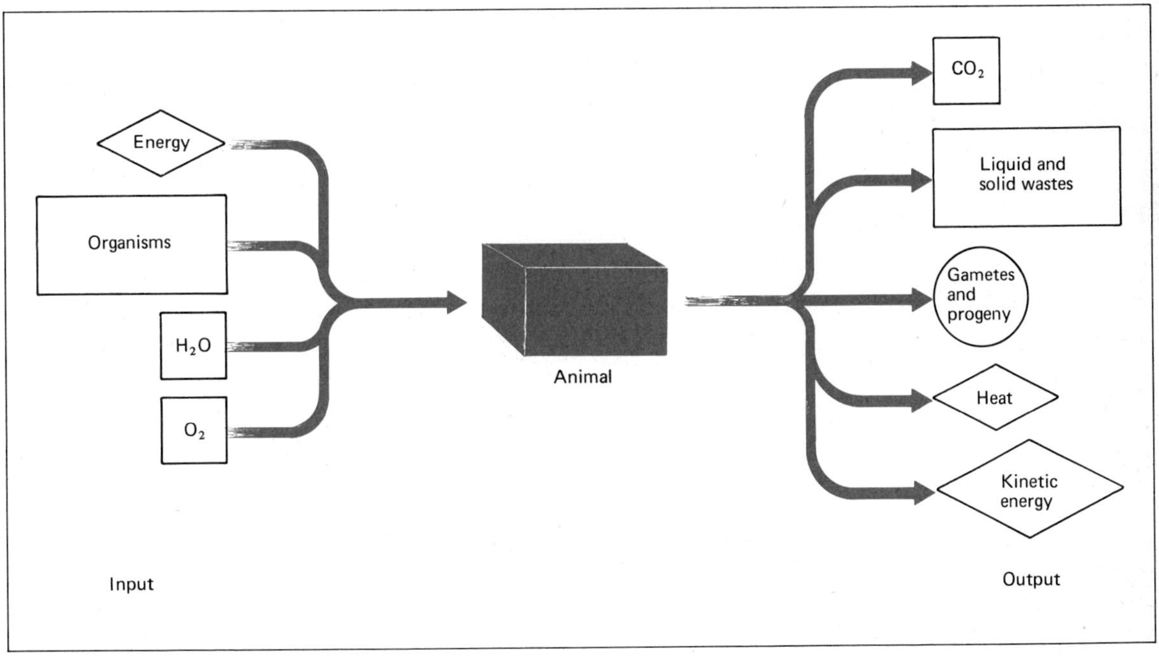

Input

Energy

Organisms

H_2O

O_2

Animal

CO_2

Liquid and solid wastes

Gametes and progeny

Heat

Kinetic energy

Output

suggest that some kind of oxidation, or burning, is going on in the bodies of animals, that energy is released from its potential state in the molecules of food, and that this energy is used to power the animal's processes.

It is interesting that the source of carbon compounds for both airplanes and animals turns out to be the same: the bodies, or the remains of bodies, of organisms. A silverfish feeding on the pages of a book and an airplane consuming gasoline are both being energized by the remains of long-dead organisms. Crude oil, from which hydrocarbon fuels are refined, is called a fossil fuel because of its origin from organisms which died millions of years ago. The pages of this book are mostly cellulose, the structural carbon compound from trees, artificially fossilized by man in the process of papermaking.

Sources of Fuel for Airplane and Organism

The airplane-organism analogy at first seems to break down completely when plants are considered (Fig. 3-4). Plants require no carbon compounds of organic origin on the input side, although they do need CO_2. On the output side, they mostly eliminate oxygen (O_2), although the production of some CO_2 can be detected, especially at night. In the first chapter, it was mentioned that plants utilize CO_2, which they extract from the atmosphere. At this point it should be clear why. The question can be attacked systematically:

1   The bodies of organisms, both plants and animals, are composed largely of complex carbon compounds.

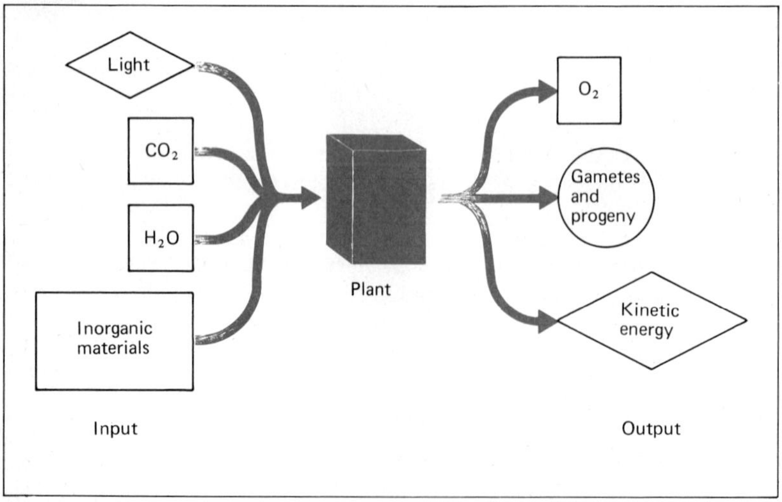

Fig. 3-4 Inputs and outputs of a daytime plant shown as a black box.

2 Plants require light but do not require or consume the bodies of other organisms. They are not observed to take in complex carbon compounds in any form.

3 Therefore, the complex carbon compounds in plants must be produced inside the plants, and the carbon for the synthesis of these compounds must enter the plant as an inorganic molecule—CO_2.

ANIMALS CONSUME THE BODIES OF OTHER ANIMALS OR OF PLANTS AS SOURCES OF COMPLEX CARBON COMPOUNDS. PLANTS USE LIGHT ENERGY AND CO_2 TO MANUFACTURE SUCH COMPLEX COMPOUNDS.

Output of Organisms

Included in the output of organisms are various forms of energy quite similar to the output of an airplane. There is kinetic energy in the mechanical state and expressed as movement of organisms or their parts. Aside from sound and heat energy, there are also forms of electromagnetic energy, such as light and electric fields. There may be other, undetected energy outputs.

The output of matter by organisms can be categorized as (1) waste materials; (2) organized materials used in construction (spider webs), reproduction (egg cases), or offense/defense (poisons); (3) reproductive products such as eggs, sperm, and offspring; and (4) growth, that is, increase in size. These last two categories of output, reproductive and growth materials, reveal another weakness in the airplane-organism or machine-organism analogy. It is true that some relatively simple machines have been designed to reproduce themselves, but inasmuch as machine progeny lack the high degree of complexity characteristic of the progeny of organisms, the distinction is still significant. And no airplane, unlike certain organisms, gradually enlarges itself and then reproduces.

ORGANISMS AND MACHINES HAVE SOME OUTPUTS WHICH ARE SIMILAR, BUT COMPLEX MACHINES HAVE NEITHER REPRODUCTIVE PRODUCTS OR GROWTH AS OUTPUTS.

THE SYSTEMS WITHIN THE ORGANISM

If one looks closely at an airplane, it is clear that it is made up of thousands of subunits: spars which support the wing, a "skin" composed of sheets of metal riveted to supporting structures, an intricate array of articulated parts which make up the engine, many wires, light bulbs, transistors, etc. Organisms, of course, also are revealed on close examination to be composed of subunits: organs such as livers and eyes, tissues such as muscles and nerves, and a wide variety of cells and molecules.

Basically, there are two ways to construct an airplane. The commonest way is to go to nonairplane raw materials, such as iron, bauxite, and petroleum, and by mining, smelting, and otherwise processing them, create airplane raw materials—steel for crankshafts, aluminum for spars, plastics for insulation, and so on. This is similar to the commonest way in which organisms are made. Plants take nonorganic raw materials such as water and CO_2 and use them to produce organic compounds.

There is a second way that an airplane can be constructed. Usable parts can be obtained from an airplane junkyard and used to assemble a new airplane. Before a part can be used in an airplane, it must be separated from the surrounding junk. Basically, that's how animals are constructed. They scrounge parts (subunits) from organic junkyards (the chewed up remains of other organisms). Before an animal can use a part, it must be separated from the surrounding junk in a process called digestion.

Synthesis and Digestion

Plants have a system for using light energy for synthesizing complex carbon molecules out of CO_2 and other inorganic materials. Animals also synthesize complex carbon molecules from simple ones, but in animals another preliminary step is required. Animals take in complex carbon compounds, break them up into smaller units, and then reassemble these into themselves—making self from nonself (Fig. 3–5). The latter

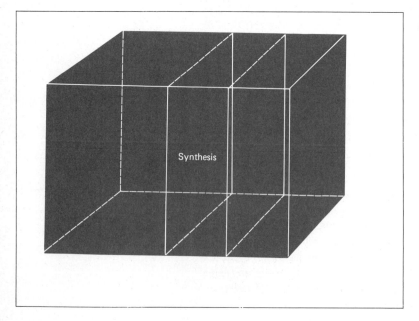

Fig. 3–5 Inside the black box: synthesis systems.

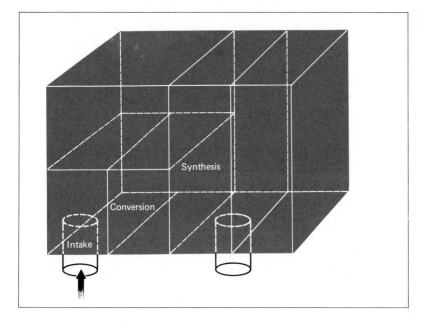

Fig. 3-6 Inside the black box: intake and conversion systems added.

step, assembling the pieces, is **synthesis;** the former step is a kind of conversion step, called **digestion.** When we are discussing the contents of an animal black box, then, a conversion system should be assigned some space. Space should also be allocated to intake systems such as pores and mouths (Fig. 3-6).

PLANTS ARE ABLE TO BUILD UP COMPLEX CARBON COMPOUNDS FROM CO_2 AND INORGANIC MOLECULES. ANIMALS BREAK DOWN COMPLEX CARBON COMPOUNDS INTO THEIR BASIC SUBUNITS AND, FROM THESE, SYNTHESIZE THE OTHER COMPLEX COMPOUNDS THEY REQUIRE.

Reception, Perception, and Effector Systems

Space for other systems can be assigned merely on the basis of our experience. Both plants and animals react to their environments. The large, compound flower of a sunflower plant and the leaves of many other plants follow the sun across the sky during the daylight hours. Nearly all the complex behavior of animals is a reaction to environmental conditions. For instance, the bill of the adult herring gull has a red spot near its tip. Upon seeing this red spot (a stick with a red spot works just as well), the nestling gulls open their mouths wide and beg to be fed. The red spot is part of the nestling's environment. When they perceive this pattern, the young birds respond in a predictable way.

These observations of the sunflower and the gulls imply the existence of three systems inside the black boxes, systems which may overlap. The first is a sensory reception system capable, for instance, of recording the position of the sun and the color red. But mere reception is useless if the incoming information is not organized and "understood," that is, **perceived.** Therefore, a perceptual system must exist as well. A third system still is required if organisms are to **react** to the environment. This is the **effector,** or response system. It is not enough to *see* danger (sensory reception). It is also necessary to understand that it *is* danger (perception). Finally, to survive, the organism must *avoid* the dangerous part of his environment (response).

ORGANISMS REACT TO CHANGES IN THEIR ENVIRONMENT. SUCH RESPONSES REQUIRE A RECEPTOR SYSTEM, A PERCEPTUAL SYSTEM, AND AN EFFECTOR SYSTEM.

These systems are not restricted to organisms. An airplane has sensory organs, such as the altimeter, airspeed indicator, and radio. And planes equipped with automatic pilots and landing devices perceive (organize) the data fed to them by the various kinds of sensory apparatus. Finally, the effector system of a plane is made up of its engine controls and the movable surfaces of its airfoils: the rudder, elevators, and ailerons. An automatic pilot is able to operate these effectors and fly the airplane.

Putting all these systems together into a model individual (Fig. 3–7)

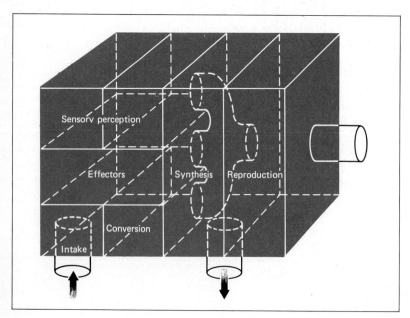

Fig. 3–7 Inside the black box: all systems shown.

gives us some idea of the kinds of systems to expect inside the black boxes. Since these boxes usually behave as organized wholes, we might also suspect the existence of communication or integrating systems.

So far, the machine-organism analogy has held up fairly well. The only serious flaws are the matters of reproduction, growth, and complexity. For instance, the complexity of a modern oil refinery, one that is automated and computer-controlled, is probably about the same as the complexity of the simplest bacterial cell. But the oil refinery cannot reproduce or enlarge itself.

Probing more deeply into the analogy may uncover more differences. So far, the economies—the input and output—of a machine and of organisms have been discussed. Matter and energy enter these systems; they are transformed inside the system, and what emerges as the output of these systems is much changed from what entered. But, as we have seen, organisms are not merely passive matter-energy transformers. They *behave* as well. They stop, they start, they move, they hunt, they mate, they flower, they communicate, and some even read and write. The question that will now be investigated is whether the behavior of machines differs from that of organisms. The behavior of a bird and that of an airplane will be our models.

To begin with, the act of taking off shows some very revealing distinctions. A bird, when "so inclined," simply springs into the air and flies off. Airplanes cannot be "so inclined"; a plane does not have the equipment—the mind—that is the prerequisite for being "inclined." When a plane takes off, an extrinsic human agent is responsible. The human being decides to take off; he starts the engine and manipulates the controls in such a way that the plane taxis to the proper place and becomes airborne at the proper time. The same distinction applies to the behavior of planes and birds when obtaining energy. When a bird requires energy, it seeks, catches, and consumes what food is necessary for its maintenance. Airplanes, by themselves, have no such behavior patterns. Again, a white-crowned sparrow migrating south in the fall will avoid flying into bad weather, will attempt to escape from other dangers such as hawks, and will change its course when about to collide with a mountain.

Airplanes do none of these things by themselves, although some military aircraft can be programmed to avoid terrain and antiaircraft missiles. This is not to say that planes could not decide to perform these acts without a pilot; eventually systems involving timing devices, radar, inertial navigation, computers, and automatic controls could be commonly

THE CONTROL OF SYSTEMS: HOMEOSTASIS

Behavior of Organisms

used to do just these operations. Avoidance behavior, in other words, *can* be intrinsic in planes.

This brings us to the essential point. Unlike the machine, the bird comes equipped with systems that ultimately maintain it and ensure its survival in the face of a very complex and sometimes inhospitable environment. These control systems are called **homeostatic** systems; their function is to maintain a steady state called **homeostasis**. A bird's homeostatic system will prevent overheating, overcooling, overfeeding, starvation, overactivity, lethargy, oversexuality, and sexlessness.

IN ADDITION TO BEING MATTER-ENERGY TRANSFORMERS, ORGA-NISMS BEHAVE IN RESPONSE TO ENVIRONMENTAL CHANGE. BE-HAVIOR IS MEDIATED BY CONTROL SYSTEMS WHICH MAINTAIN A STEADY-STATE CALLED HOMEOSTASIS.

Feedback Control

Figure 3–8 is a generalized diagram showing how homeostatic systems work. Birds offer many examples, but an example perhaps more familiar to you is human body temperature, which is usually held at 37°C. If your body temperature begins to fall, one of the responses is shivering, and the intensity of shivering is a function of the departure of the temperature from normal; that is, the lower the temperature, the more one shivers. The shivering response—rapid contraction and relaxation of muscles—produces heat and tends to offset the falling body temperature.

On the other hand, if the body temperature rises above the normal, possibly as a result of physical exertion or high environmental temperatures, perspiration increases. This cools the body because of the heat removed from the skin by the evaporating water. Usually the rate of perspiration increases with body temperature. If the departure from normality is in one direction, one response occurs; if the departure is in the other direction, the other response occurs. In both cases, the intensity of the response depends on the extent of the departure from normality.

This sort of system is called a **feedback system.** In order for it to operate, information about how the organism is responding must be fed back to the control center. A thermostat controlling a furnace operates in the same way. If the temperature of the house is 65° and you set the thermostat for 72°, the furnace will be turned on. When the temperature of the room reaches 72°, this information is fed back into the system and the furnace is turned off.

The distinction between homeostasis in machines and that in organisms is one of degree rather than one of kind, quantitative rather than qualitative. With the rate of modern technological improvements (systems en-

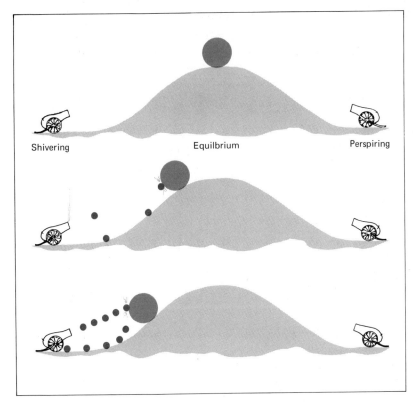

Shivering   Equilbrium   Perspiring

Fig. 3–8 Generalized diagram to show the working of a homeostatic system, the regulation of human body temperature. Shivering and perspiring are responses to a change from equilibrium (37°C). As a person becomes colder, he begins to shiver as shown by cannon firing; the farther down the large ball falls, the more cannon balls are fired to push it back up. Thus the intensity of the control response is a function of the departure from equilibrium.

gineering) proceeding as it is, the "homeostatic gap" between machines and organisms is being closed very rapidly. For example, the elegance and success of modern rocketry are due largely to the development of highly refined homeostatic, or feedback, controls.

In a very real sense, the only limitation to building a truly autonomous machine—an airplane, for instance—is the engineering problem of miniaturization, that is, the problem of making instruments small enough and light enough so they can be packaged conveniently and efficiently. Organisms have solved this problem. Even very small bats, for instance, carry a sonar system, including the generator to produce sound signals, receiving antennae (ears), and message-analyzing system (brain). All this apparatus would fit easily into a matchbox (Fig. 3–9).

ORGANISMS ACHIEVE HOMEOSTASIS BY MEANS OF FEEDBACK SYSTEMS, IN WHICH THE DIRECTION AND DEGREE OF RESPONSE ARE FED BACK INTO A CONTROL SYSTEM WHICH DETERMINES THE CORRECT RESPONSE.

Fig. 3–9 Generalized drawing of the head of a bat showing location of three components of its sonar system.

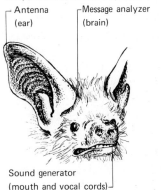

Antenna
(ear)

Message analyzer
(brain)

Sound generator
(mouth and vocal cords)

GROWTH AND DEVELOPMENT

An important characteristic of multicellular plants and animals is their capacity for growth and development. Multicellular organisms begin their life as a fertilized egg, or zygote. In the course of time and many cell divisions, a multicellular organism is produced, consisting of a great many different cell types, tissues, organs, and systems of organs, specialized for a variety of purposes. In a sense, the period during which a machine is being built is analogous to the period of growth and development of organisms. Construction of a machine requires, in addition to input of matter and energy, a maker of machines. Organisms, in a sense, construct themselves, using an input of energy and matter and following the genetic code of their DNA.

When a herring gull hatches, it is much smaller than an adult and it is covered with a soft gray down. Three years later, the gull has its adult plumage and size. From an egg smaller than a pinhead, the peppered moth goes through the stages of larva, pupa, and adult, each of them very different from the others. Except for the deterioration and aging that machines undergo, machines do not change with time in predictable ways. (A possible exception may be the computers that are capable of learning from experience. Some computers, for example, have been programmed to improve their chess game.) Therefore, the machine-organism analogy is really inappropriate in relation to growth and development.

ORGANISMS CHANGE IN TIME IN PREDICTABLE WAYS IN THE COURSE OF GROWTH AND DEVELOPMENT. THIS IS ACCOMPLISHED UNDER THE CONTROL OF THE GENETIC MATERIAL IN THE NUCLEUS OF EVERY CELL AND WITH INTERACTION WITH THE ENVIRONMENT.

SUMMARY

By making analogies between organisms and machines, one may be led to significant questions about how organisms function. In organisms, as in machines, many different transformations of energy occur. In a machine such as an airplane, potential energy in its hydrocarbon fuel is released as heat and mechanical energy when that fuel is combined with oxygen. Organisms also combine oxygen and carbon-containing compounds, releasing potential energy as heat and kinetic energy. Animals consume the bodies of other animals or of plants as a source of complex carbon compounds. Plants use light energy, carbon dioxide, and inorganic molecules to manufacture such complex compounds. Animals break down complex carbon compounds into their basic subunits and, from these, synthesize other complex compounds.

Organisms react to changes in their environment in various ways. These responses maintain the organism in a steady-state equilibrium. Such responses require a receptor system, a perceptual system, and an effector system. Maintenance of a steady state is called homeostasis. Organisms achieve homeostasis by means of feedback systems, in which the direction and degree of response are fed back into the control system, which determines the correct response.

Unlike most machines, organisms grow and develop. In the course of growth and development, organisms change through time in predictable ways. Growth and development are accomplished under the control of the genetic material in the nucleus. Throughout development, interaction with the environment takes place and plays an important role in determining the phenotype.

Elements are classes of chemical substances; they cannot be subdivided by using chemical techniques. Water, if subjected to an electric current, may be decomposed into hydrogen and oxygen. It is therefore, by definition, not an element. On the other hand, both hydrogen and oxygen resist decomposition by heating, electrical currents, or indeed any other treatment devised by chemists. They are elements.

The basic units of elements are atoms. Figure 3–1–1 shows two different representations of atoms. Remember that these are *representations,* not pictures of atoms. No one has ever seen an atom, and no one ever will. We can say this with certainty because atoms are so small, having a diameter on the order of 0.00000001 cm. This is roughly one four-thousandth of the wavelength of violet light, that visible light with the shortest wavelength. The ability to "see" things is a function of their size relative to a wavelength of light; for instance, two objects cannot be discriminated under a microscope if they are separated by a distance less than the wavelength of the light under which they are being viewed. We will never see atoms directly, although we may eventually be able to see pictures of them made by an electron microscope, which "views" them with beams of electrons that have extremely short wavelengths.

Look at the structure of the atom in Fig. 3–1–1, which is an atom of the element helium. Notice that it has a central **nucleus** (not to be confused with the nucleus of a cell!), around which orbit two electrons. The vast majority of the mass (for our purposes, this is the same as weight) of an atom resides in its nucleus, but the vast majority of its volume is occupied by electrons. Their relative proportions are not

BOX 3–1. ELEMENTS, ATOMS, AND MOLECULES

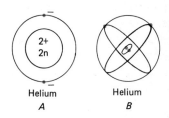

Helium
A

Helium
B

Fig. 3–1–1 Above, two representations of an atom of the element helium; below, ionic bonding of an atom of sodium and an atom of chloride, in which both sodium and chlorine have stable outer shells with eight electrons.

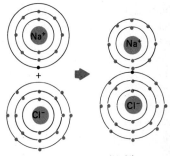

Sodium chloride (NaCl)

indicated in our figure. The nucleus, which has over 99.9 percent of the mass of the atom, has a diameter about one ten-thousandth that of the entire atom. Two kinds of particles make up the nucleus: **protons,** bearing a positive electrical charge, and **neutrons,** which have no charge. A great deal of energy binds together the particles of the atomic nucleus. Indeed, it is the release of only a small part of this energy in the splitting or fusion of certain nuclei which powers nuclear atomic reactions and hydrogen bombs.

The positive charge of the nucleus is exactly balanced by the negative charge of the electrons. Each bears a single negative charge, and the number of electrons is exactly the same as the number of protons. Thus an atom is electrically neutral. It was at one time thought that electrons traveled in fixed orbits around the nucleus, much as planets circle the sun, and our conventional diagrams show them that way. Actually, the positions of the electrons can be determined only in a statistical sense since it is impossible simultaneously to predict both the position and the velocity of a single electron. Furthermore, electrons confined within an atom by the electrical "pull" of the nucleus take on certain wavelike characteristics and can exist at one of a number of discrete energy levels. These energy levels may be thought of as a succession of shells at increasing distances from the nucleus, each successive shell representing a higher energy level. Each is actually a volume of space in which one may expect to find electrons at that energy level. The further the shell is from the nucleus, the more energy is contained in its electrons. As you will see later, it is the electron configuration of an atom—its number of electrons and their probabilistic distribution in shells—which gives elements the chemical properties which are of interest to us.

A table of biologically important elements and their chemical abbreviations follows. As you doubtless know, these abbreviations also stand for atoms of the elements in chemical formulas. For instance, H_2O stands for two atoms of hydrogen combined with one of oxygen to make a molecule of water. Don't concern yourself if you do not remember about molecules; they will be reviewed for you below.

Boron	B	Hydrogen	H	Oxygen	O		**Biologically Important Elements**
Calcium	Ca	Iodine	I	Phosphorus	P	
Carbon	C	Iron	Fe	Potassium	K	
Chlorine	Cl	Magnesium	Mg	Sodium	Na	
Cobalt	Co	Manganese	Mn	Sulfur	S	
Copper	Cu	Molybdenum	Mo	Zinc	Zn	
Fluorine	F	Nitrogen	N			

A molecule is the smallest subunit of a substance which can exist independently and still retain the properties of that substance. A water molecule consists, as we said, of two hydrogen atoms and one oxygen atom linked together. Remove any one of the three atoms, and the remainder is no longer water. In general, molecules may be thought of as units built up of atoms, but in some cases atom and molecule are identical. For instance, a single iron atom is also a molecule; it can exist independently and has the properties of iron. A single hydrogen atom is not a molecule because it does not normally exist in isolation. A flask of hydrogen gas is made up of H_2 molecules, units consisting of two hydrogen atoms joined together.

An all-important aspect of molecules is the nature of the bonds that hold them together. In general, atoms are bonded together by a process of transfer or sharing of electrons, which brings all the participating atoms to a state of maximum stability. In *ionic* bonding, electrons are transferred to achieve this stability. Consider, for instance, at atom of sodium and an atom of chlorine. Like most (but not all) atoms, they are in their most stable configuration when they have eight electrons in their outer shells. When the sodium gives up an electron, it becomes positively charged (it now has more protons than electrons), and when the chlorine accepts an electron, it becomes negatively charged. Opposite charges attract, and an electrical force holds the atoms together as a molecule of sodium chloride, NaCl (Fig. 3-1-1).

Ionic bonding, however, does not ordinarily produce molecules in the classical sense. Ionic compounds (salts) in the solid state have a crystalline structure in which the atoms are regularly spaced (Fig. 3-1-2). As you can see in the figure, each atom of one kind has as its nearest neighbors six atoms of the other kind. It is not possible to segregate specific pairs as *the* molecules. Similar difficulties occur in describing solutions.

The other major type of bonding, *covalent bonding,* comes closer to the intuitive notion of a chemical bond. In general, covalent bonds do hold together groups of atoms in discrete molecules. The large carbon-based organic molecules which are characteristic of living systems are held together in large part by covalent bonds. In these bonds, atoms do not reach a stable electron configuration by one atom donating one or more electrons to another. Instead, stability is achieved by the *sharing* of electrons. Consider, for example, the small inorganic molecule methane (Fig. 3-1-3). It consists of one carbon atom covalently bonded to four hydrogen atoms; its formula is CH_4. Carbon has four electrons in its outer shell; it needs four more to attain stability.

Molecules

Fig. 3-1-2 Crystalline structure of sodium chloride. Atoms of sodium and chlorine are spaced to cubical crystals.

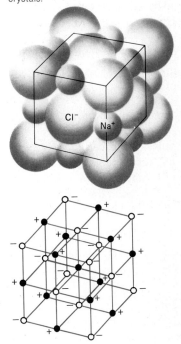

Hydrogen, on the other hand, can attain stability either by losing its one electron (as in our example of ionic bonding) or by gaining a second electron. In the covalent bonding to form methane, each of the five atoms can attain stability by sharing. There are four electrons in the outer ring of carbon. In total, the 5 atoms need 16 electrons in their outer rings to achieve stability (8 for the carbon, 2 for each of 4 hydrogens). Stability is achieved by the five atoms sharing eight electrons, which travel freely around the new molecule.

Conventionally, in organic chemistry, the methane molecule is represented by a structural formula in two dimensions, as shown in Fig. 3–1–3.

Each straight line connecting atoms indicates a *shared pair of electrons.*

Studies of the geometry of the methane molecule have shown there to be a carbon atom at the center of an equilateral tetrahedron with the hydrogen atoms at its points. An equilateral tetrahedron is a figure with four equal sides and four apices (Fig. 3–1–4).

Sometimes atoms in organic molecules will share two pairs of electrons, as in ethylene, C_2H_4, or three pairs as in carbon monoxide, CO:

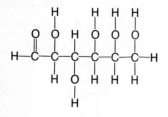

Ethylene

C≡O
Carbon monoxide

The chemical properties of carbon, which we cannot explore in detail here, make it ideal for the construction of large, complex molecules. The carbon atom's most important quality in this respect is its ability to form stable covalent bonds with other carbon atoms. This may result in molecules containing long straight chains of carbon atoms, as in the sugar glucose:

Fig. 3–1–3 Two representations of a molecule of the gas methane: *left,* the sharing of electrons in covalent bonding is shown by paired dots; *right,* each straight line indicates a shared pair of electrons.

Fig. 3–1–4 The geometry of a methane molecule is based upon the bonding properties of carbon which place the hydrogen atoms in the positions of the apices of an equilateral tetrahedron.

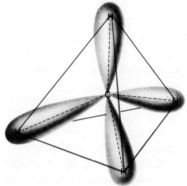

or the fatty acid stearic acid:

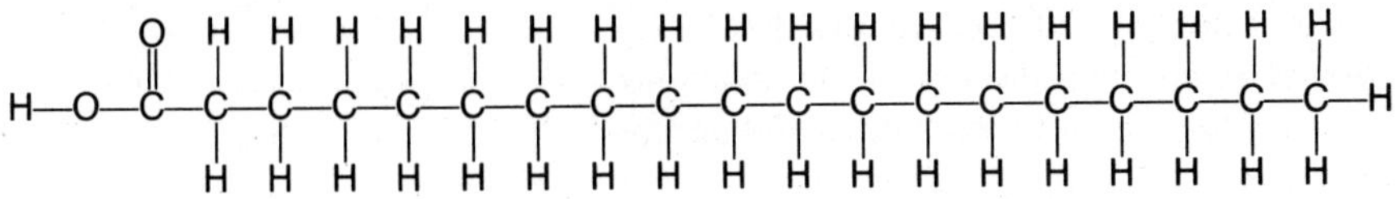

Sometimes carbons may be joined in a ring, as in benzene:

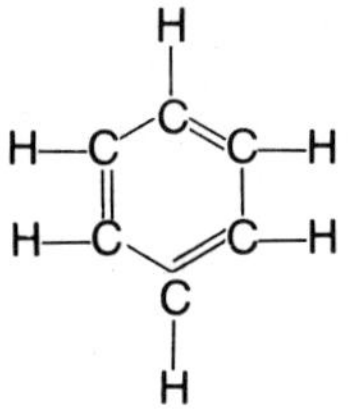

or in more complex configurations, as in the amino acid tryptophan:

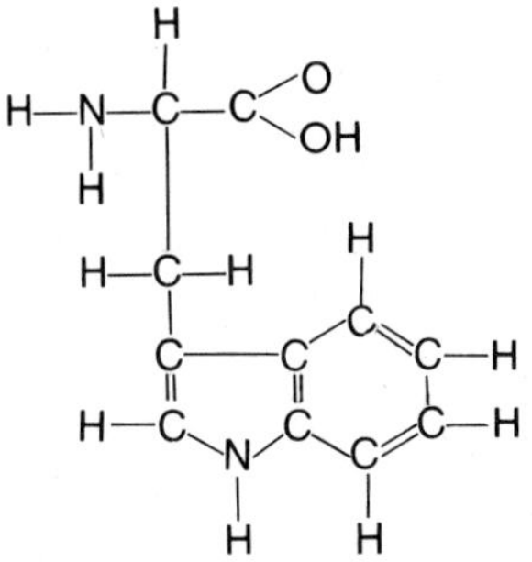

Often the structural formula is simplified; either not all the bonds are shown, or carbon atoms may be omitted, as in these representations of tryptophan and benzene:

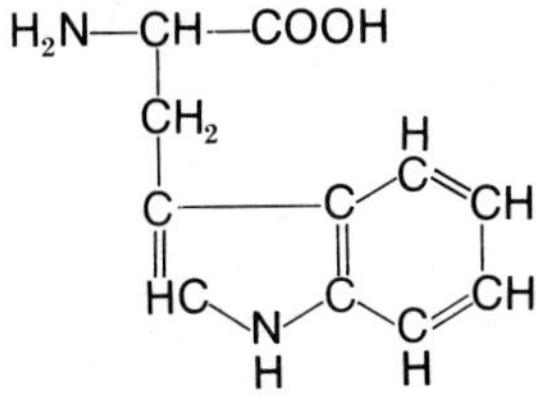

Tryptophan

Benzene

You must remember that there are many conventional ways of indicating the structure of molecules.

There is an almost endless variety of organic compounds, which function in myriad ways in living systems. As you will see, those we have illustrated here are among the smaller of these compounds. Fortunately, at least for our ease of understanding, the larger ones are made up of repeating subunits. These are the **macromolecules,** the giant molecules which are characteristic of living systems.

One additional kind of bond needs mention here, the hydrogen bond. Hydrogen atoms often form a weak bridge between nitrogen and oxygen atoms in proteins and DNA and, as you will see, are critical to the structure and functioning of these molecules. The most important thing for you to remember about hydrogen bonds is that they are easily formed and easily broken. Indeed, only about one-tenth the energy is required to break a hydrogen bond as to break an average covalent bond.

SUPPLEMENTARY READINGS
Books

Baker, J. J. W., and G. E. Allen: "Matter, Energy and Life," Addison-Wesley, Reading, Mass., 1965.

Henderson, L. G.: "The Fitness of the Environment," MacMillan, New York, 1913.

Kennedy, D.: "The Living Cell," Readings from the Scientific American, Freeman, San Francisco, 1965.

Lehninger, A. L.: "Bioenergetics: The Molecular Basis of Biological Energy Transformations," W. A. Benjamin, New York, 1965.

Articles

Buswell, A. M., and W. H. Rodebush: Water, *Sci. Amer.* **194**, no. 3 (1956).

CHAPTER 4

SYNTHESIS: MAKING SELF FROM NONSELF

Organisms of all kinds carry out three basic kinds of processes: growth, maintenance, and reproduction. Each of these requires that an organism synthesize new molecules, using energy and matter from the environment. In this chapter we will consider how the synthesis of molecules takes place in cells and how energy is utilized.

With only very minor exceptions, all the energy used by organisms is solar energy. (The laws of thermodynamics are discussed in Box 4–1.) As we have discussed earlier, the radiant energy of the sun is absorbed by pigments in green plants and some prokaryotes and converted into the chemical energy that binds atoms together into molecules. Plants that are able to use solar energy directly and to use CO_2 as their only carbon source are known as **autotrophs** (Greek: "self-feeders"). Other organisms are able to use solar energy only in the form of chemical energy, and they require organic compounds as a source of carbon. They obtain these compounds, of course, by eating autotrophs or by eating organisms that have eaten autotrophs. The energy stored in the molecules of the food organisms is made available to the eater by the complex chemical process of **oxidation,** a kind of slow burning with the release of energy. Organisms (mostly animals and microorganisms) that are unable to utilize solar energy directly and that require organic carbon are known as **heterotrophs** (Greek: "other-feeders").

SOURCES OF ENERGY AND MATTER

The process by which the sun's energy is bound by green plants is known as **photosynthesis.** The process by which both autotrophs and heterotrophs release that energy from the chemical bonds in which it is stored is called **respiration.** Respiration goes on in the cells of all organisms— including the photosynthetic cells of plants. Green plants essentially make their own food and then utilize its energy by burning it in the process of respiration. Fortunately for animals, green plants collectively trap more energy than is required for their own growth and reproduction; the excess is used for the growth and reproduction of heterotrophs.

But synthesis of molecules requires more than energy: it requires matter of many kinds. Living systems require a variety of chemical elements: carbon, oxygen, hydrogen, nitrogen, sulfur, phosphorus, calcium, potassium, and sodium (to name a few of the most important). The primary source of all but carbon and oxygen is the uptake of these materials in watery solution by the roots of plants. The basic reservoir of carbon and oxygen is the gas carbon dioxide, both free in the atmosphere and dissolved in fresh water and in the oceans. The paths along which matter and energy move in biological communities will be discussed in some detail in Chap. 14.

All the synthetic activities of organisms are controlled by proteins called **enzymes.** In order to understand how synthesis takes place, it is necessary to understand how enzymes are produced and how they act. Enzymes, like other proteins, are produced in the cell under the direction of molecules called **nucleic acids,** which make up the genes. Our detailed discussion of synthesis properly starts with a discussion of the nature of genes and how they control the production of enzymes.

Nucleic Acids

In 1868, three years after the publication of Mendel's discoveries, Friedrich Miescher began to study the chemistry of the nuclei of cells. He assumed that the nucleus was the cellular organelle that contained the hereditary information. With remarkable skill and foresight, he was able to extract and study a component of the nucleus of salmon sperm, a substance he called nuclein, which was rich in phosphorus. Nuclein is known today as deoxyribonucleic acid (DNA). Miescher's discovery is amazing since it was not until 1876 that Oscar Hertwig demonstrated that fertilization involved the fusion of the nucleus of a sperm with that of an egg cell.

DNA AS HEREDITARY MATERIAL. HERSHEY-CHASE EXPERIMENT There is now so much evidence that DNA is the hereditary material that this has become a foundation principle of biology. One of the most original of the many demonstrations of this was an experiment performed in 1952 by Hershey and Chase. They used two kinds of organisms in their experiment: bacteria and bacterial viruses. The latter are usually called bacteriophages or, simply, phages. Phages, like other viruses, are minute infectious agents that invade the cells of their hosts and use the cellular machinery of the host cells to make copies of themselves. The final stage of such an infection is the rupture of the host cell, called **lysis,** and the release of large numbers of new viruses. Viral structure is the epitome of economy. Figure 4–1 shows a phage that infects the colon bacterium *Escherichia coli,* which inhabits the intestines of all healthy

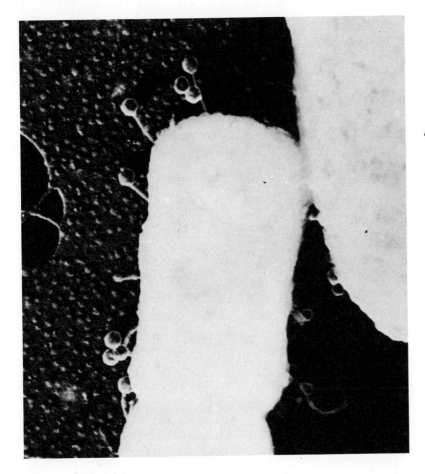

human beings. It consists of an outer protein layer, with a large head and complex tail. DNA is packed into the head; it is injected into an *E. coli* cell by a musclelike contraction of part of the tail when the phage attaches to the surface of a cell by its tail fibers. If only DNA is injected into the host cell and very little or no protein enters, then DNA must be the material that carries the information determining the reproduction of the phage. That is, DNA must be the genetic material of the virus. On the other hand, if significant amounts of protein enter the host, the protein cannot be ruled out as a bearer of the information. The challenge was to label the protein and the DNA with different tags which could be detected and distinguished inside the infected bacteria.

Hershey and Chase used a technique involving the marking of molecules by the use of **radioisotopes**. Isotopes are alternative forms of an element;

Fig. 4-1 Bacteriophage: (*A*) electron micrograph of a colon bacterium to which particles of bacteriophage are attached; (*B*) reproduction of a bacteriophage—injection of the DNA followed by lysis of the host cell after multiplication of the virus. (*A, courtesy of T. F. Anderson, E. L. Wollman, and F. Jacob.*)

they have the same number of protons as the element but differ in the number of neutrons. Some isotopes are very unstable and break down radioactively; they can be detected by appropriate instruments, and molecules containing these isotopes are said to be **labeled.** Hershey and Chase cultured bacteria on growth media that contained either the radioactive isotope of phosphorus, ^{32}P, or that of sulfur, ^{35}S (Fig. 4–2). It was known that DNA contains phosphorus but no sulfur. Furthermore, protein contains sulfur but no phosphorus. Therefore, bacteria grown on a medium containing ^{32}P would have their DNA, but not their protein, labeled by the isotope. In the same way, bacteria grown on the ^{35}S medium would have label-free DNA, but their protein would be labeled.

Each of these cultures was then infected with phages, which reproduced themselves, using the bacterial synthetic apparatus to make new DNA and protein molecules containing whatever radioisotopes were in the growth medium. Phages that reproduced in bacteria in ^{32}P medium incorporated radioactive phosphorus into their DNA, and those that reproduced in bacteria in ^{35}S medium incorporated radioactive sulfur into their protein. The next step was to infect unlabeled and previously uninfected bacteria with one or the other kind of phage. Bacteria infected with ^{35}S-labeled phages had little or no radioactivity inside their cells, whereas those

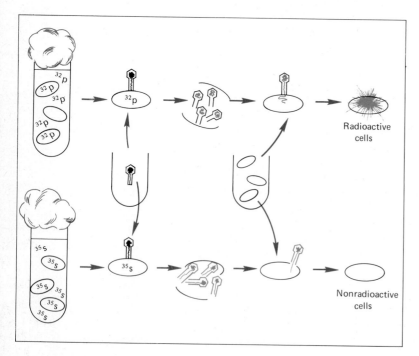

Fig. 4–2 Diagram of the Hershey-Chase experiment.

infected with ^{32}P-labeled phages were very radioactive. Furthermore, the new phages released from these infected cells contained ^{32}P but not ^{35}S. This experiment showed that no important amount of protein is transmitted from one generation of phages to another and that, therefore, the genetic material must be DNA.

STRUCTURE OF DNA   The molecule of DNA is an extremely long one and is therefore called a **macromolecule.** Because it is made up of many subunits, or building blocks, called **monomers,** it is a **polymer.** The monomer building blocks of which DNA is made are called **nucleotides.** Each nucleotide is itself made up of three subunits: a phosphate group, a sugar with five carbons (deoxyribose), and a structure called an organic base (Fig. 4-3). There are four bases found in DNA nucleotides: two are purines and two are pyrimidines. The purines are adenine (A) and guanine (G). The pyrimidines are cytosine (C) and thymine (T). Another important kind of nucleic acid, which we shall discuss later, is ribonucleic acid (RNA). The monomers of this compound differ from those of DNA in having a different sugar (ribose) and in having the base uracil (U) instead of thymine.

DNA may be partially described as a polynucleotide chain, with the nucleotides linked to one another in linear fashion via the phosphate groups, as shown in Fig. 4-4. The backbone of this macromolecule is a repeating series of sugar–phosphate–sugar–phosphate–, with either a purine or a pyrimidine base attached to each sugar group. But there is more than this to be said about DNA—it has a very special three-dimensional structure. One clue to working out this structure was the consistent quantitative relationship found among the bases in the DNA of any given species. The ratio of adenine to thymine and that of guanine to cytosine are always 1. That is, A and T always occur in equivalent amounts, as do G and C. The relative amounts of AT and GC vary widely from species to species, but the equivalence within each pair is a rule. Another clue to the three-dimensional structure of DNA came from x-ray diffraction studies—a technique used to analyze regularly repeating atomic patterns. With x-ray diffraction analysis, one can measure the spacing of atoms in a molecule. M. F. H. Wilkins and Rosalind Franklin were able to determine that the distance between the bases is relatively small and, furthermore, that the molecule consists of two or more helical (spiral-shaped) polynucleotide chains.

One of the most important scientific accomplishments of the century was the proposal of a model for DNA by J. D. Watson and F. H. C. Crick in 1953 which accounted for the base-ratio and x-ray diffraction observations. The Watson-Crick model of DNA (Fig. 4-5) proposes that DNA

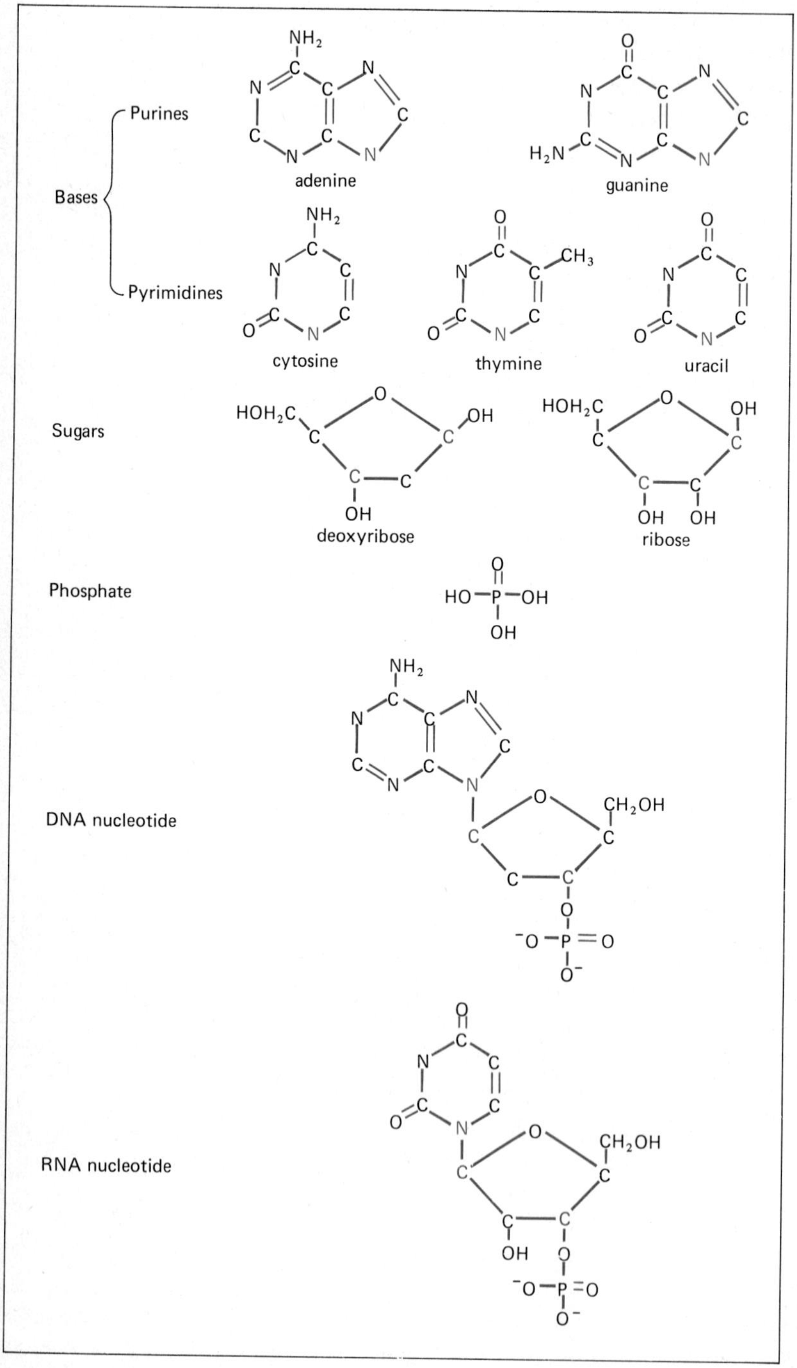

Fig. 4-3 The constituents of nucleic acids. (For clarity, hydrogens are omitted from carbon atoms in the rings.)

is made up of two polynucleotide chains oriented in opposite directions and wound around each other to form a double helix. The bases are on the inside, forming the steps of a spiral staircase. The most crucial feature of the model, and the one that accounts for the equivalence of purine and pyrimidine bases, concerns the kind of chemical bonding between the juxtaposed bases. The proximity of these bases permits special kinds of weak bonds called hydrogen bonds to join them (see Box 3–1). These bonds can form only between A and T and between G and C without upsetting the helical structure of the molecule. Thus a specific purine base *has* to be opposite a specific pyrimidine base.

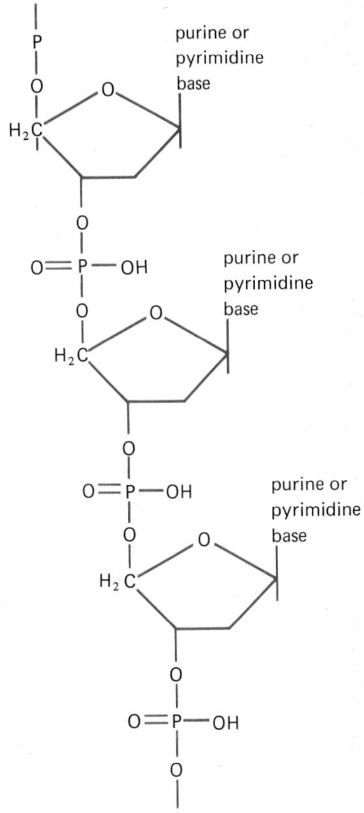

This model for the structure of DNA includes all the requirements for a chemical that stores and transmits the hereditary messages. These requirements are fourfold; **replicability, information-storage, capacity, and mutability.**

DUAL TEMPLATE REPLICATION OF DNA   The genetic material must have the property of replicability, or the ability to copy itself, so that the hereditary information can be faithfully passed on to daughter cells in mitotic divisions and to gametes in meiotic division. The DNA molecule is perfectly fitted to this task. The key to understanding replication of DNA is the **complementarity** of the four bases. Adenine is always the complement of thymine, and guanine of cytosine. Because of this base-pair specificity, the sequence of bases along one chain determines exactly the sequence along the sister chain. This suggests that each chain by itself can act as a template, or model, for the reconstitution of the double chain as long as a pool of nucleotides is available to form hydrogen bonds with complementary bases of the template chain. This scheme is outlined in Fig. 4–6.

Fig. 4-4   A DNA polynucleotide chain.

It is now generally believed that the unwinding of the two template polynucleotide chains occurs simultaneously with the synthesis of the two daughter chains. The unwinding alone is a formidable problem. Even the smallest DNA chains, those of phages (which contain one-thousandth the amount of DNA contained in a bacterial chromosome), have 1,000 turns. The energetics of the unwinding of such long chains is not a serious problem, but the mechanics is mind-boggling.

Many lines of evidence tend to confirm the Watson-Crick model of DNA structure and the dual-template replicating hypothesis. One of the most elegant verifications was that done by M. Meselson and F. Stahl in 1958 (Fig. 4–7). The colon bacterium *Escherischia coli* was the organism used for the experiments. (Much of our knowledge of molecular genetics has been obtained through experiments on this organism.) The bacteria were cultured on a medium containing nitrogen in the form of its heavy

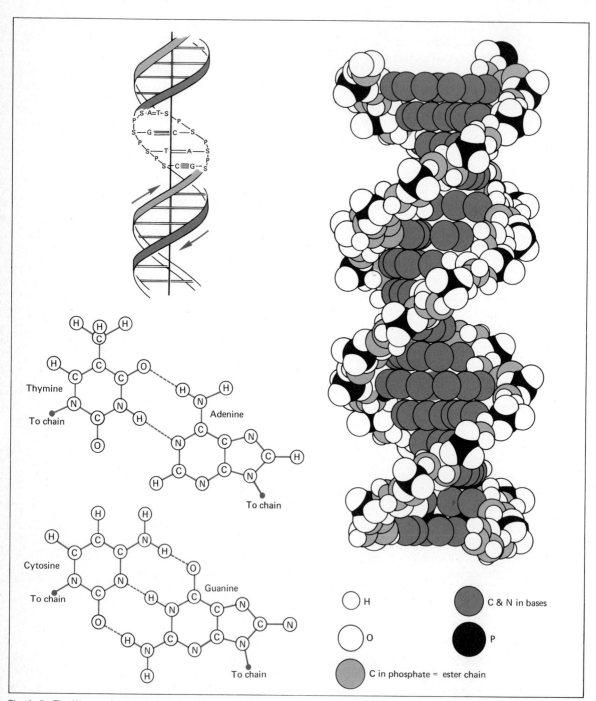

Fig. 4-5 The Watson-Crick model of the structure of DNA.

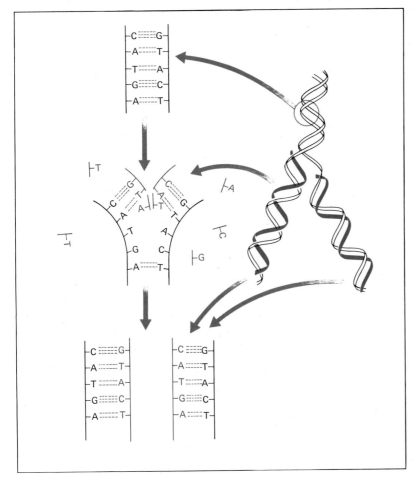

Fig. 4-6 Mode of template replication of DNA.

isotope ^{15}N, the atoms of which contain eight neutrons instead of seven, as in the common isotope ^{14}N. Molecules containing ^{15}N are denser than those containing ^{14}N. After several generations, all the DNA extracted from the *E. coli* was labeled with this isotope and therefore denser than DNA containing the normal isotope. The culture medium then was changed to one containing ^{14}N. Samples of bacterial cells were taken periodically for a number of generations thereafter, and the DNA was extracted for determination of its density.

As shown in Fig. 4-7, the DNA of the *E. coli* grown on the ^{14}N medium for one generation had a density exactly intermediate between the densities of the pure ^{14}N and pure ^{15}N DNAs. That is, after a second generation, the DNA was found to be of two kinds: half was the hybrid-density

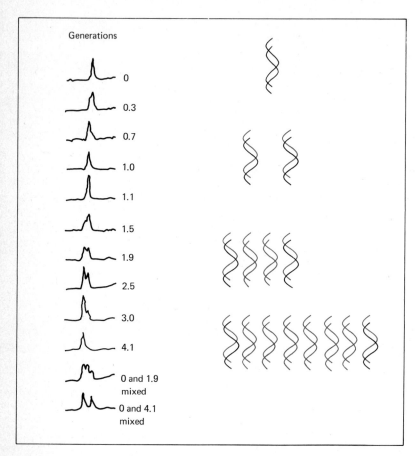

Generations

0

0.3

0.7

1.0

1.1

1.5

1.9

2.5

3.0

4.1

0 and 1.9 mixed

0 and 4.1 mixed

Fig. 4-7 Diagram of the Messelson-Stahl experiment.

type and half had the density of pure ^{14}N DNA. After another doubling of the cell number (three generations of cells), there were still the same two kinds of DNA, but there was three times as much of the ^{14}N DNA as of the hybrid DNA.

As indicated in Fig. 4-7, the results are most easily interpreted in terms of the Watson-Crick model. If each strand of the double-stranded DNA acts as a template, it can be seen that the newly replicated DNA will have one old strand containing ^{15}N and one new strand polymerized from monomers (nucleotides) in the new ^{14}N medium. After two generations, only half of the DNA molecules will retain the original ^{15}N strains; after three generations, only one-fourth.

INFORMATION STORAGE IN DNA The second function of genetic material is information storage. Information can be stored in both "real"

and symbolic ways. A specimen of a fish in a bottle is a "real" representation of the information contained in the animal. The word "fish" is a
symbolic representation of the animal—a convention used to communi-
cate the idea of fish. At still another level, the symbols $\cdot - \cdot$, $\cdot \cdot$, $\cdot \cdot \cdot$,
and $\cdot \cdot \cdot \cdot$ are representations of the letters f, i, s, and h in Morse code.
The Morse code consists of only two basic symbols: dots and dashes.
Our alphabet code has 26. The genetic code has four—the four bases.
It hardly seems possible that the information for a whole human, for example, could be packed into the nucleus of the zygote, but Crick has made
a reassuring calculation. If the base pairs of DNA corresponded to the
dots and dashes of the Morse code, the DNA in the nucleus of a human
cell would be sufficient to encode 1,000 large textbooks—a fine example
of organic microminiaturization. Perhaps a more impressive fact is
that the DNA that codes all the hereditary variation of the entire human
population would fit into three aspirin tablets!

DNA MOLECULES ARE VERY LARGE MOLECULES CONSISTING OF
TWO POLYNUCLEOTIDE CHAINS WOUND ABOUT ONE ANOTHER TO
FORM A DOUBLE HELIX. EACH CHAIN CONSISTS OF A LINEAR AR-
RANGEMENT OF FOUR DIFFERENT KINDS OF NUCLEOTIDES, AND
THE SEQUENCE OF NUCLEOTIDES IN ONE DETERMINES THAT IN THE
OTHER. THE NATURE AND STRUCTURE OF DNA ACCOUNT FOR ITS
REPLICABILITY AND ITS INFORMATION-STORAGE ABILITY.

Now that you have been introduced to the information-carrying molecule,
let us turn to protein molecules. It is the structure of protein molecules
which is specified by the information in the DNA. The essentials of protein structure must be understood in order to comprehend how proteins
are synthesized "under the direction" of DNA.

Proteins

Proteins, whether found in bacteria, green plants, or man, are linear
polymers made up of a series of different nitrogen-containing acids, called
amino acids (Fig. 4-8). Amino acids are the monomers of proteins,
just as nucleotides are the monomers of nucleic acids. There are 20 kinds
of amino acids commonly found in proteins, but a few others turn up
occasionally. The structure of amino acids is somewhat analogous to the
structure of nucleotides. Both have a "business piece," (4 kinds in the
nucleotides of DNA, roughly 20 kinds for proteins). Both also have two
different "fitting ends," which join together to form the backbone of the
molecules. In amino acids, one end has an amino group ($-NH_2$) attached
to a carbon; the other end has a carboxyl group ($-COOH$). The chemical
groups that are attached to the same carbon atom that bears the amino
group are the "business parts," the so-called R groups.

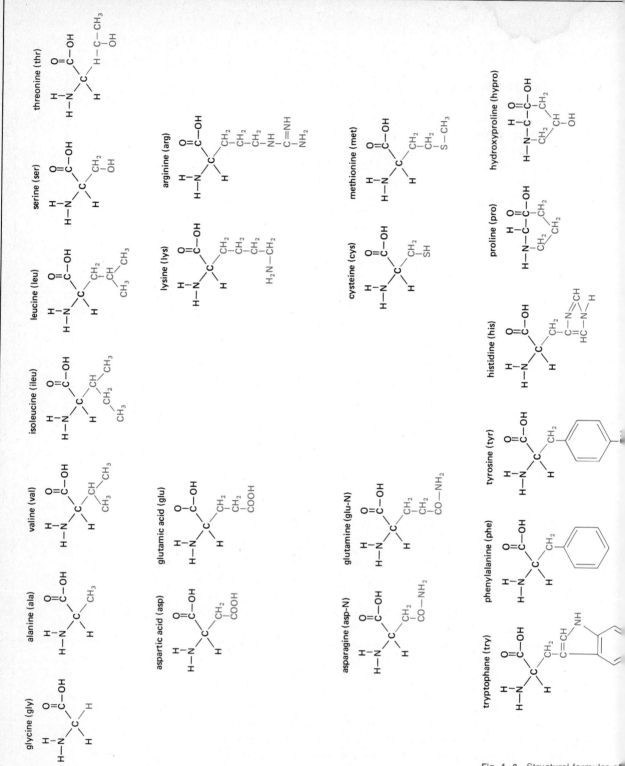

Fig. 4-8 Structural formulas of important amino acids.

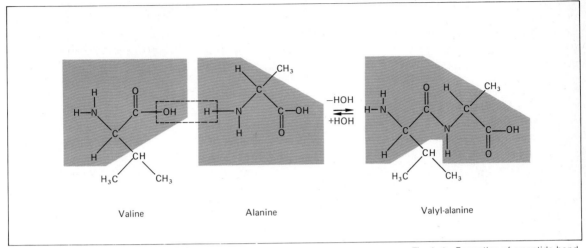

Valine   Alanine   Valyl-alanine

Fig. 4–9 Formation of a peptide bond. *(From "How Proteins Start" by Brian F. C. Clark and Kjeld A. Marcker.* *Copyright © 1968 by Scientific American, Inc. All rights reserved.)*

The amino acid molecules, the monomers of proteins, are polymerized by joining the $-NH_2$ group of one amino acid with the $-COOH$ group of another. This forms what is known as a peptide bond (Fig. 4–9). In the process, a molecule of water is formed. The products of this reaction, then, are a molecule of water and a **dipeptide,** composed of the two **amino acid residues**—what is left over after part of a water molecule is given off. During the synthesis of a protein molecule, this process is repeated serially until a complete polymer, a **polypeptide** chain, is synthesized.

PROTEIN STRUCTURE Protein structure can be inspected and discussed at four levels: (1) the sequence of amino acids in a chain; (2) the spatial relations of adjacent amino acids; (3) the folding of the polypeptide chain; and (4) the interactions of different polypeptide chains. These are referred to as the primary, secondary, tertiary, and quaternary structures, respectively.

The sequence of amino acid residues is the **primary structure** of a protein molecule. As late as the post-World War II era, protein chemists as a group were still not convinced that proteins had a specially ordered sequence of amino acids. In fact, it was only with the introduction into protein chemistry of the technique known as paper chromatography in 1941 that it became possible to separate all the amino acids of a protein and to demonstrate that the percentage composition for the amino acids was a constant in any given protein (Fig. 4–10). At this point, a program was begun by Frederick Sanger to determine whether the sequence of amino acids in a particular protein, insulin, was predictable or random. The

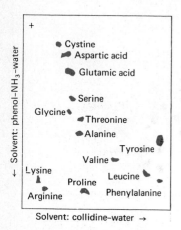

Solvent: phenol–NH₃–water

- Cystine
- Aspartic acid
- Glutamic acid
- Serine
- Glycine
- Threonine
- Alanine
- Tyrosine
- Valine
- Lysine
- Proline
- Leucine
- Arginine
- Phenylalanine

Solvent: collidine–water →

Fig. 4–10 Use of paper chromatography to determine the amino acid composition of a protein. Different amino acids move at different rates when solvent is applied to the edge of a piece of filter paper on which a protein sample is placed. After one solvent has caused separation along one axis, a second is applied along the axis at right angles to the first.

complete answer was not to come for 10 years. The primary structure of insulin, a protein hormone important in the regulation of sugar metabolism, was determined in 1954 (Fig. 4–11). Since then, the primary structures of many other proteins have been determined.

The second aspect of protein structure is the spatial relationship between the adjacent amino acid residues, the **secondary structure.** From x-ray diffraction analysis it was discovered that a common shape of the –CCN–CCN–CCN– backbone is that of a kind of spiral called an α-helix (alpha helix). The shape of the α-helix is partially determined by hydrogen bonds between the –NH and –CO groups of every amino acid and the third residue beyond it (Fig. 4–12). Fibrous proteins, such as those that make up skin, muscle, and hair, are composed of numerous interwoven α-helices.

However, most metabolically functional (as opposed to structurally functional) protein molecules are **globular** rather than elongate. The globular proteins get their shape from their patterns of folding, their **tertiary structure,** which in turn is determined by the sequence of amino acids and especially by the interactions of R groups. In other words, primary structure is an important determinant of tertiary structure. When the tertiary structure, or pattern of folding, is somehow altered irreversibly, the protein loses its biological activity as well as other of its chemical and physical properties. This change in shape and activity is called **denaturation.** Many proteins, for instance, are denatured by heating. The white of a fried egg is the protein ovalbumin denatured by heat. However, denaturation is not always irreversible. Heating the digestive enzyme trypsin to between 80 and 90°C will result in loss of activity, but activity is restored if the solution is cooled to 37°C.

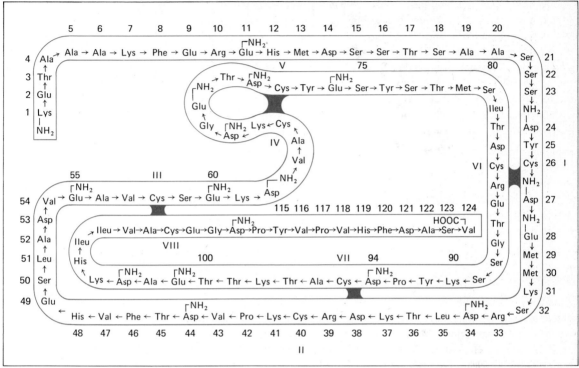

Fig. 4-11 The primary structure of the protein insulin which was discovered by Frederick Sanger.

There is still another aspect of protein structure—***quaternary structure,*** which is the precise way in which two or more polypeptide chains fit together. Many enzymes and other metabolically active proteins are aggregations of two or more polypeptide chains. Normal hemoglobin, the oxygen-carrying protein of the blood, is constructed from 2 pairs of polypeptide chains, one pair of which has 141 amino acid residues in each strand and the other pair 146.

PROTEINS ARE LARGE ORGANIC MOLECULES COMPOSED OF AS MANY AS SEVERAL HUNDRED AMINO ACIDS. THERE ARE SOME 20 KINDS OF AMINO ACIDS DIFFERING IN STRUCTURE AND CHEMICAL CHARACTERISTICS.

Structure and Function of Enzymes

In biological systems, structure and function are inseparable. In fact, it is misleading to distinguish between these two manifestations of biological materials. That we do commonly separate structure and function reflects either our ignorance of the biological activity of a particular material or our tendency to categorize our perceptions into static (structural) and dynamic (functional) forms. Much of the history of biology is describable as either finding the function of some mysterious structure or, on the other hand, elucidating the structure of the components that participate in some known process. These two aspects of matter-energy are inseparable in an evolutionary sense and also in the obstacle course of life; those structures that fail to function adequately lead to the demise of their bearers. In the long run, proper functioning is the criterion for survival, and proper structure is the basis of proper functioning.

Proteins are the most versatile compounds that exist in nature. The 20 amino acids can be arranged into a great diversity of different molecules that seem capable of performing an endless array of tasks. Considering a protein molecule of 600 amino acids, the number of possible arrangements (different sequences) is 20^{600}, a number inconceivably greater than the estimated number of electrons in the universe.

THE FUNCTIONS OF PROTEINS: ENZYMES This versatility of structure is reflected in the diversity of functions that proteins perform. These functions will be mentioned as they come up in later discussions, but a few examples will be illustrative. Most of the bulk of muscles (commonly called meat), the organs of movement, are composed of the proteins actin and myosin. Two of the materials responsible for clotting of blood are the proteins thrombin and fibrinogen. Insulin, the first protein to have its amino acid sequence determined, is a chemical messenger, or hormone, that is important in carbohydrate metabolism. The protein trypsin, mentioned above, is an enzyme secreted by the pancreas. It catalyzes the splitting of other proteins into smaller units in the process of digestion. Antibodies, the materials that act on foreign materials in the body, are also proteins. It is in their functioning as enzymes, however, that the critical importance of proteins is most clearly seen.

An organism can be thought of as a thoroughly automated chemical factory. But unlike the processes of a refinery or some other factory, those of an organism occur in a tiny area and with tools and products so small that it is very difficult to appreciate the precision and complexity of the processes. The organism's most important tools, both within and outside cells, are the protein biocatalysts called enzymes. These biocatalysts serve the same function as do catalysts in nonbiological reactions: they accelerate reactions that would occur only very slowly or not at all without them. Like other catalysts, enzymes increase the rate at which a chemical equilibrium (the most stable proportions of reactants and products) can be reached, but they do not shift the equilibrium. Many reactions, even those that release energy, require a push to get them over what

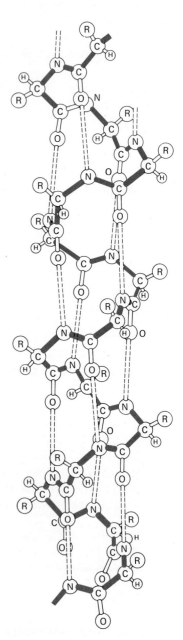

Fig. 4-12 The secondary structure of a protein showing the formation of an alpha helix (hydrogen bonds are shown by the dashed lines).

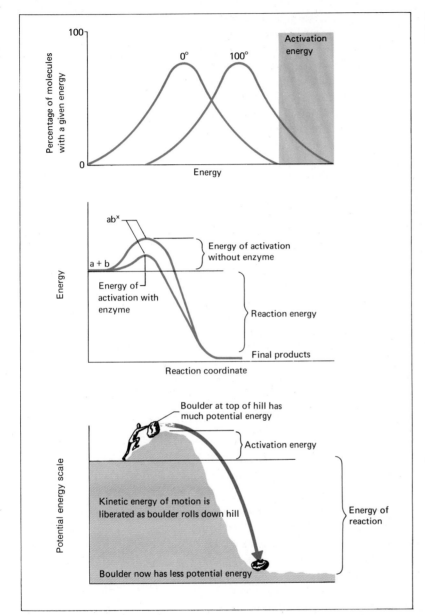

Fig. 4–13 Diagram to illustrate the concept of the activation energy barrier. (*After Simpson and Beck.*)

is called the ***activation-energy*** barrier (Fig. 4–13). A catalyst allows a reaction mixture to reach equilibrium more quickly because it lowers the ***energy of activation*** of the reacting compounds. When no catalyst is present, the amount of energy required to form the ***activated complex*** (a temporary, intermediate transition state) may be so great that spontaneous reactions are rare events. That is, few if any of the molecules are moving fast enough (have enough kinetic energy) to collide with sufficient force to react. Heating, of course, speeds up most reactions because the average kinetic energy of the molecule is increased. But there is a limit to which heating can be employed in biological systems because, for instance, proteins are denatured by heat. This is one reason why enzymes are absolutely essential to the chemical functioning of organisms. Enzymes are given a name which indicates their function, and this name usually ends in "-ase." Thus ribonuclease breaks down ribonucleic acid; tryptophan synthetase is involved in the synthesis of tryptophan. The names of the first enzymes discovered, such as trypsin, were given before the "-ase" convention was established.

ENZYMES ARE PROTEINS THAT SERVE AS CATALYSTS, INCREASING THE RATE OF BIOLOGICAL REACTIONS WITHOUT AFFECTING THE EQUILIBRIUM WHICH WOULD OCCUR IF NO ENZYME WERE PRESENT.

THE STRUCTURE AND FUNCTIONS OF ENZYMES With relatively recent improvements in microscopy has come an appreciation of the architectural complexity of cells. Before that, one view of the cell was that of a bag of enzymes in which the enzymes randomly performed their catalytic functions whenever they came in contact with their **substrate,** the compound acted upon by the enzyme. Concomitant with our increased knowledge of biochemical localization and the function of organelles within the cell has come an increasing awareness of the complexity of enzyme structure-function.

Now we can say with reasonable surety that an enzyme may have many sites of action, each of which performs a particular, somewhat independent function. These functions include (1) **localization,** or anchoring; (2) **polypeptide association;** (3) **coenzyme binding;** (4) **substrate binding;** (5) catalysis or specific activity at the **active site** or **center;** (6) recognition and binding site for **inhibitors.** This is shown diagrammatically in Fig. 4–14 for a hypothetical molecule. It is very likely that other functions will be discovered.

1 *Localization.* Many enzymes are associated with particular organelles in the cell. It is known, for instance, that some of the enzymes that function in energy metabolism (the cytochromes) are localized on the membranes of the mitochondria. One part of the enzyme must have a specific affinity for these membranes or for particular places on them.

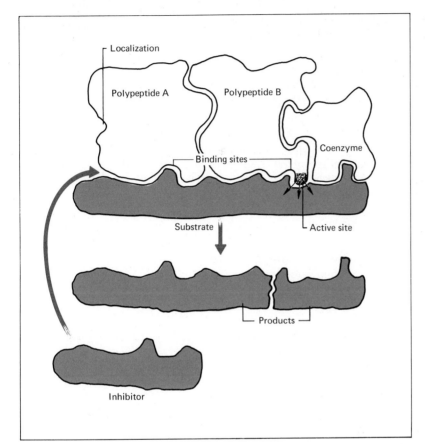

Fig. 4-14 A hypothetical enzyme molecule to show diagramatically various sites of action.

2   *Polypeptide association.*   Many enzymes are composed of two or more unlike polypeptide chains. Mention of this was made above with regard to hemoglobin. In many cases the separate chains show little or no activity when separated. The enzyme tryptophan synthetase in the colon bacterium, for instance, has two components, polypeptide A and polypeptide B, neither of which alone is more than 10 percent as active as is the complex of both. Undoubtedly there are regions on the polypeptides that function to bind the polypeptides together.

3   *Coenzyme binding.*   Many enzymes, including those that catalyze oxidative reactions (such as removing hydrogen from a molecule), will not function unless they are loosely bound to molecules called coenzymes. Either the act of coenzyme binding alters the conformation of the enzyme, enabling it to function properly, or the coenzyme actively participates in the reaction. Coenzymes have two universal

properties: First, they are not denatured by temperatures as high as 100°C, so they are nonprotein material (all proteins are denatured by such high temperatures). Second, they have relatively low molecular weights. This can be demonstrated by the technique called *dialysis,* in which a membrane sausage containing a mixture of large and small molecules is placed in water. The pores in the membrane permit the passage of small molecules into the surrounding medium but retain such large molecules as proteins and nucleic acids. When tissue extracts are repeatedly dialyzed, the oxidative enzymes in the mixture soon lose their activity due to the gradual leakage of the coenzymes. Many so-called vitamins are converted to coenzymes in the bodies of animals, including ourselves. Another group of dialyzable materials needed for the functioning of some enzymes comprises special ions, particularly divalent cations, including Ca^{2+}, Mg^{2+}, Zn^{2+}, Cu^{2+}, and Co^{2+}.

Some enzymes have nonprotein prosthetic groups. These nonproteinaceous groups are, by definition, firmly bound to the enzyme molecule and are not dialyzable. In reality, there exists a continuum, with easily removed coenzymes at one end and firmly attached prosthetic groups at the other.

4 *Substrate binding.* Recently, x-ray diffraction studies have begun to elucidate the forces that bind the substrate to the enzyme molecule, but there is good reason to believe that these sites are not precisely the same as the reaction or active sites (which probably are areas within or adjacent to the substrate binding site). The removal of a 20-amino-acid fragment from an enzyme in the pancreas that acts on ribonucleic acid, for instance, destroys the activity of the enzyme, but it is still capable of binding the substrate.

5 *Active site or center.* Many enzymes have been found to have similar amino acid sequences in the region thought to be the active site. Two intestinal proteolytic enzymes (those that catalyse the breakdown of proteins), trypsin and chymotrypsin, share the amino acid residue sequence asparagine–serine–glycine at this region. The sulfur-containing amino acid cysteine is also common in active sites, as is histidine. Many enzymes lose their activity when denatured, suggesting the possibility that the active site may be a configuration of amino acids extending across two or more polypeptide chains or across one folded chain.

6 *Inhibition.* Some enzymes combine with metabolites other than their usual substrates. This has the effect of blocking access to the enzyme or changing its shape and thus inhibiting its catalytic function. In many cases, the inhibitor is the end product of a biosynthetic pathway—a sequence of enzyme-catalyzed reactions in which each en-

zyme acts upon the product of the previous reaction. This situation can be diagramed as follows:

Reactants $\quad$ A $\rightarrow$ B $\rightarrow$ C $\rightarrow$ D
$$\uparrow \quad \uparrow \quad \uparrow$$
Enzymes $\quad$ I $\quad$ II $\quad$ III

An excess of D, by inhibiting enzyme I, shuts down the pathway that results in the synthesis of D, thereby preventing the wasteful accumulation of precursors in the same pathway. This is an example of feedback control at the enzyme level. Other regulatory systems operate at the DNA-RNA level and will be discussed shortly. The advantage of a feedback system at the enzyme level is the rapidity with which it can take effect without the regulation of protein-synthetic steps.

AN ENZYME MOLECULE HAS A DEFINITE SHAPE AND SEVERAL REGIONS OF ACTIVITY, EACH WITH A DIFFERENT FUNCTION.

Protein Synthesis

Earlier we discussed DNA, the hereditary chemical which has encoded in it the information necessary to produce and maintain an organism. Now we have discussed another group of macromolecules, the proteins, and specifically the enzymes that actually perform the work of making and maintaining an organism. How does the information stored in the sequence of nucleotides in DNA become translated into the sequence of amino acids of the enzymes?

A fairly clear picture of protein synthesis has emerged over the last few years, although there are still more than a few gaps in our knowledge. In general, what happens can be thought of as a two-step process. In the first step, the DNA acts as a template for the synthesis of ribonucleic acid (RNA). (RNA, remember, differs fron DNA in containing the sugar ribose instead of deoxyribose and in containing uracil in place of thymine.) In the second step, some of the RNA serves as a second-order template which dictates the sequence of amino acids in the resulting proteins.

SYNTHESIS OF RNA $\quad$ The first step is analogous to the replication of DNA, although only one of the two DNA strands serves as a template for RNA synthesis in any given region of the double helix. The nucleotides of the transcribed DNA strands are free to pair with the monomers of RNA (Fig. 4–15). The RNA monomers then are joined by the enzyme RNA polymerase into a single strand of RNA, in a process called **transcription.** It happens, however, that there is much more information in the DNA

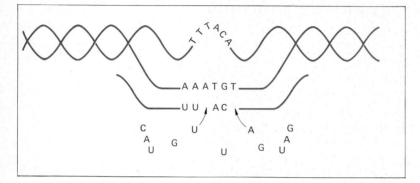

than any cell can use at any one time, so all the DNA is not transcribed at once as a large molecule of RNA. (Even if all the information were needed at one time, it would probably be impossible to handle an RNA molecule as long as the whole DNA molecule in a cell.) The DNA is transcribed in short pieces, and there are start and stop signals on the DNA to mark where these pieces are to be made.

The place on the DNA at which transcription is to begin is recognized by one of the five polypeptide chains of the RNA polymerase. This chain, σ, does not participate in the catalytic function of the enzyme but serves solely to recognize the start signals in the DNA, which are called **promotors.** Another, less well understood protein factor, ρ (which is not part of RNA polymerase), recognizes the stop signals in the DNA and terminates transcription. After transcription, the strand of RNA, now called **messenger RNA** (mRNA), moves to the sites of protein synthesis. In this strand, a segment of the DNA message (in the A, T, G, C deoxyribose language) has been transcribed into the RNA ribose language (A, U, G, C).

THE RIBOSOME Some protein synthesis takes place within the nucleus, but most occurs in the cytoplasm and specifically at the organelles called **ribosomes.** Ribosomes are usually associated with the endoplasmic reticulum, but they may also be found free. They consist of about 40 percent protein and about 60 percent ribosomal RNA. They are always made of two subunits, the smaller of which seems to have an affinity for the mRNA coming from the nucleus and bearing the transcribed DNA message. The larger subunits apparently have attachment sites for a third kind of RNA, soluble or **transfer RNA** (tRNA). The exact function of ribosomal RNA is unknown at present. It may serve temporarily as a template for the assembly of ribosomal proteins. All this is conjecture, however. Both mRNA and ribosomal RNA can be shown, experimentally, to pair with particular regions of the DNA. This suggests that both types are synthesized in the nucleus, since the pairing of RNA and DNA depends upon the similarity of nucleotide sequence.

TRANSFER RNA   As its name implies, tRNA is a carrier molecule. Each molecule carries a specific amino acid to the ribosomes and positions it where it can polymerize with the growing chain to form the correct polypeptides. There are over 20 kinds of tRNA – enough so that there can be at least one for each amino acid. All will pair with particular regions of the DNA, suggesting that they are synthesized in the nucleus. The tRNA molecules are very small compared with the other types of RNA; they have only about 70 to 80 nucleotides. When carrying their amino acids, they bind specifically to the ribosomes. But what ensures that the amino acids will be laid down in just the right sequence to synthesize the correct polypeptide?

ACTIVATING THE MOLECULES   The story is a little complex, so let us look at the components of the system first. First of all, there are more than 20 kinds of tRNA. Second, the process requires energy, and the source of this is the compound **adenosine triphosphate** (ATP), a fuel compound that is discussed in more detail in Box 4–2. Finally, there need to be 20 staging places, where each amino acid will be brought into contact with its specific kind of tRNA in the presence of ATP. This function (catalysis and recognition) is performed by a set of triple-threat amino acid activating enzymes, as shown diagrammatically in Fig. 4–16. Each enzyme must recognize and bind the one amino acid for which it is specific, the corresponding specific tRNA, and ATP. The amino acid is **activated** by reacting with the ATP molecule, yielding an amino acid adenylate and a pyrophosphate molecule. The amino acid adenylate is a very reactive high-energy compound, but the enzyme holds it until it can react with the proper tRNA to yield the charged tRNA (amino acid-tRNA complex) as well as the low-energy molecule adenosine monophosphate (Fig. 4–16).

ASSEMBLING THE AMINO ACIDS   The actual polymerization of the amino acids is thought to occur as follows. The mRNA becomes bound to a series of ribosomes, thus forming a complex known as a **polyribosome** (Fig. 4–17, Fig. 4–18). Each ribosome has two side-by-side positions: one site, P, that receives the newest tRNA bearing the amino acid-tRNA complex and a second site, A. The A site is occupied by the previous tRNA with its amino acid now bound to the end of the forming protein chain. As shown in Fig. 4–19, as a ribosome moves along the mRNA

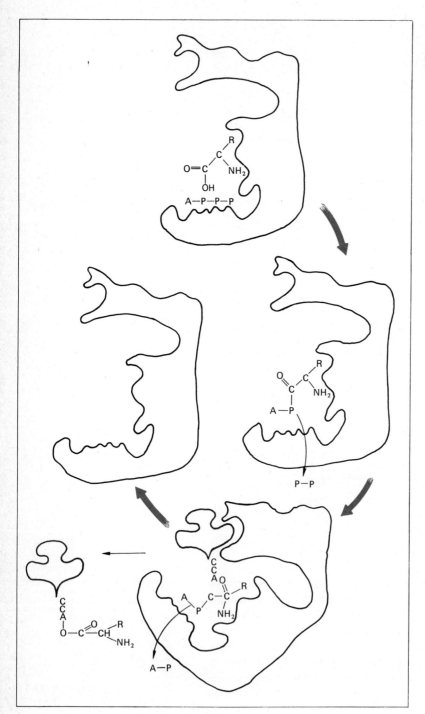

Fig. 4-16 Function of amino acid activating enzymes: the tRNA (shown in color) becomes associated with the proper amino acid and ATP to produce the charged tRNA.

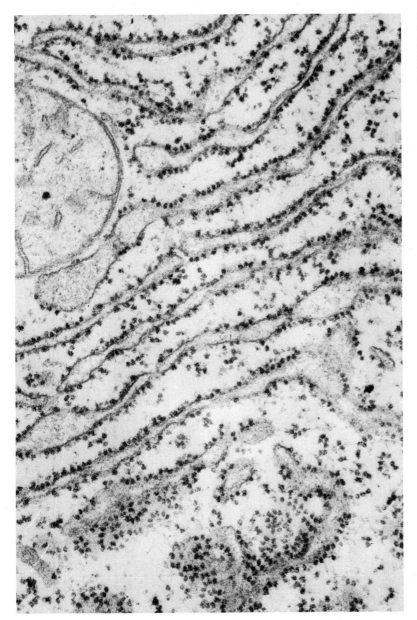

Fig. 4-17 Electron micrograph of a cell in rat liver. Membranes of the endoplasmic reticulum were cut transversely (above) and tangentially (below). Polyribosomes are thus seen in two views. (*Courtesy of G. E. Palade.*)

template, charged tRNA molecules carrying the proper sequence of nucleotides enter the ribosome P site and then move over to the A site, where they lose their amino acids and are ejected.

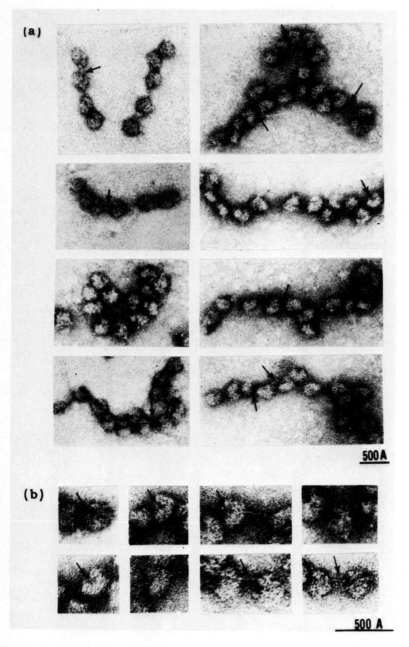

500 A

500 A

Fig. 4-18 Electron micrographs of polyribosomes. The strand mRNA between ribosomes is shown by arrows. In *b*, at higher magnification, the mRNA is clearly shown and it can be seen that the ribosomes consist of two subunits of unequal size. *(Courtesy of D. Sabatini, Y. Nonomura, and G. Blobel.)*

An important feature is the pairing of nucleotides of the tRNA with complementary nucleotides on the mRNA. The latter determines which of the 20 possible charged tRNA molecules become aligned and, thus, the order

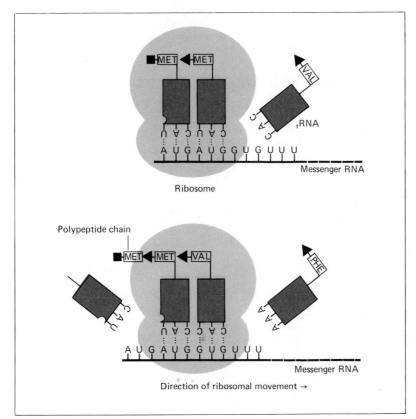

Fig. 4-19 Formation of a polypeptide chain as a ribosome moves along a strand of mRNA. The special amino acid, formyl methionine (indicated by a square), associates with AUG in the ribosome which is the beginning of a polypeptide chain. (*After Clement and Marcker.*)

in which the amino acids are polymerized by an amino acid polymerizing enzyme.

The role of the tRNA is especially critical since the accuracy of protein synthesis, that is, of the *translation* of the RNA code into the amino acid sequence, is largely dependent on the double specificity of tRNA. It must have a region that recognizes the proper amino acid activating enzyme so that it becomes charged with the correct amino acid; in addition, it must recognize a trinucleotide code word on the mRNA so that its amino acid is properly placed on the growing polypeptide chain.

PROTEINS ARE SYNTHESIZED BY THE JOINING OF AMINO ACIDS END TO END. THIS PROCESS USES THE ENERGY OF ATP AND IS CATALYZED BY ENZYMES. THE SEQUENCE OF AMINO ACIDS IS DETERMINED BY THE SEQUENCE OF NUCLEOTIDES IN ONE OF THE TWO STRANDS OF A DNA MOLECULE. A SPECIAL KIND OF RNA, MESSENGER RNA (mRNA), IS FORMED IN COMPLEMENTARY FASHION WITH A PORTION

OF DNA. THIS mRNA, BEARING THE CODE FOR AMINO ACID SE-
QUENCE, LEAVES THE NUCLEUS AND ATTACHES TO RIBOSOMES,
THE SITES OF PROTEIN SYNTHESIS. AT THE RIBOSOMES, AMINO
ACIDS, BROUGHT TO THE RIBOSOME BY TRANSFER RNA (tRNA)
MOLECULES ARE JOINED IN THE PROPER SEQUENCE DICTATED BY
THE NUCLEOTIDE SEQUENCE IN THE mRNA.

In Chap. 10, we shall discuss evidence showing that cells from different parts of plants or animals contain the DNA required to produce all the proteins found in an organism. It is clear that not all this information is used by any one cell. (See Box 4-3.) It has been demonstrated that the tadpole intestinal-cell nucleus contains the genes for producing hemoglobin since it and its descendants can control the development of an entire frog—hemoglobin and all. The nucleus does not, however, call for the production of hemoglobin when it is in an intestinal cell.

We have seen that the hereditary information is stored and replicated in the form of DNA and is dispatched to the protein synthesizing centers— the ribosomes—in the form of RNA. At the ribosomes the genetic information is translated into a totally different language, the functioning language of proteins. The information is put to work in the form of a sequence of amino acids capable of performing specific enzymatic or other functions.

The Genetic Code

It should be obvious that a nucleotide, or a sequence of nucleotides, in the DNA and RNA molecules codes for each of the 20 amino acids. "Cracking" this code was biochemistry's major *tour de force* of the sixties.

THE NUCLEOTIDE CODE   How many nucleotides make a word? That is, how many nucleotides specify an amino acid? Does just one nucleotide or does a series of two, three, four, or more nucleotides code for each amino acid? Many years before experimental results began to resolve this question, theoretical evidence suggested that a nucleotide word contained three letters. Obviously, a one-letter code would not do, since the 4 nucleotides could code for only 4 amino acids out of the 20. A two-letter code would also be insufficient, allowing only 16 (4^2) combinations. A three-letter, or triplet, code permits 64 (4^3) combinations, more than enough. Any more than three letters in the code would be uneconomical and would thus be inconsistent with the "principle of parsimony" that often seems to apply to the biological world. Naturally, this is not itself evidence that the code is triplet. Massive experimental evidence now indicates, however, that nature *has* been parsimonious and that the code does indeed consist of triplets.

If the code were overlapping, more information could be packed into a smaller number of nucleotides. As shown in Fig. 4–20, a sequence of six nucleotides would code for two amino acids if the code were a triplet, nonoverlapping code. However, if any nucleotide could be part of more than a single code word, the six nucleotides could code for four amino acids. One kind of evidence that militates against an overlapping code is the effect of mutations, changes in nucleotides. (As will be shown below, most mutations seem to be either replacements, deletions, or additions of *single* nucleotides.) Where it has been possible to investigate the effect of mutations on amino acid changes in polypeptides, it has been shown that only a single amino acid is changed by a single mutation. If the code were overlapping, one would expect that at least some mutations would affect two or more coding words and would lead to two or more amino acid changes in the corresponding protein. The tobacco mosaic virus, which produces lesions on tobacco leaves (Fig. 4–21), is a single strand of RNA (in some plant viruses, the genetic code may not be DNA) coated by a protein that contains 158 amino acids. Mutations have never been known to affect two adjacent amino acids in this polypeptide.

THE GENETIC CODE SPECIFYING AMINO ACID SEQUENCE IS WRITTEN IN THE SEQUENCE OF NUCLEOTIDES OF THE DNA. A SEQUENCE OF THREE NUCLEOTIDES SPECIFIES A PARTICULAR AMINO ACID. A CHANGE IN EVEN ONE NUCLEOTIDE MAY AFFECT PROTEIN SYNTHESIS, CAUSING A MUTATION.

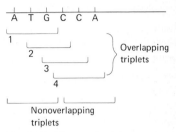

Fig. 4–20 Diagram to show overlapping and nonoverlapping triplets in a sequence of six nucleotides.

Fig. 4–21 Lesions on leaves of tobacco (*Nicotiana*) produced by the RNA virus called tobacco mosaic virus. (*Courtesy of H. Fraenkel-Conrat.*)

BREAKING THE CODE   A major breakthrough in the decoding of the nucleic acids came with the in vitro (literally, "in glass," meaning in a test container) synthesis of these compounds—the Nobel Prize-winning work begun by Ochoa and Kornberg. This system permitted the synthesis of artificial mRNAs such as polyuridylic acid, U–U–U–U–. When these man-made RNAs were added to test tubes containing ribosomes and the other necessary components of protein synthesis, polypeptides formed, which could then be purified and analyzed for their amino acid content and sequence.   Where the "messenger RNA" was polyuridylic acid, the resultant peptides were strings of phenylalanine, even when all 20 amino acids were present.   Apparently U or a string of U's codes for this specific amino acid.   Refinements and variations of this technique have produced coding words **(codons)** for all the amino acids.   A triplet code is now well documented.   The assignment of codons is shown in Table 4–1.

Another result of this approach has been the detection of what is called **degeneracy** in the code.   As you can see from the table, more than one

TABLE 4–1. The RNA Nucleotide Triplet Codes Specifying Amino Acids: A, Adenine; U, Uridine; G, Guanine; C, Cytosine.

TRIPLET	AMINO ACID	TRIPLET	AMINO ACID	TRIPLET	AMINO ACID
UUU, UUC	Phenylalanine	CGU, CGC, CGA, CGG	Arginine	GCU, GCC, GCA, GCG	Alanine
UUA, UUG, CUA, CUU, CUG, CUC	Leucine	CAU, CAC	Histidine	GGU, GGC, GGA, GGG	Glycine
		CAA, CAG	Glutamine		
UCU, UCC, UCA, UCG, AGU, AGC	Serine	AUU, AUC, AUA	Isoleucine	GAU, GAC, GAA	Aspartic acid
		AUG	Methionine	GAA, GAC	Glutamic acid
UGU, UGC	Cysteine	ACU, ACC, ACA, ACG	Threonine		
UGG	Tryptophan			UAA, UAG	Glutamine or termination signal
		AAU, AAC	Asparagine		
UAU, UAC	Tyrosine	AAA, AAG	Lysine		
CCU, CCC, CCA, CCG	Proline	GUU, GUC, GUA, GUG	Valine		

codon specifies most amino acids. One important consequence of this discovery is that many changes in nucleotides may go unnoticed by biologists since they may not have any effect on the amino acid sequence in the protein. This, of course, would result in an underestimate of mutation rate. Degeneracy might also tend to buffer the organism from the effects of mutations. For instance, a change from U–U–U to U–U–C might make little difference since the organism contains phenylalanine tRNA for both of these triplets. One should not conclude from this, however, that members of a degeneracy set, such as U–U–U and U–U–C, are identical from a functional point of view. The triplets may be serviced by tRNA molecules which are present in different combinations or which have different chemical properties. These differences quite possibly play critical roles in developmental processes.

The question naturally arises as to whether there is a difference in the code among different groups of organisms. The possibility of such "dialects" is perfectly consistent with our understanding of the evolutionary process. If all organisms evolved from a common stem population, then all organisms share a common ancestral DNA code. We know that 3 billion years of hereditary isolation has resulted in all the organic diversity in form-function that a glance out the window will demonstrate.

There is every reason to believe that the code, like every part of organisms, should evolve differences. In short, variation on a theme is the expected result.

In general, though, the code seems to be remarkably universal. This was strikingly demonstrated when the protein-making machinery of bacteria and rabbits were mixed (Fig. 4–22). mRNA and ribosomes were isolated from the reticulocyte cells (hemoglobin-synthesizing cells) of rabbits and combined in vitro with amino-acid-charged tRNA from the colon bacterium *E. coli*. In order for this system to produce recognizable protein, specifically hemoglobin, the bacterial tRNAs would have to read the rabbit mRNA exactly as do the rabbit's own tRNAs. Indeed, the hemoglobin that was isolated from this hybrid factory had exactly the same primary structure as hemoglobin made with rabbit reticulocyte mRNA.

How does the protein synthesis system know where to begin and end a polypeptide chain? It is now known that mRNA has coded into it precise starting and stopping points for polypeptides. It has also been shown that certain tRNA molecules carry a special, chemically altered amino acid, formyl methionine, to the start position. Two forms of the tRNA that carries methionine are known. Both bear the codon CAU, but only one can start a polypeptide, because it can participate in the process by which the methionine is formylated. The special amino acid is apparently able to go directly to the peptide position in the ribosome to start the polypeptide

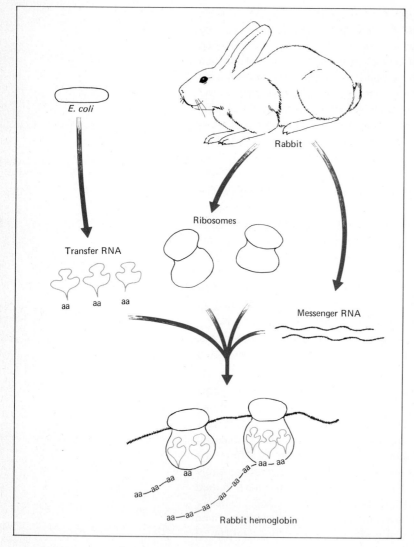

Fig. 4-22 Diagram of an experiment in which mRNA and ribosomes of rabbits were combined in vitro with amino acid–charged tRNA from *E. coli.* The hemoglobin specified by the rabbit mRNA was produced by the bacterial system.

chain (Fig. 4-19). The mRNA has the codon AUG in the start position. There are three different "stop" codons, which may occur in pairs (double stops). These stop codons are identified not by tRNA molecules but by proteins known as **release factors.**

GENETIC INFORMATION IS STORED AS TRIPLETS OF NUCLEOTIDES IN THE DNA. THE INFORMATION IS TRANSCRIBED INTO A COMPLE-MENTARY STRAND OF RNA. THIS, IN TURN, IS TRANSLATED INTO A

*POLYPEPTIDE BY THE POLYRIBOSOME COMPLEX OF RIBOSOMES
AND mRNA, AIDED BY tRNA MOLECULES AND A COTERIE OF EN-
ZYMES.*

CELLULAR ENERGETICS

Now that we have an understanding of how genetic information is stored
in the DNA and how this information is translated into the structure of
enzymes, we may turn to the enzyme-mediated synthetic processes
themselves. The first of these is photosynthesis; this process is the source
of virtually all the chemical energy available to living systems.

Photosynthesis

Two basic processes are involved in photosynthesis. The first is the
conversion of light energy into chemical energy. The other process is
the conversion of carbon dioxide to organic carbon compounds (CO_2
fixation), a conversion which is powered by energy obtained from the
first process.

THE LIGHT REACTIONS   The uniqueness of photosynthesis lies not in
the fixing of CO_2 but in the conversion of light energy to chemical energy.
This photochemical process poses two problems: (1) How is the light
energy transferred to ATP? (2) What is the source of the hydrogens that
eventually reduce the CO_2?

One of the first clues to the nature of the photosynthetic process was
the discovery of oxygen in 1774 by the English amateur chemist Joseph
Priestley. (Priestley, however, did not call it oxygen.) He found that
air which would no longer support an animal could be ameliorated by
plants, which liberate oxygen (Fig. 4–23). In contrast, animals liberate

Dead mouse   Dead plant   Live plant
Live mouse

Fig. 4–23   The simple experiment of
J. B. Priestley enclosed mice and
plants separately and together in
sealed glass containers. Only those
which were combined survived.

another gas (CO_2) but require oxygen. Shortly thereafter, the Dutch physician Jan Ingen-Housz learned by experiments that only the green parts of plants give off oxygen and only when placed in strong sunlight.

Microscopic analysis was soon to show that the green color of leaves and other structures is due to small disk-shaped bodies called **chloroplasts** (Fig. 4–24). Still later, the latter were found to contain the green pigment **chlorophyll.** (In algae, chloroplasts assume other shapes, and in blue-green algae and some photosynthetic bacteria, chloroplasts are missing altogether.) Chlorophyll appears green because it strongly reflects green light while absorbing the red and blue parts of the visible spectrum. Actually, two chlorophylls, *a* and *b,* are typically found in green plants and in green algae. Other kinds of chlorophyll occur in other kinds of algae and in photosynthetic bacteria.

When light is passed through a solution of chlorophyll, an instrument called a spectrophotometer can measure exactly how much of the light of any given wavelength is being absorbed by the solution. When the absorption in a chlorophyll solution is plotted against the wavelength, the resulting graph is called the **absorption spectrum** of chlorophyll (Fig. 4–25).

Another group of pigments, the carotenoids (from the word "carrot," since carrots owe their color to these pigments), are found associated with

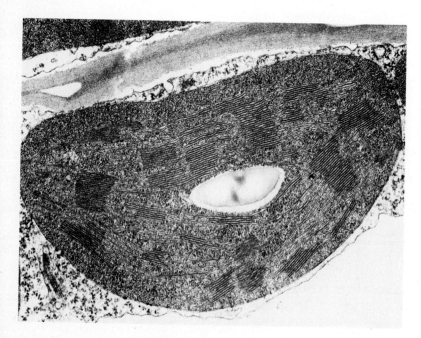

Fig. 4–24 Electron micrograph of a chloroplast from a cell of a tomato leaf. The pigment chlorophyll is associated with the darker membrane systems. × 22,000. (*Courtesy of Myron C. Ledbetter.*)

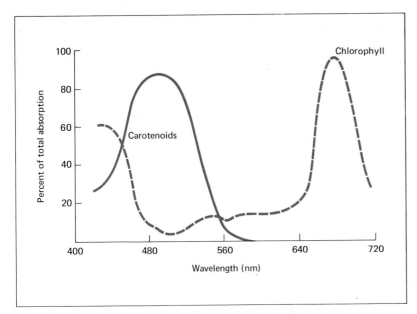

Fig. 4-25 Absorption spectrum of chlorophyll and carotenoid pigments. The amount of light at each wavelength from violet (400 nm) to red (720 nm) is shown.

chlorophylls in plants (Fig. 4-25). However, they are usually not noticed because they are masked by the chlorophylls. The decomposition of the chlorophylls in the leaves of deciduous trees in autumn unmasks the carotenoids, resulting in the annual display of reds, oranges, and yellows.

Which of these two groups, the chlorophylls or the carotenoids, is most important in photosynthesis? An ingenious experiment (Fig. 4-26) in 1881 by the German plant physiologist T. W. Engelmann answered this question. Engelmann illuminated a strand of green alga with a tiny spectrum from a special prism so that light of different wavelengths impinged on different regions of the alga filament. In the medium with the alga were motile, aerobic (oxygen-requiring) bacteria. When the light was turned on, causing the spectrum to fall on the alga, the bacteria began to congregate around certain regions of the strand—especially those regions where the red light and blue light were falling. The first conclusion is that red light and blue light result in the release of oxygen, which in turn is an indication of photosynthetic activity. The second conclusion is that it is the chlorophylls which are the active photosynthetic pigments rather than the carotenoids, since it is the former that absorb both the red and the blue. Plotting the rate of photosynthesis against the wavelength of light produces an **action spectrum** (Fig. 4-26) for the process. Note the similarity of the action spectrum of photosynthesis with the absorption spectrum of the chlorophylls. The carotenoids do seem to function in

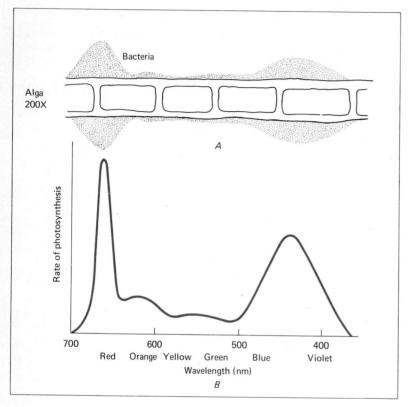

Fig. 4-26 Action spectrum of chlorophyll: (A) Engelmann's experiment which showed that motile bacteria aggregate at regions of an alga filament, illuminated by a spectrum, where oxygen is produced and hence photosynthesis is occurring; (B) The rate of photosynthesis is measured at each wavelength from red to violet.

photosynthesis, however. They are thought to transfer to the chlorophylls some of the light energy they absorb.

How is the light absorbed by chlorophyll linked to the generation of ATP and the hydrogens needed to reduce the CO_2 in the synthesis of carbohydrate? Not all the steps in these processes have yet been elucidated; nevertheless, outlines have emerged with sufficient clarity to permit some general statements. When a unit of light energy (called a quantum) is absorbed by a molecule of chlorophyll, an electron of the molecule is raised to a higher energy level; it is then in a very reactive condition. If this occurs in a chlorophyll molecule in solution, the electron quickly returns to its former energy level, releasing the energy as heat and light (called fluorescence). But in chloroplasts, the architecture of the membranes ensures the transfer of these excited electrons to an electron acceptor and then on to a cytochrome molecule. Cytochromes, which will appear again in our discussion of respiration, act as carriers in an electron bucket brigade. From the cytochrome, the electrons are passed on in a cyclic pathway to a second cytochrome acceptor and then returned to the chlorophyll a. There is also a noncyclic pathway, involving both chlorophyll a and chlorophyll b, which will be discussed below.

In the cyclic pathway, the loss of energy by the electron as it is transferred from the original acceptor back through a pair of cytochrome acceptors to the chlorophyll provides enough energy to form two or more ATP molecules (Fig. 4–27). Energy-poor molecules of adenosine diphosphate (ADP) are converted into energy-rich molecules of adenosine triphosphate (ATP) in a process of phosphorylation, the adding of a phosphate group. Phosphorylation is an energy-requiring (endergonic) reaction (see Box 4–2). In a sequence of enzyme-mediated steps, each acceptor molecule in turn is reduced by the acceptance of the electron. The electron loses energy as it moves down the chain of acceptors, and some of this energy is used to convert ADP to ATP. This is an example of a coupled reaction in which one reaction furnishes the energy to drive a second reaction. In this case, the reduction of the acceptors by the electron drives the phosphorylation of the ADP molecule. As shown in Fig. 4–27, chlorophyll can be thought of as an electron pump (or generator). The force that drives this pump is sunlight. Since the primary synthetic step of this cyclic process is the phosphorylation of ADP to make ATP, the process is usually referred to as **cyclic photophosphorylation.** This is the source of most of the ATP used in CO_2 fixation.

The noncyclic pathway is a little more complex (Fig. 4–27). Again the chlorophyll pump kicks electrons up to a high energy level. Again an electron acceptor traps them. But instead of having their energy bled for the synthesis of ATP, they are passed on to another electron acceptor, nicotinamide adenine dinucleotide phosphate (NADP), which itself becomes highly reactive. NADP is a coenzyme containing a derivative of vitamin B_2. It combines with hydrogen ions from water and is converted to $NADPH_2$. This leaves hydroxyl ions (OH^-), essentially a surplus of negative charges (electrons), with no place to go. See Boxes 4–2 and 4–5.

At this point the second chlorophyll, chlorophyll b, comes in. Light excites electrons of chlorophyll b, and they travel along the same acceptor cascade as those electrons in the cyclic path discussed above. They return to chlorophyll a, replacing the chlorophyll a electrons that took off on the noncyclic path and ended up exciting NADP. All that is left to balance things out is to replace the lost electrons of chlorophyll b. That is what the hydroxyl ions do. They donate their electrons to fill the holes left by the departed electrons of chlorophyll b. The remaining OH radicals combine: $4OH \rightarrow 2H_2O + O_2$, producing an important by-product of photosynthesis, oxygen.

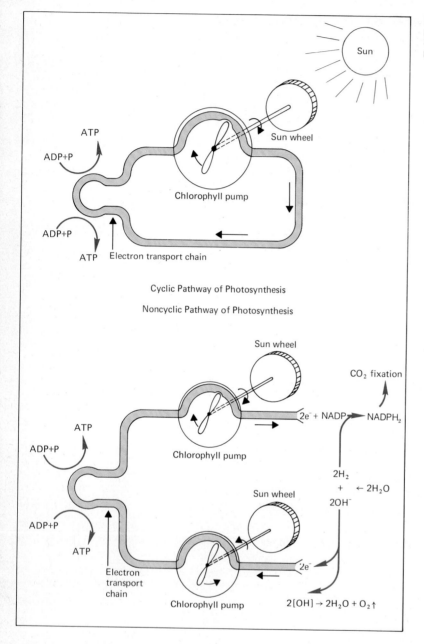

Fig. 4–27 Diagrams of the cyclic pathway of photosynthesis (*above*) and the noncyclic pathway of photosynthesis (*below*).

Both the hydrogen that eventually reduces the fixed CO_2 in the dark reactions (discussed below) and the O_2 result indirectly from the splitting of water. The process is indirect because the actual ionization of water is driven by the removing of hydrogen and hydroxyl ions from solution, thereby driving the reaction

$$H_2O \rightleftharpoons H^+ + OH^-$$

to the right.

Together, these reactions, which yield ATP, $NADPH_2$, and O_2, are called the light reactions of photosynthesis since light is required to drive them. The CO_2-fixation reactions are called the dark reactions because they can go on in the dark as long as there is sufficient ATP and NADPH to drive them.

It is worth noting that the light reactions occur in the membranes in which the chlorophyll is found, whereas the enzymes that mediate the dark reactions occur in the substance surrounding the membranes. This arrangement is similar to that found in mitochondria, those organelles that perform essentially the reverse of photosynthesis by liberating the energy in sugar molecules.

A BASIC PROCESS OF PHOTOSYNTHESIS IS THE CONVERSION OF LIGHT ENERGY TO CHEMICAL ENERGY. LIGHT ENERGY IS ABSORBED BY CHLOROPHYLL AND, IN A COMPLEX SERIES OF STEPS, IS TRANSFERRED TO ATP. IN THIS PROCESS, NADP IS REDUCED AND OXYGEN IS PRODUCED.

CARBON DIOXIDE FIXATION   Carbon dioxide fixation is not, as was once thought, limited to autotrophic organisms, and it can occur in the dark as well as in the light. The capacity to attach the carbon atom of CO_2 to other carbon atoms has been found to occur throughout the heterotrophs, although they do not fix nearly as much carbon as do plants. If both autotrophs and heterotrophs are capable of making their own carbon skeletons from the CO_2 in the atmosphere, why do not the latter become nutritionally independent of plants? If animals can absorb CO_2 through their gills, lungs, and skin, why do they not synthesize their own carbon compounds, as do plants?

A look at the CO_2-fixation reactions will suggest an answer to these questions. These reactions have been studied by the biochemist Melvin Calvin and his associates by combining the methods of organic chemistry, paper chromatography, and radioautography. Their procedure was to expose the green alga *Chlorella* in a suspension with radioactive carbon dioxide (^{14}C substituted for ^{12}C) to flashes of light. Samples of the *Chlorella* were then quickly killed, and the cell contents were separated by paper chromatography. To determine which of the compounds separated on the chromatogram were radioactive, a sheet of x-ray sensitive film was placed next to the chromatogram for a time and then developed. The areas on the chromatogram corresponding to the exposed areas on the film were then cut out, and proper analyses were performed to identify the radioactive compound at that site. It was found that the glucose on the chromatograms was radioactive after only 30 sec of photosynthesis. This means that only 30 sec is required for a CO_2 molecule to become part of the six-carbon sugar. With less time, other carbohydrates, including three-carbon and five-carbon compounds, were found to contain most of the ^{14}C. By painstaking work, which lasted many years, the intermediate compounds were identified and the circular pathway of CO_2 fixation gradually was elucidated (Fig. 4–28).

CO_2 enters the fixation cycle at the point where a five-carbon **(pentose)** acceptor molecule has been "energized" by reacting with the high-energy molecule ATP. In the fixation cycle, following a series of reactions in which two more ATPs are utilized, 2 three-carbon **(triose)** molecules are generated. Via a complex series of enzyme-catalyzed reactions (as are all these steps), five or six (on the average) of these trioses are utilized to *regenerate* the pentose molecule which originally acted as the CO_2 acceptor. The remaining triose enters into the general metabolism of the cell. Usually a molecule of glucose is synthesized from two of these trioses. During this series of reactions, hydrogens are required to reduce the CO_2. These hydrogens are provided by the hydrogen-transfer molecule $NADPH_2$.

We see then that three molecules of ATP and one of $NADPH_2$ are required merely to fix one CO_2 molecule. To fix 6 carbons, equivalent to 1 sugar

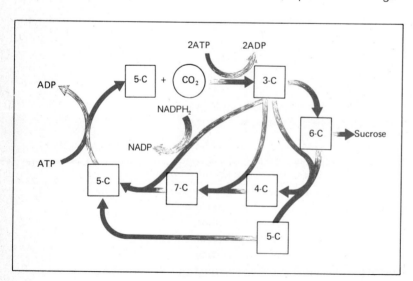

Fig. 4–28 The circular pathway of carbon dioxide fixation. (*Adapted from Kimball, BIOLOGY, 2/e, 1968, Addison-Wesley, Reading, Mass.*)

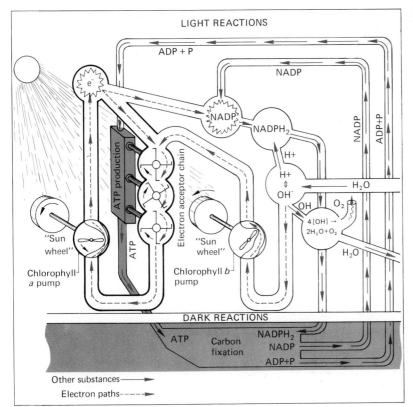

Fig. 4-29 Combined diagram showing the pathways and products of photosynthesis.

molecule, 18 molecules of ATP and 6 of NADPH$_2$ are required. Energetically, this is a rather expensive business for the cell. It would be a heavy drain on the energy stores of an organism if there were not an extremely economical way of producing ATP and reduced NADP. Such a process *is* available to organisms that can convert sunlight, the electromagnetic radiation in the visible range from the sun, to chemical energy. The pathways and products of photosynthesis are diagramed together in Fig. 4-29.

PHOTOSYNTHESIS CONSISTS OF TWO BASIC PROCESSES. ONE OF THESE—FIXATION OF CARBON DIOXIDE INTO ORGANIC MOLECULES —OCCURS IN BOTH AUTOTROPHS AND HETEROTROPHS. ENERGY OF ATP IS USED TO COMBINE C ATOMS TO BUILD COMPLEX ORGANIC SUGARS. THE OTHER PROCESS IS THE CONVERSION OF LIGHT ENERGY TO THE CHEMICAL ENERGY OF ATP.

In the last section, we saw how some organisms (higher plants, algae, and some bacteria) can exploit solar energy as an environmental resource to provide the energy for their metabolic processes, as well as all the carbon scaffolding on which they and the heterotrophs depend. We shall now discuss the fate of the carbohydrates produced in photosynthesis. (Carbohydrates are discussed further in Box 4–4.)

Carbohydrate metabolism is the hub of biochemistry for most organisms, which share nearly identical operations in this aspect of their biochemistry. Once heterotrophic microorganisms have acquired carbohydrates as part of their input, they can survive with little or no supplemental organic material (with some very important exceptions to be discussed below). Even autotrophs can survive without light—that is, without photosynthesizing sugar—as long as they are provided with a sugar source. However, some *inorganic* compounds—salts and a source of nitrogen—are universal nutritional requirements.

The central role of carbohydrate metabolism is illustrated diagrammatically in Fig. 4–30. Once sugars are photosynthesized, they are utilized both for energy and for construction material in all the other biochemical systems. Sugar, in the form of **glucose**, can be polymerized and stored as a **starch** or as a very similar compound, **glycogen**, which is found in animals (Fig. 4–31). In plants, sugar is often transported as the disac-

Carbohydrate Metabolism and Cellular Respiration

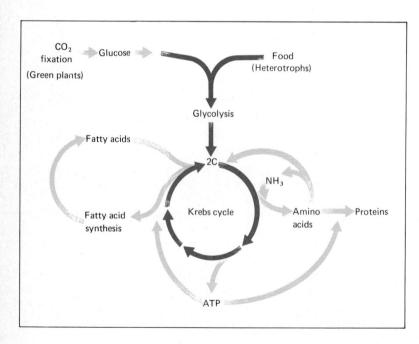

Fig. 4–30 The central role of carbohydrate metabolism (shown in color) in cellular biochemistry.

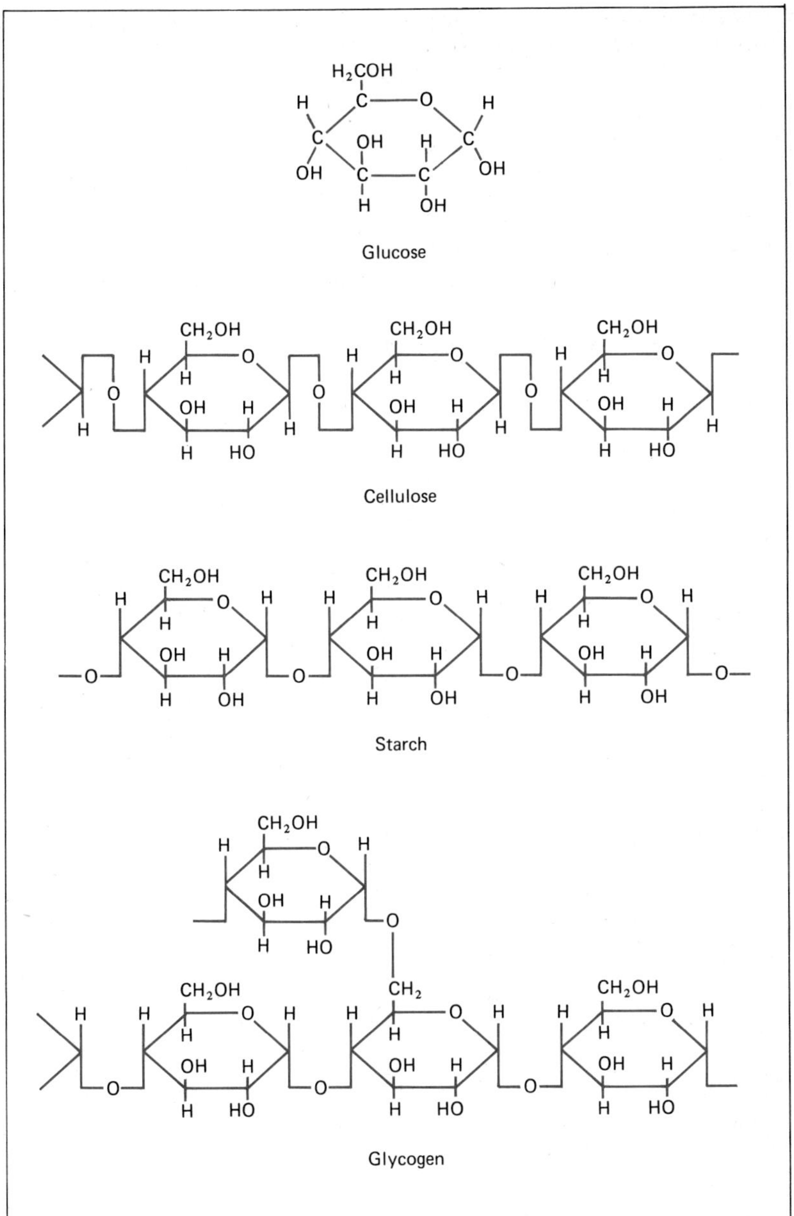

Fig. 4-31 Important carbohydrates found in cells.

charide **sucrose** (one molecule of glucose combined with one of fructose), familiar as table sugar. Other polymers of glucose are used for structural materials. The principal component of the supporting tissues of plants, for instance, is a glucose polymer, **cellulose.** Another glucose polymer, **chitin,** is a major skeletal material of insects and many other invertebrates. The carbon backbone of sugar molecules is the scaffolding from which amino acids and lipids (to be discussed below) are made.

Finally, the potential chemical energy released by the oxidative breakdown of carbohydrate is captured for the synthesis of the "high-energy" bonds of ATP in a process known as **cellular respiration.** Since the pathways of carbohydrate oxidation and ATP production are intimately related and often part of the same general process, we shall consider them together.

The process by which sugar (glucose) is converted to yield energy can be approached at various levels of complexity. In very elementary terms, the process can be stated in the form

$$C_6H_{12}O_6 + 6O_2 \rightarrow 6CO_2 + 6H_2O + \text{energy (686 kilocalories)}$$

The complete oxidation of a **mole** of glucose to CO_2 and water liberates 686,000 cal of energy, 45 percent as heat and the rest as chemical bonds. (One mole of any substance is an amount which contains the same standard number of molecules or atoms, roughly 6×10^{23}. A mole of lead is obviously much heavier than a mole of hydrogen. A calorie is the amount of energy required to raise one gram of water one degree centigrade. A kilocalorie is 1000 cal.) This formula is the same as that showing the result of burning 1 mole of glucose, except that when glucose is burned, all its bond energy is liberated as heat.

At the other extreme, a discussion of the biological process in all its complexity would require a description of 30 or so intermediate compounds, together with 35 or so enzymes, coenzymes, and accessory reactants. However, an understanding of the main processes and concepts involved can be achieved better by standing back from all this complexity and focusing on some of the prominent features of this biochemical landscape.

BOTH PLANTS AND ANIMALS UTILIZE THE CHEMICAL ENERGY STORED IN CARBOHYDRATES MANUFACTURED BY AUTOTROPHS. IN THE PROCESS OF CELLULAR RESPIRATION, CARBOHYDRATES ARE OXIDIZED AND ATP IS PRODUCED.

FERMENTATION AND GLYCOLYSIS   The first stage in the mobilization of glucose is an anaerobic (without oxygen) series of reactions and is called fermentation or glycolysis, depending on the nature of the final products. Fermentation, carried out by such microorganisms as yeast, yields organic compounds such as ethyl alcohol and carbon dioxide (and, of course, some ATP):

$$C_6H_{12}O_6 \rightarrow 2CO_2 + 2CH_3CH_2OH$$

Ethyl alcohol

(Later we will worry about the balancing of this and the following equations.)

Louis Pasteur demonstrated that microorganisms are the agents responsible for fermentation and that oxygen is not necessary for their activity. This discovery directly contradicted the then predominant view of Lavoisier that all organisms require oxygen.

In many other organisms, including ourselves, an almost identical series of reactions results in the production of pyruvic acid or lactic acid (under anaerobic conditions):

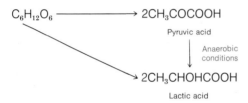

$$C_6H_{12}O_6 \longrightarrow 2CH_3COCOOH$$

Pyruvic acid

Anaerobic conditions

$$2CH_3CHOHCOOH$$

Lactic acid

In animals this series of reactions is called **glycolysis** since it usually begins with the degradation of glycogen to glucose. As shown in Fig. 4–32, the six-carbon **(hexose)** chain, after a phosphate group (P) is added to each end, is split into 2 three-carbon **(triose)** chains. These trioses are, in turn, phosphorylated with an inorganic phosphate (H_3PO_4). Eventually, these triose-diphosphates are dephosphorylated (their phosphates are removed). The phosphates, when removed, have a "high-energy" bond connecting them to the trioses. This bond energy is transferred with the phosphate group to an ADP molecule to form ATP. In the case of fermentation, the triose chain is further degraded to the two-carbon ethyl alcohol and CO_2.

In the course of these reactions, part of the bond energy of the glucose is transferred via the phosphate groups to ATP (Fig. 4–32). Remember, this bond energy, now found in glucose, was originally the energy of sunlight. After the transfer of the phosphate group on the third carbon to ADP, the phosphate group on the first carbon of the triose is enzymatically moved to the second carbon. At this point, the bond energy of the compound becomes redistributed, and it, too, becomes a "high-energy"

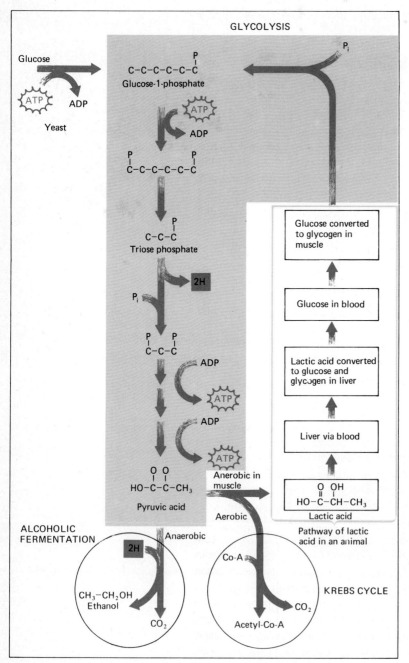

Fig. 4–32 Diagram of the reactions of glycolysis.

phosphate. Then, like the first phosphate group, it reacts with ADP to form another molecule of ATP.

The significance of this stage in glucose breakdown lies in the difference between the energy of the phosphate groups when they are attached to glucose and their energy when removed. Note that two ATPs are used to phosphorylate each glucose molecule. These are the cost, or debit side, of the process in terms of energy. On the profit side, we see that each triose molecule contributes two "high-energy" phosphate bonds to the formation of two ATPs. Since there are two of these trioses for each glucose that enters, a total of four ATPs are synthesized—a profit of two ATPs. Recall that the two extra phosphate groups were *not* derived from ATP but, rather, came from inorganic phosphate. A reorganization of the bond energies of the trioses changed these inorganic phosphates to energy-rich phosphate groups.

In alcoholic fermentation, the glucose is phosphorylated with an ATP so that the resultant equation for fermentation in yeast is

$$2 \text{ ATP} + C_6H_{12}O_6 + 2 \text{ ADP} + 2H_3PO_4 \rightarrow 2CH_3CH_2OH + 2CO_2 + 4 \text{ ATP} + 2H_2O$$

In yeast, the 2 two-carbon alcohol molecules are excreted into the medium and their energy is lost to the cell but not to the drinker. On the other hand, the lactic acid produced in anaerobic glycolysis may eventually be oxidized further, and the energy still remaining in the bonds of the triose is not lost.

THE FIRST STAGE IN THE UTILIZATION OF THE CARBOHYDRATE GLUCOSE IS CALLED FERMENTATION OR GLYCOLYSIS AND DOES NOT REQUIRE OXYGEN. FERMENTATION PRODUCES ETHYL ALCOHOL AND ATP, AND GLYCOLYSIS PRODUCES LACTIC ACID AND ATP. ONLY A PORTION OF THE ENERGY STORED IN THE GLUCOSE IS TRANSFERRED TO ATP.

THE KREBS CYCLE   The first step in cellular respiration has a net profit of two ATP molecules. The second stage in the breakdown of glucose occurs in the presence of oxygen in the mitochondria. These cellular powerhouses, as they are nicknamed, consist of a double membrane. The inner membrane has its surface area greatly expanded by infoldings called **cristae** (Fig. 4–33). Small granules occur on the surfaces of the cristae. These granules play a role in the electron-transfer systems discussed below. The double membrane, with associated macromolecules, seems to be typical of the organelles in eukaryotic cells which are involved in energy transformations. Chloroplasts, for instance, have very similar structures.

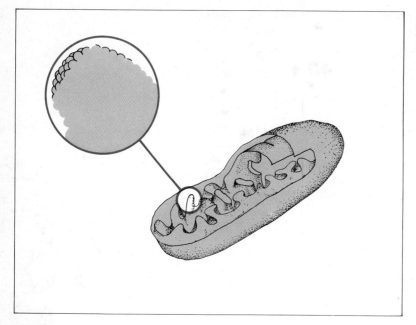

Fig. 4–33 Diagram of a mitochondrian showing the granules that play a role in the electron transfer systems.

The reactions in which the pyruvic acid resulting from fermentation or glycolysis is degraded all the way to CO_2 and water are referred to collectively as the **Krebs cycle,** named for its discoverer Sir Hans Krebs. The Krebs cycle (Fig. 4–34), also called the **tricarboxylic acid cycle,** is one example of numerous circular metabolic pathways. Above we considered one of the circular "pentose" pathways, which is important in the fixation of CO_2 in the dark reactions of photosynthesis and also in the synthesis of ribose and deoxyribose, the sugar components of nucleic acids.

A biochemical cycle, such as the Krebs cycle, can be thought of as two (or more) biochemical pathways linked together. The first pathway is a synthetic **(anabolic)** pathway, and the second is degradative **(catabolic).** Suppose that a principal energy-producing pathway in cells involved the degradative reactions six-carbon $\overset{C}{\rightarrow}$ five-carbon $\overset{C}{\rightarrow}$ four-carbon. The rate of this reaction would depend on, among other things, the supply of the six-carbon compounds. These theoretically can be synthesized by a number of different possible pathways:

> I.   3-carbon + 3-carbon → 6-carbon
> II. 4-carbon + 2-carbon → 6-carbon
> III. 5-carbon + 1-carbon → 6-carbon

If the second reaction, which utilizes the same four-carbon compound that is the end product of the degradative pathway, could be linked to the degradative chain, the whole synthetic-degradative system would continually be regenerating its own starting constituent, the six-carbon chain (Fig. 4–35). The only input required of such a circular system would be the two-carbon compound. Nothing would accumulate but the single-carbon waste products, which could be eliminated. Three enzymes could mediate the whole process: one to control the degrading of the six-carbon to the five-carbon, one to degrade the five-carbon to four-carbon, and one to combine the two-carbon with the four-carbon. The coupling of the degradative reactions with either of the other (I or III) six-carbon generating reactions would be much more complex.

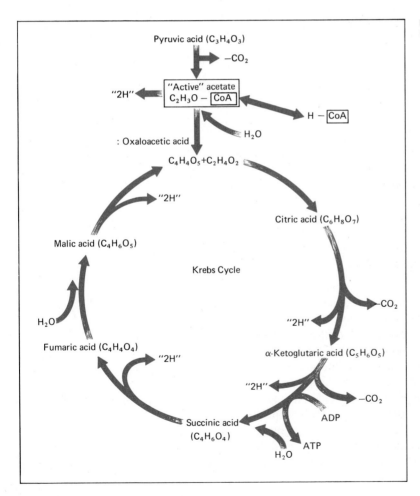

Fig. 4–34 Diagram of the Krebs cycle (tricarboxylic acid cycle).

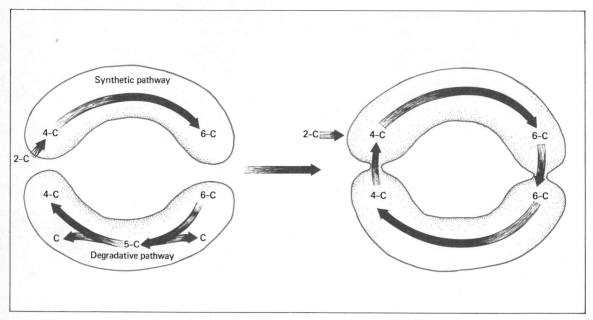

Fig. 4–35 A model of the Krebs cycle as two biochemical pathways linked together.

From the consideration of such a hypothetical case, it is not difficult to see why natural selection would favor the cycle over the less economical linear pathways. And the examination of the three alternative synthetic pathways suggests why some pathways are used over others.

The Krebs cycle is not unlike our model. As shown in Fig. 4–34, a four-carbon acid (oxaloacetic acid) condenses with a two-carbon fragment derived from pyruvic acid, the end product of glycolysis when oxygen is present. The fragment is called acetyl-coenzyme A, or **acetyl-CoA.** It also plays a part in lipid metabolism. The six-carbon compound, citric acid, then undergoes a series of enzymatically controlled rearrangements brought about by the loss and then the gain of a water molecule. The next step is a coupled oxidation-reduction reaction (see Box 4–5) with an electron acceptor NAD (oxidant) similar to NADP. The six-carbon donates two hydrogens (a pair of electrons and a pair of protons from hydrogen) to the acceptor. The electrons can be used to manufacture ATP in the electron-transfer chain. The next step is the degradative step, yielding a five-carbon molecule and a CO_2 molecule. This sets the stage for a complex series of reactions, during which there is an oxidation step (the loss of hydrogens), another loss of a CO_2, and the synthesis of ATP. The resulting four-carbon undergoes, in order: (1) the oxidative loss of two more hydrogens, (2) an addition of water, and (3) another oxidative loss of two hydrogens. The last reaction regenerates oxaloacetic acid—

the acceptor of the two-carbon fragment, which, in turn, comes from pyruvic acid.

So far in this description, very little energy in the form of ATP has actually been synthesized. Only a single ADP has been converted to ATP during one turn of the cycle, that is, during the degradation of a six-carbon to a four-carbon, plus two molecules of CO_2 and four pairs of electrons (together with their hydrogen protons). The balance of the energy remains as the potential reducing power of the electrons in $NADH_2$. This potential reducing power can lead to 3 ATPs as a result of the change of pyruvic acid to acetyl-CoA, and electrons from the Krebs cycle can produce 12 more ATP molecules.

THE SECOND STAGE IN THE UTILIZATION OF GLUCOSE TAKES PLACE IN MITOCHONDRIA AND REQUIRES OXYGEN. IT CONSISTS OF A CYCLE OF REACTIONS IN WHICH THE CARBOHYDRATE MOLECULE UNDERGOES A SERIES OF OXIDATIONS, WITH THE PRODUCTION OF CO_2 AND ATP.

THE ELECTRON-TRANSPORT SYSTEM   We have already seen that the membranes in chloroplasts contain an electron-transport chain which captures the potential energy of high-energy electrons and transfers it to "high-energy" bonds in ATP. A similar electron-transfer system in the cristae of the mitochondria produces most of the ATP in the living world (Fig. 4–36). Bacteria lack mitochondria, but they contain the characteristic elements of the electron system on infoldings of the cell membrane. Similar infoldings also behave as the analogs of chloroplasts in photosynthetic bacteria.

In the Krebs cycle, pairs of electrons (always accompanied by two hydrogen protons) from the carbon compounds pass to electron acceptors such as NAD (nicotinamide adenine dinucleotide, an electron acceptor chemically similar to NADP, which has already been discussed under photosynthesis), yielding $NADH_2$. From these acceptors, electrons then move on to a flavo-protein acceptor from which, in turn, electrons pass to a series of cytochrome acceptors, known as the cytochrome system. Cytochromes are pigmented electron-transfer molecules which are also involved in photosynthesis. They, like hemoglobin, each contain an iron atom, which in the cytochromes is essential to their electron-transfer ability. The iron atom may exist in two states, oxidized and reduced, and it alternates between these states as electrons are passed on. The last cytochromes in the chain, oxidases, are able to catalyze the transfer of the electrons and the accompanying H^+ protons to oxygen, the final electron acceptor. Water is formed as a result: $\frac{1}{2}O_2 + 2e^- + 2H^+ \rightarrow H_2O$.

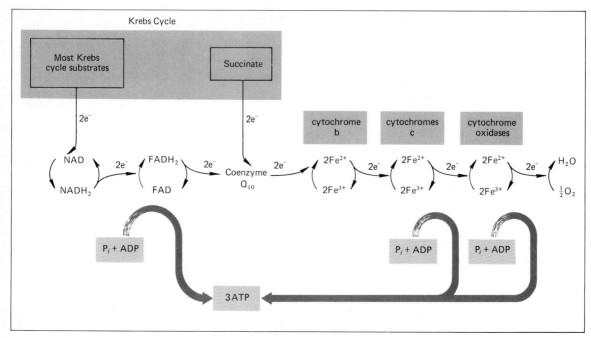

Fig. 4–36 Diagram of the electron-transport system.

It is helpful to think of the electron-transport system as a series of water-falls (Fig. 4–37). At the top of a series of falls, the water has the maximum potential energy because it can do work over the greatest distance. As it goes over the highest falls, it does work (producing heat as it strikes the rocks below or as it turns a waterwheel) and ends up with less potential energy than it started with. Finally, all its potential energy has been con-verted to kinetic energy by the time it reaches the bottom of the last fall. Analogously, the electrons, upon being removed by a dehydrogenase en-zyme from a substrate in the Krebs cycle, have the maximum potential energy. Each transfer results in the transformation of a portion of this potential energy to kinetic energy, some of which is used to drive the phosphorylation reaction $ADP + H_3PO_4 \rightarrow ATP$.

It is important to note that it is the properties of the electron-transfer system which permit energy to be captured. The electrons themselves are not changed any more than a drop of water is changed in passing over a waterfall. Acceptance of an electron by a transfer molecule flips the molecule into an activated state. It reverts to an inactivated state by giving off energy to produce ATP. Once inactivated, it is again ready to receive an electron. The chain of transfer molecules is a potential-energy gradient analogous to a waterfall. Since oxygen is required as the final electron receptor, this process is sometimes called **oxidative phosphorylation.** If the system is deprived of oxygen, it becomes "clogged" with electrons and stops phosphorylating. Furthermore, all the NAD molecules are quickly reduced by accepting hydrogens, and this prevents the Krebs cycle from functioning. Essentially the same thing happens if metabolic poisons such as cyanide are introduced. They block respiration by combining with the iron-containing group of the cytochromes. This makes it impossible for the electrons to be transferred down the respiratory chain to oxygen and thereby brings much of the metabolic machinery to a halt for lack of ATP.

THE GREATEST AMOUNT OF ATP IS PRODUCED IN THE ELECTRON-TRANSPORT SYSTEM. HIGH-ENERGY ELECTRONS FROM THE KREBS CYCLE ARE TRANSFERRED IN A COMPLEX SERIES OF STEPS TO MAKE ATP.

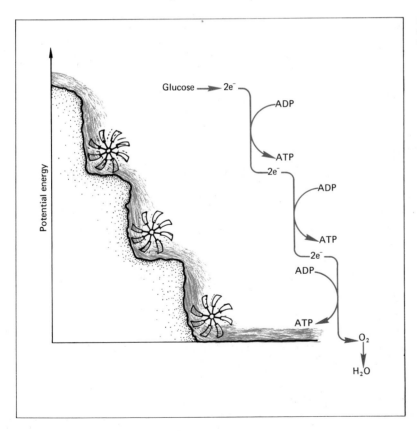

Fig. 4-37 Diagram comparing the electron transport system with a series of waterfalls generating energy.

THE PROFIT BALANCE   To recapitulate, the energy of glucose in the form of chemical bond arrangement is mobilized as a source of cellular energy by transforming it via its breakdown to the bond energy in ATP (Fig. 4–38). All in all, the complete oxidation of a single molecule of glucose yields a total of 38 of these energy-rich ATP molecules. Let us add them up. In the oxidation of glucose to pyruvate, 2 ATP molecules were made and 4 electrons led to the production of 6 more molecules. The change of pyruvic acid to acetyl-CoA led to 3 ATP molecules, and this change occurred twice for each glucose molecule. Thus 6 more ATPs were made at this stage, for a total of 14. In the Krebs cycle, 4 pairs of electrons were

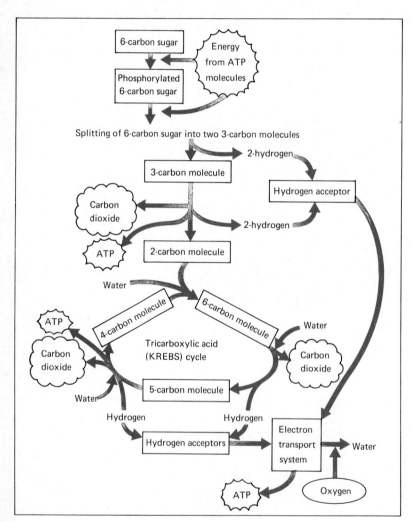

Fig. 4–38   Energy-producing reactions of a cell.

used to make 12 ATPs, and again, this must be doubled for each glucose. Thus 24 plus 14 equals 38 molecules of ATP for each molecule of glucose.

Since a mole (6×10^{23} molecules) of ATP yields about 10 kilocalories (kcal) when it is split into ADP and H_3PO_4, the oxidation of 1 mole of glucose yields about 38×10, or 380, kcal of usable energy, or about 55 percent of the total 686 kcal in a mole of sugar (glucose). The remaining 45 percent is liberated as heat. In warm-blooded animals, the heat thus produced is far from being a waste product of metabolism. On its slow passage from the respiring cells to the surface of the body, this heat maintains the high body temperature that is essential for the vigorous and dynamic life style we associate with birds and mammals.

THE PROCESS OF PHOTOSYNTHESIS USES SOLAR ENERGY TO MANUFACTURE ATP, WHICH IS USED BY CELLS TO FIX CO_2 INTO CARBOHYDRATE MOLECULES. THE PROCESS OF CELLULAR RESPIRATION TRANSFERS THE ENERGY STORED IN THE BONDS OF THE CARBOHYDRATE TO ATP, WHICH IS THE SOURCE OF CHEMICAL ENERGY FOR ALL CELLULAR ACTIVITIES.

Now that carbohydrate metabolism and cellular respiration have been discussed, let's take a brief look at how respiration is related to the metabolism of some other substances.

Other Metabolic Pathways

AMINO ACIDS   The Krebs cycle is considered central to biochemistry for reasons over and above its importance in the energetics of metabolism. Many of the carbohydrates that are reactants in the oxidative metabolism of glucose are also important intermediates in the synthesis of other cellular constituents. We have discussed above how the protein enzymes, the regulators and catalysts for all the steps of metabolism, are composed of amino acids joined by peptide bonds. The metabolism of amino acids illustrates the economy of biochemistry. When proteins are ingested, enzymes in the gut called **proteases** break the peptide linkages by adding water across these bonds (this is called **hydrolysis**—"splitting with water"). The separated amino acids can then be absorbed intact into the cells and thence into the bloodstream. In the liver, various uses are made of these amino acids. The amino group $-NH_2$ can be removed (**deamination,** Fig. 4–39), leaving an acid residue, which, in turn, can enter into the Krebs cycle to be oxidized or can be shunted into some other metabolic pathway. The excess ammonia is excreted in various ways, as will be discussed later.

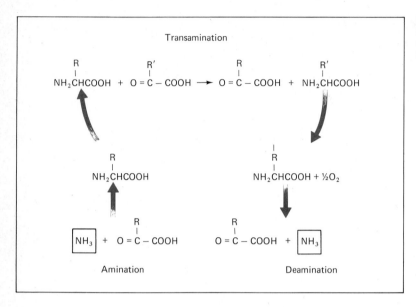

Fig. 4–39 Diagram to show transamination, amination, and deamination in the metabolism of proteins in a cell.

The reverse of this operation, the **amination** (Fig. 4–39) of keto acids from the Krebs cycle to amino acids, also is a source of amino acids for protein synthesis. The amino group from an amino acid can be enzymatically transferred to a keto acid molecule, thereby creating a new amino acid; this is called **transamination.**

CELLULAR METABOLISM INVOLVES THE SYNTHESIS AND BREAKDOWN OF AMINO ACIDS BY THE ADDITION OR REMOVAL OF AMINO GROUPS. MANY OF THE CARBOHYDRATES PRODUCED IN RESPIRATION ARE USED IN THE SYNTHESIS OF AMINO ACIDS.

LIPIDS Lipids are not a "natural" assemblage of substances; they have in common only their relative insolubility in water and similar solvents and their relative solubility in other kinds of materials such as alcohols, ethers, and other lipids.

An important group of lipids, **fats,** like carbohydrates, are composed mostly of carbon, hydrogen, and oxygen, but in fats the ratio of hydrogen to oxygen is usually much greater than 2:1. To put it another way, fats are more highly reduced than carbohydrates. **Oils,** by convention, are fats that are liquid at room temperature. Fats have two components: **fatty acids** and the alcohol **glycerol.** Most fatty acids are long chains of carbon and hydrogen with an acidic carboxyl group, $-\overset{\displaystyle O}{\overset{\|}{C}}-OH$, at one end. There usually is an even number of carbons in fatty acid molecules (see below), but odd-numbered fatty acids also occur. The long carbon chain can have the maximum number of hydrogens, in which case it is a **saturated** fatty acid, or it may have less than the maximum number and double bonds between some of the carbons, in which case it is an **unsaturated** fatty acid (Fig. 4–40).

The hydrocarbon part of the fatty acid chain is soluble in nonpolar solvents such as alcohol or acetone, and the carboxyl end is soluble in polar solvents such as water. (See Box 4–5.) This double nature of fatty acids allows them to become dissolved in both polar and nonpolar materials at once. Soaps, which are the sodium and potassium salts of fatty acids, utilize this property to **emulsify** fats. An emulsion is a suspension of minute drops of one liquid in another, such as oil in water. Usually the minute

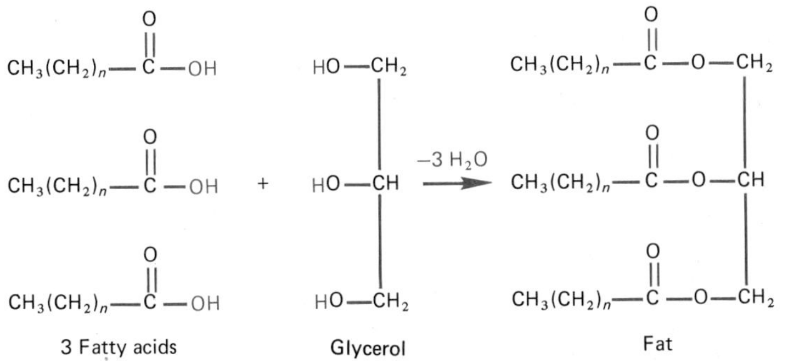

Fig. 4–40 The metabolism of lipids: *above*, some common fatty acids; *below*, formation of a fat by the combination of three fatty acids with glycerol.

Some common fatty acids:

$H-COOH$	Formic
CH_3-COOH	Acetic
CH_3-CH_2-COOH	Propionic
$CH_3-CH_2-CH_2-COOH$	Butyric
$CH_3(CH_2)_4COOH$	Caproic
$CH_3(CH_2)_5COOH$	Caprylic
$CH_3(CH_2)_{14}COOH$	Palmitic
$CH_3(CH_2)_{16}COOH$	Stearic

Formation of a fat:

3 Fatty acids + Glycerol $\xrightarrow{-3 H_2O}$ Fat

drops of the dispersed fluid will coalesce and become separated from the other liquid (like a salad dressing of oil and vinegar) unless an emulsifying agent is added. The emulsifying agent coats the oil droplets and keeps them from combining.

To make a fat molecule, three fatty acids combine with one glycerol (Fig. 4-40). In the process, three water molecules are liberated. The reverse of this, the splitting of the fat molecule with water (hydrolysis) occurs when fats are digested. First they are emulsified by salts in the bile, a secretion of the liver, and then they are split into glycerol and fatty acids by the digestive enzyme *lipase.*

There are other groups of lipids. Among these are *waxes,* in which glycerol is replaced by longer-chain alcohols, and *phospholipids,* in which one of the fatty acids is replaced by a phosphorus-containing compound. The phospholipids are major constituents of cell membranes, and they may be important in maintaining the continuity between the aqueous medium of the cell and the lipid-rich membranes of cells. *Steroids* are a complex group of lipids which also occur in eukaryotic membranes and which also act as vitamins (vitamin D) and hormones.

An important intermediary in the metabolism of fats and oils is the two-carbon fragment that joins with the four-carbon oxaloacetic acid to start an oxidative turn of the Krebs cycle. This compound is the major oxidative product of lipids. When an organism is living on stored fat (for example, a bear in hibernation or a man on a diet), the carbon chains of fatty acids are oxidized by breaking off two-carbon fragments from the carboxyl end and joining them to coenzyme A. These acetyl-CoA molecules can then enter the Krebs cycle, eventually to yield ATPs. Actually, the fats in the bodies of animals are constantly being torn down and replaced, although there may be no net change in the amount of fat in the *adipose* (fatty) tissue of the body.

You now know enough biochemistry to understand why the consumption of excessive carbohydrates causes one to deposit fat in the adipose tissues of the body. Excess glucose will be oxidized to pyruvic acid and CO_2; the pyruvic acid, in turn, will be oxidized to CO_2 and acetyl-CoA. If there is already sufficient ATP for the cells' needs, this surplus acetyl-CoA is shunted to fatty acid metabolism, in which the acetyl groups become linked together to form fatty acids (Fig. 4-40). The fatty acids in organisms usually have an even number of carbons, because both the synthesis and the oxidation of fatty acids in adipose tissues involve the two-carbon acetyl fragment.

Fats are more highly reduced (have more hydrogen relative to carbon) than carbohydrates. Therefore, they yield more hydrogens and, in turn, more ATP than a comparable weight of the less reduced carbohydrates. This is obviously why fats are favored as storage materials in animals, which move around. Plants, on the other hand, for which weight is not as much of a problem, often store food as the carbohydrate starch, for example, in potato tubers. But where space is at a premium, as in seeds, oils also are common in plants.

Oxidation of fats yields a large amount of water. In fact, sufficient "metabolic water" is yielded to provide for all the water requirements of some animals. The desert-dwelling kangaroo rat, among other vertebrates, lives mostly on oil-rich seeds. It can survive perfectly well on the water produced by the oxidation of the fats in these seeds.

LIPIDS ARE IMPORTANT CELLULAR CONSTITUENTS, PLAYING A VARIETY OF ROLES IN METABOLISM. ONE IMPORTANT ROLE IS THE STORAGE OF ENERGY. LIPIDS ARE SYNTHESIZED FROM FRAGMENTS PRODUCED IN GLYCOLYSIS.

VITAMINS   In olden days, sailors on very long voyages often would be stricken with **scurvy,** a disease which, if severe, leads to loosening of the teeth, hemorrhaging in the mucous membranes and skin, anemia, debility, and foul breath. Captain Cook, the famed eighteenth-century English explorer, saw to it that the diet of his men was supplemented with limes, as they were known to prevent scurvy from developing. Ever since, the term "limey" has been applied to British seamen. It is now known that citrus fruits contain ascorbic acid, or vitamin C, which prevents scurvy.

Since then, many other diseases have been attributed to the lack of vitamins. Vitamins, as now defined, are organic compounds necessary in small quantities but not used for construction material or as energy sources and not synthesized by the organisms in question. Vitamins often function as coenzymes or prosthetic groups for enzymes. Although there are certain "essential" amino acids that cannot be synthesized by some organisms and must be ingested, they are not vitamins because they can be used as construction material or as energy sources. Different organisms require different vitamins. Conversely, what is a vitamin for one species may be merely a "homegrown" metabolite for another. For instance, most animals synthesize their own vitamin C, although man, as well as the higher primates and some rodents, cannot.

Some vitamins are synthesized under certain environmental conditions but not under others. The steroid vitamin D, a rickets-preventing factor, is synthesized in the skin in the presence of ultraviolet light (unfiltered sunlight). Here is one of the few nonsocial advantages of a suntan. Some of the data on vitamins are summarized in Table 4–2.

Why should these vitamin deficiency diseases have evolved in many higher organisms? Autotrophs are biochemically self-supporting, as are many heterotrophic microorganisms. Presumably these deficiencies could have accumulated only if they posed no disadvantage to their carriers. As soon as they became in any way deleterious, selection would have eliminated them. One can only assume that under nearly all natural nutritional regimes, these vitamins are provided. Man's natural nutritional regime until about 11,000 years ago was hunting and gathering. Omnivorous food habits are still characteristic of human diets. All the necessary vitamins may be found in a *balanced* diet, but such a diet may be difficult to obtain with today's highly processed and calorie-overrich foods. Many people take vitamin pills to supplement their diets. If they ate a truly balanced diet, this would be unnecessary; indeed, vitamins A and D are harmful if taken in large quantities (although a single, multipurpose vitamin pill taken daily does not contain enough of either to cause harm).

The need for certain vitamins is often masked because they are produced

TABLE 4–2. Facts about Some Vitamins

NAME	FUNCTIONS	SYMPTOMS OF DEFICIENCY	IMPORTANT SOURCES
Thiamine (B_1)	Coenzyme in cellular respiration	Beriberi (muscular disorder); polyneuritis (nervous-system disorder)	Yeast, meat, unpolished cereals
Riboflavin (B_2)	Prosthetic group in cellular respiration	Membrane inflammations	Liver, dairy products, green leafy vegetables
Nicotinic acid	Precursor of enzymes in respiration	Pellagra (skin and nervous-system disorder)	Yeast, meat, milk
Folic acid	Used in nucleic acid metabolism	Anemia	Green leafy vegetables
B_{12}		Pernicious anemia	Liver
Ascorbic acid (C)	Cellular respiration; cell cement	Scurvy (loss of blood from blood vessels; connective-tissue and bone disorders)	Citrus fruits, tomatoes, green peppers
A	Precursor of pigments in vision; membranes	Night blindness; skin disorders	Dairy products, fish liver oils, carrots, tomatoes, spinach
D	Calcium and phosphate metabolism	Rickets (softening of bones and teeth)	Fish liver oils, butter (vitamin made in skin from precursors with ultraviolet irradiation)
E	Cellular respiration	Degeneration of muscles; sterility	Eggs, green leafy vegetables, vegetable and cereal oils
K	Cellular respiration	Impaired blood clotting	Leafy green vegetables (vitamin synthesized by intestinal bacteria)

by the bacteria that live in the intestines of many animals. Among these are folic acid and vitamin K. If the bacteria are killed by an antibiotic administered to combat infections, a vitamin-deficiency disease may develop.

VITAMINS ARE ORGANIC SUBSTANCES NEEDED BY PLANTS AND ANIMALS IN SMALL QUANTITIES FOR PURPOSES OTHER THAN AS ENERGY SOURCES OR STRUCTURAL MOLECULES. DIFFERENT ORGANISMS REQUIRE DIFFERENT VITAMINS, DEPENDING UPON THEIR SYNTHETIC ABILITIES, AND SOME OBTAIN NECESSARY VITAMINS FROM BACTERIA WHICH INHABIT THEIR BODIES.

SUMMARY

The genetic material of all organisms has been shown to be nucleic acid. Except for a few viruses, genetic information is coded in the molecules of deoxyribonucleic acid (DNA). DNA molecules are very large and consist of two polynucleotide chains wound about one another to form a double helix. Each chain consists of a linear arrangement of four different kinds of nucleotides, and the sequence of nucleotides in one chain determines that in the other. The nature and structure of DNA account for its replicability and its ability to store information.

The DNA code specifies the construction of proteins by the cell. Proteins are usually very large molecules consisting of some 20 kinds of amino acids arranged together in linear fashion. The chains of amino acids commonly become twisted and folded into more or less globular shape. Proteins serve many functions in cells, and one of the most important of these is as enzymes. Enzymes are catalysts that increase the rate of biological reactions without affecting the equilibrium which would occur if no enzyme were present. Each enzyme has a specific shape in which are sites of activity, each with a specific function.

Proteins are synthesized by the joining of amino acids end to end. The process is catalyzed by enzymes, and the energy of ATP is utilized. The sequence of amino acids is determined by the sequence of nucleotides in one of the two strands of a DNA molecule. A special kind of RNA, messenger RNA (mRNA), is formed in complementary fashion with a portion of DNA. This mRNA moves to the sites of protein synthesis in the cytoplasm, the ribosomes. The code for amino acid synthesis is contained in the mRNA. Amino acids are brought to the ribosome by transfer RNA (tRNA) molecules which pair with the mRNA. There they are joined in the proper sequence.

The genetic code specifying amino acid sequence is written in the sequence of nucleotides of the DNA. A sequence of three nucleotides specifies a particular amino acid, and a change in even one nucleotide may affect protein synthesis, causing a mutation. The genetic information of the DNA is *transcribed* into a complementary strand of mRNA. This, in turn, is *translated* into a polypeptide by the polyribosome complex of ribosomes and mRNA, aided by tRNA molecules and many different enzymes.

Cellular energetics involves the conversion of light energy into chemical energy and the subsequent use of this energy by cells. Photosynthesis, carried out mainly by green plants, is responsible for virtually all the energy used by plants, animals, and microorganisms. Light energy is absorbed by chlorophyll and, in a complex series of steps, is transferred to ATP. Both plants and animals are able to use ATP to fix carbon dioxide into organic molecules, which become the fuel molecules in cellular respiration. In the process of cellular respiration, carbohydrates are oxidized and ATP is produced. The greatest amount of ATP is produced in the electron-transport chain, where high-energy electrons from the Krebs cycle are transferred in a complex series of steps to make ATP.

Cellular metabolism involves the synthesis of all the varied structural and other molecules of cells. For example, amino acids are synthesized and broken down by the addition or removal of amino groups. Lipids are important cellular constituents, playing a variety of roles in metabolism —among others, in the storage of energy. Many of the carbohydrates produced in the course of cellular respiration are used in the synthesis of amino acids, lipids, and other cell constituents.

BOX 4–1. THE LAWS OF THERMODYNAMICS

The first law of thermodynamics states that in ordinary physical and chemical processes (that is, ignoring nuclear reactions), *energy is neither created nor destroyed*. Although we may think that energy is lost in a particular situation, what we actually see is energy changed from one category to another. Energy may exist as chemical energy, heat, light, electrical energy, kinetic energy (energy of motion), or potential energy (e.g., gravitational), as well as in certain other forms. You are familiar with the transformation of electrical energy into light and heat in a light bulb and the conversion of chemical energy into light and heat in a fire. The first law of thermodynamics tells us that in any physical process, only the distribution of energy from one category to another changes. The *total* amount of energy does not change.

Obviously, the first law affects how energy is treated in biological processes, such as the "capture" of the sun's energy by autotrophs and its eventual use by heterotrophs. Specifically, these processes do not create energy, not do they destroy it. Energy is ***transformed*** as it moves through biological systems (and all other systems, for that matter), and the details of these transformations are affected by the second law of thermodynamics. The second law states, in essence, that *the disorder of the universe is always increasing.* The second law requires a little more explanation than the first.

Suppose that you put a salt-permeable membrane across a water-filled container and dissolve some salt in the water on one side. Gradually, the salt ions (see Box 4–5) will cross the membrane until there are roughly equal numbers on each side. The salt–no-salt arrangement (order) will be replaced by a chaotic random assortment of ions scattered throughout the container (Fig. 4–1–1).

Similarly, if you put ice in one side of the container and hot water in the other, this order would disappear and you would be left with warm water.

Finally, if you leave a dead cat out on your lawn, you will find that most of its carefully ordered structure, with the help of bacteria and certain insects, rapidly is lost (Fig. 4–1–2).

In none of these cases would you expect the process spontaneously to reverse. The reason for this is that there has been a loss of ***useful energy.*** Useful energy is energy that can do work. There is useful

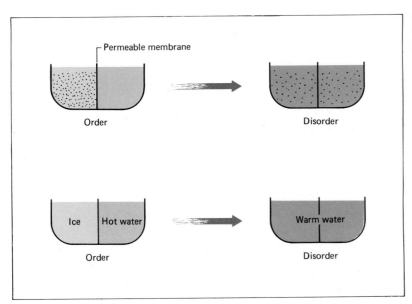

Fig. 4–1–1 Diagram of the operation of the second law of thermodynamics: *above,* salt ions become evenly distributed across a permeable membrane; *below,* temperature becomes even on both sides of a membrane.

Permeable membrane

Order

Disorder

Ice | Hot water

Order

Warm water

Disorder

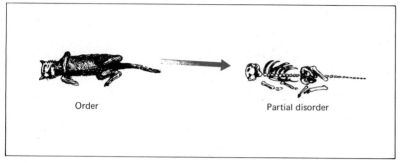

Order   Partial disorder

energy in the *difference* between the salty and nonsalty sides of the first container, and if the membrane were semipermeable (that is, if water could pass through but not the salt), you could make this energy do work. Similarly, hot water in cool surroundings can do work. For example, it can cause air to rise above it and a tiny parachute could be lifted by the moving air. Similarly, the body of the dead cat contained a great deal of useful energy in its chemical bonds. This energy was utilized by the heterotrophic decomposer organisms which disposed of the body.

The second law tells us that all spontaneous processes go in the same direction—toward an equilibrium of greatest disorder, or randomness. Physicists and chemists have a name for this disorder. They call it **entropy.** The second law calls for a continuous increase of entropy in the universe. Now, *local* reversals of the increase of entropy are possible. For instance, with a pump and filter, we could arrange a redistribution of the salt to restore some of the lost order to the system. But this local *decrease* in entropy would inevitably be accompanied by an *increase elsewhere.* If the pump motor were electric, that increase would occur at such places as in the burning of fuel at the electrical generating plant and in the production of heat by friction in the bearings of the motor.

The **production of heat** is the common way in which entropy is increased in biological processes. Heat, with its random, chaotic movement of molecules, is a form of energy which can do work only when temperature differentials exist. The water in an "unheated" swimming pool, on a day when the temperature is 34°F, contains a great deal of heat, much more than in a pan of boiling water sitting by the pool. Yet you can do no work with the heat in the pool because there is no temperature differential between the pool and its surroundings. You can do work with the heat in the pan since it is hotter than the air around it. It has **useful energy.** In a sense, the heat in the pan is concentrated and, in moving away from concentration, can do work. The heat in the pool is at the same concentration as that of the air and "has no place to go." When biological processes lead to the generation of heat, the useful energy in the heat is usually dissipated without doing useful work—and once gone, it is gone forever. In part, this is because the heat is at quite low temperatures as compared with those of boiling water or sunlight falling on a leaf. This, then, is the answer to the question of what happens to the sun's energy in living systems. Its useful energy decreases as heat is dissipated by living systems. We shall follow the pathways of heat loss in Chap. 14. The energy is still there, but it has been degraded from a useful to a relatively useless form.

In practice, all biological processes create some additional entropy, and in all biological transfers of energy, considerable useful energy is lost. For instance, as you will see, in the basic process by which cells extract energy from the chemical bonds of food molecules, almost half of the useful energy is lost. The laws of thermodynamics tell us why we need a continual input of energy to maintain ourselves, why we must eat much more than a pound of food in order to gain a pound, and why the total weight of plants on the face of the earth will always be much greater than the total weight of the plant eaters, which will, in turn, always be much greater than the total weight of flesh eaters. Since the laws of thermodynamics tell us that everything that goes on on the planet that uses energy also produces heat, we must consider how man's activities affect the heat balance of the earth. In using larger and larger amounts of energy, man is also producing more and more heat. The consequences of this will be discussed in Chap. 16.

All matter is doing work or has the potential for doing work. We describe this capacity as **energy,** and we recognize that energy can take on many different forms. A brick lying on a table has a potential energy, for it could do work by falling. A moving mass has kinetic energy, and the energy of a stream of moving electrons is electrical energy. The individual molecules in a gas have kinetic energy by virtue of their continuous random motion, and we recognize this as heat since their kinetic energies are proportional to the measured temperature of the gas.

All molecules are in motion, and their motion can be broken into several parts, each of which has a particular kind of energy. They have a translational energy because they are generally moving relative to

BOX 4-2. ENERGY, ATP, AND NAD

some outside reference; they have rotational energy because they are constantly spinning on their various axes; and they have vibrational energy because the bonds which connect their atoms to one another are more like springs than sticks and so these atoms are constantly changing position relative to one another. They also have an electronic energy because of the positions of the electrons within them which form the chemical bonds between atoms. The electrons associated with a pair of carbon nuclei are subjected to forces different from those associated with a carbon and a nitrogen nucleus; the bond energy between carbon and carbon is greater than the bond energy between carbon and nitrogen.

The energy of a molecule can be changed in various ways. For example, the molecule may be struck by a photon (light wave) that is vibrating with just the right frequency (which means that it has just enough energy) to make the molecule rotate faster or vibrate faster or shift some of its electrons into a slightly higher energy level. In this case, the molecule will absorb the light photon and increase its energy. Some molecules can also gain or lose certain electrons easily. (The molecule is then different, but it often retains the same name.) A molecule which *loses* electrons is oxidized, and one that *gains* electrons is reduced.

A chemical reaction is a process in which a different set of chemical bonds is formed. The energy of the molecules one begins with—the reactants—must then be redistributed in some way to form the new molecules—the products. Thus every chemical reaction entails the loss or gain of some energy. If energy is lost, the reaction is **exergonic** *(ex-,* "out," and *ergon,* "work"; energy is going out), whereas if energy is gained, the reaction is **endergonic** *(end-,* "in"). Such reactions are analogous to moving downhill and uphill, respectively. It should be obvious that no reaction runs uphill spontaneously. However, many of the reactions needed to synthesize the complex organic molecules of an organism are endergonic. The only way to make these molecules is to *drive* their synthesis by carrying out some other reaction that liberates enough energy and *coupling* that reaction to the synthetic reaction. (You drive your car uphill by burning gasoline and coupling the release of energy to the turning of the drive mechanism.) Cells use the following strategy to make molecule A into molecule B, where B has more energy than A: They use a compound XY whose breakdown into $X + Y$ liberates more than enough energy to make A into B. They couple the two reactions by making $AX + Y$ and then making AX into $B + X$.

Several molecules in cells play the role of *XY*, but one of them stands out above the others in importance. It is **adenosine triphosphate (ATP),** the adenine-ribose nucleotide found in RNA, but with a string of three phosphates instead of one. It happens that the release of the terminal phosphate, to make adenosine diphosphate (ADP), or of the last two phosphates, to make adenosine monophosphate (AMP), liberates enough energy to drive many synthetic reactions. ATP and some other nucleotides have been selected by evolution as the monetary units of energy exchange in cells—that is, like dollars, they are earned in many processes and used in many others. (They are sometimes referred to as "high-energy" compounds, but this is something of a misnomer; the energy released is not vastly greater than that released by other compounds.) The transfer of the terminal phosphate from ATP is called **phosphorylation;** this phosphate is the *X* of the example given above, where *Y* is the remainder of the molecule ADP. Thus, to make *A* into *B*, the cell might carry out

$$A + ATP \rightarrow A - PO_4 + ADP \rightarrow B + PO_4$$

The energy of ATP can be used in other ways, but the basic idea is the same.

How does the cell obtain its ATP in order to drive these reactions? In **phototrophs,** the cell makes use of the first method described above for getting energy—it absorbs the energy of light. The mechanism in which light is trapped in chlorophyll is described in this chapter. This light energy generates a stream of electrons with enough energy to phosphorylate ADP with an inorganic phosphate to make ATP. On the other hand, in **chemotrophs,** energy is obtained in the second way described above: reduced compounds are taken from the environment and oxidized, and the energetic electrons thus obtained are again used to phosphorylate ADP into ATP.

In both phototrophs and chemotrophs, another nucleotide, **nicotinamide-adenine dinucleotide (NAD),** is used in oxidation-reduction reactions; its reduced form is designated $NADH_2$. (The related compound NADP or $NADPH_2$ has an extra phosphate on the adenosine ribose.) NAD is used to oxidize many of the organic compounds used by heterotrophs. The resulting $NADH_2$ can be used to reduce other compounds in synthetic reactions, or the electrons stored there can be used to generate ATP molecules. $NADPH_2$ is made in photosynthesis to store the electrons necessary for reducing CO_2 to organic compounds.

Clues to the mechanisms which regulate the synthesis of proteins come largely from experiments with bacteria. These show that the regulation, or switching off or on of protein synthesis, is primarily at the stage of transcription from DNA into mRNA. Bacteria demonstrate two closely related control phenomena. Many enzymes are said to be **inducible** because they are made only in the presence of small molecules called **inducers.** For example, the enzyme β-galactosidase is responsible for the first step in the metabolism of lactose and other similar sugars. The enzyme is made only when one of these sugars is present; therefore, lactose serves as an inducer. Other enzymes are said to be **repressible** because they are always made except when certain small molecules known as **corepressors** are present. For example, the enzymes which synthesize the amino acid tryptophan are usually present in bacteria. If, however, there is ever an excess of tryptophan for any reason, the tryptophan acts as a corepressor and stops the synthesis of additional enzymes. This is another kind of feedback control, similar to the feedback inhibition described earlier. Feedback inhibition, you will recall, operates at the enzyme level to stop the synthesis of the end product of a metabolic pathway. Repression operates at the DNA-RNA level to stop the synthesis of more of the enzymes which catalyze the pathway.

It is always observed that a group of enzymes with closely related function are simultaneously induced or repressed. When the genes which carry the information for these related enzymes are mapped, it is often found that they lie very close to one another. For example, there are five enzymes which specifically synthesize tryptophan, and all five usually form a small cluster in the DNA. These clusters of tightly linked genes which are induced or repressed at the same time are called **operons.** Since the genes carry the information for the structure of enzymes, they are called **structural genes** to distinguish them from regulatory genes. Each operon has three kinds of regulatory elements, a regulator gene (R), a promoter (P) region, and an operator (O) region. The latter two elements are not genes since they do not code for proteins.

The regulator carries the information for a protein known as the **repressor.** This is the protein which has some affinity for one of the inducer or corepressor molecules, as shown in Fig. 4–4–1. At one end of every operon there is an operator, to which an active repressor protein can bind. In an inducible operon, the repressor will usually be attached to the operator, so the operon is turned "off" and no mRNA can be synthesized there. (Notice that the operator is between the promotor site and the rest of the operon, so it simply blocks the progress of the RNA polymerase along the DNA and prevents mRNA synthesis from the structural genes.) However, if an inducer enters the cell, it binds to the repressor protein and inactivates it. The repressor can no longer bind

BOX 4–3. REGULATION OF PROTEIN
SYNTHESIS

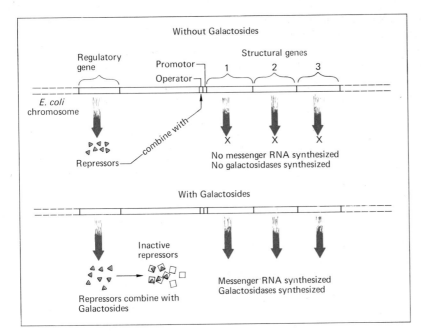

Fig. 4-4-1 A model of regulation of protein synthesis based on the galactosidase (an enzyme) operon in *Escherischia coli*. The presence or absence of the substrate determines whether or not transcription will occur.

to the operator, and the operon is turned "on" and RNA polymerase begins transcribing mRNA.

In a repressible operon, the principal difference is that the repressor protein functions differently. The repressor usually has no affinity for the operator. Therefore, messengers are usually made, and the enzymes coded by the structural genes are synthesized. However, if the level of corepressor builds up in the cell, some corepressor will bind to the repressor protein and activate it. Then it binds to the operator and turns off the synthesis of mRNA from this operon.

Remember that these schemes are based on studies of bacterial cells. Similar phenomena probably occur in eukaryotic cells, including those of higher organisms, but there are probably much more complex control mechanisms in these cells that are superimposed on operon controls.

Carbohydrates are organic molecules made up of carbon, hydrogen, and oxygen. In many of them, there are two hydrogen molecules for every oxygen molecule, the same ratio as in water. Their name derives from "hydrate of carbon," that is, a carbon-water combination.

BOX 4-4. CARBOHYDRATES

Simple sugars, which are made up of single molecular units, are known as **monosaccharides.** These are generally classified as to the number of carbon atoms they contain: three in a **triose** sugar, four in a **tetrose,** five in a **pentose,** six in a **hexose,** etc. Glucose is an important hexose sugar. Its structural formula is given in Box 3–1. Glucose and another hexose sugar can be united into a disaccharide, a dimer (two-part polymer) called sucrose, common table sugar:

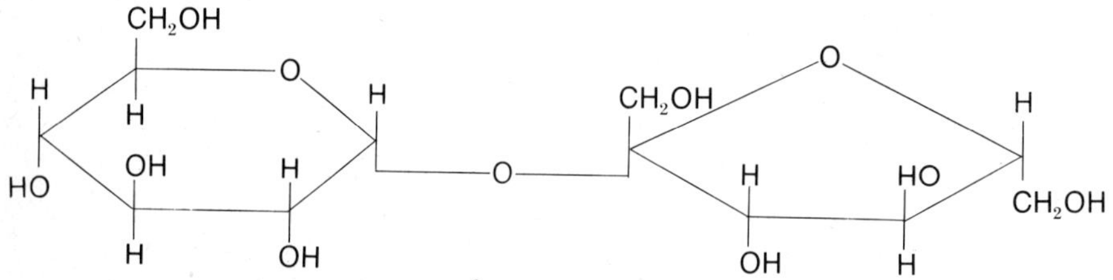

Sucrose

This representation of sucrose gives some idea of the three-dimensional configuration of the molecule. It is viewed at an angle from above the plane of the two rings.

Polysaccharides are polymers made up of three or more monosaccharide units. The macromolecule of starch is a polymer made up of some 200 to 1,000 glucose monomers joined in a series by bonds known as α-glycoside linkages. Each monomer is in the same relative position to each other, as shown in this skeleton diagram (linkages between carbons of successive monomers are colored):

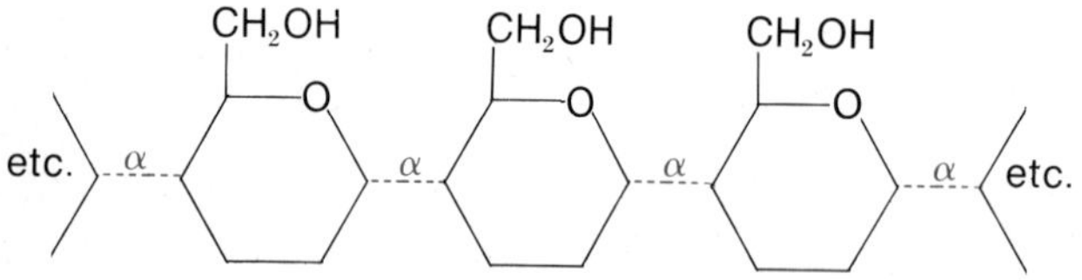

Starch

Cellulose has a similar structure, with alternate monomers flipped, and the linkages are β-glycoside linkages:

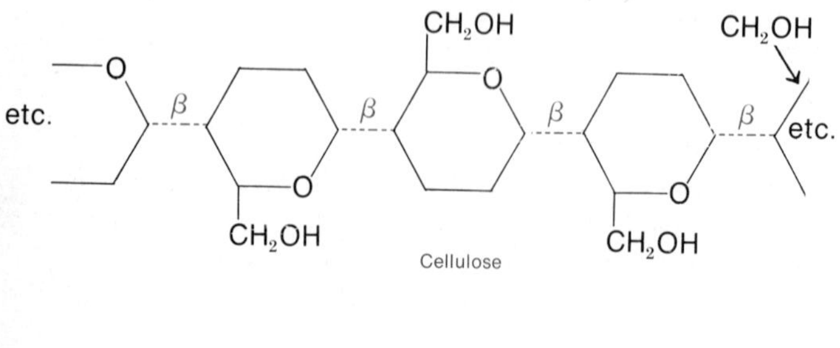

Cellulose

Glycogen is, in essence, animal starch. It is also a high polymer of glucose, containing as many as 500 monomers. It has a more branched structure than starch, which may show various degrees of branching.

The arrangement of chains of these polymers is extremely important to their functioning. In starch, the chains radiate out from a center of formation. It is not structurally strong and is easily digested by enzymes possessed by animals. Cellulose, on the other hand, occurs as bundles of microfibrils (tiny fibers). It is extremely strong; in plants, cell walls are able to withstand internal osmotic pressures (see Chap. 5) of up to 100 pounds per square inch (psi). Few animals possess enzymes which can break down cellulose.

BOX 4-5. SOME ADDITIONAL CHEMICAL IDEAS
Ionization: Ions and Radicals

An electrically charged atom is called an *ion.* You will recall from Box 3-1 that when sodium and chlorine combine into NaCl, sodium donates an electron to chlorine. . Thus the sodium atom becomes positively charged, having lost a particle of negative charge, and the chlorine atom becomes negatively charged. The electrical attraction between them is called an ionic bond because it is the attraction between two ions.

Molecules of many materials separate into two ions when they are dissolved in water. Water molecules intrude between the charged halves of the molecule, reducing the attraction between them. These charged atoms are very important in biological processes. For instance, the distribution of sodium ions (Na^+) and potassium ions (K^+) is critical to the generation of nerve impulses.

In many substances, two or more atoms take on an overall charge when ionization occurs. Such charged groups of atoms are most exactly called *radicals,* although they are often referred to simply as ions.

For instance, sodium hydroxide, NaOH, dissociates into ions as follows:

$$NaOH \rightarrow Na^+ + OH^-$$

The OH^- is known as the hydroxyl radical. It is a combination of an oxygen and a hydrogen atom with an overall negative charge. The hydroxyl ion shows a special attraction for hydrogen ions, that is, for protons (hydrogen ions are hydrogen atoms stripped of their electron —they are naked protons). A substance which dissociates to form ions attractive to protons, especially hydroxyl radicals, is known as a **base.**

Consider what happens when H_2SO_4 dissociates:

$$H_2SO_4 \rightarrow 2H^+ + SO_4^=$$

Notice the two negative charges on the SO_4, the sulfate radical, to balance the two protons. Substances which are able to give up hydrogen ions readily are called **acids,** and as you probably know, H_2SO_4 is sulfuric acid.

Water itself is a most interesting compound. As you know, it is essential to life; indeed, it is difficult to conceive of any kind of life evolving in the absence of this unique material. If you are interested in a detailed discussion of its properties in relation to life, you will be rewarded by consulting Henderson's book, "Fitness of the Environment," which is listed in the Supplementary Readings section.

Water itself dissociates slightly:

$$H_2O \rightleftharpoons H^+ + OH^-$$

or, more rarely,

$$2H_2O \rightleftharpoons H_3O^+ + OH^-$$

The double arrows indicate that the ions are constantly recombining into molecules as other molecules dissociate. One of the interesting things about water is that under the proper circumstances, it may function either as a proton acceptor or as a proton donor. It plays the role of acceptor (base) much more readily than that of donor (acid).

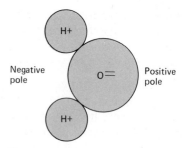

Negative pole

Positive pole

Fig. 4–5–1 A molecule of water is polarized, with the positive charges toward one side, the negative charges toward the other.

Water also displays an important characteristic shared by many other molecules, that of **polarity.** A polar molecule is one in which one end or part is positive (or negative) in relation to another. The geometry of the molecule (which is neutral overall) permits this separation of charges. This can be seen clearly in the water molecule, which has a V shape (Fig. 4–5–1).

Polarity

Notice how the negative charges are concentrated on the right side and the positive on the left. Polarity tends to bring molecules together in specific configurations, which often prove critical to their functioning. The structure of cell membranes and the shape of proteins are in part due to polar attractions among molecules and parts of molecules.

Polar attractions are electrical forces of the same sort as occur in ionic bonds, but they are considerably weaker. Hydrogen bonds are formed by the simultaneous polar attraction of a proton of one molecule for a negatively charged nitrogen or oxygen atom on the same or a different molecule.

Oxidation, in spite of its name, means simply loss of electrons. When our sodium atom donated an electron to the chlorine atom (Box 3–1), it was oxidized—even though there was no oxygen involved. Similarly, **reduction** means gaining of electrons. The chlorine atom on the sodium chloride molecule was **reduced.** Substances are often oxidized by the addition of oxygen (which accepts electrons) and reduced by the addition of hydrogen (which donates electrons). Oxidation involves a loss of electrons for the atom or molecule oxidized. Reduction, the other side of the coin of oxidation, involves a gain in electrons.

Oxidation and Reduction

SUPPLEMENTARY READINGS
Books

Conn, E. E., and P. K. Stumpf; "Outlines of Biochemistry," Wiley, New York, 1966.
Henderson, L. J.: "Fitness of the Environment," MacMillan, New York, 1913.
Loewy, A. G., and P. Siekevitz: "Cell Structure and Function," Holt, New York, 1969.
Rabinowitch, E., and Govindjee: "Photosynthesis," Wiley, New York, 1969.
Watson, J.D.: "Molecular Biology of the Gene," W. A. Benjamin, New York, 1965.

Articles

Kornberg, A.: The Synthesis of DNA, *Sci. Amer.,* **219,** no. 4, Offprint 1124 (1968).
Nomura, M.: Ribosomes, *Sci. Amer.,* **221,** no. 4, Offprint 1157 (1969).
Yanofsky, C.: Gene Structure and Protein Structure, *Sci. Amer.,* **216,** no. 5, Offprint 1074 (1967).
Goodenough, V. W., and R. P. Levine: The Genetic Activity of Mitochondria and Chloroplasts, *Sci. Amer.,* **223,** no. 5, Offprint 1203 (1970).

CHAPTER 5

WATER: PROBLEMS, PRINCIPLES, AND STRATEGIES

Life has existed on dry land for almost 500 million years. A 3,000-million-year sojourn in the sea preceding the invasion of the land left its mark: All life is water-based. Most of the weight and volume of living cells and tissues are water. Many cellular constituents are water-soluble and are able to move from one site to another. Water is essential as a hydrogen contributor in photosynthesis (except in some bacteria that can use compounds such as H_2S). Hydrolytic reactions depend on water. In short, life is wet. This chapter will survey the physiological and ecological strategies of "water relations" in plants and animals—how water is obtained, how it is retained, the nature of its role in general homeostasis.

Organisms must continuously spend energy to maintain order. The second law of thermodynamics says essentially that the maintenance of an ordered arrangement of parts (such as can be seen in a cell or a whole plant or animal) requires expenditure of energy (see Box 4-1). Natural physical processes reduce order; they all increase randomness. It is as though nature were perversely, if mindlessly, antilife. As we said in Chap. 3, order is not the only attribute of living systems, but it is a universal one.

In this chapter, we are particularly concerned with water. First, then, we will look at the processes that lead to randomness in aqueous (water-based) systems. We will see what processes and structures have evolved to counteract the entropy-producing forces in aqueous organisms.

DIFFUSION AND OSMOSIS

If a sugar cube is placed at the bottom of a glass of water, it will gradually dissolve and disappear into solution. Given enough time, the sugar molecules will become distributed evenly throughout the fluid. A movement of sugar molecules from a region of high density (the cube) to a region of lower density will have occurred. The phenomenon is called **diffusion.**

If allowed to proceed for a sufficient length of time, diffusion inevitably results in an equal distribution of molecules. Most people do not find it intuitively obvious why molecules, if free to do so, will always move to an area of lower concentration. Perhaps it will help if you remember that molecules in a liquid or gas state are in constant motion and that a molecule will move with constant velocity and direction until some force acts on it. Such a force comes from the collisions of molecules. Although molecules in a liquid move with great speed, rarely do they go any appreciable distance before striking another; their motion is essentially random. As shown in Fig. 5–1, an imaginary plane through a container with a sugar cube in the bottom separates a lower region of high sugar concentration from an upper region of less sugar. Some of the molecules are within "single-move" distance from the plane, and more are below than above. Because the sugar molecules move randomly, about half will move up in the next instant, going upward through the plane if they were below it and going even higher if they were above it. The other half will move down, going through the plane downward or moving even lower. As you can see, this results in more molecules above the plane. As long as molecules are being fed from below by the sugar cube, this process of net upward movement will continue. This is a statistical phenomenon, in that the movement of any particular molecule at any time is unpredictable but it is still possible to make predictions about a population of molecules. It can be said that, on the average, the sugar molecules will move up until there is an equal concentration of sugar throughout the glass because collisions will tend to propel them away from the region of high density. Organisms do not rely on diffusion to bring them the materials they require. It is simply too slow. Nevertheless, a special case of diffusion, called **osmosis,** is exceedingly important in living systems.

Diffusion

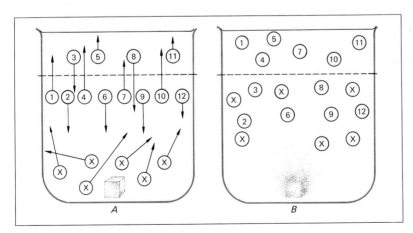

Fig. 5–1 Diffusion of sugar molecules from a sugar cube in a container of water. There are eight sugar molecules directly below an imaginary plane in (A) and only four above. An instant later, in (B), of the original twelve molecules there are six above and six below, but the concentration below remains higher because of the arrival of more sugar molecules (X's) from the sugar cube. Until the cube is completely dissolved, this random process of diffusion will continue to increase the number of molecules above the imaginary plane. When the cube is dissolved, diffusion will have resulted in a uniform distribution of molecules throughout the container.

THE PROCESS OF DIFFUSION RESULTS IN THE MOLECULES OF A DISSOLVED SUBSTANCE BEING UNIFORMLY DISTRIBUTED IN THE SOLVENT.

Osmosis

If a prune is placed in water, the skin will permit the entrance of water while impeding the exit of the sugars and other materials inside. This process continues until the prune is so swollen that a break occurs in the skin. We will examine this phenomenon with a somewhat simpler system. If a sugar cube is dropped into one side of a glass of water vertically divided with a cellophane membrane impermeable to sugar molecules (Fig. 5–2), an interesting thing occurs. The water level on the sugarless side is lowered, thus indicating a traffic of water from the pure-water side to the sugar-water side. Similarly, if some green leaves, such as spinach or lettuce, are placed in sugar solution overnight, in the morning they will be wilted due to a loss of water from their cells, whereas leaves placed in pure water will still be crisp. Microscopic examination of the cells in these two sets of leaves will show that those immersed in the sugar-water solution have **plasmolyzed,** that is, shrunk away from the cell walls because of a loss of water from the central vacuole (Fig. 5–3).

These two phenomena, plasmolysis and the movement of water toward the sugar-water side of the glass, are both examples of the process called **osmosis.** Osmosis is merely diffusion (strictly speaking, of any solvent, but usually in biology referring to water) across a **differentially permeable membrane,** such as cellophane or a cell membrane (Fig. 5–4). (Details of membrane structure and function are discussed in Box 5–1.) A differentially permeable membrane is one that permits the passage of

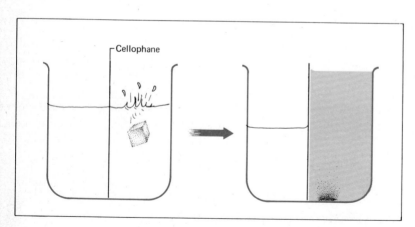

Cellophane

Fig. 5–2 Movement of water through a membrane (differentially permeable) permeable to water but not sugar, toward the side with dissolved sugar.

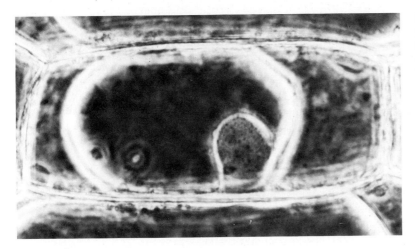

Fig. 5-3 A cell from the epidermis of an onion bulb scale which has been placed in concentrated salt solution. Water moves out and the cell shrinks within its cell wall. Such a cell is said to be plasmolysized. × 1,600. (*Jeroboam, N.Y.*)

some kinds of molecules but restricts the passage of others. Water molecules are very small and can easily pass through the cellophane or cell membranes. Larger molecules, such as sugars, proteins, and lipids, will pass through more slowly or not at all.

What determines in which direction the traffic of water will be? Recall that a material will diffuse from a region of high concentration to one of

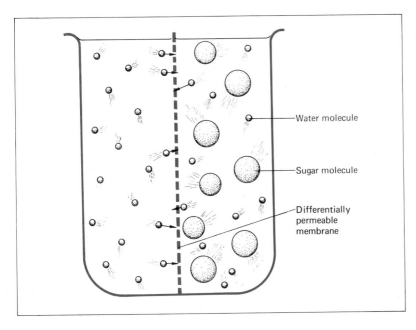

Fig. 5-4 Diagram to show the effects of a differentially permeable membrane: water molecules can move freely through the pores of the membrane but the larger sugar molecules cannot.

Water molecule

Sugar molecule

Differentially permeable membrane

low concentration. Distilled water has a water concentration of 100 percent; that is, all the molecules in such a system are water molecules. On the other hand, a sugar solution or the contents of a cell vacuole are less than pure water; some of the volume is occupied by other kinds of molecules. Therefore, the water will move toward the relatively impure side—from a region of high concentration of water to a region of lower water concentration. Only the water will diffuse through the membrane since it is impermeable (by definition) to the other materials. At any instant, more water molecules on the left side in Fig. 5–4 will be moving toward the pores in the membrane. Of course, the same *proportion* of water molecules will be moving toward the pores from both sides of the membrane, but as there are more water molecules on the left, there will be a higher absolute number passing from left to right than from right to left (see Fig. 5–4). The actual physics of the water movement is more complex than that presented here. For one thing, the kind of membrane determines the means of transport—water may actually flow through a membrane. The membrane may also be differentially permeable not because of pore size but because it can dissolve one kind of molecule in it more easily than another. We also have not taken into account the collision effects of nonwater molecules on one side of the membrane.

How long will this process go on? If the membrane is perfect—that is, if water and only water can pass through it—and if no counterforce intervenes, all the water will move through the membrane at a rate proportional to the difference in concentration of the two sides. The typical laboratory display of osmosis, however, provides such a counterforce. The display diagramed in Fig. 5–5 demonstrates water rising up a tube filled with a sugar solution. In this setup, the water level in the tube will rise until the hydrostatic pressure of the column of water (pressure = mass × height/unit area) equals the pressure resulting from the concentration differential. The pressure created by such a concentration differential is called **osmotic pressure.**

If two solutions have the same osmotic pressure, they are said to be **isotonic** to one another (*iso*, "same, and *tonic*, "relating to tension"). If solution A has a higher osmotic pressure than solution B, then A is said to be **hypertonic** (*hyper*, "higher") to B and B is said to be **hypotonic** (*hypo*, "lower") to A.

OSMOSIS IN BIOLOGICAL SYSTEMS IS THE DIFFUSION OF WATER ACROSS A DIFFERENTIALLY PERMEABLE MEMBRANE. WATER MOLECULES CAN EASILY PASS THROUGH SUCH A MEMBRANE, BUT LARGER MOLECULES CAN PASS ONLY SLOWLY OR NOT AT ALL. IF THE CONCENTRATION OF WATER IN A SOLUTION OF LARGE MOLECULES IS

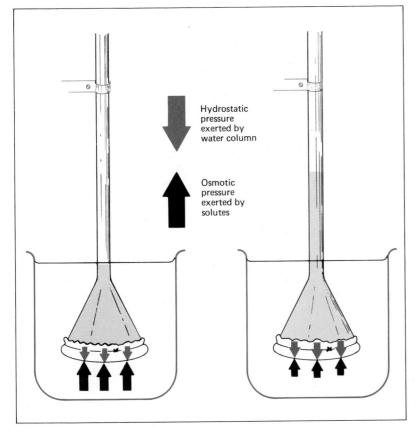

Hydrostatic
pressure
exerted by
water column

Osmotic
pressure
exerted by
solutes

Fig. 5-5 Osmosis demonstration using a vertical tube bounded by a differentially permeable membrane and containing a sugar solution. Water enters the tube until the osmotic pressure equals the hydrostatic pressure.

DIFFERENT ON EITHER SIDE OF A DIFFERENTIALLY PERMEABLE MEMBRANE, OSMOSIS WILL OCCUR UNTIL THE CONCENTRATIONS ARE THE SAME ON BOTH SIDES OR UNTIL SOME COUNTERFORCE CAUSES IT TO STOP.

Diffusion randomizes molecular arrangements—it tends to equalize concentrations of things. Osmosis is a special case of diffusion—it tends to equalize concentrations of water across membranes. Were it not for the ability of cells to counteract osmotic forces, by expelling water, for example, they would quickly swell and burst and their contents would scatter (Fig. 5-6). Yet this destructive side of osmosis is only part of the picture. These processes, osmosis in particular, can be made to work for organisms as well as against them. In this section we will show how plants do

WATER REGULATION AND MOVEMENT IN PLANTS
The Importance of Being Turgid

just this. Osmotic pressure is the muscle and, partly, the skeleton of plants. What animals do by contraction of muscles and the rigidity of bone, plants do in part by osmosis and its effects.

TURGOR   One important thing to keep in mind about terrestrial and freshwater plants is that they are hypertonic (saltier) to their environment. As a result, they are constantly tending to absorb water by osmosis, usually through their roots. For freshwater animals, this constant inflow of water poses a major physiological problem, and they would burst if it were not for a considerable expenditure of energy used in pumping out excess water against the osmotic gradient (Fig. 5–6). Plants, however, survive perfectly well in dilute environments without wasting their energy resources. This is because plant cells are contained in rigid boxes, the cell walls. As we discussed above, water will keep diffusing indefinitely into a hypertonic medium unless the osmotic pressure is opposed by an opposite pressure.

In Fig. 5–5, we saw that the opposite pressure can be the force (weight) of the column of fluid in the tube. The cell walls perform the same function

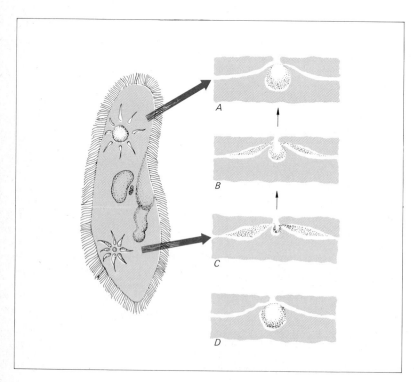

Fig. 5–6   The protozoan *Paramecium* lives in fresh water. Since its concentration of solutes is greater than that of the environment, water moves into the cell. Specialized vacuoles, called contractile vacuoles, accumulate excess water and, by contracting, force it outside the cell. This requires use of energy by the *Paramecium* to compensate for osmosis.

in plants. As water enters the cell (and thence the vacuole, which by volume makes up the bulk of the cell contents), the cell presses more tightly against the cell walls. This is analogous to the pressure of an inner tube against the inside of an automobile tire as air is being pumped in. Eventually, the cell wall, like the tire, exerts a back pressure, called **turgor pressure,** on the cell that is equal to the osmotic pressure—about 10 atmospheres, or 147 psi (about five times normal tire pressure). This alone is striking evidence for the strength of cell walls.

THE CONTENTS OF PLANT CELLS ARE HYPERTONIC TO THEIR EN-VIRONMENT. THEREFORE, WATER TENDS TO MOVE INTO PLANT CELLS, CAUSING THEM TO SWELL. THEY SWELL UNTIL RESTRAINED BY THEIR CELLULOSE CELL WALL, WHICH EXERTS A COUNTER-FORCE CALLED TURGOR PRESSURE.

TURGOR AND STOMATA   The regulatory role of turgor in plants is perhaps best illustrated by the **stomata** (sing. **stoma**), the tiny openings that modulate the passage of materials between the photosynthesizing surfaces, usually leaf cells, and the surrounding atmosphere. (See Box 5-2.) Both the upper and lower surfaces of leaves are covered with continuous layers of flattened cells, making up the epidermis (Fig. 5-7). A waxy covering (cuticle) of the epidermis protects against excessive water loss. Dispersed among the other epidermal cells, especially on the lower epidermis, are pairs of sausage-shaped **guard cells,** which surround and control the stomata (Fig. 5-8). The size of the stomatal opening determines the rate at which gases pass through the opening between the guard cells. By regulating the size of the stomatal openings, the guard cells control (1) water loss and (2) the rate of photosynthesis —crucial functions in the integration of the plant's activities. We will discuss water regulation first.

Plants often lose tremendous quantities of water, a phenomenon called **transpiration.** As much as 100 gal per day is transpired from large trees. Normally, this loss can be made up by increasing the water input from the water conducting cells (see below), but excessive water loss causes the cells of the leaf to lose turgor and the leaf wilts. If it persists for long, damage to the plant will result. As this is happening, however, a compensatory process begins to operate. The guard cells lose turgor, too. In the turgid condition, these cells are spread apart, permitting the air spaces of the leaf free access to the outside. But as they become flaccid, they close the stoma, thus effectively halting further transpiration except for the minute amount transpired through the waxy cuticle of the epidermis.

Water loss is not the only stimulus to which the guard cells respond. When

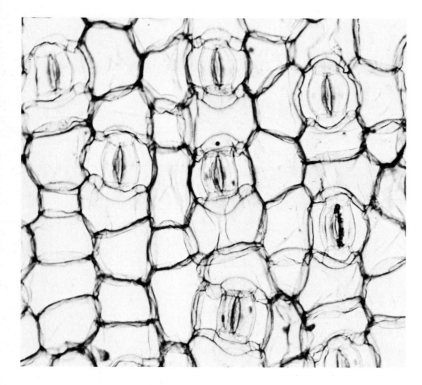

Fig. 5-7 The lower epidermis of a leaf of a dayflower (*Commelina*), in face view, showing the epidermal cells with interspersed stomata. The kidney-shaped cells are the guard cells of the stomata. × 220. (*Courtesy of Philip Feinberg.*)

placed in strong illumination, a closed stoma may open in less than 1 min. At night, stomata are closed, preventing water loss. How might light affect the opening of a stoma, given that turgor pressure controls the shape of the guard cells? First, the turgor pressure depends on the amount of water in the cell. (The more water, the more pressure and the wider the stoma is opened.) Second, water will flow into a cell (by osmosis) if the concentration of solutes in the guard cells increases. So, presumably, light somehow causes an increase in the concentration of solute molecules. The obvious mechanism for bringing about this increase is photosynthetic production of sugar molecules in the presence of light. Guard cells have chloroplasts, unlike the other epidermal cells. Actually, the guard cell response is too fast for this. Apparently, light triggers other changes that quickly bring about an increase in the concentration of solutes and raise the osmotic pressure of the cell.

CHANGES IN TURGOR OF THE GUARD CELLS OF STOMATA IN THE LEAVES OF PLANTS CAUSE THE STOMATA TO OPEN OR CLOSE. THIS REGULATES THE UPTAKE AND LOSS OF GASES, INCLUDING OXYGEN, CARBON DIOXIDE, AND WATER VAPOR. THEREFORE,

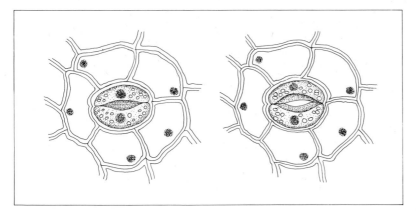

Fig. 5-8 Diagram to show the operation of guard cells: *left*, stoma closed; *right*, stoma open.

CHANGES IN THE STOMATAL OPENINGS REGULATE THE LOSS OF WATER BY PLANTS (TRANSPIRATION) AND THE RATE OF PHOTOSYNTHESIS.

TURGOR AND GROWTH   Turgor pressure plays many other roles in the life of a plant. It is now known, for instance, that turgor pressure is essential for plant growth. A major component of growth in plants is cell elongation. Turgor pressure pushes on the cell walls from the inside, stretching them and causing them to grow. Do you think a wilted plant will grow? Some plant cells expand in all directions under turgor pressure, but others elongate without expanding laterally. This differential growth is a result of differences in the structure of the cell walls and is affected by the hormone auxin (discussed in Chap. 7). The walls are made up of cellulose microfibrils with a very specific orientation, as well as other carbohydrates. The orientation of the microfibrils often permits stretching along one axis but not along the other. Other reactions are important in addition to turgor pressure. This is clear because growth is rapidly stopped by substances which inhibit cellular metabolism. Auxin presumably is capable of modifying the metabolism of the cell to permit more or less stretching under the force of turgor pressure.

A MAJOR COMPONENT OF GROWTH IN PLANTS IS CELL ELONGATION. UNDER THE INFLUENCE OF AUXIN, THE CELL WALL BECOMES MODIFIED AND ENLARGEMENT OCCURS AS TURGOR SWELLS THE CELL. THE SIZE AND SHAPE OF CELLS ARE DETERMINED BY THE ARRANGEMENT OF FIBRILS OF CELLULOSE IN THE CELL WALL.

In the preceding section we saw how an osmotic effect, turgor, gives rigidity to such structures as stems and leaves, controls the loss of water

Water Intake and Transport in Land Plants

by transpiration, and is the force behind plant growth. In this section the processes of water intake and distribution will be explained. Multicellular land plants are faced with a variety of problems. For one thing, they must be in two kinds of environment at the same time. They must take up water and dissolved minerals found primarily in the soil, where there is no light. On the other hand, they must also be exposed to light and the atmosphere in order for photosynthesis to be carried out (Fig. 5–9). The most abundant land plants have evolved root systems, which are primarily anchoring and absorbing structures. They have also developed the leafy shoot, which is specialized for holding the photosynthetic appendages up into the light, usually in the air but also in water (Fig. 5–9). It is almost as if the plant were two cooperating organisms: a heterotrophic one, living an underground existence which supplies water and minerals to the autotrophic one, which supplies organic molecules (food) to the former. The questions now are: (1) How does the water, with its load of precious minerals, get into the roots? (2) How is it carried to the highest leaves of trees, where it is needed for growth, photosynthesis, and turgor support?

The root system of vascular plants is its absorbing and supporting system (Fig. 5–10). The lateral extent of the root system usually is greater than that of the aerial branches. An oak tree some 40 ft high may have a taproot going down 15 ft into the soil. The lateral roots may extend 60 ft, however. The root system is usually very extensively branched and, if all the myriad branches are measured, usually turns out to be astonishingly long. For example, a plant of rye grass was carefully measured by one investigator. It was 20 in. high and had 20 shoots, with 480 leaves. The total surface area of the aboveground parts was about 52 sq ft. It is difficult to believe, but the much-branched diffuse root system measured a total of 2 million ft, or 380 miles! The total surface area of the root system was 2,500 sq ft.

You can see that the root system has an enormous surface area for absorbing water and dissolved materials in the soil. Absorption takes place through specialized cells called **root hairs** (Fig. 5–11), which are formed some distance behind the tip of the root (growth of the root is discussed in Chap. 10). Each root hair (Fig. 5–12) is a long, tubelike extension of a single cell of the root. It grows out among the soil particles where absorption takes place. The root-hair cell is hypertonic to the soil solution, so that water diffuses into the cell. From the root hair, the absorbed water and minerals diffuse from cell to cell of the root until they reach the vascular tissue in the center of the root (Fig. 5–13).

THE ROOT SYSTEM OF VASCULAR LAND PLANTS IS AN ABSORBING-SUPPORTIVE SYSTEM. ROOTS TAKE UP WATER AND DISSOLVED MA-

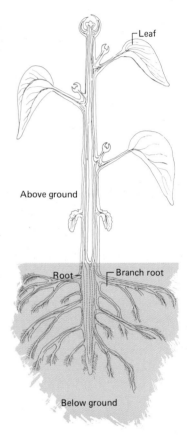

Above ground

Below ground

Fig. 5–9 Diagram of the basic structure of the plant, showing photosynthetic leaves, transporting stem, and anchoring and absorbing roots.

Fig. 5–10 Comparison of the root systems of a grass and a shrub. Grasses have a dense, much-branched root system. (*SCS–United States Department of Agriculture.*)

TERIALS FROM THE SOIL SOLUTION AND BEGIN THEIR TRANSPORT TO OTHER PARTS OF THE PLANT.

Once water and minerals are in the root, the plant must move them to where they are needed for photosynthesis, growth, and all the other processes in the aboveground portion of the plant. The tissues which carry on these functions are called **vascular tissues.** There are two main kinds of vascular tissues: **xylem,** which functions primarily in conducting water upward, and **phloem,** which conducts photosynthetic products (carbohydrates) downward. There is nothing in vascular plants really comparable to the multipurpose circulatory system found in many animals.

XYLEM   In this chapter we will examine only the xylem. Its structure will provide hints as to how it works. Xylem is composed of living cells, called parenchyma, as well as elongate, relatively thick-walled cells which are dead when functionally mature. It is the dead cells that include the chief water conduits.

Water-conducting cells of xylem are of two kinds: ***tracheids*** and ***vessel***

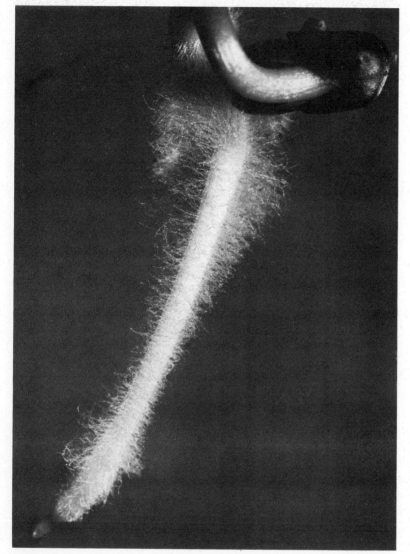

Fig. 5-11 A radish seedling, showing the abundant, absorptive root hairs, each of which is an outgrowth of a single cell. (*Lester V. Bergman and Associates, Inc.*)

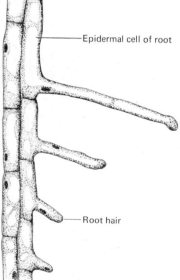

Epidermal cell of root

Root hair

Fig. 5-12 Formation of root hairs as extensions of epidermal cells.

elements. (These, plus various kinds of fibers, make up wood.) In the development of a tracheid—the only kind of water-conducting cell in the more primitive vascular plants, such as ferns and conifers—the cell elongates greatly and develops slanting end walls, which overlap the ends of other tracheids above and below. As with most plant cells, the relatively thin cellulose primary wall has numerous tiny pores, called *pit fields,* through which plasmodesmata (cytoplasmic bridges) pass into adjacent cells. As the tracheid develops, a thick secondary cell wall is laid down inside the primary cell wall, except where there are pit fields. The result is a series of openings, called **pits,** in the secondary wall.

The tracheid is functionally mature when the cytoplasm and nucleus disappear. The cell then is an empty tube whose secondary cell wall is perforated by pits, which are opposite similar pits in contiguous cells. Note, however, that there are no open connections between adjacent cells. Each cell still has its primary cell wall, and the middle lamella (intercellular cement) holds them together.

Vessel elements of the xylem usually are not as elongate as are the tracheids. In fact, the most specialized vessel elements may be broader than long. Vessel elements may have sloping end walls, but the more specialized have transverse end walls. Development of a vessel element is similar to that of a tracheid, except that the secondary-cell-wall material is not laid down where the end walls of adjacent elements overlap. In these areas also, the primary cell walls and middle lamella disappear, along with the protoplast of the cells. Thus, as you can see in Fig. 5–14, there is an open connection between vessel elements. A vertical column of vessel elements is called a **vessel.** Vessels are analogous to a pipeline constructed of individual segments of pipe. They are efficient water-

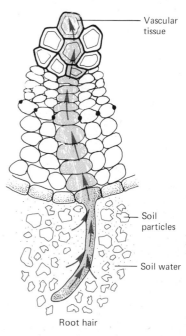

Vascular tissue

Soil particles

Soil water

Root hair

Fig. 5–13 Diagram to show movement of water–cell to cell–from root hair to vascular tissue in the center of the root.

Fig. 5–14 A longitudinal section of a portion of wood (xylem) of the silver maple (*Acer*). Running vertically are the narrow fibers and the larger vessels. Some vessel elements are sectioned to show the pits on their walls (center); some show the absence of end walls (left and right). (The elliptical groups are composed of ray cells which conduct horizontally.) ×130. (*Courtesy of Philip Feinberg.*)

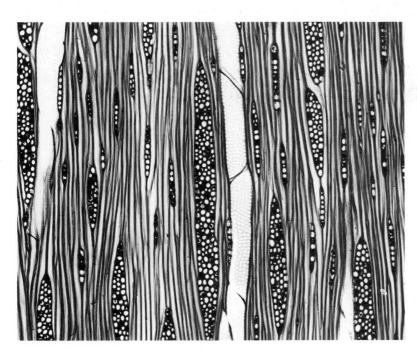

conducting tubes and may range in length from a few feet to perhaps 50 ft.

The aboveground stems of plants are designed to expose the principal photosynthetic organs, the leaves, to the maximum amount of light and to conduct water to them from the roots. In most habitats, the taller a plant, the more light it receives and therefore the more energy. As a plant increases in height, however, water conduction becomes more difficult. It is also necessary for the stem to become stronger. The tissue xylem serves the function both of water conduction and of strengthening the stem.

In most of the gymnosperms, tracheids both conduct water and add mechanical strength. In their xylem, cells called fibers have lost the ability to conduct water and develop very thick walls. Angiosperms may have both tracheids and vessels, but the latter usually predominate. Vessels are very highly specialized for water conduction, and their thin walls do not strengthen the stem. Instead, very specialized fibers have evolved which carry out this function.

WATER IS CONDUCTED, PRIMARILY UPWARD, IN THE VASCULAR TISSUE CALLED XYLEM. THE CELLS OF XYLEM ARE DEAD WHEN FUNCTIONALLY MATURE AND ARE ANALOGOUS TO A PIPELINE.

WATER MOVEMENT   Although the details of arrangement of xylem are different in root and stem, xylem is continuous from young roots, through the stem, to the photosynthetic layers of the leaves (Fig. 5–15). Thus, water absorbed through the root hairs may travel several hundred feet to the leaves at the top of a large tree. The problem then arises: What are the forces responsible for such movement? Many hypotheses have been advanced, and they have been hotly disputed. Since the tracheids and vessels are very narrow and since water will rise in very small tubes due to the cohesion of the water molecules and to the adhesion of the water molecules to the walls, this phenomenon, called **capillarity,** has been proposed to account for the movement. However, the distances involved are simply too great. Nor is the principal path of movement through the cell walls themselves. If the cavity of the cells is blocked, water conduction does not take place. It was once thought that "root pressure" accounted for the movement of water in a tall tree. The stumps of some plants exude water at a considerable rate, but experiments have shown that the rate of transport is much too low to balance water loss at typical rates of transpiration. On the other hand, cut stems will transport water for days without roots.

Since so much water is being evaporated from the leaves, it is tempting

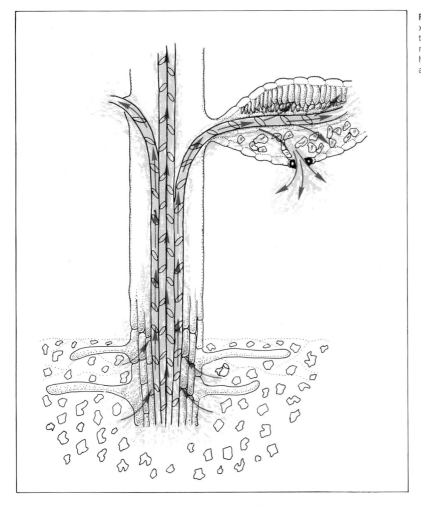

Fig. 5-15 Diagram showing that xylem is continuous from the root, through the stem, to the leaves; the movement of water is into the root hairs, through the root, stem, and leaf, and out the stomata.

to think of some sort of "suction" being applied from above. This would imply that atmospheric pressure pushed the water up from below. You will remember, however, that atmospheric pressure can push water up only 32 ft in height at sea level. Clearly, this cannot account for movement in tall trees. The answer lies in the properties of water and the behavior of water in leaf cells.

Water molecules tend to cohere very strongly. That is why drops form instead of films or sheets. If you were able to pull up a single water molecule, you would drag along other molecules of water. This is what happens in the xylem, where tracheids and vessel elements isolate "strands" of water several microns in diameter. The moist photosynthetic cells of the interior of the leaf evaporate water into the intercellular spaces and eventually out of the leaf. This leaves them water-deficient with respect to adjacent cells, and therefore diffusion occurs along this concentration gradient. Such a gradient extends to the dead but water-filled tracheids and vessels, and water is removed from them. Since the water molecules cohere and resist being "torn apart," water movement will take place up the stem.

Thus, although there is no suction from above, water is "pulled" to the top of the tree as water evaporates from the leaves. You may think of a vascular land plant as having an essentially continuous flow of water, which becomes very rapid when the stomata are open and the plant is in sunlight. Dissolved minerals are carried along in this stream. Lateral movement of water and solutes occurs by diffusion, an extremely slow process. By eliminating the protoplast and providing pits or perforations in tracheids and vessels, plants have speeded up vertical water movement tremendously. A plant probably uses, in metabolism, less than 5 percent of the water taken up. Most of it is transpired, leaving the dissolved minerals behind.

WATER MOVES UPWARD THROUGH THE XYLEM BECAUSE IT IS CONTINUALLY BEING REMOVED FROM THE LEAVES IN TRANSPIRATION. WATER MOLECULES COHERE STRONGLY AND ARE BEING ADDED AT THE ROOTS. SINCE THE XYLEM AND ITS INCLUDED WATER ARE CONTINUOUS FROM ROOT TO LEAF, LOSS OF WATER FROM LEAVES RESULTS IN AN UPWARD MOVEMENT.

THE ECOLOGY OF STAYING TURGID   Next to photosynthesis, the processes associated with water balance, especially transpiration, are the most important ecological activities of plants. This is reflected in plant evolution. The control of transpiration has been a major force in the evolutionary diversification of vascular land plants (ferns and seed plants). A corollary to this is that it is possible to predict the productivity and kinds of plants to be found in any region from data on transpiration and temperature (Fig. 5–16). Transpiration also has an effect on climate near the ground. Before examining these matters in detail, we will briefly survey the factors affecting transpiration.

FACTORS AFFECTING TRANSPIRATION   Temperature is the way we measure the kinetic energy of molecules. As heat is added to a system, the average velocity of the rebounding molecules in it increases. If the system is a liquid or gas, these higher velocities are reflected by higher rates of diffusion. Accordingly, a leaf transpires more water on a hot day than on a cold day because more molecules escape through the cuticle and stomata. At high temperatures, transpiration may exceed water absorption from the roots, resulting in water deprivation and wilting. For this reason, the danger of forest fires and brush fires is more severe on hot days.

The rate of transpiration is proportional to the difference in vapor content between the saturated air within the leaf and the surrounding atmosphere. Therefore, humidity and wind are important factors. Air circulation increases water loss because it prevents a blanket of water vapor from developing around transpiring leaves. When there is little or no wind, transpiration is decreased by these blankets of moisture. Wind plus high temperature is a dangerous combination for plants not having mechanisms to prevent water loss.

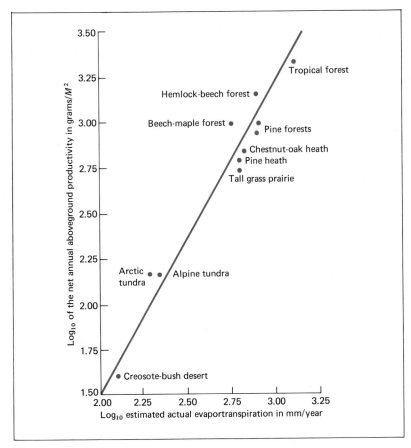

Fig. 5-16 Graph showing that photosynthetic production of plants in various habitats is proportional to the amount of water available and the temperature (evapotranspiration is the water lost from the soil by evaporation, as well as transpiration). (*Adapted from the AMERICAN NATURALIST from Vol. 102, 1968, p. 72, from an article by M. L. Rosenzweig, The University of Chicago Press.*)

So long as the roots absorb water at the rate at which it is transpired from the leaves, a plant suffers no water deficit. In midsummer, a corn plant transpires about 3 or 4 quarts per day without ill effects, but if soil moisture is not readily available, the cells of the plant lose their turgor and wilting results. Water may adhere very tightly to soil particles, and the less water in the soil, the more difficult it is for the root system to acquire it. Much of a plant's root structure and water-absorbing capacity are temporarily destroyed by transplanting, and more water than usual must be provided to compensate for this impairment of the absorption system.

As discussed above, light triggers a series of responses that open the stomata by increasing their turgor. Since carbon atoms in the form of CO_2 are often the limiting factor in photosynthesis, it is obviously advantageous for a plant to keep its stomata open whenever there is sufficient illumination for photosynthesis.

TRANSPIRATION, THE LOSS OF WATER FROM PLANTS, HAS IMPORTANT ECOLOGICAL EFFECTS IN ADDITION TO REGULATING THE UPTAKE OF WATER AND MINERALS AND THE RATE OF PHOTOSYNTHESIS.

FACTORS REDUCING WATER LOSS If a water deficit develops and the stomata close, photosynthesis and growth essentially cease. The plant soon dies if these conditions persist. Yet desert plants grow with only a few inches of precipitation annually, so one is compelled to seek explanations for drought resistance. Briefly, there are two main ways for a plant to resist desiccation: one is to minimize transpiration, and the other is to maximize absorption. Among the modifications in the first group are a reduction in the number of stomata, such as in the cacti. Another device is to have the stomata sunken in pits. The humid air accumulating in the pits retards transpiration when it is windy. The leaves of drought-resistant grasses can reduce their surface area by rolling up when a water deficit develops. The rolling is accomplished by rows of specialized thin-walled cells, which lose turgor and collapse while the normal epidermal cells are still turgid. Two other widespread modifications are a small leaf surface area and a thickening of the cuticle. During long periods of drought, many perennial trees and shrubs shed their leaves. A person from the Temperate Zone is often surprised to see forest trees in the drier tropics devoid of leaves in the warmest season.

Another group of modifications to arid conditions involves the roots. The roots of some legumes, such as acacia and mesquite, can penetrate many meters into the soil. They usually occur in washes, where underground water is relatively close to the surface. Other plants, such as cacti, produce a shallow, fibrous mat of roots that take advantage of very small amounts of precipitation. The typical restriction of the larger desert cacti and shrubs to rocky hillsides is the result of the greater penetration of precipitation into loose, gravelly soils. Much less water percolates into the more compact, finer soils of the flatlands.

Most of the water that is absorbed by the roots of a plant is transpired. Just as the evaporation of sweat from our skin cools our bodies, so does the transpiration of water from a tree cool its leaves. On a warm day this cooling may lower the foliage temperature by as much as 1 or 2°. This form of natural air conditioning is one of the reasons a forest seems so pleasant on a hot day. Extensive forests can affect local climate as well. Unfortunately, the world's forest lands are decreasing at an alarming rate due to the pressure of man, who exploits them for forest products and agricultural land. Only a quarter of the original forests in the United States remain. The cutting of woodlands in East Africa and northeastern Brazil is thought to be responsible for the increase in temperature throughout these regions. This warming trend may, in turn, cause further deterioration of these ecologically brittle parts of the tropics.

LAND PLANTS HAVE EVOLVED MANY MEANS OF REDUCING WATER LOSS. THESE RANGE FROM STOMATAL MECHANISMS OR LOSS OF LEAVES TO MODIFICATION OF THE ROOT SYSTEM.

WATER AND SOLUTE BALANCE IN ANIMALS

Water makes up the greatest fraction of our bodies—about 65 percent. Jellyfish are 95 percent water, and some insects go as low as 45 percent. Most of the cellular chemical reactions proceed in aqueous solution, water being an excellent solvent for most organic chemcials. Water also was the environment in which the first organisms evolved. Most organisms die if their tissues become very dehydrated, but there are many strategies for surviving drought. Animals inhabiting temporary ponds usually have waterproof eggs or cases, cocoons, and so on. Among these are certain "annual" fishes in Africa and South America that survive in completely dried mud as eggs and hatch ("instant fish") when immersed in water. One of the few groups that actually tolerates desiccation of its tissues is the rotifers that inhabit moss. Dry, wrinkled rotifers can survive up to 4 years. This dormancy is not complete cessation of metabolism, however, since they die if deprived of oxygen.

Internal Salt Concentration

Another essential component of the internal environment is the salt concentration. Figure 5-17 shows the salt concentration in the tissues of aquatic and terrestrial animals. Note that most organisms have about 1 percent salt (exceptions are marine invertebrates and sharks and their

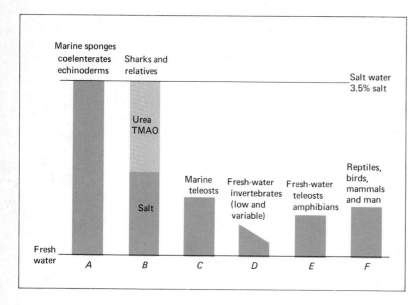

Fig. 5-17 Salt concentration in the tissues of aquatic and terrestrial animals. (Teleosts are the bony fishes.) *(Adapted from Knut Schmidt-Nielsen, ANIMAL PHYSIOLOGY, 2nd ed., © 1964, p. 54. By permission of Prentice-Hall, Inc., Englewood Cliffs, N. J.)*

relatives). Very slight changes in the concentration of these ions (mostly Na^+, K^+, Ca^{2+}, Mg^{2+}, Cl^-, HCO_3^-, PO_4^{3-}, SO_4^{2-}) can cause shock and death in higher animals and invertebrates that are never exposed to significant salinity changes. On the other hand, some animals, such as the common mussel *(Mytilus edulis)* that inhabits estuaries or bays in which the salinity of the water fluctuates with the amount of runoff of fresh water from the land, are able to withstand great changes in the salt concentrations of their body fluids. In the North Sea, the water and the mussel's body fluids contain 3 percent salt, and in some bays, they are as low as 0.5 percent salt. This is adaptation by **conformity.** Strictly marine crabs, such as the spider crab *Maja,* cannot tolerate great salinity changes. Changes in body-fluid salt levels such as occur in the common mussel are fatal to *Maja.* Other invertebrates adapt by **regulation** of the salt concentration in their tissues (Fig. 5-18). Certain shore crabs *(Hemigrapsus, Carcinus)* that venture into estuaries can retain salt in their body fluids at low salinities but conform to abnormally high salinities. Energy is required to retain salts in the body at low environmental salinities. (The crab's exertion is measured by oxygen consumption.) Energy is also required to remove the excess water that enters by osmosis. The gills, skin, or gut of most aquatic arthropods and vertebrates have specialized pumping cells that perform this osmotic work. Later, we will look at some organs of osmoregulation.

As suggested in this discussion, the regulation of water in animals cannot realistically be separated from the regulation of the chemicals dissolved in water. In the following section these homeostatic variables will

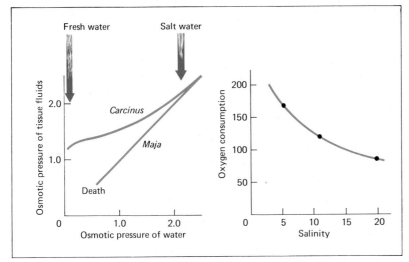

Fig. 5–18 Regulation of salt concentration in crabs: *left,* strictly marine crabs such as the spider crab (*Maja*) are not able to control salt concentration. The shore crab (*Carcinus*) conserves its body fluid salts in water at low salinity; *right, Carcinus* expends energy to maintain salts in the body at low salinities (as measured by oxygen consumption). *(Adapted from C. Ladd Prossor and Frank A. Brown, COMPARATIVE ANIMAL PHYSIOLOGY, 2/e, 1961, W. B. Saunders Company, Philadelphia.)*

be considered together. First, a brief detour into the problem of nitrogen excretion will help to broaden your understanding of the web of inter-relations in water and mineral homeostasis.

IMPORTANT IN THE MAINTENANCE OF HOMEOSTASIS IS PRESERVING THE BALANCE OF SALTS IN THE TISSUES AND CELLS OF ORGANISMS. THESE SALTS ARE DISSOLVED IN WATER, WHICH MAKES UP THE HIGHEST PERCENTAGE OF THE BODY, AND THEREFORE THE REGU-LATION OF WATER CONTENT AND THE REGULATION OF SALT CON-CENTRATION GO HAND IN HAND.

Nitrogen Excretion

Nitrogen in the reduced (hydrogenated), or amino ($-NH_2$), form is an essential part of amino acids, the monomers of proteins. If these amino acids are not present in sufficient amounts in the diet of most animals, signs of protein deficiency soon appear, a disease that in humans is known as **kwashiorkor** (Fig. 5–19). It is a little ironic, therefore, that the amino group is highly toxic if liberated as ammonia, NH_3. One part per 20,000 of ammonia is lethal in mammals. This **deamination** reaction usually proceeds in the presence of a **deaminase** enzyme and oxygen:

$$\tfrac{1}{2}O_2 + R-C\!\!\underset{COOH}{\overset{NH_3}{\diagup\diagdown}} \rightleftharpoons NH_3 + R-C\!\!\underset{COOH}{\overset{O}{\diagup\diagdown}}$$

Animals, including man, do not store amino acids or the amino groups, a fact in itself worth noting since it suggests that protein deficiency was rarely a problem to man in a "state of nature" as a hunter and gatherer.

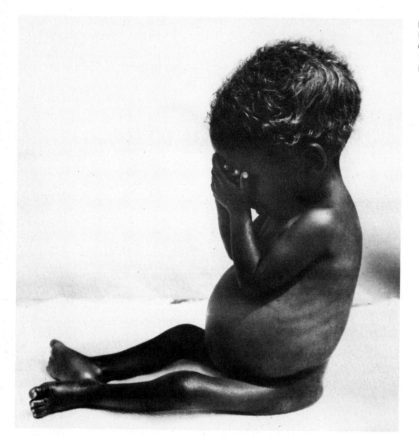

Fig. 5-19 A child suffering from kwashiorkor demonstrates the characteristic swollen belly and thin legs. Kwashiorkor is caused by severe protein deficiency. (*Jeroboam, N. Y.*)

Furthermore, few animals except marine invertebrates appear to secrete excess nitrogen in the form of amino acids—probably because this would be a profligate waste of the potential energy of these organic molecules. Some animals, because of their circumstances (see below), can secrete the excess nitrogen as ammonia without any problem. But there are other nitrogen-containing excretory products, especially urea, trimethylamine oxide, and uric acid (Fig. 5-20). The synthesis of urea is a rather complex affair. Two ATPs are used in the synthesis of each urea molecule. There is thus an appreciable cost to making urea, but it is worth bearing in some cases because urea is much less toxic than ammonia. Little is known about the synthesis of trimethylamine oxide. Uric acid is less toxic than either ammonia or urea, partly because it is so insoluble in body fluids. Its cost, however, in terms of ATP is five or six times that of the urea. Why should there be such a range of nitrogenous excretory products? It has nothing to do with evolutionary advancement; the reptiles, for example,

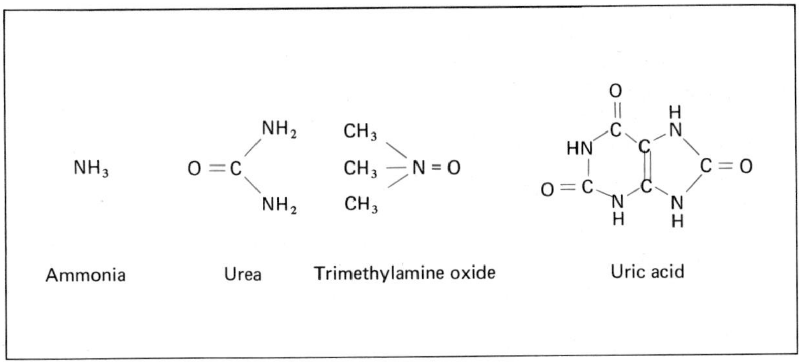

Fig. 5-20 Structural formulas of various nitrogen-containing excretory products.

use all three. The answer lies in the ecology of the individual species, particularly the water relations.

PROTEINS ARE THE BASIC STRUCTURAL COMPONENT OF CELLS, AND THEY PLAY A VARIETY OF OTHER ROLES AS WELL. PROTEINS ARE CONSTANTLY BEING SYNTHESIZED AND BROKEN DOWN, AND THERE-FORE CELLS HAVE THE PROBLEM OF DEALING WITH PROTEIN CON-STITUENTS, ESPECIALLY NITROGEN. MECHANISMS HAVE EVOLVED FOR EXCRETING EXCESS NITROGEN IN WAYS NOT HARMFUL TO THE ORGANISM.

MARINE INVERTEBRATES   Of all animals, marine invertebrates have the simplest osmotic problems. Because their body fluids are isotonic with seawater, they neither gain nor lose water. The simplest groups, including sponges, coelenterates, and echinoderms, don't even have to spend ATP in actively transporting salts because their body fluids have the same proportions of ions as does seawater. Many of the marine molluscs, however, and marine crustaceans have an ionic composition different from that of seawater and so must expend some energy combating diffusion and osmosis (Fig. 5-21). Marine forms have an unlimited volume of water at their disposal. Hence ammonia, because it is so soluble, will never reach dangerous concentrations in their body fluid because it will always tend to diffuse away (in the direction of lower concentra-tions) across membranes at the body surface. As a general rule, animals living in water employ this frugal form of nitrogen balance.

Nitrogen, Salt, and Water: Problems and Solutions

MARINE FISHES   It surprises most people to learn that marine bony fishes must drink large quantities of seawater to keep from drying out. Figure 5-17 shows that their body fluids are only 1 percent salt and are therefore hypotonic to seawater. They lose water rapidly by osmosis but, unlike camels, cannot find an oasis with fresh water to replenish their continuously dehydrating bodies. Seawater is the only source of water, so they must drink it, but this creates another problem. The gut epithelium through which the water is absorbed also allows the salts to diffuse through. If the salt concentration of a fish's body fluid rises much above normal, it dies. Calamity is averted by special salt-secreting cells in the gills. Of course, they must employ energy-consuming active-transport processes for this. (See Box 5–1.) The gills also account for most of the nitrogen excretion, two-thirds of which is ammonia, as might be predicted for an aquatic animal.

Sharks, rays, and other cartilaginous fishes (together called elasmobranchs) have evolved a physiological solution that obviates the expenditure of energy in secreting salts against an osmotic gradient. Their principal nitrogenous waste product is urea, but unlike most other urea-secreting animals the kidneys of elasmobranchs retain an extraordinary amount (2.0 to 2.5 percent) in the blood, enough so that the body fluids are slightly hypertonic to seawater. Primitive freshwater elasmobranchs in the Amazon River retain only a trace of urea in their blood. This is one example of why it is impossible to separate a discussion of nitrogen excretion and water balance—the two are interdependent.

FRESHWATER ANIMALS Osmotically speaking, freshwater organisms are semipermeable bags of salt in a hypotonic medium. Water tends to enter by osmosis. Actually, this is overstating the problem because organisms erect osmotic shields. These are impermeable waxy cuticles in insects and crustaceans, scales in fish, and shells in molluscs. All these barriers restrict osmotic transfer to a limited surface, such as the gills. But though this minimizes the problem, it does not solve it. Furthermore, the body surface cannot prevent some loss of salts due to diffusion. This twin problem of water gain and salt loss is solved by the secretion of large amounts of very dilute urine on the one hand and, on the other hand, absorption of salts by cells at the body surface or lining the gut. Both require large energy expenditures because they are working against concentration gradients (Fig. 5–17), suggesting how expensive it is in terms of energy to be a freshwater animal and possibly why there are not more different kinds of them.

The larvae of mosquitoes and other aquatic insects absorb salts through the anal papillae. Most freshwater fish absorb salt through the gills—the reverse of the direction of salt movement in salt water. Since the tissues of freshwater animals are constantly being flushed by water, they can and usually do secrete most of their nitrogen as ammonia. Some

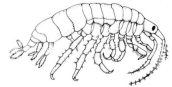

Amphipod (*Gammarus*)

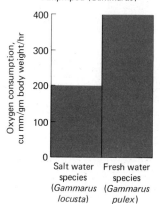

Fig. 5–21 Oxygen consumption in two closely related arthropod species. The freshwater species must do twice the osmotic work of the saltwater species. (*Adapted from R. O. Barnes, INVERTEBRATE ZOOLOGY, 2/e, 1968, W. B. Saunders Company, Philadelphia.*)

exceptions are aquatic insects, who retain the ancestral terrestrial biochemical trait of uric acid secretion.

TERRESTRIAL ANIMALS   There is one big advantage to being on land or at least to breathing air: oxygen is abundant and accessible. Hence, the metabolic rate of cells can increase many times over. One of the disadvantages is a relative shortage of water in which to disperse ammonia.

Vertebrates and insects are eminently successful terrestrial groups, and their complete independence of water came to both groups when they evolved a waterproof egg. Why is the egg rather than the adult the critical stage? It is simply that the relatively large adult animal has the sense organs and navigational equipment to find water when it becomes too dry. Furthermore, the adult has less surface area relative to its volume, so it tends to dry out more slowly, and many land animals, such as the land crab, have a waterproof cuticle. The eggs of most, if not all, aquatic animals if placed on land, would dry out long before development of a waterproof hatchling was complete. Nevertheless, waterproof eggshells have not really been the evolutionary hurdle; paradoxically, it is the problems created by them that have been the obstacle to invasion of the land by most marine animals. One of these problems is getting enough oxygen through a waterproof shell. Another is the problem of nitrogenous waste. Ammonia is certainly out of the question; it is too poisonous. So is urea, although it is less poisonous than ammonia. The answer in both insects and vertebrates is uric acid, a relatively insoluble compound that crystallizes out of solution where it accumulates in the organism. Almost no water is needed if uric acid is the excretory material. As we stated above, its disadvantage is its cost in terms of ATP.

Insects, birds, and reptiles, with few exceptions, all secrete mostly uric acid. Some aquatic reptiles (Fig. 5-22) have reverted to the less costly ammonia and urea excretion. Egg-laying (oviparous) mammals, such as platypuses, excrete uric acid; adults of the live-bearing (viviparous) mammals also still do to some extent, although urea is their principal nitrogenous waste. The developing embryo mammal makes only urea since it can be disposed of quickly and safely in the mother's circulation.

MAMMALS IN VERY DRY PLACES   Although some mammals, including man, have a small amount of uric acid in their urine, they have lost the ability of the reptiles to divert most of the waste nitrogen into this inert material. Since urea can be secreted only in solution, mammals have no

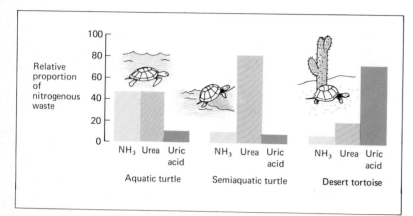

way to dehydrate their urine completely and water is inevitably lost, a grave danger to the desert and saltwater dweller. A true desert mammal, the kangaroo rat of the American Southwest, has been studied in some detail. One of the ways in which this animal conserves water is by having no sweat glands. Kangaroo rats can get by without these by leaving their deep burrow only at night, when it is cool. But water is still lost in the breath, the feces, and the urine—in fact, more water than is in the food. Cellular respiration makes up for the deficit. Most of their food consists of lipid-rich seeds. Obviously, the higher the ratio of hydrogen to other atoms in the food, the more respiratory water can be made when these hydrogens are removed in respiration and combined with oxygen in the electron-transport system of the mitochondria. These seeds thus supply enough hydrogen to maintain the animal's water balance, but a kangaroo rat will be driven to drink if kept on a low-fat or high-protein diet.

Finally, the kangaroo rat has a kidney that surpasses all others in its urine-concentrating ability, thereby minimizing water loss by way of its urine. The strength of their kidneys is demonstrated by the kangaroo rat's ability to drink seawater, but to appreciate this it is necessary to know how a kidney works.

THE MODE OF NITROGEN EXCRETION AND WATER REGULATION IN A PARTICULAR GROUP OF ORGANISMS DEPENDS UPON THE ECOLOGY OF THE SPECIES. MARINE INVERTEBRATES EXCRETE AMMONIA. MARINE FISHES ARE HYPOTONIC TO SEAWATER AND MUST THEREFORE CONSUME WATER IN LARGE QUANTITIES AND EXCRETE THE DISSOLVED SALTS. FRESHWATER ANIMALS TEND TO GAIN WATER AND LOSE SALTS AND MUST EXPEND ENERGY TO COUNTERACT THIS. TERRESTRIAL ANIMALS EXCRETE NITROGEN IN LESS TOXIC

SUBSTANCES, SUCH AS UREA OR URIC ACID. THE WATERPROOF EGG OF LAND ANIMALS PRESENTS SPECIAL PROBLEMS OF OXYGEN UP-TAKE AND NITROGEN EXCRETION. DESERT MAMMALS THAT DO NOT DRINK WATER ARE SUPPLIED WITH THIS NECESSARY SUBSTANCE BY CELLULAR RESPIRATION.

KIDNEY FORM AND FUNCTION   The kidneys of vertebrates are paired compact organs of chemical homeostasis. Their function is to maintain the *status quo* of the body fluids. Hence, it is doing the kidneys a dis-service to consider them as organs for nitrogen excretion alone, because excess nitrogen is only one of a myriad of compounds that they remove from the blood.

The structure and function of kidneys may be exemplified by the kidneys of mammals. Impure blood enters the kidney via a blood vessel, the renal artery, and purified blood leaves in a vein (Fig. 5–23). The concen-trated waste that has been removed from the blood is emptied into a cavity, the **pelvis,** inside the kidney and then passes through a duct called the **ureter** to the **bladder,** which is drained periodically through the **urethra,** a duct that leads directly to the body surface (Fig. 5–23).

The renal artery branches into smaller arteries in the outside layer, or **cortex,** of the kidney. Next, the arterioles conduct the blood into compact, ball-like capillary knots, the **glomeruli** (sing. **glomerulus).** The capil-laries join, and blood leaves the glomerulus in a single small vein. A little cup of tissue called the Bowman's capsule is wrapped around each glomerulus, and any fluid leaving the capillaries passes into the capsule through the first wall and thence into a series of tubules (where some ma-terials enter and others leave), finally reaching the pelvis. The blood vessel leaving the glomerulus branches into a second capillary network that invests the tubule leading away from the Bowman's capsule. The capillaries again reunite and enter a branch of the renal vein.

Briefly, the system works as follows: The blood entering the glomerulus is under the considerable pressure exerted by the pumping action of the heart. The walls of the glomerular capillaries, like all capillary walls, are permeable to water and small molecules. Therefore, the filtrate that is forced out of the glomerulus by hydrostatic pressure (like water through a weak place in a hose) has essentially the same composition as that of blood but lacks the blood cells and larger molecules, including proteins such as albumin and the globulins. The volume of raw glomerular filtrate in man is about 180 quarts per day. This first filtrate contains many precious substances, including glucose, amino acids, and inorganic ions. Obviously, nearly all of these must be withdrawn from the glomerular filtrate. This takes energy, and the kidney tubules use a large amount.

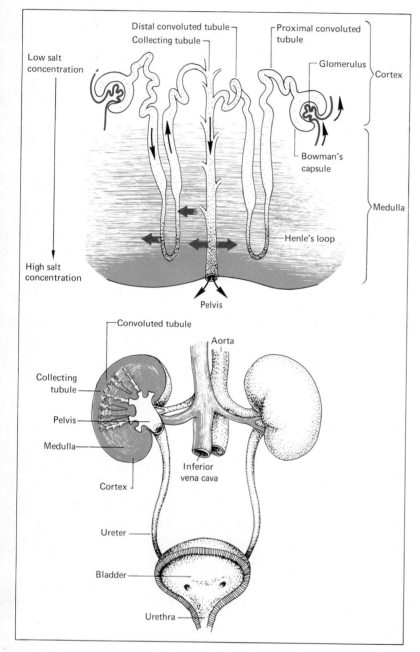

Fig. 5-23 Structure of a mammalian kidney: *below*, general structure of kidney and associated structures; *above*, detailed view: shading indicates the concentration of salt in the intercellular spaces surrounding the tubules; arrows show the direction of water movement. As the filtrate passes through Henle's loop, it first loses water, then regains it by osmosis; the final concentration step occurs as the filtrate passes through the collecting tube into the renal pelvis.

Much of the energy is used for the secretion of sodium from **Henle's loop,** a hairpin section of the tubule. The salt concentration is increased from the cortex (outer part of the kidney) to the medulla on the inside. As the filtrate flows down the collecting tube (Fig. 5–23) into the region of medulla where the salt concentration is high, the water is forced back out of the filtrate, or urine, by osmosis. The water reenters the blood through the capillary network closely surrounding the tubule. Most of the glucose, amino acids, and salts are removed in the proximal convoluted tubule. The cells making up the walls of the tubules are not strictly one-way, however. They are capable of doing some excreting themselves, including the active elimination from the filtrate of foreign materials such as penicillin.

The master chemical regulator of the body is the distal convoluted tubule. One of its activities is to adjust the acidity of the blood, adding or subtracting hydrogen ions as required. Another function of this region of the tubule is water regulation. Certain cells in the brain are sensitive to the water content of the blood. If the water content falls, these cells stimulate other cells in the pituitary (hypophysis) of the brain to release a hormone, *antidiuretic hormone* (ADH) (the purified form is called vasopressin). Upon reaching the kidney via the blood, ADH alters the cells of the distal convoluted tubule and collecting ducts so that they absorb more water, tending to replace water to the body and increasing the concentration of solutes, mostly urea and salts, in the urine. Blood that is too dilute, such as could result from drinking a lot of fluids, reduces the output of ADH and causes the secretion of more water and a dilute urine.

*KIDNEYS ARE VERTEBRATE ORGANS WHICH MAINTAIN THE HOMEO-
STASIS OF THE CIRCULATORY SYSTEM AND THUS OF THE REST OF
THE BODY. BLOOD IS FILTERED IN THE KIDNEY AND IS CHANGED
CHEMICALLY TO MAINTAIN THE PROPER CONCENTRATION OF SALTS
AND WATER AND TO REMOVE WASTES.*

SUMMARY

The process of diffusion results in the molecules of a dissolved substance being uniformly distributed in the solvent. A biologically very important kind of diffusion is osmosis. Osmosis is the diffusion of water across a differentially permeable membrane, one that is relatively highly permeable to water but relatively impermeable to dissolved molecules, such as proteins and carbohydrates, of biological systems. If the concentration of water in a solution of larger molecules is different on either side of a differentially permeable membrane, osmosis will occur until the concentration is the same on both sides or until some counterpressure causes it to stop.

The contents of plant cells are hypertonic to their environment. Therefore, water tends to move into plant cells, causing them to swell. The cytoplasm swells until it presses tightly against the tough cellulose cell wall, which exerts a counterforce called turgor pressure. Changes in turgor of the guard cells of stomata in the leaves of plants cause the stomata to open or close. This regulates the uptake and loss of gases, such as oxygen, carbon dioxide, and water vapor. Therefore, changes in the stomatal openings regulate the loss of water by plants (transpiration) and the rate of photosynthesis. A major component of growth in plants is cell elongation. Under the influence of auxin, the cell wall becomes modified and enlargement occurs as turgor swells the cell. The size and shape of cells are determined by the arrangement of fibrils of cellulose in the cell wall.

The root system of vascular land plants is an absorbing and supportive system. Roots take up water and dissolved materials from the soil solution through root hairs, which provide an enormous surface for absorption. Water diffuses from cell to cell in the root until it reaches the xylem. Water-conducting cells of the xylem are dead when functionally mature. They are analogous to a series of pipes connecting all parts of the plant. Water moves upward through the xylem because it is continuously removed from the leaves in transpiration. Water molecules cohere strongly and are being added at the roots. Since the xylem and its included water are continuous from root to leaf, loss of water from the leaves results in an upward movement. Transpiration, the loss of water from plants, has important ecological effects in addition to regulating the uptake of water and the rate of photosynthesis. Land plants have evolved many means of reducing water loss. These range from stomatal modifications or loss of leaves to modifications of the root system.

Animals have the problem of maintaining the proper balance of water and dissolved salts in their cells and body fluids. The regulation of water content and the regulation of salt concentration go hand in hand. Proteins are constantly being synthesized and broken down, and therefore cells have the problem of dealing with protein constituents, especially nitrogen. Mechanisms have evolved for excreting excess nitrogen in ways not harmful to the organism.

The mode of nitrogen excretion and water regulation in a particular group of organisms depends upon the ecology of the species. Marine invertebrates excrete ammonia, which is toxic in low concentrations but can diffuse away in the water. Marine fishes are hypotonic to seawater and must therefore consume water in large quantities and excrete the dissolved salts. Freshwater animals tend to gain water and lose salts and must expend energy to counteract this tendency. Terrestrial animals excrete nitrogen in forms less toxic than ammonia, such as uric acid or urea. The waterproof egg of land animals presents special problems of oxygen uptake and nitrogen excretion. These are solved by special mechanisms in the egg.

A cell is a system with a high degree of order. This order is both structural and temporal. The macromolecules and organelles are arranged in regular, predictable patterns in space; they also interact with each other in regular, predictable patterns in time. The events of mitosis are a graphic example of spatial and temporal order in a cell. This degree of order does not happen by chance. In fact, if cells were not equipped to control the forces that promote disorder and to constantly repair minor damage, their distinctive features could not persist. One of these forces is the randomizing effect of diffusion. Diffusion and its close relative, osmosis, tend to prevent the establishment and maintenance of high concentrations of ions and molecules in places where they are required. In the absence of a barrier against diffusion, cells could not concentrate large amounts of materials such as nutrients, nor could they selectively eliminate other materials against concentration gradients.

We must begin by admitting that it is not known exactly how cells protect themselves against the inflow of undesirable materials by diffusion or how they concentrate desirable materials where such a process is opposed by diffusion. On the other hand, we do know what the principal structures involved in these processes are. They are the cellular membranes.

The form of the membrane is typical of many membranes both on the outside surfaces of cells and within cells. There appear to be two layers separated by a thin space. The entire structure has a thickness of about 75 angstroms (Å). Such a structure is often called the unit membrane. A number of models have been proposed for the structure of such membranes.

Cell membranes behave as barriers in many ways. The central vacuole of plant cells is bounded by a membrane and may contain toxic materials. The escape of these substances has been shown in some cases to kill the cells. Many of the cellular organelles, such as mitochondria and chloroplasts, have membrane surfaces that prevent the loss of soluble molecules.

The real problem, though, is not these physical barrier characteristics of membranes but the ability of membranes to move materials in directions they ought not to go by random processes such as diffusion. Probably all cells are capable of moving solutes "uphill" against a

BOX 5–1. MEMBRANES, BARRIERS, AND TRANSPORT

concentration gradient. The colon bacterium *E. coli,* for example, can concentrate the sugar galactose inside the cell until it is at least 2,700 times more concentrated than the medium. The cells lining the intestine remove sugars and amino acids from the dilute contents of the intestine and concentrate them. Inorganic ions, too, are moved about in improbable ways. Cells control many important variables in their internal environment by the transport of ions. Cells are prevented from becoming dangerously acid by the exchange of K ions for H ions. Osmotic pressure is regulated by controlling the concentration of ions, primarily Na and K. As will be discussed in Chap. 7, this property of the limiting cell membrane affects its electrical properties and is essential for the propagation of an impulse in cells of the nervous system.

A common denominator of all these properties that depend on the control of inorganic ion concentrations is the uphill secretion of Na ions. Typically, Na ions are 10 or 20 times as concentrated on the outside of a cell membrane as on the inside. Either the pores in the membrane are too small for sodium or the sodium must be removed as fast as it enters. The former depends on the barrier property; the latter, on some active property that would require energy. Many lines of evidence lead to the conclusion that an energy-requiring "sodium pump" is constantly at work in cell membranes. Poisons, such as cyanide, that halt the production of ATP cause a sudden influx of Na

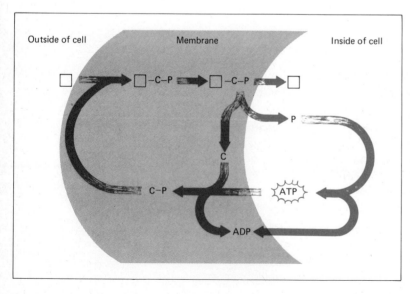

Fig. 5-1-1 A model of active transport in a cell membrane. The transported molecule is picked up by an activated (phosphorylated) carrier molecule (C-P) and moved to the inner surface of the membrane where it is released along with the phosphate. ATP provides the energy to phosphorylate and moves the carrier and its load. The ATP is regenerated by the normal processes of cell metabolism.

ions. Without ATP in the membrane, diffusion takes over and Na ions surge through the membrane.

The uphill transport, or **active transport,** of other inorganic as well as organic molecules is energy-requiring, but we have nothing but educated guesses about the actual workings of these pumps. Most hypotheses invoke a "carrier" molecule of some kind. The model in Fig. 5–1–1 is a simple diagrammatic example of the ideas now current.

The leaf is the principal photosynthetic organ of higher plants. As such, the total area that is exposed to sunlight is crucial. Leaves are not and need not be thick because photosynthetically active wavelengths of light are strongly absorbed by just a few microns of chlorophyll. In addition to the general shape of leaves, the internal structure is a good example of functional engineering. The principal photosynthetic tissue appears to be the **palisade parenchyma**—long, vertically oriented cells with many chloroplasts (as shown in Fig. 5–2–1). This shape and arrangement seem to be an especially efficient light trap, since light loss is reduced to a minimum. An extensive system of intercellular spaces provides for efficient gas exchange.

Below the palisade layer lies the **spongy parenchyma**—irregularly shaped cells with large air spaces between them and with fewer and smaller chloroplasts. In general, these cells are horizontally oriented,

BOX 5–2. THE STRUCTURE OF LEAVES

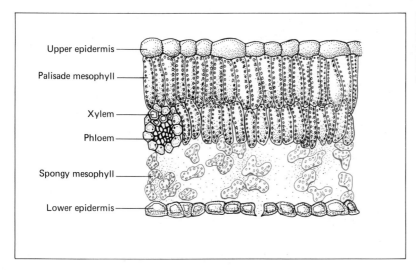

Upper epidermis

Palisade mesophyll

Xylem

Phloem

Spongy mesophyll

Lower epidermis

Fig. 5–2–1 Diagram to show the internal structure of a leaf. Photosynthesis occurs in the chloroplast-containing cells of the palisade and spongy mesophyll layers. Gas exchange is through stomata on the lower surface.

thus providing, when turgid, much of the leaf's horizontal support. They function something like the struts in an airplane wing. In addition to providing support, the large air spaces of the spongy layer permit rapid diffusion of CO_2 upward from the stomata to the palisade cells, which have an even greater air space among them. Together, the palisade and spongy parenchyma are termed the **mesophyll.** The internal air space of the mesophyll is enormous. One investigator calculated the total external surface of all the leaves of a tree to be 4,200 sq ft. Measurements and calculations of the internal surface of the mesophyll cells indicated that there were 55,000 sq ft of surface across which gas exchange for photosynthesis takes place.

The **vascular tissues,** branching repeatedly in the leaves, provide water (from the xylem) and remove excess sugar, which is transported (in the phloem) to sugar-deficient or storage tissues.

SUPPLEMENTARY READINGS
Books

Baldwin, E.: "Dynamic Aspects of Biochemistry," 4th ed., Cambridge University Press, New York, 1964.
Cannon, W. B.: "The Wisdom of the Body," Norton, New York, 1932.
Pitts, R. F.: "Physiology of the Kidney and Body Fluids," Year Book of Medical Publications, Chicago, 1968.
Polts, W. T. W., and P. Guyneth: "Osmotic and Ionic Regulation in Animals," Pergamon, Oxford, 1964.
Slayter, R. D.: "Plant-water Relationships," Academic, New York, 1967.
Sutcliffe, J.: "Plants and Water," St. Martin's, New York, 1968.

Articles

Schmidt-Nielsen, K.: Salt Glands, *Sci. Amer.* **200,** no. 1, Offprint 1118 (1959).
Smith, H. W.: The Kidney, *Sci. Amer.* **188,** no. 1, Offprint 37 (1953).
Zimmerman, M. H.: How Sap Moves in Trees, *Sci. Amer.* **208,** no. 3, Offprint 154 (1963).

CHAPTER 6

FOOD, GAS EXCHANGE, CIRCULATION

Contemporary biologists are discovering that one of the most profitable ways to look at problems of survival or adaptation is in terms of **energy.** It is impossible to describe all the ways in which animals have evolved to obtain and process the food input because there are virtually as many ways to do this as there are kinds of animals. Instead of attempting to describe all the energy-input strategies, we will discuss four examples.

FEEDING IN ANIMALS
Filter Feeding: The Bivalve

In quantitative terms, more food is caught by filtering water than by any other method. The sea occupies 70 percent of the earth's surface, and 70 percent or more of the biosphere's photosynthesis is carried out by microscopic marine **phytoplankton**—mostly diatoms and dinoflagellates (Fig. 6-1). These small organisms, and the slightly larger **zooplankton** which graze on them, form the foundation of the food webs of the oceans. In fresh water, too, phytoplankton are often the principal plants. Most aquatic animals—from the zooplankton to the largest existing animals, the baleen whales—employ a host of devices to filter out smaller organisms from the thin soup that is a pond, lake, or ocean. Copepods and other small crustaceans which make up the majority of the phytoplankton grazers, literally filter their way through the water by creating currents with their many bristle-bearing appendages. These currents force suspended particles into ciliary nets which propel the particles toward the mouth. The baleen whales catch their staple food, euphausid shrimp or "krill," by locating concentrations of these and then gulping large amounts of water, which they force back out of their mouths through the fringes of their whalebone plates. The plates are made of keratin, a horny protein. In these whales, true teeth are found only in the fetus.

Many invertebrates have evolved very elegant structures to obtain food in this rather passive manner. As an example, let us consider the gills of clams, mussels, oysters, and their relatives. These evolved from small, simple structures, used exclusively for gas exchange, into very elaborate plankton-filtering and transport organs.

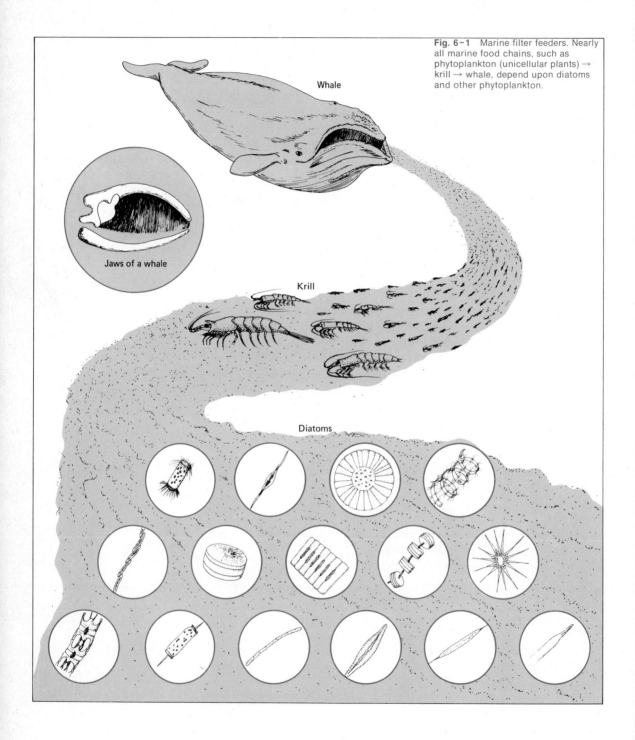

Fig. 6-1 Marine filter feeders. Nearly all marine food chains, such as phytoplankton (unicellular plants) → krill → whale, depend upon diatoms and other phytoplankton.

Whale

Jaws of a whale

Krill

Diatoms

Each gill is W-shaped in the end view (Fig. 6–2). The vertical filaments which make up the gill hang down into the mantle cavity and are fused at the ends. They extend from the mouth, at one end of the animal, well around to the other side. The mantle cavity is thus divided into a large inhalant outer chamber, into which water enters between the valves, and the smaller, inner, exhalant chamber. To reach the latter, the water must pass between the filaments. Food and other particles are filtered in the process.

By the time the water reaches the exhalant side of the gills, it is divested of its load of suspended particles, including small organisms. These are caught in strands of mucus and are already on their way to the mouth. Microscopic examination of the moving particles shows that they are being borne along in many directions by rows of cilia. The traffic pattern of the gill is quite complex. As shown in Fig. 6–3, each filament bears three kinds of ciliary rows: laterals (not visible), laterofrontals, and frontals. The main function of the lateral cilia is to propel water through the gills. A flow of up to 8 gal per hour has been measured in the American oyster at 24°C. As the inhalant water passes between the filaments, it is strained by the interlocking laterofrontal cilia, which fringe the exposed edges of the filaments and toss particles onto the mucous belt being pushed along by the small frontal cilia. These ciliary tracts move particles toward either the free edge of the gill or the base. A certain amount of sorting on the basis of particle size is accomplished on the gill surface,

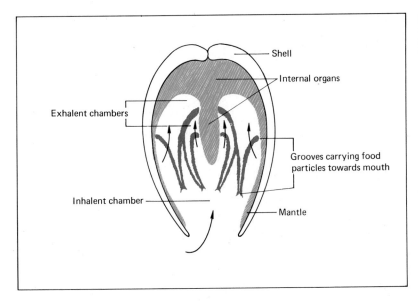

Fig. 6–2 Diagram of the gills and water flow in a bivalve mollusc, such as an oyster. The arrows show the direction of water movement.

Shell

Internal organs

Exhalent chambers

Grooves carrying food particles towards mouth

Inhalent chamber

Mantle

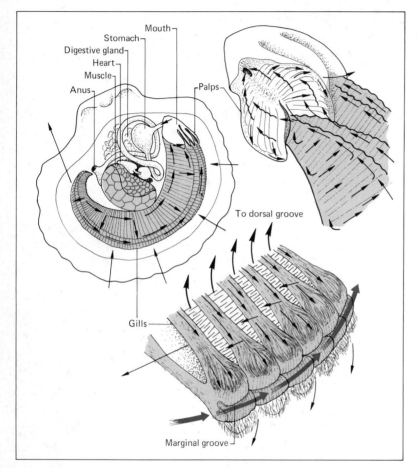

Mouth

Stomach

Digestive gland

Heart

Muscle

Anus

Palps

To dorsal groove

Gills

Marginal groove

Fig. 6-3 Filter feeding in an oyster. Particles are strained out of water passing through the gills. The arrows show the direction of water movement. *(Adapted from Martin Wells, LOWER ANIMALS. Copyright McGraw-Hill Book Company, 1968. Used by permission of McGraw-Hill Book Company and Weidenfeld & Nicolson Publishing Company, Ltd.)*

unwanted particles being pushed off the free edge. The final result is that a trapped food particle ends up moving toward the mouth in the five food grooves, one at each of the two free ends and one at each of the three attached ends of the gill.

A final sorting step is carried out by the ciliated undulating surface of the palps. One surprising fact that is emerging about molluscan filter feeders is the discrimination of the palps. Not only do they reject particles which are just a few microns too large, but certain kinds of organisms, including bacteria, are apparently selectively rejected also. All mucus feeders take in dissolved substances from the surrounding water. Some of these—for example, amino acids—may be very important in their physiological economy.

In the meantime, undesirable material (pseudofeces) cast off by the gills and palps accumulates in the inhalant side of the mantle cavity. So, on occasion, the oyster forcibly contracts its large muscle (what we eat in the scallop), ejecting the contents of the inhalant chamber. This behavior in clams probably accounts for most of the little water spouts to be seen on mud and sand flats at low tide.

Food enters the stomach embedded in a string of mucus. Here the string is coiled and agitated by a rotating rod called the crystalline style (Fig. 6-4). The end of the style continuously dissolves in the acidic stomach environment and, in so doing, liberates small amounts of a powerful amylase—a starch-digesting enzyme. Other enzymes enter the stomach from the digestive gland. Sorting of the food particles, a process begun on the gills, continues in the stomach and in a diverticulum or caecum. Coarse, indigestible particles are fed into a ciliated channel leading to the intestine. There they are fashioned into fecal pellets.

Digestion occurs in two steps. Preliminary digestion is carried out in the stomach by enzymes from the crystalline style and the digestive gland. Because this step takes place in a cavity rather than in cells, it

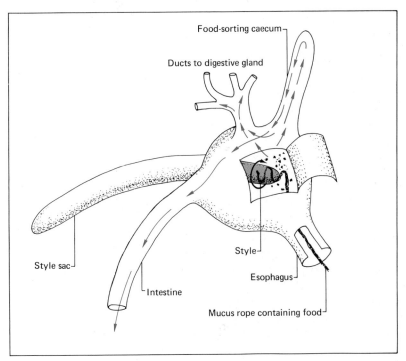

Food-sorting caecum

Ducts to digestive gland

Style sac

Style

Esophagus

Intestine

Mucus rope containing food

Fig. 6-4 Diagrammatic representation of the stomach of an oyster. The flap exposes the end of the rotating crystalline style. Long arrows show path of large, undigestible particles; small arrows show path of digestible food.

is called extracellular digestion. Large particles are engulfed by amoeba-like cells entering and leaving through the stomach lining. Smaller particles enter the digestive gland through small apertures in the stomach. There digestion is completed intracellularly, that is, within cells.

MANY ANIMALS OBTAIN THEIR FOOD BY INGESTING LARGE QUANTITIES OF WATER AND PASSING IT THROUGH A FILTER WHICH STRAINS OUT SMALL ORGANISMS.

Anyone would guess that a cow—or a deer, for that matter—is digesting its own food as it stands out in the field. But anyone would be mostly wrong. We say "mostly" because even though ruminants (cattle, deer, antelopes, sheep, and goats) do have the normal mammalian digestive organs and enzymes, other organisms are actually responsible for processing most of the ruminant's food, that is, the plant cell wall, the major component of which is the polysaccharide cellulose.

These organisms are bacteria and ciliate protozoans which are examples of **symbionts** (from **symbiosis,** "living together"). They change the coarse cellulose-laden pulp to edible products—simple organic acids and fatty acids. The cow or deer in the meadow is in reality an entire community, as shown in Fig. 6–5.

These processes of ruminant digestion occur in greatly modified and enlarged parts of the esophagus (Fig. 6–6). The first of these, and the largest, is the **rumen.** Swallowed food enters this compartment, which contains large numbers of bacteria and protozoans. No digestive enzymes are secreted into this fermentation chamber by the host, but there is a significant amount of absorption of the simpler organic acids, products of the metabolism of the bacteria. When grass is swallowed, the rumen fashions it into compact balls of cud, which are subsequently returned to the mouth for further mastication (chewing). On returning to the rumen for the second time, the pulpy food is subjected to bacterial digestion and assimilation. Especially important is the action of the **cellulase** secreted by the bacteria, which digests resistant cellulose into its component sugar molecules. So efficient is this form of symbiotic arrangement between a ruminant and its guests that cattle remain perfectly healthy for an indefinite period of time on a diet of newsprint (almost pure cellulose) and ammonium salts, such as are used for inorganic nitrogen fertilizers. (The nitrogen-containing salt is necessary to synthesize amino acids, the building blocks of protein.) But grass is cheaper than newspapers and fertilizer as a source of both cellulose and nitrogen, so it is unlikely that this form of pastoralism will ever be economic.

Herbivores: The Cow and Other Ruminants

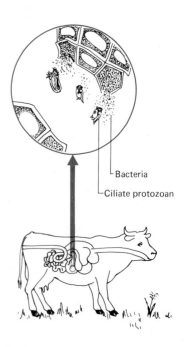

Bacteria
Ciliate protozoan

Fig. 6–5 In the rumen of a cow, bacteria produce enzymes which digest plant cells. Some of the bacteria are in turn injested by ciliate protozoa.

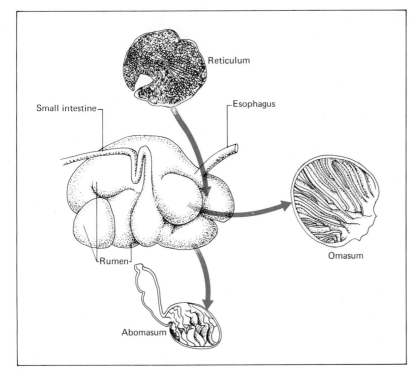

Fig. 6-6 The ''stomach'' of a cow showing sections of the reticulum, omasum, and abomasum. *(Adapted from J. E. Morton, GUTS, 1967, St. Martin's Press, Inc., Macmillan & Co., Ltd.)*

Meanwhile, the ciliate protozoans graze on the bacteria and are, in turn, digested and absorbed in the next compartments, the **reticulum** and the **omasum.** The latter is a small compartment looking something like an open book, the partially separated pages of which strain the food as it passes into the **abomasum,** or true stomach. Along with the typical gastric enzymes, a calf's abomasum produces the milk-coagulating enzyme **rennin,** from which rennet dessert is made.

Hares and rabbits (as well as some rodents) employ another strategy to get the most out of the vegetation they eat. They also have a large bacterial fermentation chamber, the **caecum,** but this appendage to the gut is located *behind* the small intestine, the principal absorptive region (Fig. 6-7). This means that the products of bacterial digestion and metabolism cannot be absorbed by the rabbit. To overcome this anatomical embarrassment, these mammals produce two kinds of feces; the small hard droppings we are used to seeing and a larger, less compact pellet from the caecum. The latter, formed only at night, is eaten directly from the anus and has a second trip through the gut. This is analogous to a cow regurgitating and chewing its cud. An experiment has shown that

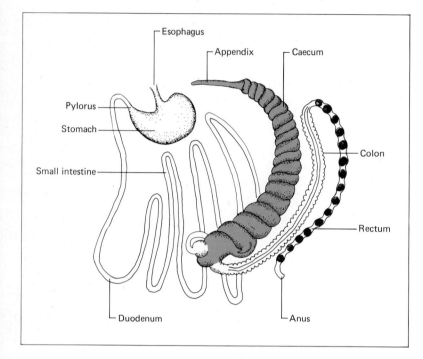

Fig. 6-7 Digestive tract of the European hare (domestic rabbit). Note the large caecum at the junction of the small intestine and the colon. *(Adapted from J. E. Morton, GUTS, 1967, St. Martin's Press, Inc., Macmillan & Co., Ltd.)*

these pellets contain essential metabolites (probably bacterial). If rabbits are restrained in such a way that they cannot reach their anus, they die in two or three weeks.

One of the advantages of these herbivore digestive systems is that these animals can consume large quantities of a coarse food when it is safe and convenient, then chew it again or recycle it later in the safety of a retreat or burrow. In this way, exposure to predators is significantly reduced. The relative indigestibility of grasses and leaves is also compensated for by the length of the intestine. Herbivores usually have much longer intestines than carnivores, as discussed below.

Among the invertebrates, termites eat wood in large quantities but are unable themselves to digest it. Digestion is performed by ciliated protozoans, which live in the termite gut and are able to produce the needed enzyme, cellulase, to break down plant tissue. Cellulose is not the only substance requiring digestion by symbiotic organisms. In at least one kind of leech, the so-called medicinal leech, the blood upon which the animal feeds is digested by a particular gut bacterium. This bacterium predominates in the gut of the leech because it produces an antibiotic that destroys other species.

A very interesting set of relationships concerning nutrition is found in Africa. There, birds are found which eat beeswax. These birds are called honey guides because, after locating a hive, they attract a larger, honey-eating mammal, the honey badger, and guide it to the hive, which the badger is strong enough to tear open. The honey guide then consumes the beeswax—which, however, is digested by bacteria and yeast living in its intestines.

HERBIVORES HAVE THE SPECIAL PROBLEM OF DEALING WITH THE MAJOR COMPONENT OF THEIR FOOD, WHICH IS CELLULOSE. CELLULOSE IS ACTUALLY DIGESTED BY SYMBIOTIC ORGANISMS—BACTERIA OR PROTOZOANS—WHICH LIVE IN THE DIGESTIVE TRACT OF THE HERBIVORE. IN OTHER ANIMALS, THE INTESTINAL MICROORGANISMS ARE ALSO RESPONSIBLE FOR DIGESTION.

Seed Eaters: The Pigeon

Birds have been around for at least 150 million years, but seed-eating birds are evolutionary newcomers because seeds are a relatively new invention of plants. It has only been within the last 100 million years that seed-bearing plants have become abundant. During this time, the great proliferation of sparrows, crossbills, cockatoos, and other seed and nut specialists has occurred (Fig. 6-8).

In contrast to the seed eaters which crack seeds with stout, powerful bills, others such as fowl, quail, and pigeons depend on the gizzard, a specialized region of the stomach, for cracking and masticating seeds (Fig. 6-9). In these birds that eat seeds and other coarse plant material, the gizzard, a reptilian legacy, has thick muscular walls and a horny lining with ridges. What a cow accomplishes with its molars, these birds do

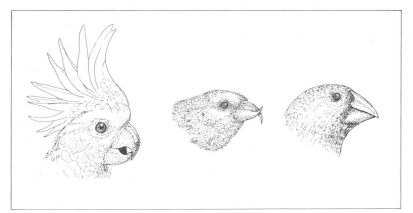

Fig. 6-8 A variety of seed-eating birds: left, the cockatoo, which has a powerful seed- and nut-crushing bill; center, the crossbill, whose scissorslike bill is used to extract seeds from the cones of conifers; right, the large-billed ground finch of the Galapagos Islands which, like most seed-eating birds, has a short, stout bill for cracking seeds.

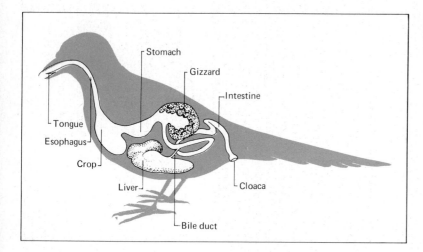

Fig. 6-9 Semidiagrammatic view of the intestinal tract and liver of a pigeon, showing the muscular gizzard and the crop.

with the muscular action of the gizzard, often assisted by stones the bird has swallowed. (In addition to masticating tough foods, the gizzard of predaceous birds—for example, hawks and owls—acts like a sink trap, holding indigestible material such as bones, fur, and feathers and compacting it into pellets to be regurgitated later.)

Pigeons, fowl, and their relatives and a few other birds have an outpocketing of the esophagus, the *crop,* that serves primarily as a food-storage organ. The crop plays only a minor role in digestion (in contrast to the rumen in cows, also a region of the esophagus). Most birds feed their young on animal material. Even most seed eaters during the breeding season switch largely to insects, ensuring a diet rich in protein to the rapidly growing juveniles. Pigeons and their relatives are unique in that during the breeding season the crop produces a white substance known as "pigeon milk," which is rich in fats and proteins. The milk is produced by a proliferation and sloughing of the epithelial cells lining the crop. This material, mixed with the food in the crop, is regurgitated into the gaping mouths of the young. Crops may serve other functions as well. For example, the crop of the greater prairie chicken is the resonating chamber for the booming call the males produce.

SEED-EATING BIRDS USE THE GIZZARD, A SPECIALIZED REGION OF THE STOMACH, FOR GRINDING UP THEIR FOOD.

Omnivores: Man

Unlike such aquatic animals as the clam, the sea anemones, and the angler fish (Fig. 6-10), very few terrestrial animals can lie in wait expect-

Fig. 6-10 The angler fish waits for prey fishes, attracted by the lurelike appendage on its head, to get close enough to be snapped up. (*Courtesy of Jim Annan.*)

ing food to fall into their mouths. Most large animals, including mammals, must actively seek and sometimes pursue their nourishment. Carnivores rely on stealth, strength, and speed, and they have poison, sharp claws, beaks, and teeth to capture and subdue their prey. Herbivores perhaps have less dramatic means of food getting, but they are equally, if not more, diverse in their structures and behavior.

Omnivores, such as man, occupy an intermediate station, both behaviorally and anatomically. Humans, like swine, eat just about anything. Such a compromise does have its disadvantages—we cannot easily subsist on a diet of hard grasses, as cattle do, nor can our jaws crush the bones of large antelopes for their marrow, as can a hyena. But what we lack in anatomical equipment, we make up for in culturally inherited techniques for processing otherwise difficult foods with heat, leaching and fermentation, grinding, crushing, and pounding.

THE MOUTH AND ASSOCIATED ORGANS Food enters our body via the mouth. The first structures to attack solid food are the teeth. The sharp **incisors** and **canines** are employed to bite and tear at an edible item, but the actual processing is often begun with the grinding and crushing action of the **premolars** and **molars** (Fig. 6–11). The last molar (the "wisdom tooth") is disappearing in human populations, and its occurrence, like that of all vestigial structures, is highly variable. Some individuals have a perfectly normal set, and others lack them altogether. The molars of carnivores are specialized for cutting by a scissorlike action of the upper and lower jaws, little or no chewing being necessary for muscle (meat) since it is highly vulnerable to attack by digestive enzymes. At the other extreme, herbaceous material would pass through the gut of an animal partly or wholly undigested if the tough cell walls were not ruptured so as to release their nutritious contents and expose more cells to digestive enzymes. Hence, the molars of herbivores usually have a large grinding surface (Fig. 6–12). Man has both kinds of teeth, as shown in Fig. 6–11. In many animals, including fish, amphibians, and reptiles, lost teeth are continuously replaced, whereas in mammals the juvenile, or "milk," teeth are typically replaced only once.

The **tongue** serves a number of important functions besides being essential for normal speech (Fig. 6–13). It tastes the food, thus determining its palatability; it transports the food from one set of teeth to another for processing; and it forms the food into a ball in preparation for its being swallowed through the **pharynx** to the **esophagus.** Another function of the tongue (in conjunction with the teeth) is to mix the food with the secretions of the three sets of **salivary glands.** The secretions of these glands (saliva) in man contain mucus to moisten and lubricate the food, various salts, and carbohydrate-splitting enzymes (amylase and galactosidase) for hydrolysing starch to glucose and maltose. Salivary amylase is also secreted by apes, pigs, and many rodents but not by horses, cats, or dogs. The latter group are grass eaters and meat eaters—such food has little starch. The total daily output of the salivary glands in man ranges between 1 and 1.5 liters. This is a drop in the bucket compared with the 50 to 100 liters in cattle.

OMNIVORES EAT BOTH PLANT AND ANIMAL MATERIALS. THEIR TEETH ARE RELATIVELY UNSPECIALIZED AND SERVE FOR BITING AND TEARING AS WELL AS GRINDING. DIGESTION BEGINS DURING THE PRO-

THE DIGESTIVE SYSTEM

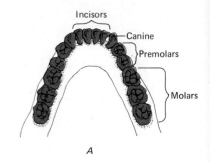

A

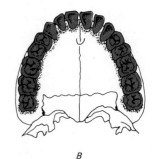

B

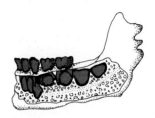

C

Fig. 6–11 Human teeth: (A) adult lower jaw; (B) adult upper jaw; (C) milk teeth in lower jaw of a child with the permanent teeth in the gums below. (*Adapted with permission from the Ciba Collection of Medical Illustrations, by Frank H. Netter, M. D.*)

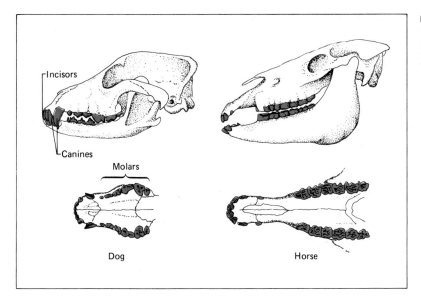

CESS OF MASTICATION THROUGH THE ACTION OF SECRETIONS OF THE SALIVARY GLANDS, WHICH ALSO MOISTEN THE FOOD.

THE ESOPHAGUS AND STOMACH When swallowing, the tongue forces the ball of food back and down into the pharynx, where it enters the straight tube leading to the stomach, the **esophagus,** normally closed at the top by a sphincter (Fig. 6-13). Where the esophagus enters the stomach, there is a junction, not a true sphincter. This is also normally closed, preventing food from backing up from the stomach into the esophagus. Food is moved down the esophagus (as well as through the stomach and intestine) by waves of muscular contractions called **peristalsis,** and these are controlled by the autonomic nervous system (Chap. 9). Only mucus is secreted by the esophagus.

The stomach, most of which is under the ribs on the left side, is a flexible bag. The combined muscular and enzymatic actions of the stomach convert the food ball into a milky semiliquid, called **chyme,** made up of small food particles acceptable for further digestion in the intestine. Powerful waves of contraction, passing over the stomach every 20 sec, churn and mix the contents with the gastric juices secreted by glands in the stomach wall. These glands secrete mucus, enzymes (the most important of which is **pepsin**), and **hydrochloric acid.**

The protease pepsin splits proteins into shorter peptide chains. It also is the milk-curdling enzyme in humans. The enzyme is inactive at normal

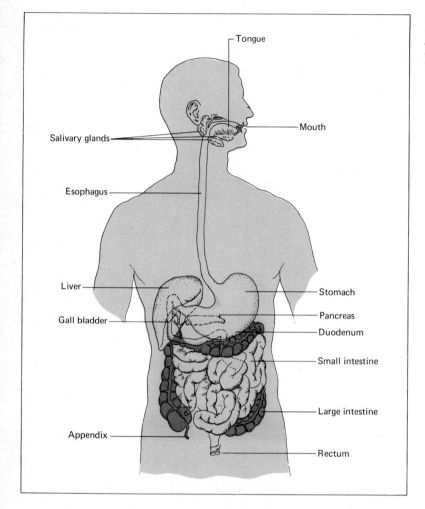

Fig. 6-13 Diagrammatic view of the digestive tract in man. *(Adapted from A. C. Guyton, BASIC HUMAN PHYSIOLOGY, 1971, W. B. Saunders Company, Philadelphia.)*

physiological pH (7.0) and is most active in the pH range of 1.5 to 2.5. In the stomach, the pH is brought within this range by the secretion of large amounts of hydrochloric acid from the gastric glands. Actually, the gastric glands do not secrete pepsin itself but produce, rather, a slightly larger protein, **pepsinogen** *(gen,* "source"). In the acid environment of the stomach, several small groups of amino acids are split off, one of which exposes the active site of the enzyme.

Hydrochloric acid not only activates pepsinogen to pepsin but also exerts its own direct digestive functions by breaking down fibers of connective tissue and muscle. It prevents bacterial growth in the stomach and the next region of the gut, the **duodenum.** Finally, it stimulates the secretion of the digestive hormones pancreomyzin and secretin, discussed in Chap. 7. Why then doesn't the stomach digest itself? After all, the stomach, like much of its contents, is made mostly of protein, which should fall easy victim to the ravages of pepsin and HCl. The explanation requires no special knowledge of biochemistry: The cell lining of the stomach (the mucosa, Fig. 6–14) secretes a double-layered barrier of mucus. In man, this layer is 1.0 to 1.5 mm thick and is neutral or alkaline, so that HCl

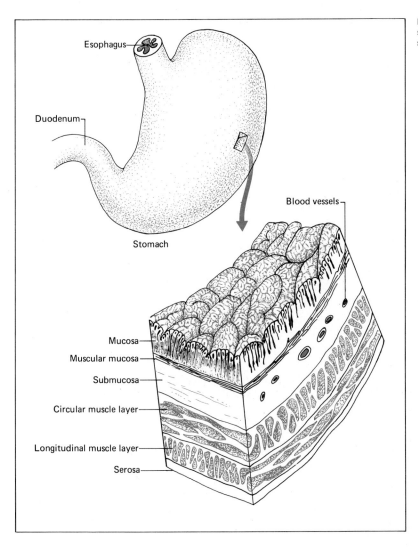

Fig. 6–14 Diagram of the human stomach showing the structure of the stomach wall.

Esophagus

Duodenum

Blood vessels

Stomach

Mucosa

Muscular mucosa

Submucosa

Circular muscle layer

Longitudinal muscle layer

Serosa

diffusing through it is neutralized before it reaches the underlying, delicate mucosa cells. Pepsin is neutralized as well because it is inactive at higher pHs. The mucous layer also provides lubrication.

FROM THE MOUTH, THE MASS OF FOOD IS MOVED BY MUSCULAR CONTRACTIONS CALLED PERISTALSIS THROUGH THE VARIOUS SPECIALIZED REGIONS OF THE DIGESTIVE TRACT. IN THE STOMACH, HYDROCHLORIC ACID AND PEPSIN CARRY OUT FURTHER DIGESTION AS THE FOOD IS CONVERTED INTO SEMILIQUID CHYME.

THE INTESTINES The next stage of digestion occurs in the first section of the intestines, the **duodenum.** Chyme enters the duodenum in very small amounts. Usually, about 1 to 4 ml passes through the pyloric sphincter between the stomach and duodenum with each of the three-per-minute peristaltic waves. Large pieces of food are rejected by the pyloric channel and squeezed back to the main body of the stomach for more digestion.

The intestine of man and other mammals is the site of most **absorption** of digested food. It usually averages seven or eight times the length of the body for mammals (over 20 ft in man). In large herbivores, the ratio is much greater. The reasons for this are twofold: (1) Plant material is more difficult to digest and more intestinal surface is required; (2) In terms of the familiar surface-area-to-volume relationship, an animal that increases in length by a factor of 2 increases its volume by the cube of this factor —about eight times. Hence, a mere doubling of intestine length would not provide the digestive and absorptive surface to feed the eightfold increase in bulk.

Upon entering the upper part of the intestine, or duodenum, the acidic chyme contains fat and partly digested protein and carbohydrates. The presence of food in the stomach and the passage of the chyme through the duodenum trigger nervous and hormonal networks that signal the pancreas and liver to begin secreting products which carry on digestion to completion. The liver secretion, called **bile,** passes from the **gall bladder** (a storage vessel) through the bile duct to a common opening with the duct from the pancreas. The most important bile constituents are bile salts. These emulsify the fats and oils of the chyme, thus exposing the lipid molecules to the action of fat-splitting enzymes called **lipases.** Lipases are secreted by the pancreas but activated by the bile salts.

The enzymes secreted by the intestine and those coming into it from the pancreas work best at neutral pH. The acidic chyme is neutralized by **bicarbonates** in the presence of pancreatic juice. The pancreatic juice also contains an amazing battery of enzymes, many of which are inactive until they are activated in the duodenum in a manner analogous to the activation of pepsinogen. Some of the more important enzymes include the proteases: trypsin, chymotrypsin, and carboxypolypeptidase.

Glands in the walls of the intestine itself secrete still another group of enzymes as well as mucus and salts. Among the enzymes are an amylase, peptidases that finally free single amino acids, and lipases that complete the liberation of fatty acids. These produce monomers that are finally absorbed by the cells of the intestinal epithelium, the surface layer of cells of the intestinal lining called the mucosa. As though enzymatic attack were not enough, the chyme is also mixed, kneaded, and thoroughly homogenized by muscular movement, exposing it repeatedly to the absorptive lining. Waves of peristaltic contractions gradually push the contents to the large intestine, or **colon.**

As we said earlier, survival is largely a matter of doing the right things so as to acquire enough energy to stay alive and reproduce. Eating and digesting food are, however, only the first steps. In a very real sense, food is not inside an animal so long as it is still in the intestine. The true barriers that must be transversed are the cell membranes of the intestinal epithelial cells—cells derived embryonically from the appropriately named endoderm, or "inner skin" (as discussed in Chap. 10). Even with active transport, a man or any other vertebrate would quickly starve with merely a smooth, tubular intestine. Several mechanisms have evolved to increase the absorptive surface (Fig. 6–15). The most obvious of these are folds of the mucosa, which triple the surface area. Secondly, the fingerlike **villi** (sing. **villus**), which are 0.5 to 1.5 mm long, further increase the surface by a factor of 10. Thirdly, electron micrographs of the exposed surfaces of epitehlial cells covering the villi reveal closely packed, 1.0-μm-long projections called **microvilli**—about 800 per cell—increasing the absorptive surface again by a factor of 20 to 24. Multiplied together, these three anatomical tactics increase the absorptive surface by a factor of about 600!

The small intestine joins the large intestine, or **colon,** in the lower right side of the abdomen of most humans. Near this point, a blind space called a **caecum** projects from the colon. In many mammals, the caecum houses a large culture of microorganisms, which assist in the digestion of cellulose as discussed above; but in man both the caecum and the troublesome **appendix,** which projects from it, have no such function. The colon ascends on the right side, crosses the upper abdomen, and descends on the left side, where it terminates as the rectum. The **rectum** communicates to the outside through the **anus,** closed by a strong sphincter muscle usually under voluntary control.

Digestion in the strict sense ends in the small intestine, but essential

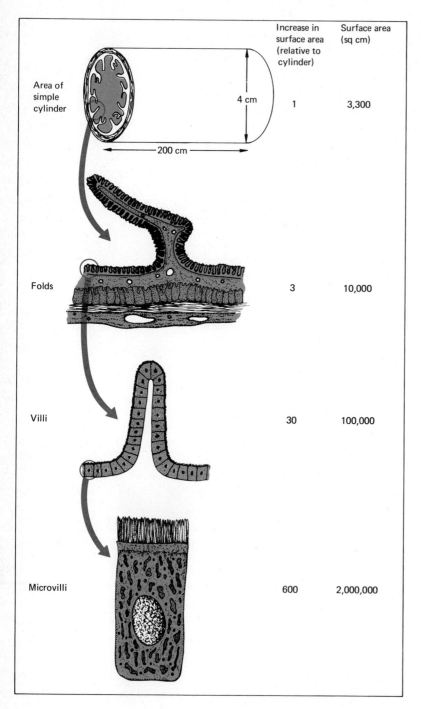

	Increase in surface area (relative to cylinder)	Surface area (sq cm)
Area of simple cylinder	1	3,300
Folds	3	10,000
Villi	30	100,000
Microvilli	600	2,000,000

Fig. 6–15 The surface area of the intestine is greatly increased by three anatomical devices: folds, villi, and microvilli.

absorption continues in the colon. The colon absorbs about 300 ml of water from the 400 to 500 ml of intestinal chyme that enter daily. This is not much, compared with the 8 liters absorbed by the small intestine. Disease causing frequent and watery defecations (diarrhea) can, however, result in serious losses of water and salts, especially potassium. It is this that is often the cause of death in the disease cholera. Most of the dry weight of feces is undigested food, especially plant cell-wall (cellulose) fragments. From 10 to 35 percent consists of microorganisms, including bacteria, yeasts, and fungi. Many animals, including man, depend on their colonic bacteria (collectively called a flora) for the production of essential nutrients, including vitamin K, essential in the blood-clotting reactions, and vitamin B_{12}. For this reason, orally administered antibiotics can have serious consequences if the intestinal flora is altered or depleted.

Gas in the colon is called **flatus** and is about 50 percent nitrogen and 40 percent carbon dioxide; the rest—methane, hydrogen, and hydrogen sulfide—as well as the carbon dioxide, are products of bacterial fermentation. It has been shown that with a diet high in beans, the average volume of expelled flatus increases from about 17 to 203 cu cm per hour, confirming the popularly held belief.

Food particles ingested during a single meal may be defecated over a period of 1 to 8 days. Part of this variation is due to the great range of normal frequency of bowel movements. Some people defecate after every meal, whereas other perfectly healthy people do so only once a week. There is no evidence that with constipation (discomfort caused by failure to defecate), persons absorb any toxic products from the feces. The color of feces is due to the breakdown products of hemoglobin in the bile. The familiar odor comes from complex alcohols, which are products of bacterial decomposition in the colon.

IN THE INTESTINES, DIGESTION IS CARRIED STILL FURTHER AND ABSORPTION BEGINS. THE SURFACE OF THE INTESTINAL WALL IS ENLARGED SO THAT THE ABSORPTIVE SURFACE IS VERY GREAT. MICROORGANISMS LIVING IN THE COLON PRODUCE IN THEIR METABOLISM ESSENTIAL NUTRIENTS FOR MAN. THE INTESTINES REMOVE LARGE AMOUNTS OF WATER FROM THE CHYME, AND THE UNDIGESTED RESIDUE EVENTUALLY IS EXPELLED AS FECES.

We have said that acquiring sufficient energy is the essential challenge of an organism's life. As you will recall from Chap. 4, the ultimate source of an animal's energy lies in the chemical bonds of organic molecules.

VENTILATION IN ANIMALS

The source of the organic molecules is food, or more specifically, the amino acids, sugars, and fatty acids released from large complex molecules by the process of digestion. Burning or oxidation of these molecules in the presence of oxygen (aerobic respiration) is the most efficient way of liberating their potential energy and synthesizing the biochemical fuel ATP.

A small organism such as a protozoan has no difficulty in obtaining the oxygen required for the synthesis of ATP. The oxygen simply diffuses in across the few hundred microns of cytoplasm between the mitochondrion in the cell and the surrounding medium. CO_2, a waste product of ATP synthesis, diffuses out in the same manner. A serious problem had to be overcome, however, before larger organisms could evolve. The larger the animal, the less surface area it has relative to its volume. In other words, bigger animals have proportionally less surface for gas (O_2 and CO_2) exchange and greater distances between the medium and the inside of the organism. The strategies evolved to overcome this barrier to large size are diverse and fascinating. They involve two systems: one to increase surface area for gas exchange and another to transport the gases around the body. Actually, this separation does not always apply—some animals, such as the insects, partially combine the two systems. Before describing the solutions, however, you must know something about the diversity of metabolic styles, the differences in the rates at which animals conduct their daily metabolic business.

OXYGEN IS ESSENTIAL FOR CELLULAR METABOLISM AND MUST BE OBTAINED FROM THE ENVIRONMENT BY DIFFUSION. SMALL ANIMALS CAN ABSORB OXYGEN THROUGH THE BODY SURFACE. LARGER ONES REQUIRE SPECIAL MECHANISMS FOR TAKING IN OXYGEN AND TRANSPORTING IT TO EVERY CELL.

How Much Oxygen Is Required?

The rate at which materials pass through the bodies of organisms and liberate their energy is referred to as the **metabolic rate.** In biochemical terms, metabolic rate is the amount of energy released by ATP going to ADP and H_3PO_4 per unit time. Many factors influence metabolic rate, as discussed below. The simplest way to measure the metabolic rate is to determine how much food is eaten. This can give misleading results if the animal is finicky about the food you have provided and is living on its own fat or other stored material. At the other extreme, it may so relish its diet that some of its intake is stored as fat. A more practical, if less direct, measure of metabolic rate is oxygen consumption, and in fact this is usually the way it is determined (Fig. 6–16).

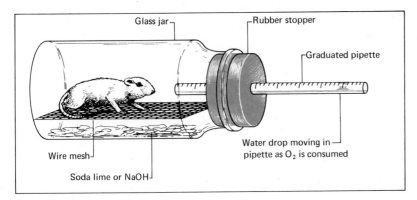

Glass jar ─ · ─ Rubber stopper

┌ Graduated pipette

Water drop moving in ─
pipette as O_2 is consumed

Wire mesh ┘

Soda lime or NaOH ┘

Fig. 6-16 A simple measure of metabolic rate. The rat inhales oxygen and exhales carbon dioxide which is absorbed by the soda lime. A water drop in the graduated pipette moves toward the stopper as oxygen is consumed. (*After Keeton.*)

Metabolic rates have a range of about two orders of magnitude. A fish needs 10 times the oxygen per unit weight of body tissue as does a sea anemone; and a mouse uses four times as much as a fish. Closely related animals such as the octopus and squid can have very disparate metabolic rates if their behavior differs. Squid are continuous swimmers and with this exertion require about four times as much oxygen as does the less active octopus, which usually sits in its lair or crawls about on the bottom. Finally, oxygen consumption appears to be inversely related to body size, as shown in Fig. 6-17. This is most noticeable in warm-blooded animals, such as birds and mammals, for the following reasons. Much of the warm-blooded animal's energy is used to maintain the body temperature at a constant and relatively high level. If we now involve the familiar surface-area-to-volume relation, we can predict that a small body would have to use more energy per unit of body weight than a large one to maintain its temperature above that of the environment. The reason is that it has relatively greater surface from which to lose heat by radiation and conduction.

THE METABOLIC RATE OF ORGANISMS IS THE RATE AT WHICH THEY USE ENERGY. SINCE OXYGEN IS REQUIRED FOR ENERGY USE, ITS CONSUMPTION IS A MEASURE OF RATE OF METABOLISM.

Sources of Oxygen

Both animals and plants obtain their oxygen either from the atmosphere or from water in which it has been dissolved. Oxygen makes up about 21 percent of the air, the balance being nitrogen, water vapor, trace amounts of carbon dioxide (0.03 percent), and inert gases, such as helium, neon, argon, xenon, and krypton. In contrast to the abundance of oxygen in the atmosphere, its concentration in water is low and depends on water

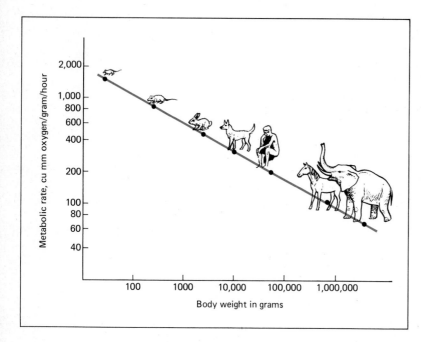

Fig. 6-17 Relationship between body weight and oxygen consumption. (*Adapted from Knut Schmidt-Nielsen, ANIMAL PHYSIOLOGY, 2nd ed., © 1964, p. 34. By permission of Prentice-Hall, Inc., Englewood Cliffs, N. J.*)

temperature. At 0°C, 1 ml of oxygen is dissolved in 100 ml of water. The solubility of gases decreases with increasing temperature, so that at 15°C, there is 0.7 ml of oxygen per 100 ml of water and at 37°C (the body temperature of humans), there is 0.5 ml of oxygen per 100 ml of water— exactly half that at 0°C. If this were the best of all possible worlds for aquatic animals, the solubility of gases would increase, instead of decrease, with the temperature.

The advantages of air breathing should be obvious. At sea level, the weight of air is much less than that of water, and therefore air requires less energy to move than water. And to get 1 cu ml of oxygen requires moving 5 cu ml of air, compared with 143 cu ml of water. Another disadvantage of the aquatic medium stems from the rate of diffusion of oxygen, which is 3 million times slower in water than in air. Obviously, air breathing has many advantages, and this is borne out by the tenacity with which animals hold onto it once it has evolved, even if they return to an aquatic habitat. The adults of all aquatic insects, reptiles (turtles, lizards), birds (penguins), and mammals (whales) are still air breathers, as were their ancestors millions of years ago.

OXYGEN MAY BE OBTAINED FROM WATER OR THE ATMOSPHERE, ALTHOUGH OBTAINING IT FROM THE AIR REQUIRES LESS ENERGY

*AND GREATER QUANTITIES CAN BE OBTAINED MORE RAPIDLY
THAN FROM WATER.*

The term "*respiration*" once meant the exchange of gases between the
organism and its environment; the organs that actively participated in
this exchange were called **respiratory organs.** As often happens with
words, the term has evolved a new meaning. "Respiration" now means the
cellular events leading to the production of energy-rich ATP from compounds such as glucose. In this process, oxygen is combined with hydrogen ions to produce water. CO_2 is released as another end product of
carbohydrate metabolism. To avoid semantic confusion, we shall employ
the word "ventilation" for all those processes involving gas exchanges
across cell membranes at the body surface. As stated before, microorganisms, including protozoans, need no special organs for ventilation
because no point in their body is far from the surrounding medium.
Diffusion over such short distances, such as a few hundred microns,
is rapid enough to propel enough oxygen in and to remove enough CO_2
along their respective concentration gradients.

Even larger animals such as flatworms can avoid investing in special
structures by ensuring that no part of the body is far from the body surface.
Animals that go through life at a leisurely pace have a low metabolic rate
and may not require special structures for gas exchange. Sea anemones
have no special structures, although their tentacles increase their surface
area and undoubtedly serve in gas exchange. The moist skin of frogs and
toads is a ventilation organ; and some lungless salamanders depend on
it entirely, having lost all vestiges of functional lungs. As might be expected, these salamanders lead a sedentary life and require relatively
little oxygen. But even in these lethargic animals, gas exchange is increased by the penetration of blood vessels much closer to the surface
of the skin than in other vertebrates.

What about larger or more active animals? Recall how mammals have
solved the problem of the need for more and more absorptive area in the
intestine by a series of projections or invaginations into the intestinal
cavity—microvilli on top of villi on top of folds. Where the relatively
unmodified body surface does not provide enough area for the exchange
of gases, the problem has been solved in essentially the same way: by
the outpocketing or inpocketing of the body wall. Let's examine the advantages and disadvantages of both approaches in the major habitats,
water and air.

VENTILATION IN WATER   The great weight of water has certainly been a
factor in the evolution of structures for ventilation. If you have ever had

Structures for Ventilation

to carry a bucket of water very far, you know what an effort it is. Moreover, a great deal of water must be moved over the gas-exchange surface because the oxygen concentration in water is much less than in air. Compare the mechanical difficulties to be overcome in an invaginated structure such as the simple lung with those in evaginated structures such as simple gills (Fig. 6–18). Muscular activity of some kind would be necessary to force water into the lung and to force it out again. The amount of force necessary to push the water in and out would depend on the surface area and how the channels branched, that is, the resistance to flow in proportion to the surface area. Compare, for example, the force necessary to squeeze water out of an ordinary balloon with that required for a balloon filled with pieces of sponge. Much more energy would be required to empty and refill a spongy lung with water than with air. When these difficulties are added to the difference in oxygen content between water and air, absence of "water lungs" in animals is not surprising.

Perhaps the best solution for an aquatic animal would be to sit in a moving current of water, which would bathe its gills or the outpocketings of its body surface. No effort would be required, and this is, in fact, the solution of many organisms, especially those inhabiting streams and shorelines where waves sweep over the substratum. The plumelike gills of many tube-dwelling marine worms serve the dual purpose of gas exchange and food gathering. Other worms bear segmented gills along their sides (Fig. 6–19).

Some means of actively moving the water is necessary in cases in which a passive approach to gas exchange is insufficient. Larger size, low oxygen tension, or higher metabolic rate—or a combination of these—

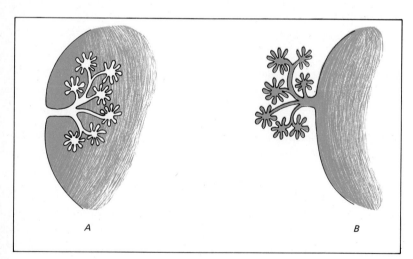

A   B

Fig. 6–18 Comparison of (*A*), an invaginated ventilation structure, and (*B*), an evaginated ventilation structure.

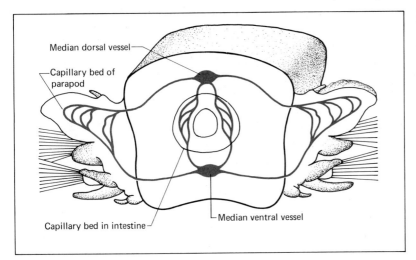

Median dorsal vessel

Capillary bed of parapod

Capillary bed in intestine

Median ventral vessel

Fig. 6-19 Diagrammatic section through a segmented marine worm. Blood is aerated in the capillary bed of the segmented appendages (parapods) and then circulated to the body cavity.

would call for such mobility. Some animals create currents that circulate fresh water across the gills. We have seen how the cilia on the gill filaments of molluscs can move many gallons of water every hour. The gills of some animals are attached to the appendages or, as in crayfish, lobsters, and crabs, project from the walls of the thorax into a chamber created by the overhanging carapace. Water enters the gill chamber from under the free edges and is propelled forward and out anteriorly by bailers—flattened plates or appendages near the mouth (Fig. 6-20). In the vertebrates, **gill slits** develop early in development on both sides of the throat or pharynx (Fig. 6-21). These persist to adulthood in fishes, as well as in the lower vertebrates such as lampreys and hagfish. They permit a circulation of water from the mouth to the outside of the body.

In sharks and rays, there are five or six gill slits, each of which opens directly to the outside. Tufts of red, blood-filled gill filaments, separated by gill arches, are attached, and through these run large blood vessels. The exchange of oxygen and carbon dioxide is enhanced by the circulation of blood through the filaments in a flow direction opposite to that of the water—a process known as **countercurrent exchange.** Because gases diffuse relatively slowly in water, the amount of exchange through the gill membranes between the blood and the water will depend on the amount flowing past a given blood volume. When the blood and water are both flowing in the same direction, the exchange will be less than when they flow in opposite directions (Fig. 6-22).

A simple system of valves and pumps circulates water past the gills. Muscles lower the floor of the mouth and simultaneously expand the walls

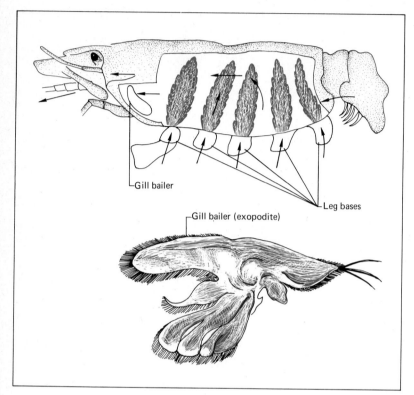

Fig. 6-20 Diagrams showing path of water, through the gill chamber of a small crustacean. Arrows show the path of water moved by the specialized exopodite or gill bailer. *(Adapted from R. O. Barnes, INVERTEBRATE ZOOLOGY, 2/e, 1968, W. B. Saunders Company, Philadelphia.)*

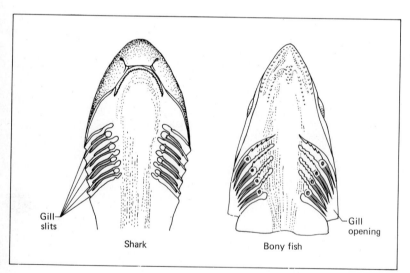

Fig. 6-21 Comparison between the gills of a shark and a bony fish. *(Used by permission of Hill and Wang, A division of Farrar, Straus & Giroux, Inc. from A HISTORY OF FISHES by J. R. Norman, illustrated by W. P. C. Tenison.)*

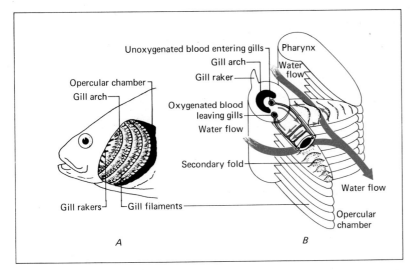

Unoxygenated blood entering gills
Pharynx
Gill arch
Water flow
Gill raker
Opercular chamber
Gill arch
Oxygenated blood leaving gills
Water flow
Secondary fold
Water flow
Opercular chamber
Gill rakers
Gill filaments
A
B

Fig. 6-22 Diagram to show current flow in the gills of a fish. Arrows show that blood flow and water flow are in opposite directions. *(Adapted from Claude Villée, et al.; GENERAL ZOOLOGY, 3/e, 1968, W. B. Saunders Company, Philadelphia.)*

of the pharynx, while flaps of skin close tightly over the gill slits. Water rushes in through the mouth and the spiracles, which are modified gill slits just behind the eye. Ventilation is then accomplished by closing the mouth, narrowing the pharynx, and raising the floor of the mouth. The water pressure forces open the gill-slit covers and closes a check valve in the spiracles. In bony fish, the gills hang freely from the arches into chambers covered by a single rigid flap called the operculum, which opens posteriorly and which pumps water past the gills. Furthermore, the bony fish does not need to close its mouth all the way during the expiratory phase because water, tending to flow forward, pushes together flaps of soft tissue which serve as a valve. Energy, as we have seen, is a precious commodity, and this is a device for conserving it. Some constantly swimming fish, such as mackerel, have no muscular control over gill ventilation and depend for gas exchange on the constantly flowing current in their open mouths. They will quickly suffocate if restrained.

VENTILATION IN WATER IS ACCOMPLISHED BY STRUCTURES CALLED GILLS. THESE ORGANS PROVIDE EXPANDED SURFACE FOR DIFFUSION AND EXTEND INTO THE WATERY ENVIRONMENT OR ARE ARRANGED IN SUCH A WAY THAT WATER CAN BE FORCED OVER THEM. GILLS ARE USUALLY RICHLY SUPPLIED WITH BLOOD VESSELS. OXYGEN DIFFUSES INTO THE BLOOD, AND CARBON DIOXIDE DIFFUSES OUT.

VENTILATION IN AIR Evaginated structures such as gills are not satisfactory in air. For one thing, the delicate filaments collapse when unsupported by the dense medium of water. For another, evaporative water loss from all the exposed surface would quickly dehydrate an animal in dry air. Everything else being equal, an animal needs much less gas-exchange surface in air because of the much higher (about 25 times as much) concentration of oxygen and the high rate of diffusion of oxygen in air.

Snails are animals with relatively slow metabolism that take advantage of fast diffusion of oxygen in air. They have a mantle cavity with many blood vessels and a restricted air passage to the outside. Snails, such as the common garden snail *Helix,* inspire by opening the air passage while lowering the muscular floor of the mantle cavity, or "lung." The passage then closes, and gas exchange occurs across the moist membranes of the mantle. Stale air is expired by raising the floor of the cavity and opening the passage. The loss of moisture by diffusion is kept at tolerable levels by this arrangement.

Compared with snails, much more oxygen is required for cellular respiration by insects, especially flying ones, some of which have metabolic rates higher than those of birds. The exchange surface is brought into intimate contact with every part of the body by the **tracheoles,** the terminal branches of tubes called **tracheae** (sing. trachea). These open to the outside via paired spiracles on the sides of the body. As in the stomata on leaves of plants, the size of the spiracular opening is controlled by a valve; but unlike the arrangement in plants, it is operated by muscles. The opening is bounded by hairs that prevent dust from clogging the tracheae. Active organs are invaded by a rich rootlike mass of minute tracheoles, sometimes less than 1 μm in diameter, which may even enter individual cells. Exchange of air in a tracheal system is often enhanced by rhythmic movements of the abdomen. Large internal air sacs increase the air circulation in some insects.

Some aquatic insects, including beetles, bugs, and wasps, have evolved a remarkable elaboration of spiracular ventilation. Very fine waterproofed hairs extend over much of the ventral surface, retaining a thin layer of trapped air next to the body. One would think that the oxygen would soon be depleted from this air layer. However, as the oxygen concentration falls, it tends to be replaced by diffusion from the surrounding water, the rate of inward diffusion being inversely proportional to the concentration in the bubble.

TERRESTRIAL ANIMALS USUALLY TAKE UP OXYGEN AND EXPEL CARBON DIOXIDE BY THE USE OF VENTILATION ORGANS CALLED

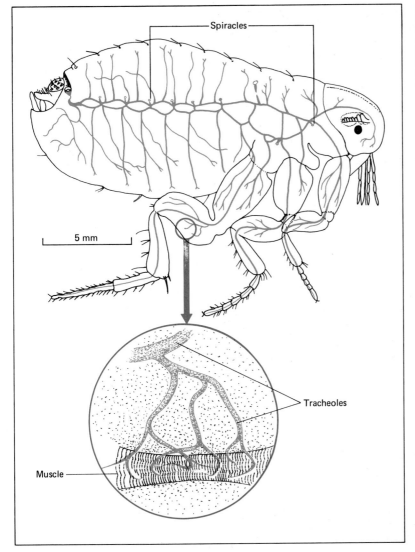

Fig. 6–23 Diagrammatic view of the tracheal system of a flea which shows the association of tracheales with muscle fibers in the leg. *(Adapted from Vincent P. Wigglesworth, THE PRINCIPLES OF INSECT PHYSIOLOGY, 6/e, 1965, Associated Book Publishers, Ltd.)*

LUNGS OR BY A SYSTEM OF FINE TUBES, CALLED TRACHEAE, WHICH EXTEND TO EVERY PART OF THE BODY. VENTILATION IN AIR REQUIRES MUCH LESS GAS-EXCHANGE SURFACE THAN DOES VENTILATION IN WATER.

VENTILATION SYSTEMS IN THE HUMAN: VARIATIONS ON A THEME

The insect tracheal systems ultimately depend on diffusion for the movement of gases from the larger tracheoles to the smaller and finally into the cells. A more efficient system is required by large, active animals—one that combines a large surface area for gas exchange (the ventilation system) with a delivery or circulatory system for transporting the gases to and from the cells. In order to discuss this system, we must divide what is, in reality, a unitary functional complex into two sections. You should, nonetheless, keep in mind that ventilation and circulation are inseparable structurally and functionally. The first part to be described is the complex of structures and functions that move the gases to the site of gas exchange. Following this, we will examine the circulatory systems evolved to transport the gases as well as other materials. The human will be our model.

Our discussion will follow the path air follows to the lung. Air generally enters the nostrils, where it is cleaned, warmed, and humidified by passing over the moist mucous membranes lining the nasal cavity (Fig. 6–24). The skull bones that line the nasal cavity have ledges, or **turbinates,** on them which increase the surface area of the nose and cause eddies and rebounding of the air as it is inhaled. The swirling air is slowed down in its passage over and around the turbinates, and particles of matter larger than 5 μm or so tend to drop out and become trapped in the mucus. A wide-spectrum bacteriocide (lysozyme) and another agent with antiviral capability are secreted by the epithelial cells lining the nasal cavity and are a first line of defense against infection and disease. The turbinates of some mammals are much more complex than those of man, looking like an impossible maze in cross section. By this device, both the "air-conditioning" and olfactory surfaces are increased.

Mammals, including man, do not normally breathe through their mouths. When overheated, however, panting may lower the body temperature by evaporation from the moist surfaces of the mouth and tongue. The food and air passages meet in the throat, or **pharynx** (Fig. 6–24). The air passage is called the **trachea.** The epiglottis closes over the uppermost part of the trachea—the specialized **larynx**—whenever something is swallowed. In ruminants and other animals that swallow a great deal of liquified food, the larynx has a raised opening, permitting it to remain open for the passage of air at the same time food is flowing past on both sides into the esophagus. Within the larynx of man, on each side of the air passage, are the two **vocal cords.** The tension on, and gap between, these elastic cords is controlled by muscles of the larynx, and the flow of air by them causes them to vibrate, generating sound of a particular pitch. The formation of words requires the coordinated use of these, plus the cheeks, tongue, jaws, lips, and **soft palate,** the flap of tissue forming the back of the roof of the mouth.

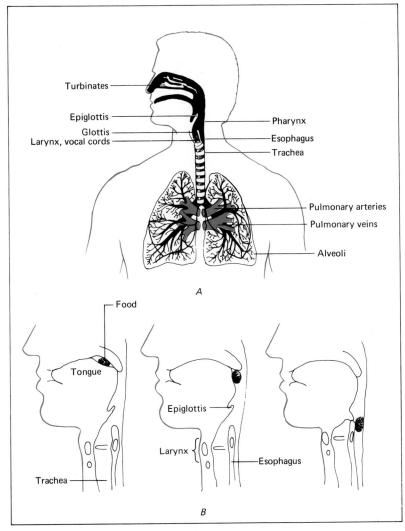

Turbinates

Epiglottis

Glottis

Larynx, vocal cords

Pharynx

Esophagus

Trachea

Pulmonary arteries

Pulmonary veins

Alveoli

A

Food

Tongue

Epiglottis

Larynx

Esophagus

Trachea

B

Fig. 6-24 Diagrammatic section of human to show the ventilation system and movements required for swallowing. During breathing, air moves through the pharynx into the trachea; during swallowing, the epiglottis closes the trachea and the food mass moves into the esophagus. *(From BIOLOGY: ITS PRINCIPLES AND IMPLICATIONS, Second Edition, by Garrett Hardin. W. H. Freeman and Company. Copyright © 1966.)*

The rest of the trachea and the two bronchi that branch from it are rigid tubes reinforced by rings of cartilage. Exhalation in some animals, such as horses, is more complete because the bore of the trachea can be decreased by muscles in each ring. Thus more of the residual, stale air in the air passages can be removed. The two bronchi divide repreatedly into **bronchioles** and finally into the alveolar ducts. Blind sacs, **alveoli** (sing. **alveolus),** that surround the alveolar ducts are the ultimate sub-divisions of the ventilation system and are the main gas-exchange

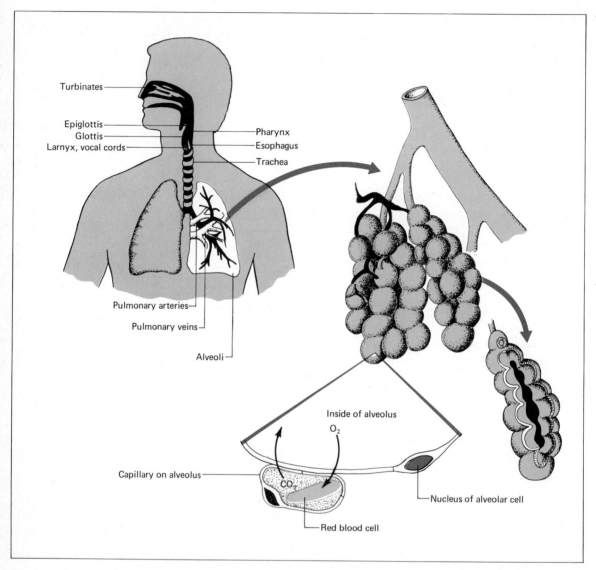

Turbinates

Epiglottis
Glottis
Larnyx, vocal cords

Pharynx
Esophagus
Trachea

Pulmonary arteries

Pulmonary veins

Alveoli

Inside of alveolus
O_2

Capillary on alveolus

CO_2

Nucleus of alveolar cell

Red blood cell

Fig. 6–25 The ventilation system in man, showing lungs, alveoli, and gas exchange in capillary. (*Adapted from A. C. Guyton, BASIC HUMAN PHYSIOLOGY, 1971, W. B. Saunders Company, Philadelphia.*)

surfaces (Fig. 6–25). Air does not blow in and out of the alveoli with each breath but exchanges by diffusion with the air in the terminal bronchioles. During forced expiration, however, the latter can squeeze out all their air by contraction of the circular muscles in which they are wrapped. The membrane separating the air in the alveoli from the blood of the capillaries that bathe the lungs is only 1 to 4 μm thick. In man, there are about

700×10^6 alveoli, and each is about 100 μm across. With this information and simple algebra, you can calculate the available gas-exchange surface of the lung: about 50 to 100 sq yd (the floor space of a good-sized room). Samples of alveolar air contain 14 percent oxygen and about 5 percent CO_2.

THE LUNGS IN MAN ARE SUPPLIED BY AIR THROUGH THE TRACHEA, BRONCHI, AND BRONCHIOLES. THE AIR IS WARMED AND FILTERED IN THE NASAL PASSAGES. GAS EXCHANGE TAKES PLACE IN BLIND SACS CALLED ALVEOLI.

BREATHING The process of moving air in and out of the ventilation system is called breathing. Everyone is aware that the chest expands when air enters the lungs, and it is natural to think of the lungs expanding and pushing out the rib cage, especially since we cannot feel the location of the muscles involved, as we can feel a leg or arm muscle when it contracts. The fact is that the lungs cannot expand by themselves, although when expanded, they are sufficiently elastic to shrink back to their original size like a deflating balloon.

Two groups of muscles cause inhalation: (1) The ribs are pulled up and out by muscles, thus increasing the volume of the rib cage (Fig. 6-26). (2) A domed partition called the **diaphragm** seals off and isolates the rib cage from the abdomen below. At the center of the diaphragm is a large central tendon; it is surrounded by a muscular periphery which is attached to the rib cage so that when its muscle fibers contract, it must necessarily move down against the abdomen. It can travel about 8 cm. Simultaneously, the abdominal wall ("stomach") muscles relax, allowing the

Fig. 6-26 Diagram to show inflation of lungs by expansion of chest cavity through muscular contractions which increase the volume of the rib cage and lower the diaphragm. (*From BIOLOGY: ITS PRINCIPLES AND IMPLICATIONS, Second Edition, by Garrett Hardin. W. H. Freeman and Company. Copyright © 1966.*)

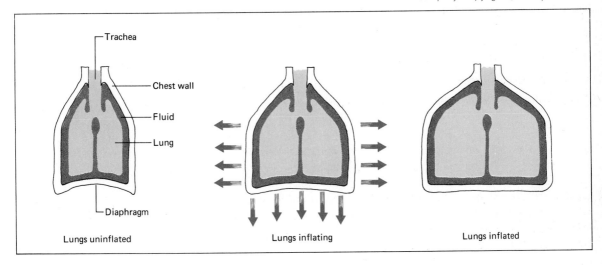

Trachea

Chest wall

Fluid

Lung

Diaphragm

Lungs uninflated   Lungs inflating   Lungs inflated

abdomen to protrude. As the chest cavity expands and begins to pull away from the lungs, a partial vacuum is created, pulling the lungs outward and in turn decreasing the air pressure in them. Air rushes in through the trachea until the pressure inside the lungs equals atmospheric pressure. This is inhaling, or the **inspirational** phase of ventilation. Exhaling, or **expiration,** is passive. The muscles relax, and the elasticity of the lungs brings them, the rib cage, and the diaphragm back to the original shape.

A resting person moves about 500 ml of air in and out with each of the 12 to 15 breaths per minute. This is called the **tidal air.** Of this 500 ml, about 140 ml is in the air passages, so that only 360 ml of the air actually reaching the alveoli is fresh. Tidal air is only a fraction of the potential volume of the lungs. Following a series of normal inhalations and exhalations, the average person can inhale about 3,000 ml if he tries. Similarly, he can exhale about 1,500 ml. The total, or 4,500 ml, is called **vital capacity.**

BREATHING IS ACCOMPLISHED BY RAISING THE RIB CAGE AND LOWERING THE DIAPHRAGM, WHICH RESULTS IN LOWERED AIR PRESSURE IN THE LUNGS, CAUSING AIR TO ENTER. RELAXATION OF THE CHEST MUSCLES AND THE ABDOMEN RESULTS IN AIR BEING EXPELLED.

CLEANING AND PROTECTION A large supply of oxygen and a means of dispelling large amounts of CO_2 are necessities for an active animal such as man. And even though we are provided with a reasonable margin of safety in surface area for exchange of gases, any significant loss of surface would seriously jeopardize life. We have already mentioned two mechanisms for protecting the lungs from disease and obstruction: the antibacterial and antiviral secretions in the mucus of the nasal passages and the turbinates, which cause the air to drop its load of larger particles. There are at least four others.

Hairs guard the entrances to the nose, filtering out the largest objects. Particles in the 2-to-5-μm range are usually trapped in the trachea and bronchi by mucus. Only very small objects—less than 1 μm or so—reach the alveoli.

Cilia clean out the noncollapsible channels—the nasal passages, the sinuses, and all the tracheobronchial tree. In the most anterior regions of the nasal cavity, cilia (at least in rats) propel mucus and entrapped particles to the nostrils. All the other ciliary currents (in the posterior nasal passages and the tracheobronchial tree) bear a one-way traffic to the pharynx, where the mucus accumulates until it is swallowed.

Further down in the collapsible bronchioles, particles are squeezed out into the bronchi, where ciliary currents take over. In the alveoli, wandering white blood cells (phagocytes) ingest lodged particles, but they cannot keep up with the rain of air contaminants in the urban environment—as can be seen by comparing the lungs of a city dweller with those of a rural inhabitant.

More violent means of cleaning also exist. In the event that a large object is accidently inhaled, a reflex system triggers spasmodic contractions of all the ventilation musculature, forcefully expelling air and the foreign material (at a velocity of up to 70 mph) during the spasmatic *cough.* Objects in the nose trigger a similar reflex, a *sneeze,* but in this case the soft palate partially seals the rear of the mouth, so that much of the gust of air is forced out the nose.

AIR IS FILTERED IN THE NOSE AND NASAL PASSAGES. THE NASAL PASSAGES, AS WELL AS THE AIR PASSAGES TO THE LUNGS, ARE PROVIDED WITH MUCUS-SECRETING CELLS AND CELLS WITH CILIA. FOREIGN PARTICLES BECOME TRAPPED IN THE MUCUS, WHICH IS THEN MOVED TO THE NOSTRILS OR MOUTH.

REGULATION OF VENTILATION The amount of exchange across the alveolar membrane depends on how often and how deeply one breathes. The muscles responsible for breathing movements are controlled by nerves from the brain's medulla, an integration center for many of the involuntary physiological processes. The cells of the medulla could theoretically monitor either the CO_2 (actually carbonic acid, the dissolved form of CO_2) or the O_2 concentration in the blood flowing past them. The British biologist J. S. Haldane determined which was being monitored by breathing different mixtures of gas and measuring the effects. He found that increasing the CO_2 tension in the lungs only 0.25 percent was enough to cause a doubling of the breathing rate, permitting more CO_2 to leave the blood.

If one voluntarily breathes very deeply for a time, or **hyperventilates,** it is possible to lose consciousness because the pH of the blood shifts to the alkaline due to the loss of carbonic acid as CO_2 from the lungs. Skin divers often hyperventilate by taking a number of deep breaths just before diving, thus reducing the blood's CO_2 content below normal. Consequently, they don't feel compelled to breathe so soon. As one might expect, diving mammals, such as seals, are relatively insensitive to increased CO_2 loads. Mammals also have an oxygen-monitoring system. Sensory cells in the aorta and carotid arteries (see below) send impulses to the medulla, but the concentration of oxygen in the blood must decrease considerably before increased breathing is stimulated.

BREATHING IS CONTROLLED BY THE BRAIN, WHICH MONITORS THE AMOUNT OF CARBON DIOXIDE IN THE BLOOD. EVEN SLIGHT IN- CREASES IN CARBON DIOXIDE LEAD TO AN INCREASE IN THE RATE OF BREATHING.

At this point, you have a picture of how food is broken down into small energy-containing molecules which can be transported through the cells lining the intestine. You also understand how and where gas exchange takes place. In Chap. 4, you learned that in most organisms oxygen is needed by the mitochondria in cells for the synthesis of ATP from small organic molecules. To complete the picture and tie together all these loose ends, you must understand how the oxygen and the food molecules reach the cells where actual respiration occurs. In man and other verte- brates, this function is carried out by the circulatory system. The vessels that transport food and gases transport other things (such as hormones and antibodies), as will be discussed in the following chapters.

The basic solution to the problem posed by the great distances mole- cules must move is to circulate a fluid. It is convenient to think of this happening in two steps: (1) the fluid passes through the gas-exchange surfaces, where it quickly equilibrates with the medium (water or air), picking up oxygen and losing carbon dioxide; and (2) the fluid is dis- tributed to the tissues, where the directions of diffusion are reversed.

Small and structurally simple animals don't require a very elaborate circulatory system. Their metabolic rates are low, and their surface-area- to-volume ratio is high. But even the structurally simple coelenterates move the contents of the gastrovascular cavity with waves of muscular contraction and with long powerful flagella (See Box 7–3), which create water currents that distribute food and dissolved oxygen. This kind of circulation works pretty well for sluggish or saclike and essentially tubu- lar animals without specialized thickened organs, but larger and more complex animals require more sophisticated plumbing.

Rigid organisms such as crustaceans (crabs, lobsters, and their relatives) cannot squeeze fluid about their bodies by contraction of the body wall. Flagellar circulation is too slow to satisfy the oxygen and nutrient re- quirements of tissues, including the muscles of locomotory appendages. In small crustaceans, the movements of legs and internal organs are sufficient to push the blood through tissue sinuses. Sometimes there are valves that confer a directionality to the blood flow. In larger crustaceans, such as lobsters, there is a dorsal **heart,** which receives blood from the gills and pumps it through arteries that branch repeatedly and penetrate throughout the body (Fig. 6–27). Eventually, the blood empties into cav- ities, where it comes into direct contact with cells. These cavities, in

CIRCULATORY SYSTEMS IN ANIMALS
Transport by the Circulatory System

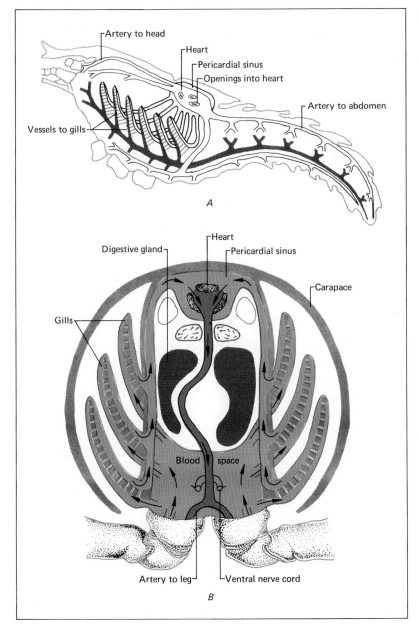

Artery to head

Heart

Pericardial sinus

Openings into heart

Artery to abdomen

Vessels to gills

A

Digestive gland

Heart

Pericardial sinus

Carapace

Gills

Blood space

Artery to leg

Ventral nerve cord

B

Fig. 6-27 Diagram of an open circulatory system in a crustacean. *(Adapted from ANIMALS WITHOUT BACKBONES by R. Buschbaum, The University of Chicago Press, 1938.)*

turn, finally empty into one large ventral sinus. Vessels carry blood from the sinus to the gills. Booster hearts, located in strategic places, may assist the flow, but directional check valves, movements of the appendages, and peristalsis in the gut are also important. The advantage of an **open circulation,** such as this, is that the blood comes into direct contact with the cells of tissues. The disadvantage is its leisurely pace.

Insects, such as flies and bees, have an even more open system than do crustaceans, the heart being the only blood vessel. It functions to pump blood into the head, and from there the blood meanders back and through sinuses, eventually returning to openings in the heart. Why should such active animals have such a poorly developed circulatory system compared with the aquatic crustaceans? Recall that insects depend on the tracheal system for the circulation of gases, so they don't require a complex circulatory system, as do aquatic arthropods.

GASES AND MOLECULES DERIVED FROM FOOD OR CELLULAR METABOLISM ARE TRANSPORTED, IN MOST MULTICELLULAR ANIMALS, IN A CIRCULATORY SYSTEM. MANY INVERTEBRATES HAVE A RELATIVELY SIMPLE, OPEN CIRCULATORY SYSTEM. THE HEART PUMPS BLOOD FROM THE GILLS THROUGHOUT THE BODY TO CAVITIES WHERE IT CONTACTS THE CELLS. EVENTUALLY THE BLOOD IS RETURNED TO THE GILLS THROUGH SINUSES.

The Circulatory System of Vertebrates

The basic vertebrate circulation pattern is closed, as shown in Fig. 6–28. The blood throughout its entire circuit never comes into direct contact with tissues; instead, diffusion of water, gases, and other small molecules takes place through the very thin walls of the smallest vessels, the capillaries. One of the advantages of a closed circulation, as already stated, is its rate of flow. Relatively high blood pressures can be maintained as long as the vessels are not too distensible. In a normal adult, for instance, the blood pressure in the arm fluctuates between 120 and 80 mm of mercury. The variation is due to the rhythmic fluctuation in the pumping action of the heart. With each beat or contraction of the left ventricle, a quantity of blood is forced into the arteries. This causes a temporary increase in the pressure as measured in the arm. The greatest pressure at any point is called the **systolic** pressure; the lowest pressure, the **diastolic pressure,** occurs when the ventricle is relaxed. The usual shorthand for blood pressure data is 120/80 mmHg, meaning that the systolic pressure is equivalent to 120 mm of mercury (about one-sixth of an atmosphere) and the diastolic pressure to 80 mm of mercury. In rodents such as rats, hamsters, and guinea pigs, the systolic and diastolic pressures are about 77/47. In the heart of a lobster, it is only 13/1. Blood pressure diminishes away from the heart due to frictional loss and increase in area—in mammals, the total cross-sectional area of capillaries is 800 times that of the large aorta leaving the heart. The bore of most capil-

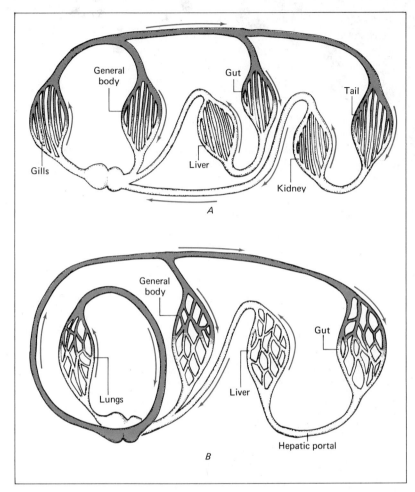

Fig. 6-28 Generalized circulatory systems in lower vertebrates (fishes) (*A*) and in birds and mammals (*B*). The circulation is closed and blood never leaves the vessels, although there is diffusion of materials across the thin capillary walls. In the birds and mammals, the blood makes two passes through the heart, first to the lungs and then to the body. In fishes, the blood goes directly to the body from the gills. (*Adapted from Alfred S. Romer, THE VERTEBRATE BODY, 3/e, 1962, W. B. Saunders Company, Philadelphia.*)

laries is just large enough (less than 10 μm in diameter) to allow the blood cells to flow through, so there is every opportunity for exchange of materials between the blood and the surrounding cells and intercellular lymph fluids. Another advantage of a closed circulatory system is the potential for fine control. Capillaries have tiny rings of muscles at their branch points. These are under the control of the ***autonomic nervous system*** (Chap. 9), and in a resting muscle, for instance, less than 5 percent of the capillaries may be open. Box 6-1 discusses circulation in man.

VERTEBRATES HAVE A CLOSED CIRCULATORY SYSTEM IN WHICH

BLOOD DOES NOT COME INTO DIRECT CONTACT WITH TISSUES.
DIFFUSION TAKES PLACE ACROSS THE WALLS OF THE CAPILLARIES.
CLOSED CIRCULATORY SYSTEMS CAN MAINTAIN A HIGH RATE OF
FLOW AND PROVIDE FOR FINE CONTROL OF CIRCULATION.

THE TRANSPORT OF OXYGEN: THE ROLE OF HEMOGLOBIN   One might reasonably think that just a solution of salts circulating through gills or lungs and around the body would deliver enough oxygen to the tissues. But in fact, very little oxygen dissolves in such a solution (about 0.3 ml per 100 m of solution), and a mammal needs about 65 times this much. This problem has been solved many times in many groups of animals by the evolution of oxygen-carrier molecules. Vertebrates (certain Antarctic fish excepted) and many other groups of animals have colored blood, and the pigments responsible for the color have a very high affinity for oxygen. In man and other vertebrates, the pigment is **hemoglobin,** a complex protein molecule containing an iron-porphyrin (heme). Some invertebrates, including sea cucumbers, many annelids, small crustaceans, insects, and molluscs, also employ hemoglobin. Other pigments, some with copper or vanadium instead of iron, are found scattered through many phyla.

What are the requirements for an efficient oxygen carrier? First, its affinity for oxygen must be great enough to pick up oxygen at the gas-exchange membranes of the ventilation organ. Second, and equally important, it must release oxygen in the tissues where it is in short supply. In other words, the carrier has to have the capacity to respond to oxygen concentration, as well as to pick up and transport the gas.

Until recently, it was believed that vertebrates required so much hemoglobin that if each molecule were in solution, it would make the blood too syrupy to flow. This was thought to be the reason given for the packaging of hemoglobin in special cells called **red blood cells.** Now it has been shown that the viscosity of blood that has had its red blood cells disrupted in such a way as to free all the hemoglobin into solution is actually *less* than the viscosity of normal blood. Why then red blood cells? The probable reason is that the osmotic pressure of mammalian blood would triple if the hemoglobin were in solution and would therefore tend to remove great quantities of water from the tissues. To counteract this, a mammal would have to reduce its hemoglobin content, consequently lowering its rate of oxygen use, that is, its metabolic rate. Alternatively, it might greatly increase its blood pressure until this about equaled the osmotic pressure. This would be analogous to increasing the height of the fluid in Fig. 5–5. The latter would put a great strain on the heart. It is interesting to note in this regard that there are many invertebrates that do carry their blood pigments in solution rather than in cells, but these pigments circulate as very large aggregates. Osmotically, these aggregates behave as one molecule instead of many.

In air, the partial pressure of oxygen is 155 mmHg, whereas the partial pressure of oxygen in human lungs is about 100 mmHg. To be most effective, a pigment should be fully saturated at the partial pressure which is highest in the organism (at the gas-exchange surface) and should release a large fraction of the oxygen in the tissues. In man, the partial pressure of tissues is between 5 and 30 mmHg, a point at which hemoglobin is about 50 percent saturated. This relation is conveniently expressed by a graph of the partial pressure of oxygen against the percentage of the hemoglobin that is loaded—that is, the percentage of saturation (Fig. 6-29). Man's saturation or dissociation curve is typical of mammals. At about 100 mmHg, the oxygen partial pressure in the lungs, virtually every iron atom of hemoglobin has an oxygen molecule bound loosely to it; we see that it is fully saturated. At lower oxygen pressures, hemoglobin releases the oxygen into solution. Hemoglobin releases all its oxygen in tissues where there is no oxygen.

Examine the dissociation curve and ask yourself the following question: In which direction would the curve shift if hemoglobin had a lower affinity for oxygen? Now pick an arbitrary oxygen pressure, say 40 mmHg. If, at that pressure, hemoglobin carried less oxygen, say 60 percent, the

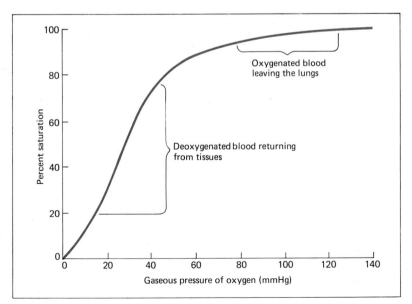

Fig. 6-29 Relationship between the gaseous pressure of oxygen and the degree of saturation of the pigment hemoglobin.

curve would have to pass through a lower point. If you do this two or three more times, you will see that a curve drawn through these new points would be to the right of the original. Now consider another question: If carbon dioxide had an effect on the affinity of hemoglobin for oxygen, in which direction ought it to shift the curve? Again, the answer is to the right. If a tissue is metabolizing at a high rate—the leg muscles of a runner, for instance—much carbon dioxide will enter the capillaries and a correspondingly great oxygen debt will develop in the muscle. This is, in fact, what happens.

Carbon dioxide facilitates the delivery of oxygen by shifting the dissociation curve to the right. In the lungs, carbon dioxide is lost from the blood and there is a corresponding shift to the left, facilitating the uptake of oxygen from the alveolar air. The carbon dioxide shift is known as the Bohr effect.

Other predictions about the dissociation curves tend to be verified as well. As discussed in Box 6-2, the "parasitic" relationship between fetus and mother is correlated with profound changes in circulation, especially in the fetus. What would you predict about the hemoglobin of the fetus? The blood of the fetus circulating through the villi of the

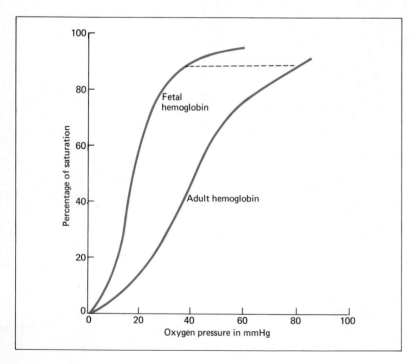

Fig. 6-30 Graph showing the difference between fetal hemoglobin and adult hemoglobin with respect to affinity for oxygen.

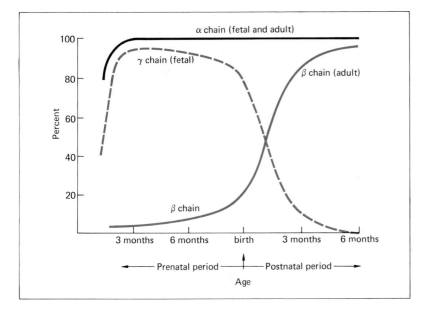

Fig. 6-31 Graph showing the change in synthesis of gamma (fetal) and beta (adult) polypeptide chains which, with alpha chains, make up hemoglobin. *(Adapted with permission of The Macmillan Company from GENETICS by Monroe W. Strickberger. Copyright © 1968 by The Macmillan Company.)*

placenta is relatively unsaturated. The mother's placental blood is freshly oxygenated, so there is a tendency for oxygen to diffuse from the maternal to the fetal blood. If fetal hemoglobin had a greater affinity for oxygen, especially at low partial pressures, the diffusion would be accelerated. In fact, fetal hemoglobin does differ from adult hemoglobin and in just this way (Fig. 6-30). In place of the two adult beta chains, there are two gamma chains with a different amino acid sequence. At about the time of birth, fetal hemoglobin stops being synthesized and the rate of beta-chain synthesis takes a sharp rise (Fig. 6-31). Besides their occurrence in mammals, fetal hemoglobins also occur in birds (egg hemoglobin) and in larval fish and frogs.

Analogous shifts occur in many animals at critical times during life, especially when the animal changes environments. For instance, one would expect to find that frogs have a dissociation curve to the right of their tadpoles (Fig. 6-32). The tadpoles, being aquatic, must extract their oxygen from water, so it is to be expected that their hemoglobin would have a higher affinity for oxygen. Some small animals living in oxygen-rich habitats have very little hemoglobin but have the ability to make it if the oxygen pressure drops. The water flea *Daphnia* is pale in oxygen-rich water and red in oxygen-poor water.

Myoglobin, a muscle pigment very similar to hemoglobin, plays a physiological role something like that of fetal hemoglobin in many animals. It

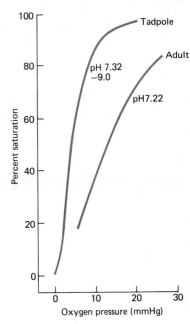

Fig. 6-32 Graph showing the difference between frog hemoglobin and tadpole hemoglobin with respect to affinity for oxygen.

has a higher affinity for oxygen than does hemoglobin, so it can "steal" oxygen from blood and store it until needed by respiratory enzymes. The efficiency of myoglobin as a transfer agent was impressively demonstrated by an experiment in which myoglobin was placed on a thin filter between two solutions of differing oxygen concentrations. The oxygen diffused several hundred times faster through this filter than through one without myoglobin.

OXYGEN, IN VERTEBRATES, IS CARRIED IN THE BLOOD IN SPECIALIZED CELLS CONTAINING THE PIGMENT HEMOGLOBIN. HEMOGLOBIN PICKS UP OXYGEN IN TISSUE WHERE THE CONCENTRATION OF OXYGEN IS HIGH (THE LUNGS). IT GIVES UP OXYGEN IN TISSUE WHERE THE CONCENTRATION OF OXYGEN IS LOW AND THAT OF CARBON DIOXIDE IS HIGH. OTHER PIGMENTS, SUCH AS FETAL HEMOGLOBIN AND MYOGLOBIN, PLAY MORE SPECIALIZED ROLES.

CARBON MONOXIDE AND NITRITE POISONING An increasingly prominent air pollutant is carbon monoxide, a toxic component of internal-combustion engines, tobacco smoke, and coal gas. Hemoglobin has about 300 times more affinity for carbon monoxide than for oxygen; and only 0.1 percent carbon monoxide in the air will combine with most of a person's hemoglobin, and this can lead to death by oxygen starvation.

Very recently, another chemical has been found to be affecting the ability of hemoglobin to carry oxygen in large numbers of people in certain areas. Where high-intensity agriculture is being practiced and where there are imputs of large amounts of nitrate fertilizers, the nitrogen is dissolved in water and often ends up in lakes and streams and in the water table. In many parts of the country, a significant amount of drinking water is obtained from these belowground reservoirs. High levels are showing up in wells in agricultural areas in many parts of the United States.

Nitrate itself is relatively innocuous, but bacteria in the colon of infants have the ability to convert nitrate to nitrite. Bacteria may also change nitrate to nitrite in baby food. Nitrite enters the circulation, where it combines with hemoglobin to form methemoglobin. Methemoglobin has a much lower affinity for oxygen than does hemoglobin, and respiratory distress and suffocation can occur. This new disease is becoming a serious health problem in the agricultural valleys of California and other places in the Southwest and Midwest.

HEMOGLOBIN HAS A MUCH GREATER AFFINITY FOR CARBON MONOXIDE THAN FOR OXYGEN, AND THEREFORE VERY LOW CONCEN-

TRATIONS OF THE FORMER CAN REPLACE MOST OF THE OXYGEN IN THE BLOODSTREAM, LEADING TO DEATH. NITRITES CHANGE HEMOGLOBIN TO A FORM WHICH HAS A MUCH LOWER AFFINITY FOR OXYGEN, AND THEY ARE THEREFORE TOXIC.

THE TRANSPORT OF CARBON DIOXIDE   Theoretically, blood should become saturated with CO_2 gas at 2.5 ml of CO_2 per 100 ml of blood, but blood returning to the heart from tissues actually holds up to 55 or 60 ml of CO_2 per 100 ml of blood. The reason for this is the chemical reaction of CO_2 with water forming carbonic acid, which is very soluble in water:

$$H_2O + CO_2 \rightleftharpoons H_2CO_3$$

Normally this is such a slow reaction that it wouldn't contribute very much to the CO_2 carrying capacity of blood, but it is accelerated by an enzyme in red blood cells called **carbonic anhydrase.** Most of the carbonic acid dissociates to hydrogen ions (H^+) and bicarbonate ions (HCO_3^-). The body's tissues are very sensitive to even slight changes in pH, so the hydrogen ions thus formed in the red blood cells represent a grave threat. Hemoglobin, however, combines with H^+:

$$H^+ \, Hb \rightleftharpoons HHb$$

preventing the blood from becoming too acid. In the lungs, nearly all the hemoglobin becomes loaded with oxygen, and oxygenated hemoglobin ($HHbO_2$) tends to liberate its H^+. The result is the formation of more carbonic acid and, in turn, more CO_2 and H_2O. Hence, the CO_2 concentration of the blood is given a boost in just the right place to facilitate its elimination by diffusion into the alveolar air.

CARBON DIOXIDE PRODUCED IN CELLULAR RESPIRATION IS TRANSPORTED TO THE LUNGS AS CARBONIC ACID DISSOLVED IN THE BLOOD. IN THE LUNGS, CARBON DIOXIDE DIFFUSES FROM THE BLOOD INTO THE ALVEOLI AND IS EXPELLED.

TRANSPORT IN PLANTS

Photosynthesis occurs chiefly in the leaves, and much of the plant—the roots and other tissues—leads a heterotrophic existence. The products of photosynthesis must somehow be conducted to these areas. In the vascular land plants, the tissue called **phloem** carries out this function. We will consider here only the type of phloem found in most of the flowering plants Fig. 6-33. Like xylem, described in Chap. 5, phloem is a complex tissue containing many different cell types. The conducting cells themselves, however, are called **sieve-tube elements.** A vertical file of sieve-tube elements is called a **sieve tube.**

In the development of a sieve-tube element, the mother cell divides to produce two daughters. These will develop along very different lines. The sieve-tube element itself is an elongate cell of relatively large diameter. The primary cell wall of a sieve-tube element is somewhat thickened, but no secondary wall is produced. At either end, there are specialized aggregations of pits. Such groups of pits form sieve plates. The cytoplasms of sieve-tube elements are connected through these pits. The structure of a sieve tube is shown in Fig. 6-34. At maturity, the cytoplasm of the sieve-tube cell changes in various ways, and the nucleus disappears. The other daughter cell remains relatively very small. It has a dense cytoplasm and a nucleus and is closely associated with the sieve-tube element by pits on its lateral walls. Presumably, the nucleus of this **companion cell,** as it is called, provides the necessary nuclear functions for both.

Although sieve tubes are vertical rows of cells connected by pits, you can see that they are very different from vessels. Vessels are *empty* tubes. Sieve-tube elements retain their cytoplasm. Photosynthetic products moving through the phloem must pass through these cells. The details of phloem transport are not well understood. By the use of radioactive tracers, the rate of flow of sugar in sieve tubes can be measured. Observed rates of flow up to 100 cm per hour are much faster than could be expected by diffusion from cell to cell.

Unfortunately, there is not complete agreement as to how phloem functions. It is possible that the contents of sieve-tube elements flow across the sieve plates from one cell to the next. This hypothesis is difficult to prove or disprove. It can be shown that there is a gradient of sugar in sieve tubes. In the roots, the sugar, originally manufactured in the leaf cells, is removed from the sieve tubes by nonphotosynthetic cells, which use it in their metabolism. There will, therefore, be a decrease in turgor pressure in such root cells. In the leaves, where sugar is being synthesized, phloem cells will have a greater turgor pressure. A gradient therefore is established, which will drive the sap to areas where it is being used. This will most often be downward, toward the roots. It may also be lateral or even upward. The parts of the shoot above the actively photosynthesizing leaves also will require photosynthetic products.

Much obviously remains to be discovered about transport in phloem. It should be clear, however, that transport in plants is basically very different from that in animals. Separate vascular tissues—xylem and

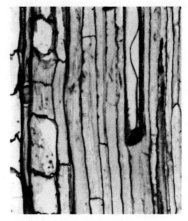

Fig. 6-33 Longitudinal section through the phloem of a squash plant (*Cucurbita*). The elongated cells in the center are sieve-tube elements with the darkly staining sieve plates. Narrower, nucleated cells are companion cells. X322. (*Jeroboam, N. Y.*)

phloem—conduct water and photosynthetic products, respectively. There is nothing comparable to the circulatory systems of most animals.

TRANSPORT OF THE PRODUCTS OF PHOTOSYNTHESIS FROM THE LEAVES TO NONPHOTOSYNTHETIC PARTS OF THE PLANT OCCURS IN THE PHLOEM. PLANTS DO NOT HAVE A CIRCULATORY SYSTEM COMPARABLE TO THAT IN ANIMALS.

Many animals obtain their food passively by ingesting large quantities of water and passing it through a filter which strains out small organisms. Most animals, however, actively seek out and eat animal or plant food, or both. Herbivores face the problem that much of the material they ingest—cellulose—cannot be digested by them directly. Symbiotic organisms—bacteria or protozoans—which live in their digestive tracts are able to digest cellulose.

Omnivores, such as man, eat both plant and animal food. Their teeth are relatively unspecialized, compared with those of herbivores and carnivores, and serve for biting and tearing as well as grinding. Digestion begins in the mouth during the process of mastication through the action of secretions of the salivary glands, which also moisten the food. From the mouth, the mass of food is moved by muscular contraction called peristalsis through the various specialized regions of the digestive tract: esophagus, stomach, duodenum, small intestine, large intestine, and rectum.

In the stomach, hydrochloric acid and pepsin carry on the digestive process begun in the mouth. The food becomes converted into a semi-liquid milky substance called chyme. Digestion is carried still further in the small intestines, with secretions from the gall bladder and pancreas. Absorption also begins in the intestines, the walls of which have modifications that greatly increase their surface area. Microorganisms in the colon, or large intestine, produce nutrients essential to man in the course of their own metabolism. In addition to food molecules, water is removed from the chyme, and the undigested residue, together with bacteria, eventually is expelled through the anus as feces.

Oxygen, essential for cellular respiration and the use of energy, must be obtained from the environment by diffusion. Similarly, carbon dioxide must be given off to the environment. Small animals can accomplish the required gas exchange through their body surfaces. Animals with greater volume in relation to surface have evolved special mechanisms for taking in oxygen and transporting it to every cell, as well as for getting rid of

SUMMARY

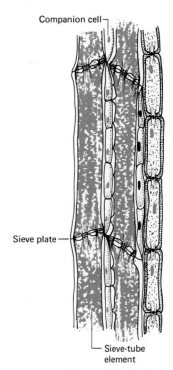

Companion cell

Sieve plate

Sieve-tube element

Fig. 6-34 Diagram of longitudinal section through phloem tissue showing sieve-tube elements with sieve plates and companion cells. (Adapted from *Relis B. Brown, GENERAL BIOLOGY. Copyright McGraw-Hill Book Company, 1970. Used by permission.)*

carbon dioxide. The metabolic rate of organisms is the rate at which they use energy. Since oxygen is required for energy use, its consumption is a measure of rate of metabolism.

Oxygen may be obtained either from water or from the atmosphere. Greater quantities can be obtained more rapidly from air than from water, and less energy is required with air as a source. "Ventilation" refers to all the processes involving gas exchanges in animals. Ventilation in water is accomplished by structures called gills. These are organs of very extensive surface which extend into the watery environment or are arranged in such a way that water can be forced over them. Gills are richly supplied with blood vessels.

Terrestrial animals basically use one of two systems of ventilation. Insects have a much-branched system of tubes called tracheae, which bring oxygen to every cell. Other animals have some sort of lung Gas exchange in man takes place in tiny blind sacs, called alveoli, in the lungs. Air is warmed and filtered in the nose and transmitted through a series of branching air passages to the alveoli. Breathing is controlled by the brain, which monitors the level of carbon dioxide in the blood.

Gases and molecules derived from food or cellular metabolism are transported, in most multicellular animals, in blood in a circulatory system. In vertebrates, oxygen is carried in the blood in specialized cells combined with the pigment hemoglobin. Hemoglobin picks up oxygen in tissues where the concentration of oxygen is high (the lungs) and gives up oxygen where the concentration of oxygen is low and that of carbon dioxide is high. Carbon dioxide is transported to the lungs as carbonic acid dissolved in the blood.

In plants, transport of the photosynthetic products from the leaves to the nonphotosynthetic parts of the plant occurs in the vascular tissue phloem. Water is conducted upward in the xylem. Gas exchange occurs across cell membranes throughout an extensive series of intercellular air spaces. There is no circulatory system in plants comparable to that of animals.

BOX 6–1. CIRCULATION IN MAN

Once it was thought that blood was formed in the intestines and transported to the limbs and other organs, where it was destroyed. The one-way scheme was demolished in favor of the round-trip model in 1628 by William Harvey in his classic work, "On the Movement of the Heart and Blood in Animals." We will begin our description of circulation in man and other mammals with blood in the aorta, the giant vessel transporting blood away from the heart to the general, of *systemic,* circulation (Fig. 6–1–1). (Any vessel carrying blood from

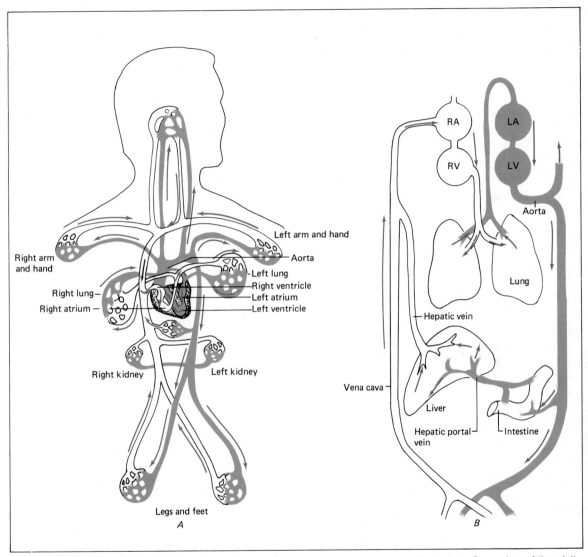

Right arm
and hand

Left arm and hand

Aorta

Left lung

Right ventricle

Left atrium

Left ventricle

Right lung

Right atrium

Right kidney

Left kidney

Legs and feet

A

RA

RV

LA

LV

Aorta

Lung

Hepatic vein

Vena cava

Liver

Hepatic portal
vein

Intestine

B

Fig. 6-1-1 Comparison of the adult
human circulation with the generalized
bird or mammal circulation. *(Adapted
from Relis B. Brown, GENERAL
BIOLOGY. Copyright McGraw-Hill
Book Company, 1970. Used by
permission.)*

the heart to an organ is called an *artery;* the aorta is the largest artery.)
Many arteries branch off from the aorta, the first ones carrying blood
to the head and upper extremities and the lower branches carrying
blood to the digestive tract, the other abdominal organs, and the lower
extremities. As arteries penetrate organs and masses of tissue, they
subdivide into smaller and smaller vessels. The smallest subdivisions

of arteries, called **arterioles,** are microscopic and branch, in turn, to the even smaller, thin-walled **capillaries.** Transport of small molecules between the blood and the surrounding tissues finally occurs through the walls of these delicate tubes.

Blood leaving the "beds" of capillaries in a tissue is collected in tiny **venules,** which join to form larger **veins.** The largest veins return the blood to the heart. Veins do not always carry blood directly to the heart. Blood collected in veins from capillary beds in the digestive tract passes through the hepatic portal vein en route to the liver. Here

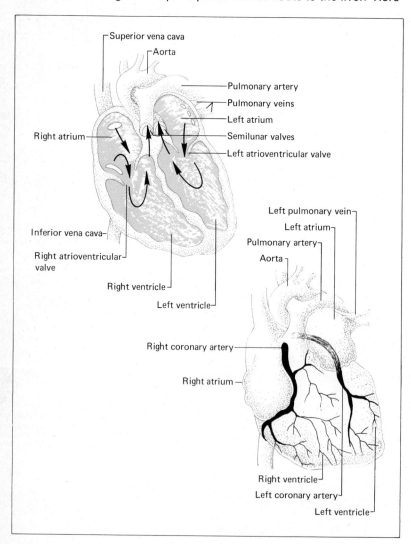

Fig. 6-1-2 Structure of the heart and the coronary circulation.

Superior vena cava

Aorta

Pulmonary artery

Pulmonary veins

Left atrium

Semilunar valves

Left atrioventricular valve

Right atrium

Inferior vena cava

Right atrioventricular valve

Right ventricle

Left ventricle

Left pulmonary vein

Left atrium

Pulmonary artery

Aorta

Right coronary artery

Right atrium

Right ventricle

Left coronary artery

Left ventricle

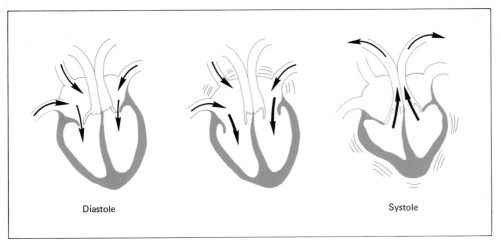

Diastole   Systole

Fig. 6 – 1 – 3   Diagram to show the
contraction of the heart. *(From
BIOLOGY: ITS PRINCIPLES AND
IMPLICATIONS, Second Edition, by
Garrett Hardin. W. H. Freeman and
Company. Copyright © 1966.)*

it enters the liver's capillary bed, where two-thirds of the glucose
and most of the other nutrients absorbed by the villi in the intestine
are removed.   Here, also, special cells ingest the bacteria that are
continually leaking through the intestinal wall. Upon coursing through
this second capillary bed since leaving the heart, the cleaned and filtered blood is collected in the hepatic vein before it finally reaches
the heart via the large **vena cava.**

The pump that keeps the blood circulating is the heart. The heart
of man, as well as those of other mammals and birds, is really two
pumping systems. One of these systems pumps newly returned blood
to the lungs through the pulmonary arteries. After the blood is oxygenated, it is returned to the heart in the pulmonary veins. The second
pumping system propels it into the aorta and throughout the body.
The first system occupies the two chambers of the right side of the
heart. Returning blood enters the thin-walled right **atrium** (pl. **atria**).
The right **ventricle** is filled by the pumping action of the right atrium.
The blood is then forced into the pulmonary circulation by the pumping
action of the right ventricle. Blood returning from the lungs enters
the left atrium and then the left ventricle, from which it is forcefully
propelled into the aorta. The muscle of the heart is supplied with
oxygenated blood by the **coronary** arteries branching off the aorta
(Fig. 6– 1– 2).

The basic mechanics of the heart are really very simple. Each chamber
is a muscle-lined sac (Fig. 6– 1– 3). When the muscle contracts, the sac
gets smaller and the contents are forced out through a hole into another chamber or an artery. The proper direction of flow is ensured by

the order in which the chambers contract and by valves that prevent
blood from flowing backward. The atria contract first, filling the
powerful ventricles. When the ventricles contract, blood is prevented
from flowing back into the atria by valves in the wall between the two

Fig. 6–1–4 Diagram to show the
extent of the lymphatic system in
an adult male.

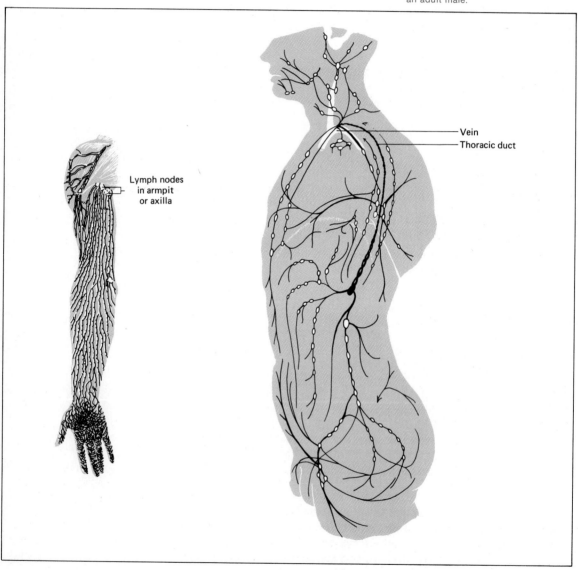

Lymph nodes
in armpit
or axilla

Vein

Thoracic duct

chambers. The sudden closing of these valves is what makes the "lub" part of the "lub-dub" sounds heard through a stethoscope. The "dub" sound is caused by the closing of valves in the pulmonary artery and the aorta. These valves prevent blood from flowing backward into the ventricles when the ventricles relax between contractions.

Man also has another, open circulatory system. More fluid leaves the capillaries than returns. This fluid, called **lymph,** bathes the cells in the body and is the medium for all the exchanges between the blood and the cells (Fig. 6–1–4). The contraction of the body's muscles ultimately squeezes the lymph into tiny **lymphatic vessels.** These unite into larger vessels, which eventually spill their contents into veins. Special lymphatics draining the intestine carry the absorbed lipids to a large vessel called the thoracic duct. An excess in lymph in tissues produces swelling and puffiness, a condition called **edema.**

The fetus in its mother's womb (uterus) is, in a sense, a parasite. It is totally dependent on the host for nutrients, oxygen, protection, and excretion. Its gut is not used for digestion, nor are its lungs used for ventilation, and very little blood flows through these organs compared with the flow immediately after birth. The circulatory system of the fetus (Fig. 6–2–1) differs from that of a baby in flow pattern as well as flow rate, and we will look at these specializations for parasitic life and how they change dramatically at birth.

In the fetus, both the intestine and the lungs, and to a large extent the liver and kidneys, are replaced by a single organ that performs all their functions. This is the placenta, a very complex structure which carries out many metabolic and regulatory functions in addition to those already mentioned. Food- and oxygen-enriched blood leaves the placenta through the **umbilical vein** (in the umbilical cord), entering the body of the fetus at the location of the future **navel,** or **umbilicus.** The umbilical vein courses directly to the liver. Most of its blood flows straight through in the **ductus venosus** without passing through capillaries. From there into the right atrium, the fetal and adult circulations are the same. Once in the right atrium, most of the blood passes directly through a hole, called the **foramen ovale,** to the left atrium, bypassing the lungs entirely. The portion that is pumped into the right ventricle is again split into two channels, some going into the pulmonary circulation but most entering another vessel unique to the fetus—the **ductus arteriosus.** The latter connects with the aorta and is the second way the fetus has of bypassing the lungs, which are collapsed at this time and offer considerable resistance to the flow of

BOX 6–2. FETAL CIRCULATION AND BIRTH

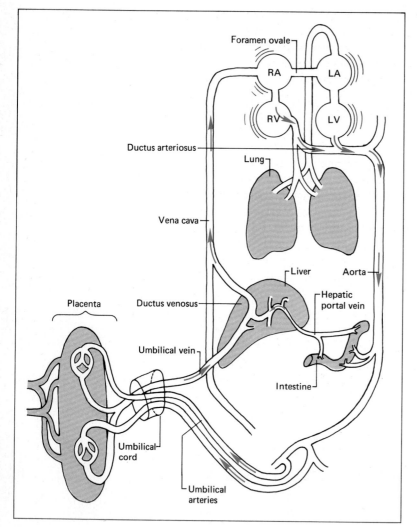

Fig. 6-2-1 Diagram of the fetal circulatory system.

blood.   Hence, most of the blood in the aorta avoids the time-con-
suming and unnecessary pulmonary shunt.

Blood enters the placenta through the umbilical arteries, which branch
off the femoral (leg) arteries just before the latter leave the abdomen.
In the placenta, the fetal blood flows into cauliflowerlike nodules
covered by villi. The nodules protrude into pools or sinuses of maternal
blood.   Oxygen and food diffuse through the membrane of the villi
and into the fetal capillaries inside.

The hemoglobin of the fetus has a higher affinity for oxygen than does that of the mother, and this accelerates the diffusion of oxygen into the villi. Carbon dioxide and other excretory products tend to diffuse into the maternal blood since they are more concentrated in the fetal blood. The maternal blood is "clean" and oxygenated since it does not pass through any capillary beds on the way to the placenta. Usually, only small molecules can pass these membranes, so that "idiosyncratic" molecules of the two individuals—proteins—are barred from transfer. This is important in protecting the fetus from rejection by the mother as a foreign tissue graft (see Chap. 11).

During or shortly after birth, the muscles of the blood vessels in the umbilicus clamp shut these vessels. The resulting oxygen deficit in the fetus apparently is the trigger for the first breath, and the lungs expand with air. Blood can then flow without obstruction through the lungs. Therefore, flow through the ductus arteriosus is stopped or even reversed. This vessel begins to occlude and is thoroughly plugged within a few weeks. Much more blood now returns via the pulmonary veins to the left atrium, for the first time raising the pressure in this chamber higher than that in the right atrium. Blood would tend to flow backward through the foramen ovale if it were not for a membranous valve that is forced over the opening by this pressure. Flow through the ductus venosus is shut off by a sphincter in its walls a few minutes after birth, completing the remarkable transformation from parasite to free organism.

SUPPLEMENTARY READINGS
Books

Guyton, A. C.: "Function of the Human Body," Saunders, Philadelphia, 1969.
Morton, J. E.: "Guts," St. Martin's, New York, 1967.
Prosser, C. L., and F. A. Brown: "Comparative Animal Physiology," Saunders, Philadelphia, 1962.
Richardson, M.: Translocation in Plants, St. Martin's, New York, 1968.

Articles

Comroe, J. H.: The Lung, *Sci. Amer.,* **214,** no. 2, Offprint 1034.
Neurath, H.: Protein-digesting Enzymes, *Sci. Amer.,* **211,** no. 6, Offprint 198 (1964).
Perutz, M. F. The Hemoglobin Molecule, *Sci. Amer.,* **211,** no. 5, Offprint 196 (1964).

CHAPTER 7

INTEGRATION I: HORMONAL SYSTEMS

In previous chapters, systems and strategies have been described more or less in isolation. You have been told about feeding and digestion in an oyster and about photosynthesis in plants—isolated functions of complex, organized, whole organisms. In reality, these functions are, of course, anything but isolated. Survival depends on integration, whether the system is a motorcycle, a cell, or a community. When parts become disengaged from the operation of the whole, the results are stripped gears, cancer, or violence. The goal of this chapter is to explain how the most complex systems in the known universe—organisms—are integrated. It will deal with the mechanisms that ensure coordination of activity.

INTEGRATION AND THE ENVIRONMENT

You might ask why we have assumed that coordination is necessary in the first place. After all, is it so hard to accept the thesis that an organism is a collection of systems functioning independent of each other—digestion, ventilation, excretion, and reproduction, for example, all functioning under one roof (or skin) but without coordination, like clocks in a clock shop? Biologists reject such a "chaos model" of organisms for many reasons. Some of these reasons are derived from simple observation; others, from experimentation. There is, however, a common denominator to these observations; it has to do with environmental change. In theory, the chaos model is fine so long as the environment is constant. A constant supply of food, oxygen, and water and perfect physical security do not demand much coordination of systems. Imagine, though, a mouse caught in the open by an owl. If it is to survive, it has to move fast and accurately. To do this, its muscles will need more oxygen, which requires a higher blood pressure and faster breathing. **Environmental change,** then, requires the integrated functioning of the circulatory, ventilation, and "movement" systems.

Many, if not most, structures play a role in the sensing of change. In fact, the entire organism can be viewed as though it were a complex machine programmed to respond properly to the vagaries of the environment. In a real sense, then, *every* system and structure plays a role in sensing and responding to change. Let's look at an example.

You are driving down a highway at 70 mph when a car swerves from the opposite lane into your path. This information reaches your eyes as changes in the pattern of arriving light waves; the change in pattern is a **stimulus** and is **perceived** as a change in course of the oncoming car. You decide instantly to turn to the left rather than brake, and a fraction of a second later your arms make the appropriate movements. If you have made the right decision, your chances of survival have been greatly enhanced.

The essence of this sequence of events is the **reception** of information about a change in the environment (stimulus); the assimilation, or understanding or **perception,** of the information; and the final relaying of appropriate "commands" to **effector** (doing) systems. In animals, there are two types of communication systems linking these three elements together. One is the nervous system—those structures whose elements receive and convey information over chains of specialized cells in times measured in milliseconds. The other is the endocrine system, based on the glands of internal secretion, by which information is transferred as chemical messengers **(hormones)** by diffusion among cells or in the circulatory system. In a slightly oversimplified sense, the nervous system is analogous to a telephone exchange, with its complex of peripheral circuits, whereas the endocrine system functions more like a loudspeaker system at a county fair. In the first case, messages enter and leave along insulated, separate lines, coming from more or less localized points and going back to specific targets. In the latter case, information is broadcast generally to everyone within hearing distance, but only those individuals to whom a particular message is relevant take any notice. Later we will see that the distinction between these two systems is not quite as clear-cut as here indicated.

INTEGRATED FUNCTIONING OF ALL SYSTEMS IS REQUIRED IF ORGANISMS ARE TO SURVIVE IN EVER-CHANGING ENVIRONMENTS. THE ORGANISM MUST BE CAPABLE OF THE RECEPTION OF INFORMATION ABOUT ENVIRONMENTAL CHANGES (STIMULI), THE "UNDERSTANDING" OF THAT INFORMATION, AND THE "COMMANDING" OF APPROPRIATE RESPONSES. IN ANIMALS, TWO SYSTEMS ARE RESPONSIBLE FOR THESE FUNCTIONS: A HIGH-SPEED, TELEPHONE-TYPE NERVOUS SYSTEM AND A SLOWER, BROADCAST-TYPE ENDOCRINE SYSTEM.

The importance of these information sensing, sending, and interpreting systems for permitting the organism to respond to changes in its environment is obvious. What is not so obvious, and is even a little paradoxical, is that there is another environment in addition to the one that is usually considered. This is the **internal environment.** The external environment of an organism consists in part of many variable conditions, such as temperature, danger of attack, salinity, pH, and humidity. Similarly, the environment of a tissue or of a cell in any large plant or animal has a certain range of conditions. Loss of function or death will occur to any tissue or cell which is kept too long under conditions which violate the limits of any of these ranges.

The great French physiologist Claude Bernard appreciated this concept of *milieu interieur* (internal environment) when, in 1878, he stated a major axiom of physiology: "The constancy of the internal environment is the necessary condition of the free life." This statement was based on his observations of the remarkable stability of internal conditions despite extreme changes in the external conditions under which the animal was kept. In this century, the American physiologist Walter B. Cannon studied the regulation of variables of the internal environment of mammals and coined the word "homeostasis" (Greek: *homoios,* "like," and *stasis,* "standing") for this kind of regulation (see Chap. 3).

The Internal Environment

Fig. 7-1 Venus's flytrap (*Dionaea muscipula*). The highly modified leaves, shaped like bear traps, close if sensitive hairs in the center of the trap are stimulated. Insects which are caught are digested by enzymes secreted by the leaf. (*Carolina Biological Supply Company.*)

The response to changes in the internal environment involves most of the same steps as are involved in the response to changes in the external environment: detection, perception, and the action of some effector organ. Conscious decision making is often not involved; nevertheless, the same structures, for the most part, carry out these functions. The nervous and endocrine systems together bring about the integration of the whole animal; they weld the separate parts into a harmonious whole. Both plants and animals show integration, but as we shall see, there are some basic differences in the integration mechanisms in these two major kingdoms.

AN ORGANISM BOTH LIVES IN AN EXTERNAL ENVIRONMENT AND HAS AN INTERNAL ENVIRONMENT. SURVIVAL DEPENDS ON ITS ABILITY TO RESPOND PROPERLY TO CHANGES IN BOTH OF THESE ENVIRONMENTS. SUCH RESPONSES DEPEND ON THE INTEGRATION OF VARIOUS SYSTEMS, A PROCESS MEDIATED BY (1) RECEPTOR SYSTEMS, WHICH MONITOR CHANGE; (2) PERCEPTUAL SYSTEMS, WHICH INTERPRET AND RELAY THE INFORMATION FROM RECEPTORS; AND (3) EFFECTOR SYSTEMS, WHICH PERFORM THE APPROPRIATE ACTIONS THAT ENHANCE THE CHANCES OF SURVIVAL.

INTEGRATIVE MECHANISMS IN PLANTS

Among large organisms there are two principal strategies for obtaining nourishment. Most animals obey the dictum "seek and ye shall find" by moving about actively in search of food. Plants, on the other hand, have adopted the "everything comes to him who waits" strategy. As long as sunlight is present, plants do not need to move rapidly and precisely, and they can be rooted in one place for their entire lives. Among the few exceptions to the generalization of insignificant movement in plants are the partly carnivorous plants. Because they inhabit soils poor in available nitrogen, they supplement their input with the nitrogen-rich bodies of the insects they manage to trap. But although these plants can move parts of their bodies rapidly, they do not, of course, move from place to place (Fig. 7–1).

The Control of Plant Growth

The relative lack of mobility in plants does not mean that plants are oblivious to environmental change or have failed to evolve integrative mechanisms. However, the coordination of plant responses to environmental changes is, in general, neither as rapid nor as complex as that of animals. As in animals, the integration of plants and their responsiveness to external stimuli are often manifested by movement—either toward or away from something. Plants can move in two ways: either slowly, by growing in a particular direction (displaying a **tropism**), or more quickly, by changing the relative turgor or rigidity of certain cells, as discussed in Chap. 5. In this section, we will consider how it is that plants grow in the "right" way. The growth of plants and the development of some plant tissues are considered in this and the following sections because they exemplify some of the general principles of integration. The outline of plant development is presented in Chap. 10.

PLANTS, UNLIKE MOST ANIMALS, DO NOT HAVE TO MOVE ABOUT IN ORDER TO OBTAIN NOURISHMENT. AS A RESULT OF A LACK OF PRESSURE FOR RAPID MOVEMENT, THEIR INTEGRATIVE SYSTEMS TEND TO BE LESS COMPLEX THAN THOSE OF ANIMALS. PLANTS MOVE IN TWO WAYS: BY GROWING IN A CERTAIN DIRECTION (DISPLAYING A TROPISM) OR BY CHANGING THE TURGOR OF THEIR CELLS DIFFERENTIALLY.

For instance, the leafy parts of a plant always grow toward a light source and the roots always grow into the soil. Let us label these processes and then delve more deeply into them. Growth directed in relation to the pull of gravity is called a **geotropism;** growth in response to the direction of illumination is called a **phototropism.** Growth in the direction of the source of a stimulus is called a *positive* tropism; growth away from the source, a *negative* tropism. There are many other kinds of tropisms in addition to geotropism and phototropism. For instance, besides displaying positive geotropism, roots also grow toward moisture—a positive hydrotropism. Plants also direct their growth response to surfaces; ivy clinging to a wall is an example of this. They also orient along heat and chemical gradients in the soil or air.

AUXIN AND PLANT GROWTH Charles Darwin and his son Francis experimented with the growing shoots of seedlings in an attempt to learn something about the mechanism of phototropism. Removal of the growing tip of a seedling obliterated the response, even though the actual bending of cells occurred in the region below the tip, the region called the zone of elongation. (In this region, the cells produced by the rapidly dividing tip region expand in length to reach their ultimate size.) The Darwins found that the response could also be obliterated by covering the tip with an opaque material even though the rest of the seedling was illuminated (Fig. 7–2). They concluded that the tip perceives the stimulus (light) and that somehow this information about the direction of illumination is passed down to the elongating cells. Presumably, the nature of the message depends on which side of the tip is illuminated, since the cells

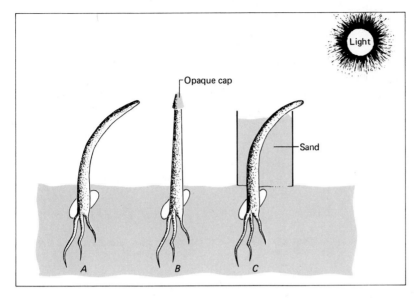

on the illuminated side elongate less than the cells on the relatively dark side, thereby causing a growth (tropic) curvature. The Danish plant physiologist Boysen-Jensen discovered that, whatever this messenger was, it could pass through gelatin but was not able to penetrate a mica sheet. Furthermore, the message was communicated down the dark side of the shoot, suggesting that the substance stimulated growth on the dark side rather than inhibited it on the lighted side.

A favorite plant for such experiments has always been the oat *(Avena)* seedling. As in other grasses, in the oat the first thing to emerge from the soil is a cylindrical protective sheath over the first leaf. This sheath is called a **coleoptile.** The coleoptile exhibits the growth (tropic) properties so far discussed.

Starting around 1930, another plant physiologist, Frits Went, began a series of investigations which led to the discovery of the nature of the message passed from the tip to the cells of the zone of elongation. First, Went found that the message could be trapped by placing the excised coleoptile tip on a small block of agar (a gelatinlike carbohydrate derived from certain seaweeds). Once the message had diffused into the agar, the block could be put on the shoot in place of the coleoptile tip. The agar block would then function like the tip, causing the resumption of elongation in the cells below. But more to the point, if the block was placed on one side of the shoot instead of over all of it, the growth was differential—more on the side under the agar block (Fig. 7–3). Further,

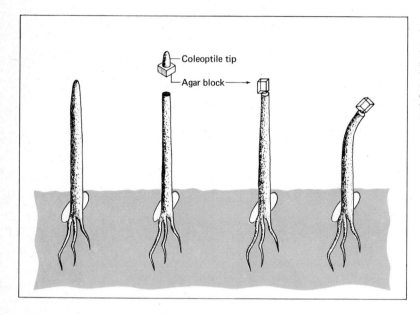

Coleoptile tip

Agar block

Fig. 7-3 Went was able to show that the message causing elongation was a diffusible substance coming from the tip of the coleoptile. After the substance diffused into an agar block, the agar was asymmetrically placed on a decapitated coleoptile. The cells of the coleoptile on the side on which the agar block rests elongated the most.

the curvature of the coleoptile was found to be proportional to the amount of message in the block. Two coleoptile tips placed on an agar block prior to placing the block on a tipless coleoptile would result in more growth than when only one tip was used. Here, then, was a way to assay the amount of message, or alternatively, the messagelike effect of substances from other sources. Such an assay, using the response of living material, is called a **bioassay.** The chemical messenger was named **auxin;** it was later to be identified as indoleacetic acid (Fig. 7-4). The term "auxin" now applies to a group of compounds with functions similar to indoleacetic acid. Among these are the weedkillers 2,4-D and 2,4,5-T, which disrupt normal growth in the concentrations used.

Auxin fits the definition of a **hormone,** that is, a chemical messenger produced in one part of an organism but having its effect elsewhere. In this case, the effect is growth, or elongation. It is now believed that light does not so much affect the production of auxin as lead to its movement to the relatively dark side.

THE DIRECTION OF GROWTH OF PLANT SHOOTS HAS BEEN DEMONSTRATED TO BE UNDER THE CONTROL OF A HORMONE CALLED AUXIN. AUXIN IS PRODUCED IN THE GROWING TIP AND MIGRATES THROUGH THE TIP AWAY FROM A LIGHT SOURCE. IT THEN DIFFUSES TO A ZONE BELOW THE TIP, CAUSING CELLS IN THAT ZONE ON THE SIDE AWAY FROM A LIGHT SOURCE TO ELONGATE. THIS, IN TURN, CAUSES THE TIP TO GROW TOWARD THE LIGHT.

Geotropism can also be explained by the migration of auxin. If a seedling is placed on its side, auxin is soon found to accumulate on the lower side. As expected, the higher concentration of auxin on the lower side of the stem causes greater elongation on the lower side and a corresponding upward curvature—a negative geotropism. It seems reasonable to expect the same response from root tissues, but this is not the case. Auxin, in the same concentration that stimulates stem elongation, apparently *inhibits* growth in elongating root cells, resulting in a downward turning of the root as a whole—a positive geotropism.

Group	Formula (Specific compound)	Function
Auxins	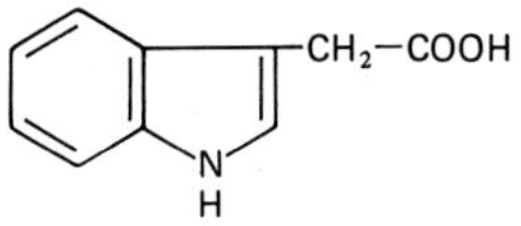 Indoleacetic acid	Cell elongation Apical dominance Phototropism and geotropism Fruit set and development
Gibberellins	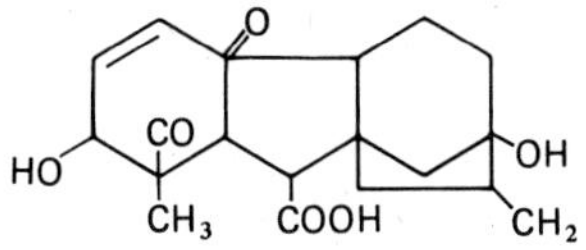 Gibberellic acid	Cell elongation Fruit set
Cytokinins	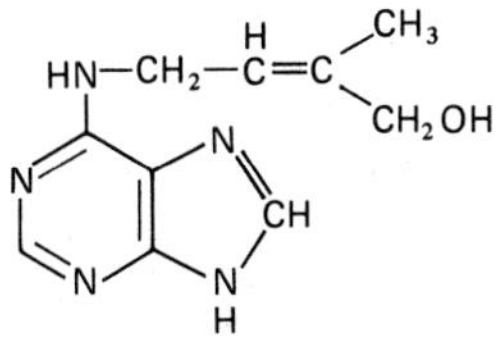 Zeatin	Cell division promotion
Abscisin	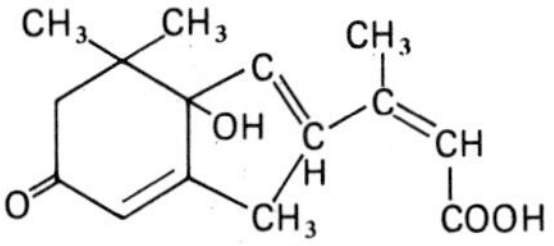 Abscisic acid	Inhibits growth, antagonistic to auxin Induces formation of winter buds Accelerates abscission
Ethylene	$CH_2{=}CH_2$ Ethylene	Fruit ripening

Fig. 7-4 Five important plant hormones and their functions.

This illustrates a most important principle about hormones: the response to a hormone depends on the "target" site, not on the chemical nature of the hormone. It is something like a radio warning of a hurricane—a listener's response depends in large part on his location relative to the seacoast, whether he is part of an organization of lifeguards, etc. Or, perhaps more appropriately, a hormone is a symbol which has different meanings to different receptors. The word "car," for example, has no intrinsic meaning; English-speaking people have endowed it by convention with the meaning "passenger-carrying vehicle," whereas French-speaking people have endowed it with an equally logical meaning, "because" or "why." Auxin, a biochemical symbol, means one thing to stem tissue but quite another to root tissue.

AUXIN ALSO CONTROLS THE NEGATIVE GEOTROPISM OF STEMS AND THE POSITIVE GEOTROPISM OF ROOTS. AS IS OFTEN THE CASE WITH ENDOCRINE MESSENGERS, THE HORMONE "MEANS" DIFFERENT THINGS TO DIFFERENT TISSUES; IT ENCOURAGES ELONGATION IN STEM CELLS AND INHIBITS IT IN ROOT CELLS.

APICAL DOMINANCE   Auxin has many other functions in the growth and development of plants. If one cuts off the terminal shoot of a growing plant, the lateral (side) buds in the axils of the leaves below the tip begin to grow rapidly, forming new branches. From this observation it can be assumed that the intact apex must exert some inhibitory influence over the lateral buds to keep them from growing and developing. This influence is called **apical dominance.** After removal of an apex, the biggest of the lateral shoots eventually becomes dominant and exerts growth inhibition over the others. Apical dominance can be mimicked by the application of auxin to the cut tip—further evidence that auxin coming from the growing apex of the plant acts as a growth inhibitor. The inhibition may be at the level of cell division because DNA synthesis is suppressed in the lateral buds by auxin from the apex. This kind of observation is very important to the understanding of the overall *form* of plants. The difference in shape between a tall and narrow fir and a stocky, looming oak may be explicable, in large part, by the degree of apical dominance exercised by the apical branch.

THE DROPPING OF LEAVES AND FRUIT   Auxin is also important in the dropping of leaves and fruit. Prior to the falling of a leaf or a ripe fruit, an abscission layer (Fig. 7–5) forms at the point on the petiole or fruit stalk where the separation will eventually occur. An abscission layer is a specialized zone of cells with distinctive morphology. Breakage

Fig. 7–5 Longitudinal section of a plant stem showing the point of attachment of a leaf and a bud. The smaller cells in the leaf base at the level of the bud make up the abscission layer. (*Jeroboam, N.Y.*)

at the abscission layer can be prevented, however. If the fruit or leaf is cut above the abscission layer and an auxin paste applied at this point, the remaining part of the stalk will stay affixed indefinitely to the stem. Here again, auxin is "read" as an inhibitory message by a receptor site—the abscission zone. In the normal course of things, leaves and developing fruit synthesize and transfer enough auxin to the abscission zone to inhibit formation of the layer. As the fruit ripens or as the leaf begins to turn, auxin production diminishes and the layer forms. For the convenience of fruit growers, citrus and other crops are often held on the trees by spraying with auxin.

Leaf drop is a fascinating topic in itself, and the "behavioral" and physiological differences between deciduous trees and evergreen trees demonstrates how these two radically different leaf types can both be perfectly successful arboreal strategies. The large, broad leaves of deciduous trees are highly efficient at trapping the maximum available light. Branches and leaves of such trees are spaced so as to have the minimum overlap. As a result of this large photosynthetic surface, deciduous trees are usually fast growers. But this large surface area can be a serious disadvantage in the winter, when the weight of accumulated snow on them could easily snap the branches. Besides, the freezing winter weather slows the metabolism of the tree to a very large extent, making the leaves superfluous. Natural selection has obviously resulted in the deciduous habit for many plants in snowy regions.

Deciduous trees are also characteristic of many tropical areas. In parts of the tropics, seasons of intense precipitation often alternate with very dry, warm seasons. Since large leaves provide a great area for water loss, it is not surprising that during the dry seasons tropical broad-leaved trees drop their leaves.

The narrow, conical shape of coniferous trees and their needlelike leaves are an alternative arboreal strategy. The surface area of each needle is small in comparison with a broad, flat leaf, but there are many of them. Branches are relatively short, and this, plus the shape of the needles, precludes the accumulation of a dangerous amount of snow. Furthermore, water loss is controlled by a thick, waxy cuticle covering the needles and by a reduction in the number of stomata. The distribution of evergreens (a common name for conifers) can be predicted from these anatomical facts. They should survive in places that have heavy accumulations of snow and in places too arid for the survival of broad-leaved trees. Hemlocks, spruces, and firs do, indeed, dominate the forest closest to the poles. Desert mountains and arid plateaus are treeless except for cypresses, junipers, and short-needled pines.

AUXIN REGULATES THE FORM AND HABIT OF PLANTS IN MANY WAYS. APICAL DOMINANCE RESULTS FROM THE INHIBITION OF LATERAL BUDS BY AUXIN PRODUCED AT THE GROWING STEM TIP. DROPPING OF LEAVES IS ALSO CONTROLLED BY AUXIN PRODUCTION, AS IS ABSCISSION OF FRUITS.

OTHER HORMONES AND HORMONAL INTERACTIONS   We have been discussing the hormonal integration of plant growth as though auxin were the only agent involved.   Actually, there are at least three other groups of plant hormones that typically act in consort with auxin: the **gibberellins,** the **cytokinins,** and a group of hormones that includes **abscisic acid** (Fig. 7–4). Like the auxins, the gibberellins are growth or elongation hormones, whereas cytokinins are agents that stimulate cell division. Gibberellins, like auxins, are synthesized in actively growing plant tissues. The two interact in such a way as to make their independent effects difficult to disengage.

Gibberellin application can dramatically change the shape of both wild and domesticated plants. Bush beans, which are a dwarf variety of the vinelike pole beans, take on the form of the latter when sprayed with gibberellin. Many dwarf plant varieties have a low endogenous production of gibberellin, and this probably accounts for their stunted growth form.

The cytokinins are a very interesting group of hormones related in chemical structure to the purine adenine. They have been found in actively growing tissues of a wide variety of plants, where they apparently stimulate cell division. Cytokinins interact with other hormones in the control of plant growth. For example, they are released during seed germination after gibberellin has led to the production of hydrolytic enzymes. Experiments with plants in tissue culture have shown that cytokinins stimulate mitosis, but only in the presence of auxin. By itself, of course, auxin results in the stimulation of cell elongation.

Auxin or auxinlike substances have a variety of other effects. An unpollinated flower soon drops from a plant as a consequence of an abscission layer that forms at its base. Why doesn't a pollinated flower absciss? As early as 1909, a substance extracted from orchid pollen was found that would initiate the formation of a fruit from the ovary. Later, it was shown that this substance was auxin and that auxin alone applied to a stigma can effect the development of a fruit and prevent its premature abscission. Such a fruit is, of course, unpollinated and has no fertile seeds. Tomatoes are now produced commercially in this way. The pollen of some plants contains gibberellins that apparently act in a similar way to achieve "fruit set." The trees producing stone fruits—peaches, apricots, plums, and cherries—as well as grapes, figs, and apples, will set fruit in the absence of pollination when sprayed with a solution of gibberellin. Developing seeds emit auxin to the surrounding tissues, thereby initiating and controlling the development of the fruit as a whole. Artificial application of auxin to flowers has become an important commercial process ensuring fruit set of all the flowers and synchrony of fruit development. Dormancy of tubers (such as Irish potatoes) can be artificially prolonged by auxin application.

In the past few years, hormones with the effect of promoting abscission were discovered. One of these, abscisic acid, or abscisin, accelerates abscission when it is applied to a stem or fruit. The highest concentrations of abscisic acid occur in fruits at the time of fruit drop. It is apparently the balance of auxin and abscisic acid that controls the formation of an abscission layer. Plant hormones that used to be called dormins and that are the agents responsible for the inhibition of growth in dormant buds were recently found to be identical to abscisic acid.

It has been known for some time that the simple molecule of the gas ethylene ($CH_2 = CH_2$) has an effect on plants. Indeed, the effect is used commercially. Many fruits can be picked green and stored under refrigeration. Shortly before they are to be marketed, they are treated with ethylene, which causes the fruits to soften and color as their starches are changed to sugars.

Ethylene is produced at higher levels in fruits than in other tissues, and its concentration increases prior to natural ripening. It undoubtedly interacts with other plant-growth regulators in controlling the formation of fruits. One of the remarkable things about ethylene is that it is biologically active on plants at concentrations of as low as 0.01 part per million (ppm) in air. Ethylene has effects on the response to wounding in plants, and there is a growing list of effects of this hormonelike substance.

THE GROWTH OF PLANTS IS INTEGRATED BY THE COORDINATED ACTION OF MANY HORMONES WITH DIFFERING EFFECTS. AUXIN AFFECTS CELL ELONGATION, CELL DIVISION, AND ABSCISSION. GIBBERELLINS AFFECT CELL ELONGATION AND MAY PRODUCE DWARFING. CYTOKININS STIMULATE CELL DIVISION. ABSCISIC ACID PROMOTES ABSCISSION AND INHIBITS GROWTH IN DORMANT BUDS. ETHYLENE HAS IMPORTANT EFFECTS ON THE RIPENING OF FRUIT, AS WELL AS OTHER EFFECTS. OFTEN IT IS DIFFICULT TO DETERMINE WHICH HORMONE IS RESPONSIBLE FOR A PARTICULAR EFFECT SINCE THEY OPERATE TOGETHER AND OFTEN PRODUCE SIMILAR CHANGES.

Have you ever noticed how poinsettias, a Christmas plant, will bloom only in the winter, whereas snapdragons will bloom whenever it is warm enough and they have adequate moisture? If you live in a desert area, have you been impressed by the sudden synchronous appearance of myriad flowers and their equally sudden disappearance? If you live in the Rocky Mountains, have you ever wondered why the monument plants almost all bloom in unison and only every other year? There are many questions to be answered about the patterns of flowering found in plants. **Flowering**

TEMPERATURE CHANGES For instance, what environmental changes (stimuli) initiate the flowering? How does the plant detect these changes and mobilize itself for the effort of flowering? In some plants it is known that temperature changes are crucial. Flowering in the early spring by plants such as lilacs and cherry trees is, of course, brought on by warmth, but the precondition for flowering is a protracted period of cold—freezing or near-freezing temperatures. The selective value of this (that is, the reason that this has been brought about by natural selection) is obvious. Often a warm spell interrupts the typically cold winter or early spring weather. If a plant began to flower in response to such unseasonal warm weather, a late frost might well destroy the buds and thereby probably kill the tree. Those individuals that, because of their particular genetic makeup, were somehow inhibited from flowering until a sufficiently long period of cold weather had passed would survive and reproduce; those that had no such inhibition would not. A carrot or a monument plant, for instance, develops only vegetative (nonreproductive) parts during its first year of growth. At the end of the first growing season, all the leaves and stems die back, leaving only the underground root, which is specialized for storage of food synthesized during the first growing season. During the second growth season, reproductive parts (flowers) are formed on the new shoots of the biennial (2-year) plants. Much of the energy for the development of the reproductive organs comes from the stored food. Flowers will not develop, however, unless the root has been exposed to cold. The whole root does not have to receive this treatment in the case of carrots—only the apical bud, which becomes the growing tip of the new shoots. As yet, the mechanism for this temperature effect has not been discovered. Perhaps the exposure to cold removes a flowering inhibitor. Alternatively, the cold may induce the formation of a flowering hormone. This seems likely since flowering can be induced in a plant that has not been exposed to cold by grafting it to one that has received the treatment. In some plants, gibberellins have the same effect, although gibberellin itself is probably not the flowering hormone but is, instead, involved in its synthesis.

PHOTOPERIODISM   Another environmental stimulus crucial to flowering in many plants is light—more precisely, the length of the day, or the **photoperiod.** The response of plants, called **photoperiodism,** was first discovered by W. W. Garner and H. H. Allard, plant physiologists with the U.S. Department of Agriculture. They found that one variety of tobacco they studied flowered only in December, unlike other varieties, which flowered in the summer. (Note that it is often the observation of an exception to a general rule that draws attention to a previously unnoticed group of phenomena.) Furthermore, the typical summer-flowering varieties of tobacco were found to flower *only* when days were long—about 14 hr or more. Originally, summer forms were called "long-day" plants and the winter-flowering types were called "short-day" plants. But this terminology was soon made obsolete by observations such as those discussed below. From this point on, we will refer to these two types as summer (long-day) and winter (short-day) plants.

If the light period (day) of a winter plant, grown in a light regime of short days and long nights, is interrupted by a short period of darkness, flowering still occurs (Fig. 7–6). But if the dark period (night) of such a plant is interrupted by exposure to light (as little as 1 min) in the middle of the dark period, flowering will be inhibited just as if the plant had been exposed to light for the entire dark period preceding the exposure. From this evidence, it would appear that some crucial process occurs during the dark period in winter plants and that if this process is interrupted by light, flowering is forestalled.

A long exposure to darkness has the opposite effect on summer plants. They produce flowers when the dark period is less than 14 hr or so. A dark period longer than this inhibits flowering. As in winter plants, a short exposure to light reverses the effects of a long dark period. Experimentally, this is demonstrated by growing the summer plant in a winter light regime. A brief exposure to light in the middle of the inhibitory dark period abolishes the inhibition, and flowers begin to develop.

In both kinds of plants, the length of uninterrupted darkness is critical. A mechanism must exist whereby winter and summer plants measure the length of the dark period; that is, they must have biological clocks. The plants are able to measure time with astonishing accuracy. A summer plant called henbane will flower when exposed to light for 10 hr and 20 min or more at 22.5°C. A light period of 10 hr or less inhibits flowering. This ability to measure time is a standard feature of biological clocks, which are common, perhaps universal, among plants and animals. The cellular mechanism by which such clocks operate is one of the least understood phenomena in biology. (See Box 7–1, **Biological Clocks.**)

Now, what can be inferred about the photoperiodic mechanism from these

Fig. 7–6 Flowering stimuli in summer and winter plants: (*A*) winter (long-night) plants are induced to flower by a long dark period; (*B*) a period of darkness during the light period has no effect; (*C*) exposure to light during the dark period inhibits flowering; (*D*) a short dark period inhibits flowering; (*E*) summer plants are induced to flower by a short dark period; (*F*) a long dark period inhibits flowering; (*G*) exposure to light in the middle of a long dark period induces flowering.

experiments? We know that for both winter and summer plants something crucial happens during the dark period and that a momentary exposure to illumination reverses this process. We will call this the "dark process," and applying the principle of parsimony, we will assume that it is the same process in both kinds of plants. In science it is a common rule of thumb to assume simplicity, that is, to test a simple hypothesis first. How can a brief exposure to light affect the dark process? To begin with, light reception in organisms always depends on pigments. Upon reception of sufficient light, the pigment undergoes some change in configuration or structure (remember that chlorophyll loses electrons). Depending on the pigment and the system of which it is a part, this transformation may initiate a train of reactions leading eventually to some other event. Let us assume that there is a pigment B which is virtually instantly changed in structure by light to pigment A:

$$B \xrightarrow{\text{Light}} A$$

and that A spontaneously reverts to B in the dark:

$$B \xleftarrow{\text{Darkness}} A$$

The verification of this scheme came with the discovery of the pigment **phytochrome** by S. B. Hendricks and H. A. Borthwick of the U. S. Department of Agriculture. For some time it had been known that a flash of orange-red light of wavelength 660 nanometers (nm) during the dark period was the most effective in either inducing flowering in summer plants or inhibiting flowering in winter plants. Further, it was discovered that a flash of far-red light of wavelength 735 nm, just beyond the limits of human vision, would reverse the effect of the 660-nm light. In fact, a plant could be illuminated alternately with 660-nm and 735-nm light any number of times, and the last treatment, regardless of which wavelength was used, was the effective one. Phytochrome exists in two forms, one more stable than the other. In reality, phytochrome was being switched from one form to the other by the light:

$$P_{660} \text{ (B)} \underset{\text{735 nm or darkness}}{\overset{\text{660 nm}}{\rightleftharpoons}} P_{735} \text{ (A)}$$

Sunlight is far richer in 660-nm light than in 735-nm light, so that during the day all the phytochrome is converted to P_{735}, that is, to the form that will absorb most strongly light of 735 nm. P_{735} is, of course, the unstable form, and it gradually reverts to P_{660} in the dark. Apparently, this spontaneous conversion is part of the clock mechanism that times the dark period. The "decision" to flower or not to flower on the part of the plant may depend, then, on how far this clock has run down (Fig. 7-7). Other reactions must be involved as well, because the spontaneous conversion

Fig. 7-7 A possible model accounting for the photoperiodic mechanism in plants. Compare with Fig. 7-6. Flower induction occurs when the amount of some compound at the end of the day is above the arrow. In winter plants light inhibits this synthesis or destroys this compound. In summer plants light induces the synthesis of this compound.

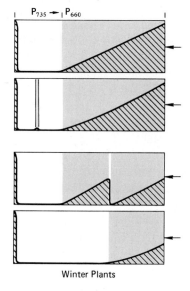

Winter Plants

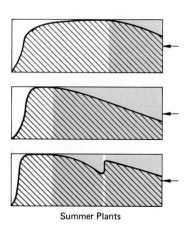

Summer Plants

of P_{735} to P_{660} takes only about 3 hr. The steps from the reception of light by phytochrome to the actual flowering are still being elucidated.

At least one chemical messenger, or hormone, is involved in these steps. The exposure of only a single leaf of a plant to the proper flower-inducing regime will lead to the development of flower primordia (the rudimentary beginnings of flowers) rather than leaf primordia in the apical and lateral buds. The hormone, called florigen, has been isolated and has been found to be transported by the phloem. In fact, its role has been substantiated by grafting experiments, as shown in Fig. 7–8.

FLOWERING IN THE ANGIOSPERMS IS CONTROLLED BY A VARIETY OF FACTORS, SUCH AS HORMONES AND PHOTOPERIOD. PHOTO-PERIOD OPERATES THROUGH A LIGHT-ABSORBING PIGMENT CALLED PHYTOCHROME, WHICH CAN EXIST IN EITHER OF TWO FORMS. PHYTOCHROME IS SWITCHED FROM ONE FORM TO ANOTHER BY SPECIFIC WAVELENGTHS OF LIGHT IN THE RED AND FAR-RED BANDS OF THE SPECTRUM. THIS CONVERSION OF PHYTOCHROME IS PART OF AN ENDOGENOUS CLOCK MECHANISM WHICH ENABLES A PLANT TO MEASURE THE LENGTH OF THE DARK PERIOD.

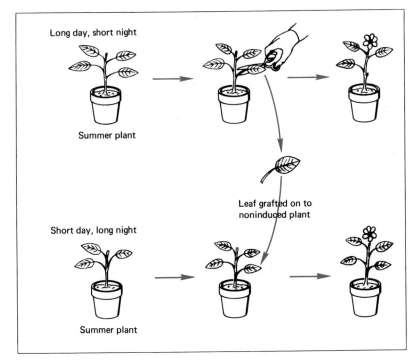

Fig. 7–8 A grafting experiment showing that a leaf from an induced plant will stimulate a noninduced plant to flower. Such a graft will also induce a winter plant (on a summer light regime) to flower.

Everyone who has turned over a board in a vacant lot or field has found that the plants growing under the board appeared rather emaciated—yellowish-white instead of green and with very long stems and little leaf development. Actually, it would be misleading to consider such a plant to be emaciated; it is merely showing an adaptive response to the absence of light. This response is called **etiolation.** The problem facing a developing seedling which happened to germinate in a dark place is to reach its source of energy before it uses up all its food reserves. The chances of reaching light are increased by unusual elongation; the retardation of leaf growth undoubtedly streamlines the lengthening stem as well as saving food. Light is required for the final step in chlorophyll synthesis, and this accounts for the pale color of the tissues. Etiolation is reversed most effectively by treatment with light of the 660-nm wavelength. Predictably, far-red light can reverse the effect of such a treatment. Phytochrome, then, is very probably the receptor for the etiolation response. The 660-nm light has three effects: (1) It is required for the last step of chlorophyll synthesis—the conversion of pale-yellow protochlorophyll to green chlorophyll. (2) It inhibits the excessive elongation of the stems. (3) It promotes the growth and development of the leaves. Light may act on stem elongation by affecting the synthesis or transport of gibberellin, but this has not been definitely determined.

Here, then, is a simple yet elegant example of how environmental change —in this case, of light intensity—can drastically alter the development and "behavior" of an organism. But the point to be emphasized is that these changes in molecules, in developmental processes, and in the shape and size of the individual are finely coordinated, so that the outcome is integrated and thoroughly functional. Here, too, our ignorance of exactly how and where the phytochrome system acts deserves emphasis.

PHYTOCHROME IS IMPORTANT IN OTHER ASPECTS OF PLANT GROWTH BESIDES FLOWERING. IN WAYS NOT COMPLETELY UNDERSTOOD, THE DEGREE OF ELONGATION OF THE STEM OF DEVELOPING SEEDLINGS, THE GROWTH OF LEAVES, AND THE DEVELOPMENT OF FUNCTIONAL CHLOROPHYLL ARE AFFECTED BY THE ABSORPTION OF ORANGE-RED LIGHT.

Certain aspects of germination (the resumption of growth by a spore or seed) will be considered in Chap. 10. A period of dormancy usually follows maturation of a seed. If this were not so, seedlings would be found at the end of the growing season when the adult plants were shedding their seeds. Why and how do you suppose this kind of dormancy evolved? Clearly, unsatisfactory conditions prevail for growth of many plants in the late fall and early winter, and the tender seedlings would not survive the severe weather. In spring, one of the environmental factors that breaks dormancy is light—specifically, 660-nm light. Certain strains of lettuce will not germinate unless they are illuminated by light of this wavelength. This means that unless these tiny seeds are at or very near the surface of the soil, they will not germinate. Light over 700 nm inhibits germination and reverses the effect of 660-nm light, thus implicating the phytochrome system. The seeds of many plants show the same responses, while those of others (the California poppy, for instance) show the opposite effect: 700 nm or more of light induces germination, whereas 660-nm light is inhibitory, thus ensuring that only buried seeds will germinate.

It may have occurred to you that the 735-nm reversal response of the phytochrome system is more of a laboratory artifact or curiosity than a response called forth under "natural" conditions. Indeed, the discovery that 735-nm light greatly accelerates the spontaneous reversal of P_{735} to the stable P_{660} form has helped to unravel the phytochrome system, but this response may not be of significance to plants under normal conditions. After all, you might ask, where in nature does far-red light occur in the absence of red light? In other words, under what natural conditions would far-red light inhibit germination?

It has been observed that some seeds fail to germinate when in the shade of other plants. The significant point here is that light that has passed through a curtain of leaves has had most of the red light filtered out, leaving nearly all the wavelengths over 700 nm at their original intensity. Therefore, the effect of even an occasional beam of unfiltered sunlight striking the seed will be constantly reversed by the far-red "shade" light. This prevents germination in a place where the seedling could receive little light that was favorable for photosynthesis.

DORMANCY IN MANY SEEDS CONTINUES UNTIL THEY ARE EXPOSED TO ORANGE-RED LIGHT. EXPOSURE TO FAR-RED LIGHT INHIBITS GERMINATION. OTHER PLANTS SHOW THE OPPOSITE RESPONSES. THESE RESPONSES SUGGEST THAT THE PHYTOCHROME SYSTEM IS INVOLVED. OFTEN THEY ARE CORRELATED WITH LIGHT CONDITIONS PREVAILING IN NATURE.

INTEGRATIVE MECHANISMS IN ANIMALS
Hormonal Integration

The earliest demonstration of hormonal integration in mammals concerned an aspect of digestion: the train of events set off by the act of swallowing food. By the time food has been transformed from digestible input to absorbable small molecules and indigestible output, a surprisingly large number of physiological events have occurred, and all in proper sequence. Movement of the food at the proper rate is accomplished by rhythmic contractions of muscles in the gut (peristalsis). Other muscles (sphincters) control openings that separate regions of the gut and permit controlled delivery. Gland cells in the intestinal lining, the mucosa, secrete their products at just the right time; the liver and pancreas secrete their digestive juices through ducts into the gut upon cue. All this and more proceeds with assembly-line efficiency. It was the discovery in 1902 of the nature of the cue received by the pancreas which began the gradual elucidation of many of the integrative mechanisms in vertebrates.

It was found that the presence of acid in the first section of the small intestine, the duodenum, stimulates the release of a chemical messenger into the blood (you will recall that stomach contents are made acid by secretion of hydrochloric acid by the cells in the stomach lining). Upon reaching the pancreas, the messenger, now referred to as the hormone **secretin,** elicits the flow of pancreatic juice through the pancreatic duct to the duodenum (Fig. 7-9). Pavlov, the famous Russian physiologist,

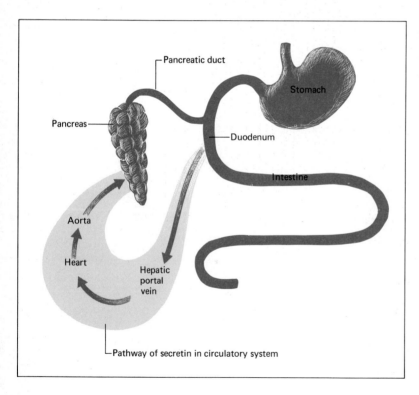

Fig. 7-9 Secretin, one of the six gastrointestinal hormones, is secreted by cells in the duodenum into the circulatory system. Upon reaching the pancreas, secretin stimulates the flow of pancreatic juice.

and his coworkers did not accept the chemical-messenger theory and quite a debate ensued. One of the critical experiments which more or less resolved the matter was the joining of the circulatory systems of two dogs and the introduction of acid into the duodenum of one of them; the flow of pancreatic juice occurred shortly in both dogs. There are a number of other gastrointestinal hormones. One of these, **gastrin,** is secreted into the blood by the gastric mucosa and induces glandular secretion in other parts of the stomach.

A **hormone** may now be more strictly defined as a chemical of a specific structure produced in a restricted area and having its effect at a distance. This definition excludes the group of messengers produced *generally* throughout the body, such as CO_2 and histamine, all of which have their effects at a distance. These are sometimes referred to as parahormones. This definition also excludes intracellular messengers such as mRNA. Even excluding these substances, there are still some intermediate cases that resist exact classification. **Thyroxin,** for instance, a major regulator of metabolic rate secreted by the thyroid gland, is thought to be synthesized in other sites as well since it is found in thyroidless animals. Adrenalin and noradrenalin are formed at nerve endings throughout the body. Other so-called hormones are elaborated in more than one place. Is there any reason to expect chemical messengers in organisms to fall into classes with sharply defined boundaries?

As stated earlier, an organism in its lifetime will usually experience three classes of change: long-term developmental changes, such as growth, puberty, flowering, and senescence; shorter-term cyclical changes, such as seasonal breeding cycles and daily activity rhythms; and relatively unpredictable changes, such as disease, trauma, digestion, and vacillations in the constancy of the internal environment. The great majority of changes in all three classes are either directly attributable to or significantly affected by integrative hormonal systems. This three-way classification of change occurring within an organism makes a convenient scaffolding on which to construct a picture of hormonal functions.

A HORMONE MAY BE STRICTLY DEFINED AS A CHEMICAL WITH A SPECIFIC STRUCTURE PRODUCED IN A RESTRICTED AREA AND HAVING ITS EFFECT AT A DISTANCE. THUS SECRETIN IS A HORMONE AND CARBON DIOXIDE IS NOT. SOME SUBSTANCES, SUCH AS THYROXIN, ARE ELABORATED AT MORE THAN ONE SITE AND DEFY EXACT CLASSIFICATION. HORMONAL SYSTEMS HELP CONTROL LONG-TERM AND SHORT-TERM CYCLIC CHANGES IN ANIMALS AS WELL AS SHORT-TERM RESPONSES TO UNPREDICTABLE ENVIRONMENTAL CHANGES.

HUMAN-GROWTH HORMONE   Human-growth hormone (somatotropin, STH) is elaborated in the anterior lobe of the pituitary gland (Fig. 7–10). It is a protein with a molecular weight of about 27,000. Bone and muscle seem to be the tissues most responsive to this hormone. The control of STH is poorly understood. It is normally active during childhood and adolescence. An undersecretion results in dwarfism, and an excess produces gigantism.

It has a stimulating effect on all epithelial cells, possibly leading to a nonspecific enhancement of endocrine secretion generally. The development of the mammary glands is dependent on the presence of STH at all stages from childhood through lactation.

HORMONAL CONTROL OF METAMORPHOSIS IN INSECTS   No one would mistake a caterpillar for a butterfly. Caterpillars generally move by crawling, feed on leaves, increase in size many times over, and do not reproduce. Interposed between these two very different life forms is a transition form called a **pupa.** Superficially quiescent, the pupa is actually undergoing a very active and complete physiological and morphological overhaul, involving the nearly complete disintegration of the larval body and the development of the adult body. This process of disintegrative and reintegrative differentiation from one life stage to another is called **metamorphosis.** It is characteristic of many invertebrates and most orders of insects.

In insects, metamorphosis is controlled by a hormone balance. One hormone, **ecdysone,** is produced by a gland in the first segment of the larva behind the head. Ecdysone is responsible for molting, the formation of the pupa, and the initiation of adult development. It is balanced by a **juvenile hormone,** which prevents premature development of the adult. By appropriate manipulations—transferring glands or connecting living insects so that their blood intermingles—it is possible to produce more juvenile molts than normal or to cause young larvae to metamorphose into tiny adults.

Vertebrates, too, may undergo metamorphosis under the control of hormones. Many amphibians, some bony fishes, and the more primitive vertebrates, such as lampreys and hagfish, all undergo a metamorphosis.

HORMONAL CONTROL OF AMPHIBIAN METAMORPHOSIS   In amphibians, hormones from the thyroid gland trigger the metamorphic changes. (These hormones do not act in the same way in those fish and lower vertebrates that undergo metamorphosis.) The thyroid hormones

Hormones Affecting Long-term Integrated Change

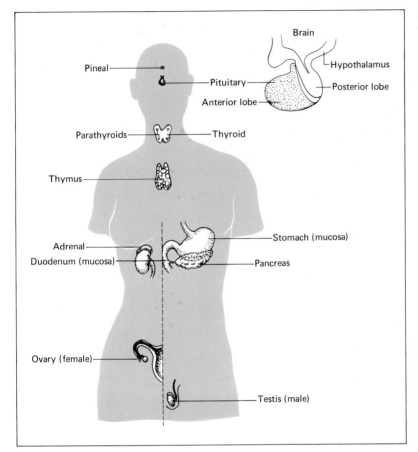

Fig. 7-10 The principal endocrine organs in man.

are amino acids containing either three (tri-iodothyronine) or four (thyroxine) atoms of iodine. Iodine alone is effective in bringing about some of the changes induced by the thyroid hormones. If the thyroid gland of a tadpole is surgically removed, the tadpole continues to grow but never metamorphoses into an adult. On the other hand, if thyroxin is administered to a young tadpole, it will soon thereafter metamorphose prematurely into a midget frog. Tadpoles are amazingly sensitive to thyroxin; concentrations of about 1 part per billion, or 10^{-9} molar, are sufficient to induce metamorphosis in leopard frog tadpoles.

Some populations of salamanders, such as those of the tiger salamander, reproduce as "larvae"; that is, most of the nonreproductive (somatic) tissues never undergo metamorphosis. The Mexican axolotl *(Ambystoma mexicanum,* Fig. 7-11) has an indefinitely prolonged larval life, as

Fig. 7-11 The Mexican axolotl has an indefinitely prolonged larval life. (*Jacana.*)

have the mud puppy *Necturus* and many other kinds of salamanders. It was once thought that this was attributable to a chronic lack of iodine, a constituent of thyroxin (such as causes simple goiter; see below), but it is now known that many "permanently" larval forms are relatively unresponsive to the induction of metamorphosis by thyroid hormone.

GROWTH IS UNDER HORMONAL REGULATION IN ANIMALS AS DIVERSE AS MAN, BUTTERFLIES, AND FROGS. HORMONES ARE INVOLVED WHEN GROWTH IS A GRADUAL TRANSFORMATION, AS IN MAN, OR WHEN THERE ARE ABRUPT TRANSITIONS BETWEEN LIFE STAGES AS IN BUTTERFLIES AND FROGS (WHICH ARE SAID TO UNDERGO A METAMORPHOSIS).

HORMONAL CONTROL OF HUMAN REPRODUCTION A woman's reproductive cells (eggs, or ova) are present in immature form in her ovaries from birth. From puberty, around the age of 13, until menopause (the cessation of the periodical bleeding called menstruation), somewhere around the age of 45 or 50, one egg cell (ovum) is brought to maturity approximately every 28 days. A single ovum is only about 0.10 mm in diameter, so small that it would be visible to the naked eye only as a tiny speck. Yet it contains, besides the DNA representing the mother's

**Hormones Affecting
Reproductive Activities in
Vertebrates**

contribution to the genetic information of a potential child, material for the nourishing of the early embryo should the ovum be fertilized.

Besides being the source of ova, the ovaries are also **endocrine** glands, that is, glands that secrete their products into the circulatory system. They secrete the female hormones, which regulate the female's menstrual cycle and the successful completion of pregnancy. One of these hormones is **estrogen** ("female producer"). During puberty, estrogen stimulates the growth and development of the reproductive organs and the expression of secondary sexual characteristics. These include the widening of the pelvis, the development of mammary glands, the characteristic distribution of body hair and body fat, and the character of the voice. The softness of skin in human females and the more rounded body contours result from the deposition of fat and are a response to a relatively high level of estrogen production from puberty to menopause.

Estrogens are produced principally by the ovarian follicles, groups of cells surrounding the developing egg, and by the corpora lutea ("yellow bodies"). The **corpora lutea** (sing. **corpus luteum**) are secretory structures which develop from the follicle cells after the ovum has been released (Fig. 7–12). The release of the ovum is known as **ovulation.** After ovulation, the corpora lutea begin to secrete the pregnancy hormone, **progesterone,** together with estrogen. Progesterone functions to prepare the uterus (womb) for implantation of the embryo by building up a lining of cells and extra blood vessels. Progesterone also reduces the sensitivity of the uterus to stimuli, such as the hormone oxytocin, which causes contractions of the uterine muscles and could lead to spontaneous abortion. During pregnancy, part of the placenta takes over the production of progesterone from the corpus luteum.

The process of ovulation and the secretion of the sex hormones estrogen and progesterone by the ovaries are themselves controlled by hormones. The controlling hormones, called gonadotropic hormones, are produced by the anterior lobe of the pituitary gland at the base of the brain. The two principal gonadotropic hormones are follicle-stimulating hormone (FSH) and luteinizing hormone (LH). These are involved in a complex feedback system with the sex hormones. For instance, progesterone inhibits the release of the pituitary hormone FSH, which stimulates ovulation. Thus, if pregnancy occurs, wastage of eggs is prevented. If conception does not occur, the corpus luteum breaks down and secretion of progesterone terminates. Then FSH again flows, and the cycle is repeated (Fig. 7–13).

The cycle of ovulation, corpus luteum formation, preparation of uterus for implantation, in the uterine wall, of the embryo, and corpus luteum re-

Fig. 7-12 Sex organs of the human female: (A) on the left are the external genitalia; on the right is a longitudinal section showing the anatomy; (B) on the left is a semidiagrammatic frontal view of the principal reproductive organs; on the right is an enlargement of the ovary and oviduct; (C) diagram of the stages in the development of an ovarian follicle, ovulation, and the development and regression of the corpus luteum.

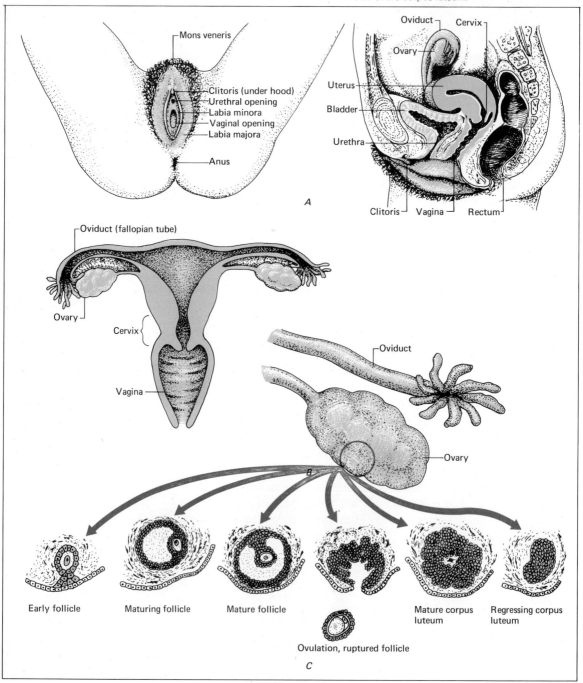

gression is known in man as the **menstrual cycle.** The cycle is normally of about 28 days' duration, and its most conspicuous event is the passage of blood through the vagina as part of the sloughing off of the lining of the uterus, which was prepared for implantation. This bleeding (actually, passage of tissue and blood), occurring roughly once a month, is known as **mensus** (Greek: "month").

THE SECONDARY SEXUAL CHARACTERISTICS AND REPRODUCTIVE CYCLE OF HUMAN FEMALES ARE CONTROLLED BY A COMPLEX HORMONAL SYSTEM BASED ON THE OVARIES AND THE PITUITARY

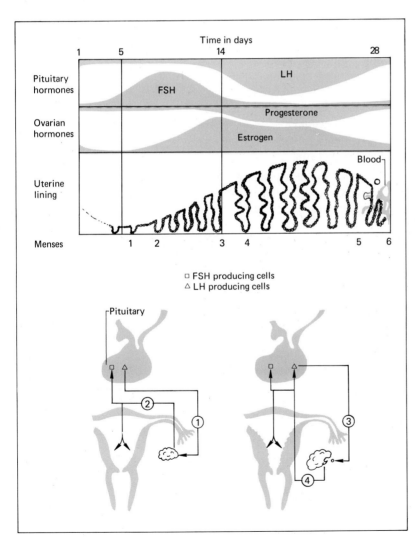

Fig. 7–13 Hormonal control of the human menstrual cycle: (1) FSH secreted by the anterior pituitary stimulates the growth and maturation of the follicle and its egg; (2) estrogen produced by the maturing follicle (a) inhibits FSH release, (b) stimulates the lining of the uterus to thicken, and (c) stimulates FSH realease; (3) LH causes release of egg from follicle (ovulation) and induces transformation of follicle into corpus luteum; (4) the corpus luteum begins secreting more progesterone and less estrogen; the progesterone conditions the uterine lining to receive a fertilized egg and prevents the maturation of another follicle by inhibiting secretion of FSH and LH; lower levels of LH cause corpus luteum to regress or atrophy and to secrete less progesterone; (5) resorption and sloughing of uterine lining (menstruation) with low progesterone levels; (6) FSH inhibition is released because of low levels of ovarian hormones—return to (1).

GLAND. THE MONTHLY MENSTRUAL FLOW IS THE MOST OBVIOUS OUTWARD SIGN OF THE CYCLICAL FUNCTIONING OF THIS SYSTEM.

In males, the gonadotropic hormones control the release of the male sex hormones from the male gonads, or **testes** (Fig. 7–14). The male sex hormones are known as androgens ("male producers"). In man and other mammals, the predominant androgen is **testosterone.** It controls the growth and development of the sex organs, especially at puberty and after. It promotes the typical male secondary sex characteristics: narrow hips, hairiness, and deepening of the voice. It also seems to promote growth in general. Testosterone is produced by special tissues in the testes, and it promotes the development of sperm cells by diffusing into the sperm-producing tissue.

The removal of the testes (castration) before puberty forestalls normal male expression of the secondary sex characteristics, including sex drive and low-pitched voice. In the Near East, castrated individuals, known as eunuchs, were employed as harem guards. And, until this century, *castrati* were used in church choirs in Italy because of their exceptionally beautiful soprano voices. Livestock are often castrated for ease of handling and because they have superior meat. Steers, geldings, and capons are castrated bulls, stallions, and roosters, respectively.

The above material is only an abbreviated outline of the complex hormonal interactions involved in human reproduction. Many different hormones are involved, and they affect each other's production and activity in ways which are only partially understood. Some of these hormones are secreted by tissues not normally thought of as endocrine glands. For instance, during pregnancy the fetal and maternal tissues that form the placenta become important, though temporary, producers of hormones that help to maintain the pregnant condition of the mother. In the absence of these hormones, the corpus luteum is not stimulated to produce progesterone and the menstrual cycle continues. Interestingly, a drop in progesterone level stops the inhibition of the hormone oxytocin (produced by the posterior part of the pituitary gland). Oxytocin may then induce uterine contractions and produce the "cramps" which so often accompany the onset of menstruation. It is still a mystery how birth contractions (labor) are induced.

IN HUMAN MALES THE DEVELOPMENT OF THE SEX ORGANS AND SECONDARY SEXUAL CHARACTERISTICS IS UNDER THE CONTROL OF HORMONES RELEASED BY THE TESTES. THERE IS NO EQUIVALENT OF THE CYCLICAL FEMALE SYSTEM IN THE MALE. THE FEMALE HORMONAL SYSTEM IS SO COMPLEX THAT IT IS ONLY PARTIALLY UNDERSTOOD.

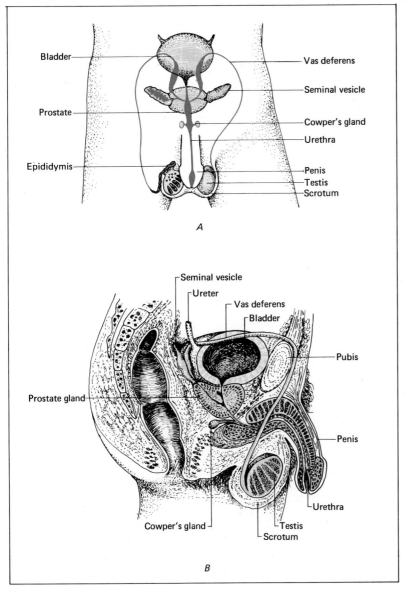

Bladder

Prostate

Epididymis

Vas deferens

Seminal vesicle

Cowper's gland

Urethra

Penis
Testis
Scrotum

A

Seminal vesicle
Ureter
Vas deferens
Bladder

Pubis

Prostate gland

Penis

Urethra

Cowper's gland
Testis
Scrotum

B

Fig. 7–14 Sex organs of the human male: (*A*) a frontal view showing the prostate gland, right testis, and right seminal vesicle in section; (*B*) a longitudinal section through the pelvic region. (*Adapted from BIOLOGICAL SCIENCES, 1/e, by William T. Keeton, illustrated by Paula DiSanto Bensadoun, by permission of W. W. Norton & Company, Inc. Copyright © 1967 by W. W. Norton & Company, Inc.*)

VERTEBRATE REPRODUCTIVE CYCLES   We have discussed the actions of the reproductively important hormones of man, and now we can extend this information into a picture of the reproductive process as a whole. Reproduction, unlike feeding, sleeping, and breathing, is not a continuous behavioral or physiological activity for most organisms. There are some partial exceptions, including man, but plants and animals for the most part restrict their reproductive activities to certain seasons, months, or even days and hours of the year. Usually, the reproductive periods are in phase with some environmental periodicity; for instance, they may occur during the spring in organisms outside the tropics. In the tropics, reproduction often commences with the beginning of the rainy season.

Clearly then, the explanation for reproductive periodicity is partly founded in ecological exigencies. To look at a trivial example, the reproduction of deer is timed so that young are born in the spring, coinciding with the emergence of a plentiful food supply for the doe and later for the fawn when it is weaned. By the ensuing winter, the fawn has become sufficiently large and self-sufficient to be able to negotiate deep snows and to maintain itself on meager food supplies. A deer born in winter would wallow helplessly in the snows, and the mother would have difficulty finding enough food. Thus deer that reproduced at an inappropriate season would leave few offspring; natural selection will automatically favor those animals that breed during the most propitious season. This gives us at least a partial answer to the question: *Why* reproductive periodicity? We shall now investigate the question: *How* reproductive periodicity? This topic is discussed here rather than in the chapter on reproduction because the timing and integrative mechanisms are largely hormonal.

Just as the breeding behavior is keyed to the most propitious season, so the production of gametes is keyed to the breeding. Under the stimulus of a hormone from the anterior pituitary (follicle-stimulation hormone), oogenesis (the production of ova) and spermatogenesis (the production of sperm) occur in waves or cycles, so that gametes are present at the time of breeding. These cycles of reproductive activity are given the general name **estrous cycles.** As the follicle, the ball of endocrine cells enclosing the ovum, matures in the ovary of a mammal, its cells secrete more and more estrogen. This, in turn, inhibits the release of the stimulating hormone from the anterior pituitary and the development of a new wave of follicles. In many mammals, the level of estrogen in the blood reaches a peak just before ovulation. At this time the female is receptive to copulation. Another pituitary hormone (luteinizing hormone) is the actual stimulus for ovulation. In many mammals, including man, ovulation is spontaneous, but in some, such as the mink, cat, and rabbit, the release of this hormone, and thus ovulation, must be induced by copulation. Ovulation in these induced ovulators is a so-called "neuroendocrine" response; the nervous and hormonal systems work together. Messages coming in on nerves inform the brain that copulation has occurred. Other nerves then stimulate the anterior pituitary to release the hormone that induces ovulation.

Man is polyestrous; that is, there is more than a single estrous cycle a year. In addition, cyclical changes of the uterine lining in response to ovarian hormones result in menstruation. Instead of being timed from the day of ovulation, as it is in most mammals, the estrous cycle of animals that menstruate is timed from the first day of menstruation. In Fig. 7–13 the events of the menstrual cycle in the human are correlated with ovarian and hormonal events. The situation in other higher primates is more complex and is affected by environmental conditions. (See also Box 7–2.)

MOST VERTEBRATES HAVE REPRODUCTIVE CYCLES WHICH RESULT IN YOUNG BEING PRODUCED AT THE MOST PROPITIOUS SEASON. THE TIMING OF THESE CYCLES IS MEDIATED BY HORMONAL SYSTEMS. IN SOME ANIMALS, THE PRODUCTION OF OVA IS DEPENDENT UPON COPULATION. THE ESTROUS CYCLE IN MAN IS LESS DEPENDENT ON EVENTS IN THE EXTERNAL ENVIRONMENT THAN THAT OF ANY OTHER ANIMAL.

MIGRATION IN BIRDS In migrating birds of the Northern Hemisphere, gonadal growth, gametogenesis, and the spring migration northward are correlated with increased day length. The southward fall migration is correlated with decreasing day length and with regression of gonads. The experimental alteration of day length will hasten or retard migratory behavior and restlessness, depending on whether the day lengths are long or short. Other variables are also important, but it seems that light acting directly on nervous tissue and indirectly through the eyes either excites or inhibits the release of gonadotropins from the pituitary. The integrative pathways that lead to migratory behavior are now being investigated, and it seems clear that migration and gonadal developments are not necessarily related.

SEXUAL DIMORPHISM IN BIRD PLUMAGE In some species of birds, the plumage difference between the sexes is under strictly genetic control; experimental alteration of the endocrine system is without effect on the plumage of the English sparrow, for instance. In other birds, control is hormonal, although both the degree and the type of hormonal control varies from species to species. In the common leghorn chicken, the removal of the male's testes does not affect his glamorous plumage. Removal of the female's ovaries, however, results in a change in her dull plumage to that typical of males. In other words, the "male" plumage is really a neutral, or sterile, plumage, whereas the "female" plumage is hormone-dependent.

Weaver finches and their relatives, such as the paradise wydah, offer an interesting case. In both sexes, the nonbreeding birds have a modest brown, henlike plumage. During the breeding season, the female's plumage is unchanged but the male develops a brilliant, showy (cocklike) plumage, even if castrated. If the proper pituitary hormone (LH) is injected into both sexes, normal and castrate, regenerating feathers develop as shown in the following table:

BIRD	PLUMAGE RESPONSE
Normal male	Cocklike
Castrated male	Cocklike
Normal female	Henlike
Castrated female	Cocklike

The LH of the male induces the skin to produce cocklike feathers in the presence or absence of male hormones. In normal females ovarian estrogen interferes with the action of LH and permits the expression of henlike characteristics. This is one of the rare instances in which a gonadotropin has a primary effect on an organ (the skin) other than the gonads.

A single example will suffice to indicate how hormones may function in a feedback system to help the organism adapt to short-term environmental changes. Suppose you are running down a road on a warm day. You are sweating profusely as part of the homeostatic control of your body temperature. As you sweat, your body loses water, and as a result, the osmotic pressure of your blood rises. This change is detected by nerve cells in the brain which have endings in the posterior part of the pituitary. The electrical activity of these **neurosecretory** cells results in stored hormone being released from the pituitary into the bloodstream. This hormone, **vasopressin,** has as its target the cells of the kidney tubules. In some way it acts to increase the rate at which the tubules reabsorb water. The functioning of this feedback loop is diagramed in Fig. 7–15.

Hormones Affecting Short-term Changes

HORMONALLY INDUCED CHANGES IN THE DEVELOPMENT OF GONADS AND GAMETOGENESIS ARE CORRELATED WITH MIGRATORY BEHAVIOR IN BIRDS, BUT THE WAY IN WHICH THAT BEHAVIOR IS ITSELF MEDIATED IS NOT CLEAR. IN SOME BIRDS, HORMONES PLAY A PROMINENT ROLE IN PRODUCING SEXUAL DIMORPHISM, WHEREAS OTHERS SHOW A DIMORPHISM UNAFFECTED BY HORMONAL BALANCE (THAT IS, UNDER STRICTLY GENETIC CONTROL). IN A MUCH SHORTER TIME SPAN, HORMONES ACT AS THE MAJOR COORDINATORS OF A FEEDBACK SYSTEM REGULATING THE WATER CONTENT OF MAMMALIAN BLOOD.

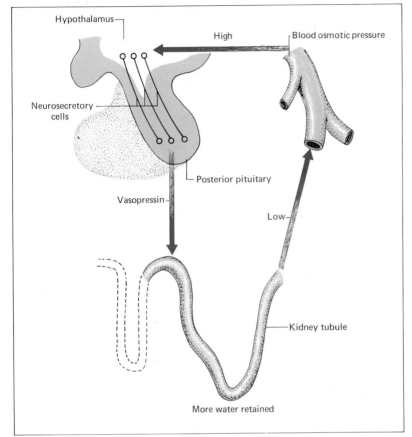

Fig. 7–15 The homeostatic system regulating the osmotic pressure of the blood. The amount of water absorbed by the kidney tubules is controlled by the amount of the hormone vasopressin secreted by the pituitary. The amount of vasopressin secreted, in turn, is determined by the osmotic pressure of the blood. If the blood becomes dilute, vasopressin levels decrease and the kidney tubule resorb less water from their lumina, thus increasing the output of urine.

MODE OF HORMONE ACTION

As discussed in Chap. 4, the activities of cells are controlled by DNA via RNA. You will recall that mitosis guarantees that each cell gets a complete set of genes, but in differentiated cells not all these genes are active. Kidney tubule cells are not busily producing hemoglobin, although they have the genetic information to do so. Young red blood cells do not produce the enzymes necessary for progesterone synthesis, even though at that stage they "know how." The set of DNA that is active in one kind of cell is not the same as that which is active in another. In each kind of cell, many, if not most, of the genes are *repressed.* The mechanism of repression and related regulation of gene activities is discussed, as far as the current state of our knowledge permits, in Chap. 10. Here we would like to look at a different question. As you have just seen, the functioning of tissues is very often controlled by hormones. And yet these tissues are made up of cells whose activities are, in a broad sense, genetically controlled. We are brought then to an interface between genetic and hormonal control—to the basic question of how hormones function. Many different suggestions have been made as to the mode of action of hormones. Vasopressin, for instance, presumably acts on the systems responsible for the transport of substances across cell membranes. Here the division between genetic and hormonal control is relatively clear. The cellular systems are produced under genetic control, and the hormone, presumably attaching to the cell membrane, alters their function. It, in some way, makes it easier or harder for the transport systems to get a substance (in the case of vasopressin, water) in or out of the cell. Other suggested modes of hormone function include several different kinds of relationship with enzymes.

There is now a growing body of evidence showing how many hormones operate directly on the genetic mechanism by regulating gene activity. For instance, estrogen has been shown to stimulate protein synthesis in the uterus of castrated animals. Within ½ hr of estrogen treatment, there is a rapid rise in the rate of RNA synthesis. If actinomycin treatment is used to inhibit this RNA synthesis, then no protein synthesis follows estrogen treatment. In full accord with the principle that hormone specificity resides in the target tissue rather than in the hormone itself, estrogen stimulates the production of different proteins in different tissues. In the uterus, many different proteins are produced; in the liver of a chicken (or a rooster), high estrogen levels induce the production of two yolk proteins necessary for egg formation. It has been shown that in the liver, estrogen acts very specifically on the protein-synthesis mechanism. One of the two yolk proteins, phosvitin, has, as almost half of its amino acid residues, those of serine. Estrogen treatment causes liver cells to produce an especially high concentration of the tRNA which handles serine.

IN SOME CASES, SUCH AS WITH VASOPRESSIN, HORMONES ALTER THE FUNCTION OF CELLULAR SYSTEMS PRODUCED UNDER GENETIC CONTROL. IN OTHERS, HORMONES DIRECTLY REGULATE GENE ACTIVITY. ESTROGENS, FOR INSTANCE, CAN INDUCE CELLS IN DIFFERENT TISSUES TO SYNTHESIZE DIFFERENT PROTEINS.

Estrogen has been shown to increase the synthesis of mRNA, appropriate tRNAs, ribosomal RNA, and an enzyme responsible for the polymerization of RNA. Ecdysone has also been shown to stimulate RNA synthesis. In the salivary glands of flies are found giant chromosomes produced by continued multiplication of the chromosomal strand until several hundred are lying adjacent to one another. In some areas the giant chromosomes show "puffs." The bands of the giant chromosomes are now known to be made up of tightly coiled DNA and proteins known as histones. The puffs are unfolded bands made up of loops of DNA and RNA polymerase and are sites of intense RNA synthesis—that is, of gene activity. New puffs appear in the chromosomes as metamorphosis begins, at the time when ecdysone is functioning. Ulrich Clever in Germany injected minute doses of ecdysone into larvae of the fly *Chironomus.* Within 30 min this produced a puff in a specific site on one chromosome. Thirty minutes later, a second puff invariably appeared at a specific site on another chromosome. The ecdysone dose necessary to "turn on" the gene or genes at these sites was so small that it consisted of no more than 100 molecules for each chromosomal thread in each nucleus.

Finally, for a third example of hormonally induced gene activity, we can turn to plants. The plant hormone gibberellin turns on the structural genes which synthesize amylase in the food-storage system of barley seeds. The barley embryo produces the gibberellin, which in turn stimulates the production of this enzyme necessary to change the stored starch into the soluble carbohydrates required by the embryo.

The precise mechanism by which hormones affect RNA synthesis is not clear. Hormones, you will recall, are not a single class of molecules but show a vast diversity of size and structure. Those which have been shown to affect the genetic apparatus of the cell run the gamut of both size and configuration. There is, of course, no reason to assume that they all act in a similar manner. Some may act directly on the chromosome, perhaps separating the DNA from its companion protein and permitting the DNA to unfold and RNA synthesis to occur. Another might facilitate the flow of RNA out of the nucleus. Still another might stabilize the relatively unstable interaction of the mRNA and the polyribosome. Whatever the exact mode of operation, these gene-hormone interactions provide a critical link between the adaptation of the organism to its environment (so often modulated by hormones) and its genetic information.

ESTROGENS HAVE BEEN SHOWN TO INCREASE THE SYNTHESIS OF ALL RNA COMPONENTS OF THE TRANSLATION APPARATUS AS WELL AS RNA POLYMERASE. ECDYSONE HAS BEEN SHOWN TO "TURN ON" GENES IN FLY CHROMOSOMES, AND GIBBERELLIN TURNS ON GENES, PRODUCING α-AMYLASE IN BARLEY SEEDS. THE EXACT MANNER (OR MANNERS) IN WHICH HORMONES AFFECT RNA SYNTHESIS REMAINS TO BE UNCOVERED.

SUMMARY

In order to survive, organisms must be able to integrate their activities. They must be able to perceive stimuli, determine appropriate responses to the stimuli, and effect those responses. Two basic integrating systems are found in higher organisms. Both plants and animals have systems based on the movement of chemical messengers known as hormones.

These systems function in a way analogous to a loudspeaker system. The hormones are produced in a restricted area and are carried throughout the organism. Responses to the hormonal message then vary from tissue to tissue. Hormonal systems are relatively slow, being dependent on diffusion and the circulation of body fluids. In animals which must move rapidly in a coordinated fashion, a nervous system roughly analogous to a telephone network has evolved. This system transmits information along specified pathways at high speed.

The growth of plants is controlled by a series of hormones, the best known of which are auxins. The distribution of auxins is controlled in part by gradients of light and gravity. Through enhancing cell elongation, the auxin distribution determines the direction of growth. Auxin regulates the form and habit of plants in many ways. Apical dominance results from the inhibition of lateral buds by auxin produced at the growing stem tip. Dropping of leaves is also controlled by auxin production, as is abscission of fruits. Plant hormones interact with each other in complex ways, and different hormones may have the same effects. Combinations of hormones also may have specific effects.

Flowering, seed germination, and growth of the seedling are all affected by light. A pigment, phytochrome, is converted from one form to another by exposure to either orange-red or far-red light. The phytochrome system is part of an internal clock mechanism that enables plants to calculate the length of dark periods to which they are exposed and to respond appropriately. Various hormones are also involved in the photoperiodic response of flowering.

In animals, growth is largely under hormonal control, both when it is gradual (as in man) and when it involves a dramatic metamorphosis (as in frogs and many insects). Animal hormones also control cyclic phenomena, such as the monthly reproductive cycle of the human female. Short-term changes in animals are also mediated by hormones. It is, for instance, a hormonal system that controls the water content of mammalian blood.

Some hormones act by modifying the function of cellular systems which have been produced under genetic control. Others intervene in the functioning of the genes themselves, in ways which have not yet been fully elucidated.

Cyclical events are a prominent feature of the natural environment. Day alternates regularly with night, tides rise and fall, wet seasons alternate with dry, and cool seasons with warm. These cycles are a result of the physical relationships of the planet earth with its moon and the sun.

BOX 7–1. BIOLOGICAL CLOCKS

Organisms must, of course, correlate their activities with these environmental changes. Moths flying in the daytime have evolved a variety of defenses against diurnal predators, but most moths avoid such predators simply by not flying except when it is dark. Many birds in temperate regions respond to the onset of fall by flying south. If they failed to recognize that onset, they would perish in the cold. Grunions, a species of small fish, swim up on the beach at extreme high tide and lay their eggs. If they did not judge the tide correctly, their eggs would be washed out to sea by a higher tide. Many flowers open synchronously with the appearance of their pollinators. If they opened too late, their chances of being pollinated would be diminished or reduced to zero. If they opened too early, their chances of being eaten or destroyed before they were pollinated would be increased.

Organisms, then, must be able to detect the "proper" time to do things. At first biologists believed that plants and animals "tell time" from the environment, much as we "tell time" from a clock. Increasing coolness or shorter days warn the bird of the onset of fall; light tells the moth that flight time is over. And, indeed, in many situations that is precisely how organisms do tell time. But in many, if not all, organisms there is an internal "clock" as well. It is, for instance, such a clock that permits animals to navigate by using the sun and stars. Since these bodies are in constant motion, navigators must have a time sense in order to keep their direction constant in an environment with changing physical reference points. After all, flying toward the sun takes you in one direction in the morning and in the opposite direction in the afternoon. Man needs a fine chronometer if he is to practice astronavigation. Birds somehow pack the equivalent of such a chronometer into their cells and tissues.

The existence of biological clocks has clearly been shown in a wide variety of laboratory experiments. The most widely studied are cyclic changes which occur with a roughly 24-hr periodicity, the so-called **circadian** (Latin: *circa,* "about," and *diem,* "day") **rhythms.** For instance, E. Bünning showed that when fruit flies *(Drosophila)* were kept under uniform conditions of temperature and humidity and were illuminated by constant dim light, they continued to show a circadian rhythm of emergence from the pupal stage.

If an organism is kept in uniform conditions for a long enough period, however, its circadian rhythm will slowly drift out of phase with the environmental cycle. The rhythm can then be resynchronized with the external cycle by an appropriate stimulus, most often a change in light. For instance, C. S. Pittendrigh was able to show that with asynchronous *Drosophila* living in total darkness, a circadian rhythm could be entrained by a single flash of light. Larvae exposed to such a flash matured into flies which emerged from their pupae at the same time of day that the flash was given.

Little is known about the physiological mechanism of biological clocks. They are, however, beginning to be a subject of great practical interest to human beings. As jet transports carry people rapidly through time zones, their biological clocks get badly out of phase with the day-night cycle. Short-term physical and psychological impairments result until the rhythm is resynchronized. The possibility of more serious long-term consequences has not been ruled out. Supersonic transports (SST) as a regular mode of travel might lead to even more serious disturbances.

BOX 7-2. THE MENSTRUAL CYCLE AND CONCEPTION

Understanding the menstrual cycle permits you now to understand two commonly used methods of contraception (prevention of conception). One is the ***rhythm method.*** This method involves abstention from sexual intercourse during the 2 days before and ½ day after ovulation. A human egg is receptive to fertilization for only 12 hr after ovulation, and the sperm survive only 48 hr. Unfortunately, the occurrence of ovulation can be detected only after the event, and not too accurately even then. When ovulation has taken place, the woman's temperature rises about half a degree. It drops again when her menstrual period begins. Predicting the time of ovulation must be done on the basis of past performance, projected from carefully kept records of previous menstrual and temperature cycles. To allow an adequate safety margin, several additional days should be included on either side of the estimated fertile period. This means abstention for a sizable part of the month, often during a period when the woman is relatively more receptive to intercourse. In addition to this serious disadvantage, about one woman in six has a menstrual cycle so irregular that the rhythm method will not work at all for her.

There is some evidence that use of the rhythm method results in a higher than normal incidence of babies with birth defects. This is believed to be due to a greater number of eggs being fertilized by "stale" sperm, several days after intercourse, or of "stale" eggs being fertilized toward the end of their life. In either case, some sort of deterioration seems to have set in, and the result is a defective child.

The ***Pill*** is the most effective contraceptive developed to date. Actually there are several kinds of pills. They are a mixture of estrogen and progestin, the latter a synthetic substance chemically resembling natural progesterone. The pills are believed to act by suppressing ovulation. A pill is taken daily for 20 or 21 days of the 28-day cycle, beginning on the fifth day after the onset of menstruation. The pills have the effect of regularizing the menstrual cycle, even in women who have never had regular cycles before. The menstrual flow will also be noticeably reduced or even occasionally suppressed altogether. About one in four or five women using the Pill have one or more side effects, many of which resemble the complaints of pregnancy. This is not surprising, since progestin, in a sense, hormonally simulates pregnancy. These effects include tenderness and swelling of the mammary glands, weight gain and retention of fluid, nausea, headaches, depression, nervousness, irritability, and vaginal bleeding.

In the course of up to 7 years of testing on large numbers of women prior to its release in the United States to the public in 1962 (and its use by millions since then), the Pill has proved to be safe for the overwhelming majority of women. It is probably safer than undergoing a pregnancy. Doctors do not prescribe the Pill for women who have histories of cancer, liver disease, or thromboembolic diseases. The Pill has shown no evidence of inducing cancer; on the contrary, it may have the opposite effect, at least for cancer of the female reproductive organs. Some kinds of liver disease are known to be aggravated by the hormones during pregnancy; the Pill seems likely to have the same effect.

Thromboembolism (blood clots blocking blood vessels) presents a more complicated picture. Research published in England in 1968 revealed that women over 35 using the Pill had a significantly higher chance of dying of thromboembolism than women of the same age not using the Pill. The risk is less than half as high in either case for women under 35. In both age groups, the risk of death from thromboembolic disease is probably about equal to that of death during pregnancy. The estrogen component of the Pill appears to be responsible for the development of these disorders. Work is now under way toward the perfection of a low-dosage progestin pill which should avoid many of the problems. Unlike the high-dosage progestin-estrogen combination, the new pill will not suppress ovulation. In spite of this, in tests it seems to be as effective as the Pill. It alters the consistency of the mucus in the entrance to the uterus, which may block the passage of sperm. It also affects the lining of the uterus, which may prevent implantation, and it may also cause the ovum to travel too rapidly down the fallopian tubes. Of course, two or all three of these factors may act in concert to prevent conception.

Other contraceptive methods have less direct connections with the

METHOD	PREGNANCY RATES/100 YEARS USE		TABLE 7-2-1
	HIGH	LOW	
No birth-control method	80	80	
Jelly or cream	38	4	
Condom	28	7	
Diaphragm and jelly	35	4	
Rhythm	38	0	
The Pill	2.7	0	
IUD	About 3		

menstrual cycle. The condom, a rubber sheath covering the penis during intercourse, simply prevents the passage of sperm. Various spermicidal jellies, foams, and creams may be introduced into the vagina to kill or immobilize the sperm. These agents may be used in conjunction with a rubber cup-shaped diaphragm fitted over the **cervix** (entrance to the uterus), giving a double barrier against sperm entrance. All these methods create problems of esthetics and convenience, but their physical side effects are essentially nil. They vary in effectiveness; most, if used properly, are superior to rhythm, but none is as effective as the Pill. Intrauterine devices (IUDs) are plastic or metal objects placed inside the uterus and left there as long as birth control is desired. The method of operation of IUDs is uncertain; presumably they prevent or disrupt implantation of the embryo. IUDs are the most convenient form of contraception: no calendars to mark, no pills to take, and no contraceptive materials to deal with. Once in place they can be forgotten. The difficulties of the IUD include bleeding, discomfort, and pelvic inflammation in some women and spontaneous (and sometimes unperceived) expulsion of the device. The latest IUDs, with the shape of a closed shield, tend to reduce these problems.

Table 7-2-1 compares the efficiency of various contraceptive methods, calculated on the basis of 100 woman-years; that is, each method is scored on the basis of how many women out of 100 using the method will become pregnant in a single year. The figures are only approximate, since various groups show wide differences in motivation toward proper use, and this has not been taken into consideration. A range of results is given for each method except IUD, where the average is shown.

Cilia and flagella—there is essentially no difference—are projections from the surfaces of cells that usually move in an oarlike manner. Sperm cells and many small animals rely on ciliary movement for locomotion. Larger animals depend on cilia to move material along the

BOX 7-3. CILIA AND FLAGELLA

surface of tissues, especially in ducts and passageways. As one biologist recently put it, "They either bring the cells into new environments or bring new environments to the cell." Wherever they occur—in plants, animals, or protozoans—their architecture is basically the same as long as they behave in the typical fashion. Where they have been modified for another purpose, such as in the detection of light in the rods and cones of the vertebrate retina, the basic structure is modified as well.

Electron micrographs of cilia (Fig. 7-3-1) reveal a common pattern. There is a core of two central filaments surrounded by nine paired filaments. The ciliary membrane, which is continuous with the cell membrane, envelopes the whole structure like the skin of a banana. Where the cilium enters the cell body, the filaments abut on a basal plate. Below this is the basal body and its rootlets extending deep into the cytoplasm. How does the cilium contract? Even so simple a structure has evaded our understanding. Isolated ciliary stalks (the cilium minus the basal body) contract in a manner reminiscent of isolated muscle fibers when ATP is added to them, proving that there is inherent mobility in ciliary protein. We still do not know what part the basal body and rootlet play in the initiation and coordination of the movement, but an intimate connection between the stalk and basal body is required for normal functioning.

But quite a bit is known about the reproduction and development of cilia. It is no coincidence that another cellular organelle, the **centriole,** has essentially the same structure as the basal body. The centrioles are small cylinders containing nine sets of filaments. A pair of them participates in cell division in animals by organizing the spindle fibers that are involved in moving the chromosomes to the new daughter cells. In the process, they reproduce themselves, so that each daughter cell has a pair. The centriole of a sperm behaves just like a basal body and somehow generates the sperm's flagellar tail. The basal body of cilia has the same reproductive and organizational ability.

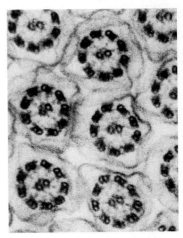

Fig. 7-3-1 Cross section showing the lateral cilia of a mussel (*Anodonta*). X75,000. (*Courtesy of I. R. Gibbons.*)

Gorbman, A., and H. A. Bern: "A Textbook of Comparative Endocrinology," Wiley, New York, 1962.

Harrison, R. J.: "Reproduction and Man," Oliver & Boyd, Edinburgh, 1967.

Turner, C. D. "General Endocrinology," Saunders, Philadelphia, 1966.

Beerman, W., and U. Clever: Chromosome Puffs, *Sci. Amer.,* **212,** no. 4, Offprint 180 (1964).

SUPPLEMENTARY READINGS

Books

Articles

Davidson, E. H.: Hormones and Genes, *Sci. Amer.,* **212,** no. 6, Offprint 1013 (1965).

Etkin, W.: How a Tadpole Becomes a Frog, *Sci. Amer.,* **214,** no. 5, Offprint 1042 (1966).

Van Overbeek, J.: The Control of Plant Growth, *Sci. Amer.,* **218,** no. 6, Offprint 1111 (1968).

Pengelley, E. T., and S. J. Asmundson: Annual Biological Clocks, *Sci. Amer.,* **224,** no. 4, Offprint 1219 (1971).

CHAPTER 8

INTEGRATION II: NERVOUS COORDINATION

Now that you have been introduced to the functioning of hormonal systems of integration in plants and animals, we can turn to the intricate high-speed integrating system found only in animals.

Hormones move from their place of origin to the target organ by diffusion or by catching a ride in the circulatory system. The interval between their release and their receipt may be measured in seconds. But in some instances, a second is too slow. Consider a sparrow attempting to escape a hawk. The difference between survival and mortality for the little bird is a matter of milliseconds, and if he is to escape the hawk, a lot has to happen in that interval. **Speed,** then, is one requirement of a nervous system. The speed with which chemicals can be moved from one part of the body to another simply is not great enough for many coordinative functions.

Our fleeing sparrow needs more than speed. To evade the hawk, its right wing must know what its left wing is doing—that is, its movements must be coordinated. Uncoordinated, convulsive movements would make it an easy target since such movements would make flight impossible and it would drop to the ground. Rapid communications are necessary, but they are not sufficient for survival. Signals must also be **accurate;** the messages must go to the correct organs and in the proper order to produce coordination. Finally, the messages must also be of the right kind. For a wing to work properly, the muscles controlling it must contract not only at just the right moment but also in just the right amount. A signal that simply commanded a muscle to "contract" would be as useless as a telegram which told you: "do something." In other words, the information coming to an organ must be **specific.** It must not only say "act," it must say how to act—that is, quickly, slowly, strongly, or delicately.

REQUIREMENTS OF A NERVOUS SYSTEM

MOST ANIMALS REQUIRE COORDINATION THAT IS FAST, ACCURATE, AND SPECIFIC; HORMONAL SYSTEMS ARE TOO SLOW FOR THIS TYPE OF COORDINATION AND WOULD NORMALLY BE INCAPABLE OF MAKING CRITICAL DISTINCTIONS BETWEEN IDENTICAL TISSUES (FOR EXAMPLE, IN ORDER FOR A SPARROW TO TURN, IDENTICAL MUSCLES IN RIGHT AND LEFT WINGS MUST RECEIVE DIFFERENT MESSAGES).

Think of the sparrow as though it were a black box. Careful observation of its behavior would yield certain facts: First, the bird senses the presence of "danger"; this is evidenced by its evasive behavior. Therefore, we can conclude from our observation that there is an input, or **sensory capability,** about the "box." Second, the evasive behavior of the box informs us that there is a response system, or **effector capability,** about the box. Symbolically, this could be represented as follows:

ELEMENTS OF NERVOUS COORDINATION

Sensory Input	Output (Effector)
System	System

In this example, the output is manifest by evasive movements. More observation (and some probing into the box) would disclose that these movements are actually exceedingly complex, involving the coordinated contraction and relaxation of dozens of muscles. Furthermore, from repeated observations of the evasive behavior of birds, one would quickly discern that evasive behavior is really a very large set of overlapping kinds of behavior. A bird when trying to escape is capable of braking, accelerating, diving (at various angles and velocities), and turning. And these movements can be accomplished in any order and in many combinations.

Success of evasive behavior must at least be based on accurate determination of the predator's velocity, size, and angle of approach. Furthermore, before orders are fed to the bird's muscles, all this information must be integrated, or "understood." Thus, we can infer that a third element must be interposed between the other two in this example of coordination:

By "integration," we mean that aspect of coordination involving judgment, calculation, or interpretation of data being fed in by the sense receptors. All these elements will be discussed in some detail below. But first we will inspect a common component of all these systems in animals: the neuron and the nerve impulse which the neuron can propagate. Roughly speaking, neurons are to nervous systems what copper wire is to telecommunication systems. A knowledge of the neuron and the nerve impulse is basic to understanding the functioning of nervous systems.

EVASIVE BEHAVIOR IN A SPARROW OBVIOUSLY REQUIRES RAPID AND PRECISE EVALUATIONS OF THE BIRD'S POSITION RELATIVE TO A PURSUER, RAPID DECISION MAKING, AND MANY PRECISE AND COORDINATED RAPID MOVEMENTS. NEURONS WITH THE ABILITY TO TRANSMIT NERVE IMPULSES ARE THE CONTROL COMPONENTS WHICH MAKE SUCH BEHAVIOR POSSIBLE.

NEURONS AND THE NERVE IMPULSE

Electrical Phenomena in Cells

The occurrence of an electrical potential, or voltage, across membranes is a widespread phenomenon. Many plant cells, and eggs and other cells of higher animals have such a **polarization** across their membranes and thus behave like miniature charged batteries. Usually the inside is negative to the outside, the potential being from 50 to 100 millivolts (mv), as shown in Fig. 8-1. This difference in charge is known as the **resting potential** in nerve cells because it appears to be an equilibrium condition requiring little or no expenditure of energy to maintain.

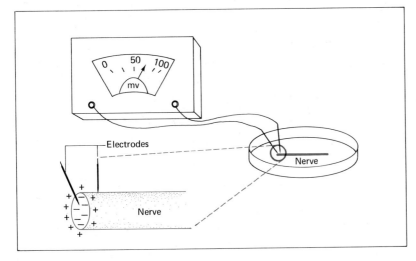

Fig. 8-1 It is possible to measure the electrical potential between the inside and outside of a cell by using extremely fine electrodes.

The **depolarization** (actually involving a momentary *reversal* of polarization) of a part of the membrane can be induced by electrical, chemical, or mechanical stimuli or by anything that can cause certain changes in the membrane. Once one region of a cell membrane is depolarized, a wave of depolarization, or excitation, is likely to spread out over the surface of the cell and even to adjacent cells.

The resting potential is a fortuitous by-product of cellular architecture and chemistry and, like many such fortuitous by-products, it has been exploited in the course of evolution by organisms in which a system of rapid communication was advantageous. It is not difficult to imagine how this evolutionary development occurred. Many cells are polarized in their normal states and are capable of conducting a wave of excitation over their surfaces when a point on the surface is depolarized. If such a cell were to become elongated, it could serve as a direct communication link between two relatively distant cells. The nervous system of animals is composed of such elongated units; these are called **neurons.**

MANY PLANT AND ANIMAL CELLS HAVE AN ELECTRICAL POTENTIAL ACROSS THE CELL MEMBRANES. THIS POLARIZATION MAY BE CHANGED BY ELECTRICAL, CHEMICAL, OR MECHANICAL STIMULI, AND A WAVE OF DEPOLARIZATION MAY BE INITIATED WHICH TRAVELS OVER THE CELL MEMBRANE IN ALL DIRECTIONS FROM THE POINT OF STIMULUS. THIS WAVE OF EXCITATION MAY PASS ON TO OTHER CELLS; IT HAS FORMED THE BASIS FOR THE DEVELOPMENT OF THE SPECIALIZED TRANSMITTING CELLS OF THE NERVOUS SYSTEM, THE NEURONS.

The Neuron and Membrane Conduction

The "wires" of the nervous system are elongated cells called neurons. Neurons are heterogeneous in form and size, but there are usually three important regions, as shown in Fig. 8–2: (1) a zone of input, or excitability, usually near or on the cell body and often ramified into **dendrites** (small branchlike projections that receive the input); (2) a zone of transmission, called the **axon,** which may be as much as several meters long and which extends from the cell body to (3) the zone of output, called the **axonal terminal,** which is specialized to transfer the excitation from itself to an input zone of one or more adjacent cells.

The actual shape and size of neurons depend on their function. Neurons that receive inputs from many other neurons may have very elaborate dendritic regions. Those that relay information to another neuron close by may be short and simple. Those that transmit information over distances—for example, from the foot of a giraffe to its spinal cord—may be longer than you.

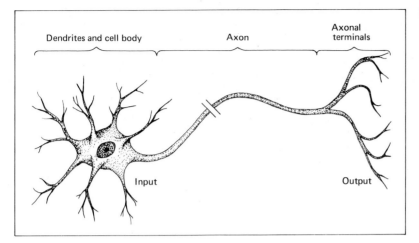

Fig. 8-2 The structure of a neuron.

How do these "wires" work? Of course, they are not really comparable to electric wires. In an electric wire, the electricity travels at a speed approaching that of light—186,000 miles per second. The fastest nerves conduct an impulse at about 0.06 mile per second, or about 100 m per second. Further, the axon of a nerve is very resistant to an actual electric current and is also very leaky to the flow of electricity. A current fed into an axon would disappear after just a few millimeters. Closer examination of the resting potential will contribute to the resolution of this problem.

RESTING POTENTIAL   Cell interiors have a chemical composition different from that of exteriors. Neurons of the squid have been studied intensively in this respect. The concentrations of potassium (K^+), sodium (Na^+), and chloride (Cl^-) ions on the inside and outside of squid axons were found to be as follows:

ION	INSIDE	OUTSIDE (BLOOD)
K^+	400	20
Cl^-	40	560
Na^+	50	440

These relative concentrations are shown diagrammatically in Fig. 8-3. Clearly, the pores in the membrane of the axon of the neuron are too small to allow the proteins or other large organic molecules easy exit. This is typical of cell membranes in general. These large organic molecules are on the whole negatively charged and occur as their ionized K salts. The question raised by this diagram is: Why don't the K^+ ions migrate out through the membrane in correspondence to the laws of diffusion; si-

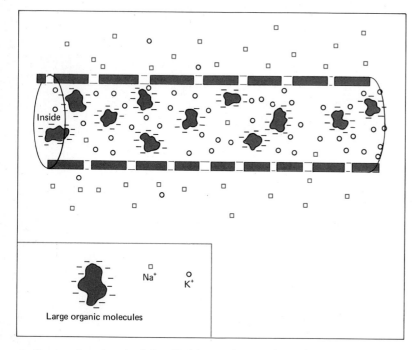

Large organic molecules

Na^+ K^+

Fig. 8-3 The distribution of ions inside and outside a neuron is shown diagrammatically. Na^+ ions are prevented from entering the cell by an energy-consuming "sodium pump" in the membrane.

multaneously, why don't the Na^+ ions migrate along their concentration gradient to take the place of the K^+ ions?

Much research has shown that the membrane does, in fact, permit the K^+ ions to exit along their concentration gradient; on the other hand, the resting membrane is almost impermeable to Na^+ ions. That is, the K^+ ions can leave, but no positive ions can come in to replace them. As a result, the outside becomes positive to the inside, and the more positive the external medium becomes, the more such positive ions as K^+ are prevented from coming through. This is because similarly charged ions repel each other just as do like poles of magnets (Fig. 8-4). Eventually an equilibrium is reached in which the diffusion force outward is balanced by the electromagnetic force inward. Usually the equilibrium is achieved when the potential difference between the two sides is about 60 mv outside positive.

This potential difference, or resting potential, could not develop across cell surfaces if sodium ions could pass freely through the cell membrane. If they did, they would compensate for the outward traffic of potassium, thereby preventing the development of a charge difference.

NEURONS COME IN A VARIETY OF SIZES AND SHAPES APPROPRIATE TO THEIR FUNCTIONS. THEIR RESTING POTENTIAL DEPENDS ON THE

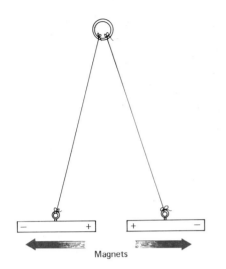

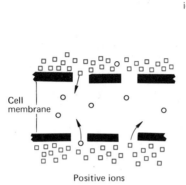

Fig. 8-4 The magnets remind us that like charges repel. K+ ions are prevented from diffusing through the membrane because the high concentration of Na+ ions on the outside acts to repel other positively charged ions.

Magnets

Positive ions

DIFFERENTIAL PERMEABILITY OF THE CELL MEMBRANE TO K+ (PO-TASSIUM) IONS, WHICH CAN PASS THROUGH IT FREELY, AND NA+ (SODIUM) IONS, WHICH CANNOT. K+ IONS, WHICH ARE HIGHLY CONCENTRATED INSIDE, MOVE OUTWARD ALONG THE K+ DIFFUSION GRADIENT, BUT NA+ IONS CANNOT MOVE ALONG THEIR INWARD GRADIENT. WHEN ENOUGH K+ IONS HAVE EMIGRATED, THE CONCENTRATION OF POSITIVE CHARGES OUTSIDE THE MEMBRANE CREATES A COUNTERACTING ELECTROMAGNETIC INWARD FORCE STOPPING FURTHER K+ MOVEMENT OUTWARD.

ACTION POTENTIALS The resting potential is an equilibrium condition; so long as the cell is alive and maintains the integrity of characteristics of the membrane, the equilibrium will be maintained. To alter the equilibrium requires the application of an outside energy source. When the membrane is excited by such an outside agency, its characteristics suddenly change and it no longer acts as a barrier to Na+ ions (Fig. 8-5). In such a case, the Na+ ions will suddenly surge inward down their concentration gradient. As the positive ions penetrate the cell, the charge differential and the resting potential disappear. Typically, there is an overshoot of about 50 mv in nerve cells, or a reversal of the polarity, before the penetration of sodium stops altogether. (What causes the cessation of Na+ infusion?)

At this stage the membrane suddenly becomes more permeable to potassium ions, and K+ ions flow out of the cell along their concentration gradient. They are no longer "held back" by the electrical forces that

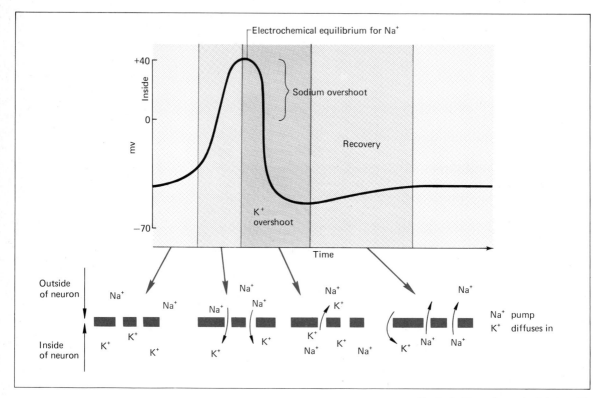

Fig. 8–5 The action potential. A rapid influx of Na⁺ ions causes a reversal of polarity across the neuron membrane. This is immediately followed by an outflux of K⁺ ions reversing the polarity again. The original ion concentrations are restored during the recovery phase.

follow the change in polarity. The outward surge of positive ions restores the negative charge to the inside. In fact, there is again a slight overshoot during this process, but final recovery is achieved quickly, during which the excess Na^+ ions are pumped out (using energy from ATP) and some K^+ ions move back in.

The whole process—the **action potential**—lasts only 5 to 10 milliseconds (msec). Most of this time the system is recovering; the spike (polarity reversal) itself lasts only about 1 msec. During the recovery period, only a very strong stimulus can initiate another action potential.

STIMULATION OF A NEURON RESULTS IN THE FORMATION OF AN ACTION POTENTIAL. THE CELL MEMBRANE BECOMES PERMEABLE TO NA⁺ IONS, WHICH FLOW IN, MAKING THE INSIDE OF THE NEURON MOMENTARILY POSITIVE WITH RELATION TO THE OUTSIDE (REVERSAL OF POLARITY). THIS IS FOLLOWED BY AN INCREASE IN THE POTASSIUM PERMEABILITY OF THE MEMBRANE AND AN OUTFLOW OF K⁺ IONS, WHICH RESTORES THE RESTING POTENTIAL. THE WHOLE

*SEQUENCE IS VERY RAPID, THE ACTION POTENTIAL LASTING ONLY
5 TO 10 MSEC.*

NERVE IMPULSE   So far, we have been discussing the action potential at a small region of the cell surface. In nerve (and other) tissues, a local depolarization spreads out rapidly over the cell membrane in the form of a wave known as the **nerve impulse.** The velocity at which the impulse is propagated depends on numerous factors, including the nature of the cell contents, the membrane, and the size of the nerve (the larger the neuron, the faster the impulse).

In the squid, the relationship between nerve size and impulse speed has been employed to control the escape reaction. When frightened, the squid can "jet away" rapidly by forcing water out of its body cavity through a funnel, which it can aim. To work efficiently, all the muscles of the body cavity wall must contract simultaneously so that the maximum water pressure is built up and the maximum velocity of escape is attained. The nerve center, or **ganglion,** that controls this reaction, in part, has very large (so-called "giant") nerve axons arising from it and going to the muscles of the body wall. These axons differ in diameter, however. If, indeed, the fibers were of the same diameter, then their conduction velocities would also be identical and the nerve impulse would reach the more distant parts of the body wall later than they reached those parts closest to the ganglion, even though the magnitude of the difference would be only a few hundredths of a second. Figure 8–6 shows how this time-lag problem has been overcome in the squid. The muscles farthest from the ganglion are innervated by the largest nerves, so that the discrepancy in distance is compensated for by the variation (gradation) in axon size.

The giant fibers of squid and worms are as large as 0.5 mm in diameter and propagate impulses of around 10 to 20 m per second. Smaller invertebrate nerves conduct at one-tenth this velocity. These velocities are fine for integration in very small animals, but large vertebrates can do much better with much smaller neurons. Some cat neurons less than 0.02 mm in diameter conduct at 100 m per second. The secret is the **myelin sheath,** a thick fatty coating around axons and interrupted by naked rings called *nodes.* In such neurons, depolarization occurs only at the nodes and the impulse jumps from one node to the next, dramatically increasing conduction velocity.

Propagation of the impulse along an axon has been compared with the burning of a fuse. The burning zone of a fuse generates enough heat to ignite the adjacent zone. Once the fuse is ignited, the burning is propagated by an infinite series of such steps. Furthermore, the rate of burning and the energy released are independent of the kind and intensity

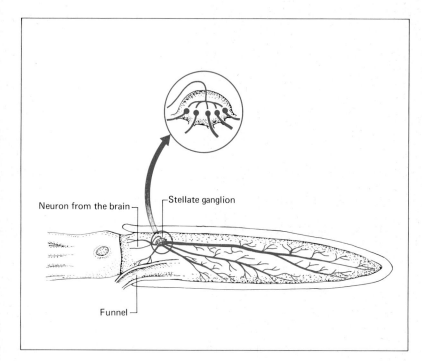

Fig. 8-6 Nerve impulses from the stellate ganglion of the squid reach all muscles in the body wall simultaneously because of the relationship between axon diameter and conduction velocity.

Neuron from the brain

Stellate ganglion

Funnel

of the stimulus (e.g., spark, match, or blow) that started it. A nerve impulse is very like this. Once an action potential has been initiated on a nerve cell, the electrochemical change "ignites" the adjacent region so that the impulse is transmitted rather than dying out. For this reason, the nerve impulse is called an **all-or-none** phenomenon; that is, once initiated, it is propagated down the axon with constant characteristics—it does not change in intensity or velocity. For a particular nerve, there is no such thing as a smaller or larger, faster or slower impulse.

Usually an impulse is initiated at the dendritic end of a nerve cell, and the impulse travels along the axon to the terminals. But if the impulse is not initiated at the end of the cell, it will travel in both directions, just as a fuse will burn in both directions if it is ignited in the middle. It is always possible for an impulse to travel "backward" on a single neuron, but the circuitry of nerve connections usually precludes reverse travel over a chain of neurons.

A precise understanding of the mechanism by which the impulse is propagated depends on some knowledge of physics, but it will suffice to say that the zone adjacent to an excited membrane is partly depolarized due to a movement of ions toward the excited zone. Like the heat of a burning fuse, the depolarization reaches sufficient proportions to alter the permeability of the adjacent membrane regions to Na^+ ions; an action potential follows, and so forth. As we noted, this type of **electrochemical** transmission of the action potential is much faster than that characteristic of hormonal systems but much slower than the electrical transmission of a telephone system.

A NERVE IMPULSE IS A WAVE OF DEPOLARIZATION SPREADING OVER THE SURFACE OF A NEURON. AN ACTION POTENTIAL ESTABLISHED AT ONE POINT STIMULATES THE FORMATION OF ACTION POTENTIALS AT ALL ADJACENT POINTS. THE RATE AT WHICH A NERVE IMPULSE TRAVELS IS A FUNCTION OF NUMEROUS FACTORS, INCLUDING THE DIAMETER OF THE NEURON AND WHETHER THE AXON IS MYELINATED. IF A NEURON IS STIMULATED IN THE MIDDLE, ACTION POTENTIALS WILL BE PROPAGATED IN BOTH DIRECTIONS FROM THE POINT OF STIMULUS, BUT THE WAY IN WHICH NEURONS ARE CONNECTED TO ONE ANOTHER MAKES THE FLOW OF INFORMATION THROUGH THE NERVOUS SYSTEM UNIDIRECTIONAL. THIS ELECTROCHEMICAL FLOW IS VERY RAPID IN COMPARISON WITH THE FLOW OF INFORMATION THROUGH A HORMONAL SYSTEM BUT VERY SLOW IN COMPARISON WITH A PURELY ELECTRICAL SYSTEM SUCH AS A TELEPHONE OR TELEGRAPH.

The Synapse

Neurons are specialized for communication, but the information itself is often useless unless it causes something to happen, that is, unless it induces an effector cell or organ to act, perhaps by secreting something or by contracting. Before effectors can act and before the impulse can be transmitted to another neuron, it must jump the gap between neurons or between neurons and other cells. The gaps between neurons are called **synapses.** The *synaptic cleft,* as the gap is called, is so small that a controversy raged for much of the nineteenth century as to whether it existed at all. In fact, the cleft is only about 20 nm wide. It is difficult to express this tiny distance in familiar terms. In molecular dimensions, it is about ten times the width of a single DNA helix.

Following the discovery of the synapse by Ramon y Cajal at the end of the nineteenth century, another battle ensued over the question of the mechanism of bridging the synapse. Some thought that synaptic transmission was electrical. They claimed that the impulse reaching the **presynaptic** membrane was strong enough to depolarize the **postsynaptic** membrane and start an impulse in the postsynaptic neuron. Others argued that the presynaptic potential acted as an electrical signal triggering the release of a chemical messenger, which diffused across the cleft and depolarized the postsynaptic membrane. As in many such controversies in biology, both sides had part of the truth. It is now thought that most synapses are of the chemical kind, but recently some electrical synapses have been found in invertebrates.

Studies of very simple animals, such as ctenophores (e.g., the sea gooseberry) and coelenterates (e.g., hydra), have suggested the theory that chemical synaptic transmission is evolutionarily an advanced form of cell communication. In these simple but successful aquatic creatures, much of the transmission is conducted not by nerves and synapses but by sheets or bands of epithelial (surface) cells.

Such nonnervous electrical transmission controls the intricate swimming movements of ctenophores (Fig. 8–7). The ctenophore might be described as a jelly-invested gut that swims by coordinated beating of eight bands of ciliary combs. Each comb along the band is formed by

Fig. 8–7 Locomotion in ctenophores is controlled by the coordinated beating of ciliary combs. The rate of beating is under the control of the apical pit organ. (*Adapted from G. A. Horridge, AMERICAN ZOOLOGIST, Vol. 5, 1965, p. 359.*)

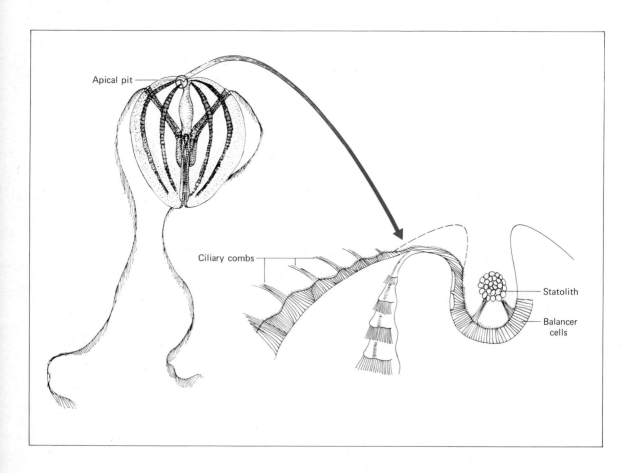

the division of large cilia. These cilia beat spontaneously, but the rate at which they beat is under the control of a single sense organ. The latter is the apical pit, which contains a small stone (a **statolith**) that rests on ciliated cells called *balancers*. If, for instance, the animal rotates clockwise, the balancers on the right must support more of the statolith's weight. The rhythmic beating of the comb plates is coupled to the balancers by ciliated grooves. Consequently, the comb plates beat more rapidly on the lower side and more slowly on the upper side of the animal, resulting in the maintenance of upright orientation. So long as the animal is upright, simultaneous waves of excitation pass down the ciliated grooves, the comb plates beat in unison, and the animal swims upward with its tentacles trailing behind, trapping its zooplankton food. All this coordination is accomplished without neurons—apparently each ciliated cell in the grooves and some of the comb row cells pass the wave of depolarization along in a fashion not unlike that of neurons.

If the ciliated groove of ctenophores is thought of as a primitive depolarizing communication channel, it is tempting to speculate that neurons were originally elongated epithelial cells not unlike these ciliated grooves.

In addition to the depolarization systems just described, coelenterates and ctenophores also have well-defined neurons, often arranged in one or more "nets." Such a net of bipolar and tripolar neurons intertwines among the bases of the epithelial and ciliated cells. Depolarization of the cells in a ctenophore net, such as might be caused by a bump, leads to immediate inhibition of the ciliary beats. These nerve nets also have excitatory functions.

The great resolution power of the electron microscope may have created more problems than it has solved by opening up so many small-scale vistas for the biologist. The functions of many structures observed in cells with this instrument are yet to be revealed. But one fortuitous discovery is the existence of small vesicles in presynaptic endings (Fig. 8–8). There is reason to believe that these vesicles contain the synaptic transmitter substance. When an impulse arrives at the terminals of an axon, the vesicles closest to the cleft eject their contents into it.

Fine Structure of the Synapse

VERY NARROW GAPS—SYNAPSES—OCCUR BETWEEN NEURONS. IT WAS LONG DEBATED WHETHER THE NERVE IMPULSE BRIDGED THE SYNAPSE ELECTRICALLY OR CHEMICALLY. BOTH KINDS OF BRIDGING ARE NOW KNOWN TO OCCUR, WITH CHEMICAL BRIDGING THOUGHT TO BE THE MOST COMMON.

The mechanism by which the transmitters bring about an action potential

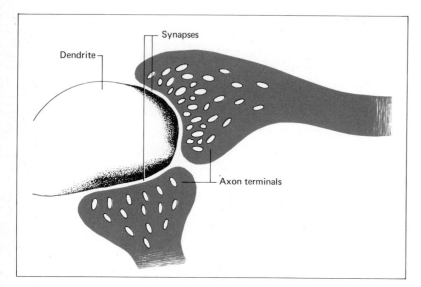

on the postsynaptic cell is only partly understood, and it is a very complex subject. From what has already been said about the action potential, it should occur to you that the transmitter substance alters the permeability of the postsynaptic membrane in such a way as to bring about a reversal of polarity; in other words, it must make the membrane more permeable to Na^+ ions. Such synapses are called *excitatory* synapses because they can induce an impulse in the postsynaptic neuron. In recent years another very fascinating kind of synapse has been discovered—the *inhibitory* synapse. The transmitter substance of inhibitory synaptic neurons affects the permeability of the postsynaptic cells in just the opposite way. Instead of increasing the permeability to Na^+ ions, which would tend to migrate into the neuron, the inhibitory synaptic transmitter enhances the permeability to K^+ ions, which tend to diffuse outward. This *increases* the polarity of the membrane by making the inside even more negative (Fig. 8–9). This increase of the potential across the postsynaptic membrane is called **hyperpolarization.**

Can you imagine what function inhibitory synapses could have in a system of neurons? It would certainly make no sense if the only kind of input a neuron had was inhibitory—there would be no need for a neuron in such a case. Actually, inhibitory synapses make sense only if they work in consort with excitatory synapses. Many neurons have more than a single axon terminating on them; some brain cells have hundreds of inputs. In such cases, what determines whether an impulse is initiated is something like the algebraic sum of the inhibitory (minus) and excitatory (plus) synaptic potentials (Fig. 8-10). This will be discussed further in relation to muscle contraction.

The chemical transmitter is not the same in all neurons. The autonomic nervous system (see Chap. 9) has at least two chemicals. Acetylcholine is the only one to be definitely identified as the synaptic chemical in the vertebrate central nervous system (see Chap. 9), but there are undoubtedly many others, including norepinephrine (related to adrenalin) and inhibitory transmitters such as GABA (gamma-aminobutyric acid).

CHEMICAL TRANSMISSION ACROSS A SYNAPSE OCCURS WHEN VESICLES ALONG THE PRESYNAPTIC MEMBRANE OF THE CLEFT DISCHARGE THEIR CONTENTS INTO THE GAP. THE TRANSMITTER SUBSTANCE THEN DEPOLARIZES THE POSTSYNAPTIC MEMBRANE (IN EXCITATORY SYNAPSES) OR MAKES IT MORE DIFFICULT TO DEPOLARIZE (IN INHIBITORY SYNAPSES). INHIBITORY SYNAPSES ARE FUNCTIONAL ONLY IN CONJUNCTION WITH EXCITATORY SYNAPSES.

RECEPTORS

In evolutionary terms, success is measured by survival and reproduction, and the diversity of organisms testifies to the variety of strategies that are successful. Each of these strategies (in a sense there are as many

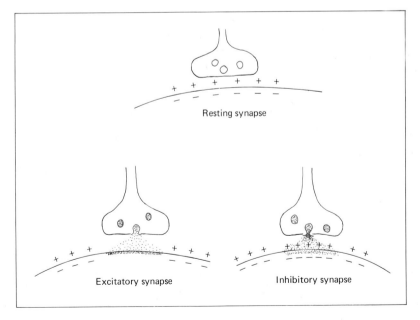

Resting synapse

Excitatory synapse

Inhibitory synapse

Fig. 8-9 Electrical properties of excitatory and inhibitory synapses. Discharge of the excitatory synaptic transmitter depolarizes the postsynaptic membrane. Discharge of the inhibitory synaptic transmitter increases the charge difference across the postsynaptic membrane.

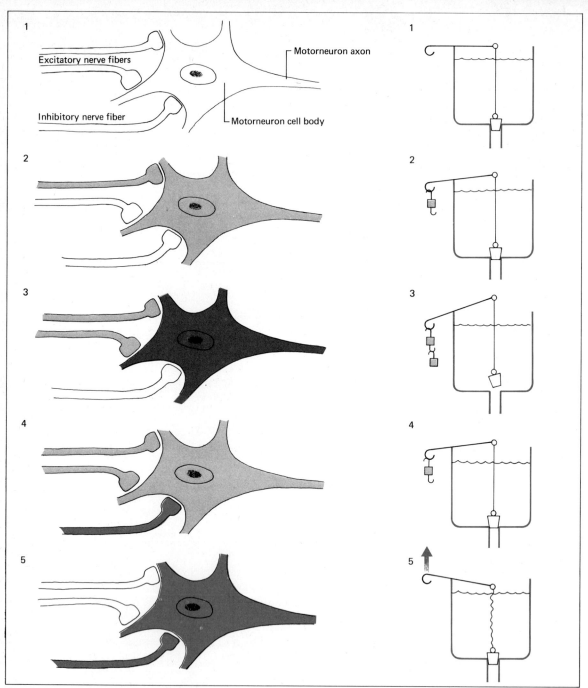

Fig. 8–10 Every neuron has a threshold for the firing of a nerve impulse, just as a certain amount of force must be applied to flush a toilet: (1) "no presynaptic impulses" is analogous to no force applied to the handle; (2) a single excitatory synaptic event is analogous to an insufficient or subthreshold force applied to the handle; (3) two excitatory synaptic events raise the motorneuron above its firing potential and are analogous to a flushing force; (4) two excitatory and one inhibitory synaptic events have the same effect as one excitatory event; (5) an inhibitory stimulus alone lowers the motorneuron's probability of firing and is analogous to an upward force on a toilet handle. (*From "The Synapse" by Sir John Eccles. Copyright © 1965 by Scientific American, Inc. All rights reserved.*)

strategies as there are kinds of organisms) includes as a crucial element the capacity to respond to external and internal changes. Such a capacity is, first of all, a matter of **detecting,** or sensing, change in the environment. Any detected change in an organism's environment, you will recall, is a stimulus. In this section, we will deal with the kinds of change detectors, or receptors, and how they respond to sensory input.

Receptors can be classified in various ways: by their location (internal or external), by their complexity (unicellular, multicellular), by the kind of input they detect. Here, receptors will be classified and discussed in terms of input.

Electromagnetic Reception

As shown in Fig. 8–11, animals can detect only a small band in the electromagnetic spectrum. The detectable radiations extend from the relatively long infrared wavelengths to the shorter ultraviolet wavelengths. Longer wavelengths do not have properties which make changes in them easily detectable by biological systems. Organisms tend to be screened

Fig. 8–11   The electromagnetic spectrum.

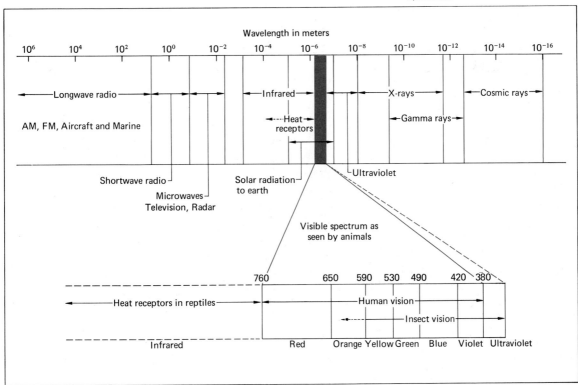

from contact with shorter wavelengths, which is just as well since their high energy content makes them very destructive to cellular machinery. Of course, it cannot be said categorically that no organisms can discriminate or detect x-rays, radio waves, etc., but there is no evidence for such sensitivities and no obvious need of them. There is, however, evidence for the detection of magnetic fields and some fish are known to be sensitive to electric fields.

Temperature Reception

Temperature is perceived by animals in two basic ways: by the detection of the infrared radiation that emanates from objects and by direct measurement of heat by physical contact. Detection of infrared radiation from warm objects is well developed in some reptiles. The rattlesnakes of the New World feed mostly on small mammals. Other reptiles are capable of detecting infrared radiation, but the rattlesnakes have developed this to a high degree. A specialized organ, the pit (Fig. 8–12), lies between the eyes and nostrils and contains a cavity lined with sensitive nerve cells capable of responding to long infrared radiation. The cells are so sensitive that objects that differ by only 0.1°C can be distinguished. Pythons and boas have similar infrared-detecting pits on their lips.

Vision

LIGHT RECEPTION: THE MOLECULAR BASIS   Light must do something in a cell, that is, it must cause some change, if it is to be detected. This is a principle applying to all sense reception, but it may not be intuitively

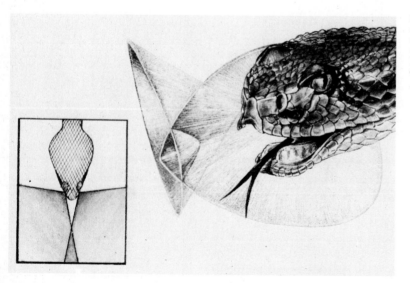

Fig. 8–12 Head of a rattlesnake showing the partially overlapping fields of reception of the heat-sensitive pits on either side of the head below and between the nostril and eye. (*Courtesy of T. H. Bullock.*)

obvious. What does light do in a cell? You will recall that one thing light can do in plants is "activate" a chlorophyll molecule so that it ejects an electron. In other words, light effects a molecular change—a photo-chemical process. In order to cause such a change, light energy must first be absorbed by a molecule; therefore, a pigment must be present. (A pigment, by definition, is a material that absorbs light.)

RECEPTORS ARE ORGANS WHICH DETECT CHANGES IN AN ORGA-NISM'S ENVIRONMENT. AS FAR AS IS KNOWN, ANIMALS CAN DETECT CHANGES IN ONLY A SMALL PORTION OF THE ELECTROMAGNETIC SPECTRUM. WITH CERTAIN EXCEPTIONS (INFRARED RADIATION, MAGNETIC AND ELECTRIC FIELDS), SENSITIVITY TO THE ELECTROMAG-NETIC SPECTRUM IS NORMALLY CALLED VISION, AND THE RADIATION DETECTED IS LIGHT. PIGMENTS, SUBSTANCES WHICH ABSORB LIGHT, FORM THE CHEMICAL BASIS OF VISION.

The eyes of vertebrates have the light-sensitive cells localized on the **retina,** a tissue at the back of the eye (Fig. 8–13). Two classes of sensory neurons are found in the retina: **rods** and **cones.** The rods are especially sensitive to light in the 500-μm region—blue-green to blue light. Toward the edges of the retina, the concentration of rods is greatest. Cones are sensitive to longer wavelengths: 560 to 620 μm—yellow to red light. Different cones have different spectral peak sensitivities, but it is not established how many types there are. It *is* known, however, that cones are the receptors involved in color vision. Cones are most dense in the **fovea centralis,** the site of the greatest visual acuity on the retina. When you stare directly at an object, the light is focused on the fovea.

Rods and cones have very different sensitivities. This can be verified by a simple experiment. On a dark night, stand outside and fix your gaze on a star. Without moving your eyes, locate a very dim star "out of the corner of your eye." Finally, look directly at the dim object. Does it seem to disappear, or does it seem to increase in brightness?

The sensitivity to light in rods and cones must be due to some change mediated by a photosensitive pigment. It is now known that light actually induces a change in shape in the pigments—the first step in a complex sequence of steps that ends in the firing of a nervous impulse.

An inability to see in dim light, night blindness, usually can be cured quickly by the administration of vitamin A, which is a carotenoid pigment responsible for the reds and oranges of tomatoes and carrots. Two forms of carotene are found in the eye of vertebrates: **retinene$_1$** and **retinene$_2$.** But these must be bound to a protein, **opsin,** before light is absorbed. Rods have a different form of opsin than cones. The distribution and

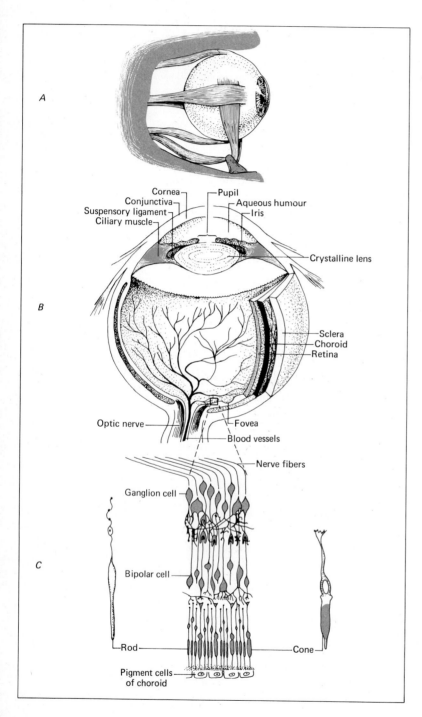

A

B

C

Cornea
Conjunctiva
Suspensory ligament
Ciliary muscle
Pupil
Aqueous humour
Iris
Crystalline lens

Sclera
Choroid
Retina

Optic nerve
Fovea
Blood vessels
Nerve fibers

Ganglion cell

Bipolar cell

Rod
Cone

Pigment cells
of choroid

Fig. 8–13 The structure of the eye: (*A*) a mammalian eye shown suspended in its orbit (socket); the three pairs of muscles perform all the positioning movements of the eyeball; (*B*) cross-sectional structure of the mammalian eye; (*C*) the retina: light enters from the top and passes through several layers of neurons before striking the photoreceptors (rods and cones). (*Adapted from Curtis, Helena, BIOLOGY, Worth Publishers, New York, 1968, pages 597 and 598.*)

	ROD OPSIN	CONE OPSIN	TABLE 8-1. DISTRIBUTION AND SENSITIVITY OF RETINENES AND OPSIN
RETINENE$_1$	Rhodopsin: λ max 500 Rods of mainly marine and terrestrial vertebrates	Idopsin: λ max 560 Cones of mainly terrestrial vertebrates	
RETINENE$_2$	Porphyropsin: λ max 522 Rods of mainly freshwater vertebrates	Cyanopsin: λ max 620 Cones of mainly freshwater vertebrates	

sensitivity of the retinenes and opsin are shown in Table 8-1. The symbol λ (the Greek letter lambda) stands for wavelength.

Actually, the maximum sensitivity of the pigments depends on the specific nature of the protein component. And since the opsins of different species may differ slightly from one another, the peak sensitivities also vary. For instance, the rhodopsin of marine fish absorb maximally at wavelengths that could be predicted from the differential penetration of light in the ocean.

TWO KINDS OF NEURONS CONTAINING VISUAL PIGMENTS ARE FOUND IN THE RETINA OF VERTEBRATE EYES: RODS AND CONES. RODS ARE SENSITIVE TO SHORT WAVELENGTHS OF LIGHT, IN THE BLUE TO BLUE-GREEN REGION OF THE SPECTRUM. CONES ARE SENSITIVE TO THE LONGER WAVELENGTHS OF YELLOW AND RED LIGHT. THE SENSITIVITY OF THE ROD AND CONE PIGMENTS DEPENDS ON THE NATURE OF THE PROTEIN TO WHICH THEY ARE BOUND.

SIMPLE PHOTORECEPTORS   An important quality of most, if not all, living systems is a sensitivity to light. The simplest protozoans, as well as fungi, algae, and higher plants, respond to light. Protozoans such as *Amoeba* and *Stentor,* which have no light receptors *per se,* respond to change in illumination by cessation of movement. The nerves themselves of echinoderms (starfish, sea urchins, etc.) seem to be generally light-sensitive. But even very simple organisms have special "organs," or *organelles,* for light reception. These often have an opaque pigment cup on one side so that light can enter from one direction only. These **ocelli** do not form an image, but they are useful in orientation.

The receptors of flatworms are hardly more complex, except that they are organs rather than organelles, since organelles are subunits of individual cells. Some ocelli are equipped with another accessory structure—a lens (Fig. 8-14). There is no evidence that ocellar lenses permit images to be formed on the sense cells. It is likely that they modify the incoming light in some way, however.

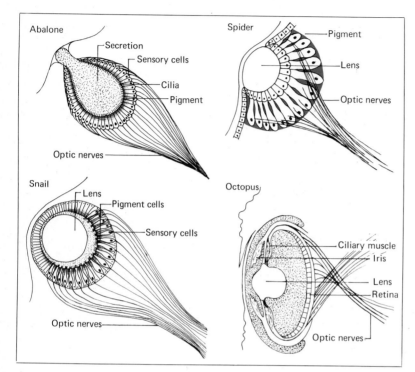

Fig. 8-14 Ocelli (simple eyes) of four invertebrates. (*After Wells.*) (*Adapted from Martin Wells, LOWER ANIMALS. Copyright McGraw-Hill Book Company, 1968. Used by permission of McGraw-Hill Book Company and Weidenfeld & Nicolson Publishing Company, Ltd.*)

Fig. 8-15 The compound eye of an insect.

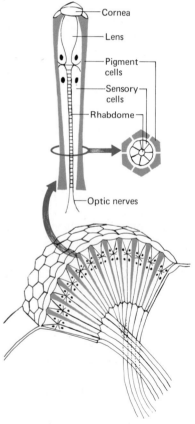

THE COMPOUND EYE   The large, faceted eyes of many arthropods are aggregations of visual elements called **ommatidia** (Fig. 8-15). Each ommatidium is a cone-shaped chamber with a cap (or cornea), a lens (crystalline cone) at the top, and a number of nerve cells at the bottom which secrete a central rod called the **rhabdome.** It is thought that the rhabdome is the site of visual pigment. Enveloping these structures is a sheath of cells, some of which are pigmented. Lateral nerve connections between the sense cells of adjacent ommatidia in insects and in the horseshoe crab can carry inhibitory signals, thus "silencing" ommatidia next to the one which is stimulated at a higher level. This could effect a sharpening of the perceived image.

Other aspects of the compound eye are worth noting. Our eyes can detect a flicker with a frequency of no more than 45 or 55 per second, whereas the eye of a blowfly is much more efficient, being able to detect a flicker of 265 per second. Light bulbs powered by 60-cycle current flicker 60 times a second—undetectable to us but easily detectable by the fly. This obviously would be advantageous for an insect with fast and erratic flight, such as a fly, since it would decrease blurring of small objects such as flowers or other insects as the fly passed rapidly over them.

In the eyes of vertebrates, compensation for or accommodation to changing light intensities is brought about by change in the size of the aperture of the iris (pupil). The same thing is accomplished in compound eyes by migration of pigment (Fig. 8–16). In bright illumination, pigment moves up around the sense cells to shade them; the pigment retracts in dim light. This response is hormonally mediated.

Insects not only are equipped with color vision but also are able to detect ultraviolet wavelengths. Ants are not so apt to move their pupae out of the sun if the ultraviolet light is filtered out. These abilities are not, however, the same thing as discrimination between wavelengths—the facility on which color vision depends. A visual receptor could be stimulated in precisely the same way by two different wavelengths of light, and thus it would be unable to distinguish between them.

The classical experiments demonstrating color vision in bees by Karl von Frisch are still a model of experimental design, although subsequent technological advancements have permitted more conclusive experiments. In addition, bees can detect the polarization of light. When light is passed through or reflected from certain substances, the normally random orientation of its electromagnetic field becomes directional; that is, the light becomes polarized. Bees are able to detect this change of orientation and use it to aid in navigation.

MOST ORGANISMS, IF NOT ALL, ARE SENSITIVE TO CHANGES IN LIGHT INTENSITY, AND EVEN VERY SIMPLE ANIMALS OFTEN HAVE SPECIALIZED PHOTORECEPTORS. THE LARGEST GROUP OF ORGANISMS, INSECTS, HAS HIGHLY DEVELOPED COMPOUND EYES, WHICH IN SOME WAYS CAN OUTPERFORM VERTEBRATE EYES.

THE VERTEBRATE EYE An intriguing question is why the compound eye reached such a stage of perfection in arthropods while a totally different visual system evolved in some worms, a few arthropods (crustaceans), squid and octopuses (cephalopods), and vertebrates. Both systems stem presumably from a simple ocellus, and both have undoubtedly evolved independently several times.

The vertebrate eye, typified by the mammalian optical system, is in broad outline very much like a camera, though only the most complex cameras (for instance, ones with built-in light meters which control the aperture of the lens and exposure time) begin to approach the eye in complexity.

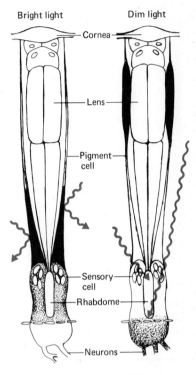

Bright light   Dim light

Cornea

Lens

Pigment cell

Sensory cell

Rhabdome

Neurons

Fig. 8–16 Migration of pigment in the compound eye of a shrimp. In bright light the rhabdome is shaded except from light perpendicular to the cornea. In dim light a rhabdome can receive light from nearly every direction.

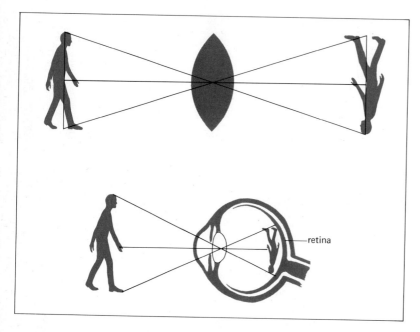

The eye (Fig. 8–13) is enveloped by a tough, lightproof membrane—the **sclera,** with its lining of black pigment. The **cornea,** a transparent saucer at the front of the eye, permits light to enter on its way to the **retina,** the light-sensitive surface. Between the retina and the cornea are several components. Behind the cornea is a transparent liquid, the **aqueous humor.** Behind this is the **iris,** a membrane in the center of which is the **pupil,** the hole or slit through which the light passes to the **crystalline lens.** Between the lens and the retina is the rear chamber of the eye, which is filled with the gelatinous **vitreous humor.** Various structures, such as eyelids, nictitating membranes, and transparent caps protect the eyes of various vertebrates from mechanical damage, keep the surface moist and clean, and in the case of lids, shut out the light.

THE LENS SYSTEM Two structures, the cornea and the crystalline lens, make up the lens system of the vertebrate eye. In a normal eye, these lenses converge the light and bring it to a focus on the retina, where a sharp image is formed. The properties of convex lenses result in an inverted image (Fig. 8–17).

There are three common abnormalities which are attributable to the lenses and which prevent light from being focused properly on the retina. **Hypermetropia** (or farsightedness) is the condition in which the focal point is behind the retina. Close objects appear to be even more out of focus or blurred than distant objects because the focal point is even farther behind the retina (Fig. 8-18). **Myopia** (nearsightedness) is the opposite of hypermetropia; the lens system converges the light too much, and the focal point is in front of the retina.

Astigmatism results from an improperly shaped lens. There is no single focal point, and when one part of the visual field is brought to focus on the retina, another part appears fuzzy (Fig. 8-18). The lenses that correct for astigmatism must be rather complex.

THE PUPIL The eyes of nocturnal animals and those which are crepuscular (active at dawn or dusk) can "open" very widely. That is, the iris can adjust the size of the pupil to varying light conditions. In bright light, the pupils of such animals contract to vertical or horizontal slits. The eyes of diurnal animals are not as light-sensitive, so the pupils do not

Fig. 8-18 (*A*) The focal point of light in normal and abnormal eyes; (*B*) lenses used to correct for hypermetropia (far-sightedness) and myopia (near-sightedness). (*Adapted from A. C. Guyton, FUNCTION OF THE HUMAN BODY, 3/e, 1969, W. B. Saunders Company, Philadelphia.*)

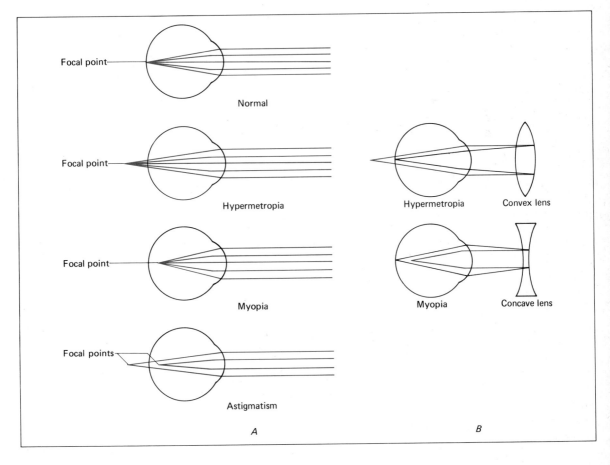

have to be able to close as far and are, therefore, typically round. In very bright light, the pupil of a man may close to a diameter of 1.5 mm, whereas in darkness it can open to 8 or 9 mm.

The iris is controlled by two sets of muscles. There is a circular set that acts as a sphincter, reducing the size of the aperture by contracting. When these muscles relax, the pupil enlarges. This can be aided by radial muscles, which are arranged like the spokes of a wheel. Light striking the retina can initiate a pupillary light reflex which closes the pupil almost instantaneously without the individual being conscious of it. In addition to reduced light, psychological factors also determine pupil size. It has been shown, for instance, that attractive visual stimuli (for instance, a picture of a beautiful woman shown to a man) trigger unconscious enlargement of the pupil.

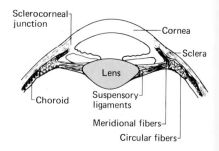

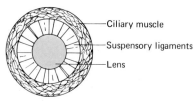

Fig. 8-19 The focusing mechanism of the eye. (*Adapted from A. C. Guyton, FUNCTION OF THE HUMAN BODY, 3/e, 1969, W. B. Saunders Company, Philadelphia.*)

ACCOMMODATION (FOCUSING) Focusing in vertebrates is accomplished by changing the shape or the position of the lens. In man the shape is changed. The mechanism consists of two rings of musculature on the internal wall of the eye—the **ciliary** muscles (Fig. 8-19). The meridional fibers release the tension on the suspensory ligaments of the lens when contracted. The circular fibers, also acting as a sphincter, reduce the tension still more. The lens is a fairly rigid structure, so that when the two sets of muscles are not contracting, the ligaments are under tension and the lens is flattened to some extent. This is why the human eye is focused for infinity when it is "relaxed." Contraction of the ciliary muscles permits the lens to assume a more ovoid shape, decreasing its focal length. In fishes, in contrast, focusing is accomplished by moving the lens in the same way focusing is done in a camera.

THE VERTEBRATE EYE RESEMBLES A CAMERA WITH AN AUTOMATIC LIGHT-METER–CONTROLLED APERTURE. IT HAS A LENS SYSTEM WHICH IS FOCUSED BY CHANGING THE SHAPE OF THE LENS OR BY CHANGING ITS POSITION (POSITION IS CHANGED IN A CAMERA) AND HAS AN IRIS DIAPHRAGM WHICH ENLARGES TO LET IN MORE LIGHT AND CONTRACTS TO EXCLUDE LIGHT (ALMOST EXACTLY LIKE A CAMERA).

THE RETINA The retina is a highly complex structure that, in vertebrates, is an outgrowth of the brain. As shown in Fig. 8-13, the retina has many layers of nerve cells. Light must pass through various layers before it actually strikes the photoreceptors (rods and cones) and is then absorbed by the pigment (melanin) behind. The eyes of albinos lack this light "sponge," and as a result, they are virtually blinded in bright light by the reflections back into the retina.

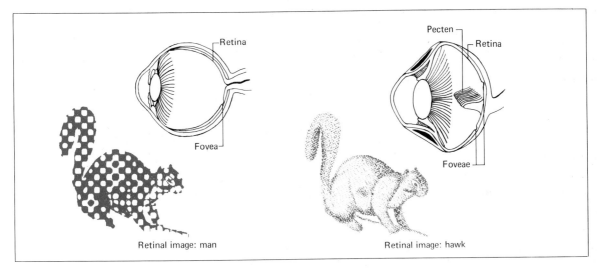

Retina

Fovea

Retinal image: man

Pecten

Retina

Foveae

Retinal image: hawk

Fig. 8-20 One of the reasons that birds-of-prey can see small animals while soaring at great heights is the great concentration of cones in their foveae (about eight times the density in a human eye). The illustration shows how this difference might improve the bird's visual resolution.

The rods can probably respond to a single quantum of light, which is the theoretical limit of sensitivity for any photoreceptor. The cones function only in bright light and are responsible for visual acuity and color perception. Unlike the rods, of which 10 to 100 connect to a single nerve fiber in man, usually no more than 2 or 3 cones are connected to 1 optic nerve. This explains the great resolution possible with cones.

The highest concentration of cones occurs at the **fovea,** the center of focus. Some birds of prey have two foveae, and the eye of a sparrow hawk (a falcon) is eight times as acute as a man's (Fig. 8-20). The ciliary muscles of birds are so placed that they can squeeze the lens for closeup vision, making the eye function alternately as a microscope for food search in the immediate vicinity and as a telescope for prey or predators.

As mentioned above, cones are the locus of color discrimination in the retina. Color vision is assumed to require at least two and probably three different color pigments, with different absorption maxima, but much still remains to be discovered.

THE VERTEBRATE RETINA CONTAINS RODS AND CONES AND IS BACKED BY A CURTAIN OF BLACK PIGMENT WHICH PREVENTS LIGHT FROM BEING REFLECTED ONTO THE PHOTOSENSITIVE CELLS. RODS ARE MORE LIGHT-SENSITIVE THAN CONES AND ARE HOOKED INTO THE OPTIC NERVE IN LARGE GROUPS, SO THAT THEY PROVIDE RELATIVELY LOW RESOLUTION. CONES ARE LESS SENSITIVE INDIVIDUALLY, BUT ONLY TWO OR THREE ARE CONNECTED TO A SINGLE OPTIC NERVE, THUS PROVIDING GREATER RESOLUTION. COLOR

*VISION IS CENTERED IN THE CONES, WHICH CONTAIN SEVERAL DIF-
FERENT PIGMENTS WITH DIFFERENT ABSORPTION PROPERTIES.*

Chemoreceptors

Just as beauty is a function of the perceptual mechanism of the beholder, so smell and taste are strictly subjective values assigned to molecular stimuli. There is nothing intrinsically "sweet" about sugar. Rather, our receptors and nerve hookups are such that molecules of a certain shape elicit the sensation we call "sweetness"—a sensation which is programmed into the nervous system. And, for sugar, it is perfectly clear why this is so. Sugar is an energy-rich compound and one that is efficiently converted to ATP. It is clearly adaptive that foods containing sugar, such as ripe fruits, nectar, and honey, be considered pleasant by many animals.

Sweet is one of the four primary taste sensations of mammals. The others are salty, bitter, and sour. These four basic flavors are detected by the **taste buds,** groups of sensory cells in the mucous membrane of the mouth. The taste buds that are sensitive to salt actually respond to electrolytes in general. The "bitter" buds respond primarily to alkaloids and other noxious plant materials. These plant biochemicals evolved as herbivore poisons, and our taste buds have developed a poison detection-avoidance system. The "acid" taste buds are pH detectors. Stimulation of these elicits a moderately pleasant sensation if the stimulation is slight, but a strong concentration of acid, such as in vinegar, is perceived as unpalatable.

It is the nose, however, that makes a gourmet. The olfactory epithelium of the nasal passages contains sense cells many times more sensitive and discriminating than the taste buds. When a cold occludes these areas with mucus, the sense of taste is said to disappear. Actually, the taste buds are functioning normally, but the important olfactory component of taste is temporarily lost.

Our world is mostly a visual one, and it is next to impossible for us to appreciate the world of a mole, earthworm, shark, dog, or moth—organisms whose olfactory senses are much more highly developed than ours. Aquatic and semiaquatic organisms are constantly "taste-smelling" their environment with receptors located on convenient points on the body surface, and the sensitivity of these senses is often remarkable. If a man were suddenly to be equipped with the chemical senses of a dog or a fish, the effect would be staggering. It might be like being able to see color after having been aware of only shades of gray all one's life. Biologists would be pleased to discover that plant identification would be infinitely easier if they had enhanced chemoreception. Almost certainly also, those confusing microorganisms under the microscope would be more readily understood if we could "taste" them, that is, readily determine their chemical characteristics. Whereas vision is paramount in primates, cats, and most birds, chemical sense is a highly important sensory window for most other animals.

THERE ARE FOUR BASIC TASTE SENSATIONS: SWEET, SOUR, SALTY, AND BITTER. THE REMAINING "TASTE" SENSATIONS ARE ACTUALLY SMELLS. MAN LIVES IN A PRIMARILY VISUAL WORLD, BUT MANY OTHER ORGANISMS HAVE MUCH MORE HIGHLY DEVELOPED CHEMO-RECEPTORS AND LIVE LARGELY IN A TASTE-SMELL WORLD.

Many animals can sniff out food underground. Dogs and pigs can be trained to find truffles (a fungus relished by gourmets) in oak forests, and the kiwi, with nostrils at the end of its bill, can sniff out worms underground. *Aedes* mosquitoes have carbon dioxide receptors which help them to locate their vertebrate prey. The parasitic wasp *Ephialtes* deposits eggs in the body cavity of the larvae of other insects, including wood-burrowing forms, which it can detect beneath the bark of trees. Certain *Euptychia* butterflies lay their eggs in the vicinity of their larval food plant. Before they lay, however, they must alight on the proper food plant, tasting it with their feet. These are examples of two of the five principal functions of chemical sensitivity: food finding and host (or oviposition-site) finding.

The other three functions are (1) sexual, (2) protective, and (3) habitat selection and territoriality. For many organisms who, unlike man, live solitary lives, finding a mate is not simply a matter of looking. During the breeding season, the spawning of some sea urchins is coordinated by chemicals released with the eggs. When these are detected by nearby males, they in turn release milky clouds of sperm. This starts a chain reaction of detection followed by spawning.

Many animals use the olfactory sense to locate their mates. Gypsy moth males can detect a female up to 2 miles away by virtue of their great sensitivity to the **pheromone** (a chemical message released into the environment) released by her scent glands. Male snakes are known to track a female by scent.

Chemical sensitivity is helpful in avoiding dangerous encounters. On the other hand, the release of a noxious chemical can often forestall an attack, a tactic well developed in skunks, stinkbugs (actually beetles), and millipedes. Many fish can distinguish different species of fish, including predators, by their odor. The subject of chemical defense will be discussed in Chap. 11.

It is well known that salmon on their return trip from the ocean invariably end up in the stream where they themselves were spawned. It is now thought that their highly sensitive noses and an excellent memory explain this feat. The fish navigate by the stars to their home river system and then by "smell" to their home branch. Blocking their noses prevents them from making the proper choice at each fork. Many other animals use the sense of smell in recognizing their home. Dogs and wolves mark out their "territory" by urinating on obvious sites such as bushes and rocks. Other dogs recognize and respect these invisible "no trespassing" signs. Social insects recognize the members of their colony by smell and will kill even their own colony-mates if they are experimentally rubbed with the smell of another colony.

The precise molecular basis of chemoreception is not known. Presumably the perceived molecules must first be dissolved in the membrane of the sense cells, and the shape of the molecule must be important. But attempts to correlate shape and odor have not been successful. Sucrose and saccharin, for instance, have very different molecular structures yet taste alike to man. The picture is complicated even further because species differ in their perceptions. Cats and cows seem to lack a sweet taste altogether, and saccharin may taste bitter to dogs. What is known about chemoreceptors is outlined in the next section.

ANIMALS USE CHEMORECEPTORS TO HELP THEM LOCATE FOOD, OVIPOSITION SITES, MATES, OR SUITABLE HABITATS. CHEMORE- CEPTORS MAY ALSO SERVE TO WARN AN ANIMAL OF DANGER. THE EXACT MOLECULAR MECHANISM OF CHEMORECEPTION IS UN- KNOWN.

MECHANISMS OF CHEMORECEPTION In vertebrates, the sense of smell depends on the contact of "odor-bearing" molecules with the specialized nerve cells, the ends of which project between the other cells of the surface of the nasal organ. This outer region of the cell is covered by a layer of mucus or other liquid. Each nerve cell is isolated from other nerve cells by special insulating cells. By recording electrical impulses of individual receptor cells, it has been possible to show that they are specialized for detecting different classes of molecules. It seems certain that the different kinds of receptor cells have different kinds of receptor molecules on their projecting ends—that is, they differ in the ultrastructure of their cell membranes. In some way, the stimulant molecule "fits" the molecules of the membrane, and this leads to an opening of the "sodium gate" and the initiation of a nerve impulse. The situation is clearly quite complex. For instance, the odor perceived may change with the concentration of the stimulating molecules. In low concentrations, one thing is "smelled"; in high concentrations, another. Apparently the distribution of receptor cells in the sensory surface is important, numerous cells showing many grades of response to a given stimulus. These various reactions are summed, in a manner not clearly understood, to give the sensation of smell. It has also been shown that the spatial pattern of the cells on the sensory surface is reflected in their terminations on the brain. The endings of those cells which respond to oil-soluble chemicals are at one end of the portion of the brain in which the sense of smell is localized; the endings of those which respond to water-soluble chemicals are at the other.

THE ABILITY TO SMELL IN VERTEBRATES ALMOST CERTAINLY DEPENDS ON THE ASSOCIATION OF "ODOR-BEARING" MOLECULES WITH DIFFERENT RECEPTOR MOLECULES ON NERVE CELLS. THE ACTUAL SENSATION OF SMELL IS SOME SORT OF SUMMATION OF RESPONSES OF DIFFERENT CELLS TO THE ODOR-BEARING MOLECULES PRESENT AND TO THEIR CONCENTRATIONS.

Taste receptors are different in detail from smell receptors. No one type of cell responds only to "bitter" substances, another only to "sweet," "salt," or "acid," the four basic taste sensations. Rather, individual cells initiate impulses when stimulated by a range of chemicals which may fall into several taste classes. The perceived taste is, like a perceived smell, a summation of messages from many cells.

Chemoreception has been studied in some detail in insects, and these studies have helped to make clear the intricacy of the interactions which must occur at the cell surface. For instance, it was discovered that two sugars, glucose and fructose, are excellent stimulators of the taste receptors on the leg of the blowfly *Phormia*. Both caused reactions in concentrations less than one-fiftieth of the concentration required of a third sugar, mannose. The presence of mannose in a mixture with fructose inhibited the stimulatory ability of the fructose. Mannose, however, was unable to impede the action of glucose. Possibly mannose and fructose combine with the same molecular sites on the cell membrane, whereas glucose combines with a separate site. On the other hand, glucose may be able to displace mannose from active sites, and fructose may not have this ability.

It should be apparent from the above discussion that the most critical mystery remaining in the area of chemoreception is that of the exact events which take place at the membrane surface of the cells. There are, of course, other important perceptual questions, but we will indicate these in our discussion of response selection and levels of neural organization below.

TASTE-RECEPTOR CELLS ARE CAPABLE OF RESPONDING TO MORE THAN ONE OF THE FOUR BASIC TASTE QUALITIES: SWEET, SALT, BITTER, AND ACID. COMPLEX MOLECULAR INTERACTIONS SEEM TO BE INVOLVED, IN ADDITION TO A FORM OF SUMMATION, IN PRODUCING THE SENSATION OF TASTE.

Fig. 8-21 *(A)* Touch, pressure, and other receptros in human skin; there is uncertainty about the function of some of these; *(B)* insects can detect strains in their exoskeleton (cuticle) because of the presence of receptors called sensilla; the sensilla shown detect stress in one direction only; compression of the cuticle along the other axes merely binds the cuticle along the sides of the sensilla but does not distort them. *(A, Adapted from Kimball, BIOLOGY, 2/e, 1968, Addison-Wesley, Reading, Mass; B, Adapted from Martin Wells, LOWER ANIMALS. Copyright McGraw-Hill Book company, 1968. Used by permission of McGraw-Hill Book Company and Weidenfeld & Nicolson Publishing Company, Ltd.)*

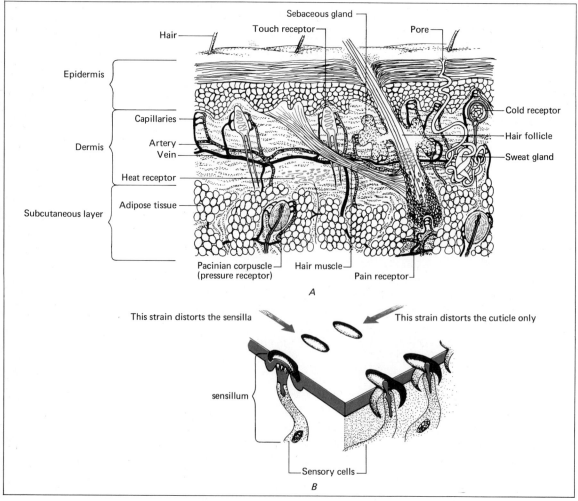

The response to a mechanical stimulus always involves a deformation of the cell membrane of a receptor cell. Pressure and stretch cause relatively long-term deformations and can be contrasted with shorter-term but repetitious deformations caused by earth-, water-, or airborne vibrations.

The receptor cells are either neurons with modified endings (Fig. 8–21 or specialized epithelial cells. When the latter are stimulated, the resulting potential triggers an impulse in an adjacent neuron. Often the receptor cells are studded with minute hairs, and it is these that probably deform the membrane when a shear force is applied to them.

PRESSURE AND STRETCH RECEPTORS Many pressure and stretch receptors have been found. Some are naked nerve endings which are modified into knobs, spirals, and other shapes. Some are encapsulated like the pacinian corpuscle (Fig. 8–21) of mammalian tendons and con-

Pressure and Movement Receptors

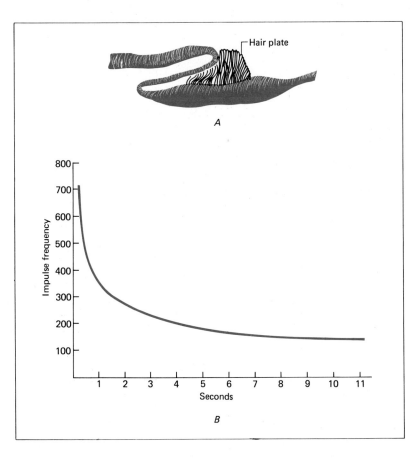

Fig. 8–22 (*A*) A hair plate at the leg joint of a cockroach (*Periplaneta*); deflexion of the hairs caused by movement in the joint increases the frequency of discharge from the neurons at the base of the hairs; (*B*) adaptation of a neuron in the above hair plate; the discharge freuqency gradually decreases following the movement. (*After V. G. Dethier: PHYSIOLOGY OF INSECT SENSES, Associated Book Publishers, 1963.*)

nective tissue. The stretch receptors of invertebrates and vertebrates are typical deformation-sensitive neurons. Insect pressure sensitivity is indirect. The cuticle of insects is more or less inflexible. Nevertheless, pressures do set up strains, and special receptors, such as the bell-shaped ones shown in Fig. 8–21, detect these. Bending of the body or an appendage alters the configuration of a joint. Hair plates (Fig. 8–22) are located in the joints to detect such movement. Like many receptors, these *habituate* quickly to stimulation (Fig. 8–22). (A good way to observe habituation is in your own heat receptors. Put your left hand in cold water and your right hand in warm water; wait a few minutes until neither feels warm or cold, and then put them both in room-temperature water. You will note the degree to which the heat receptors in the skin of each hand have adapted to the previous temperature regime.)

Another large group of pressure receptors comprises the organs that perceive gravity. Typically, these are spheres containing *statoliths,* sand-like grains that are either secreted or picked up from the environment, such as the sand grains used by crustaceans (Fig. 8–23). These spheres, or *statocysts,* are lined with sensory hairs. As the animal changes its orientation with respect to the gravitational force, the statolith moves, changing the force on the hairs with which it is in contact.

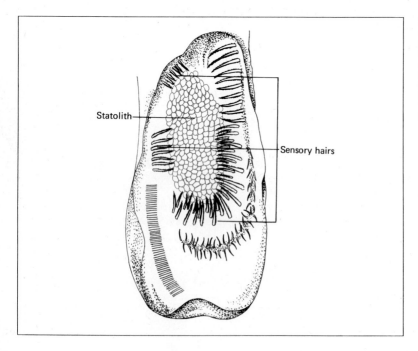

Statolith

Sensory hairs

Fig. 8–23 The statocyst at the base of an antenna of a lobster. The sand grains are bound together by a secretion from the wall of the chamber. As the animal moves, the statolith differentially stimulates the rows of sensory hairs. (*Adapted from R. O. Barnes, INVERTEBRATE ZOOLOGY, 1/e, 1963, W. B. Saunders Company, Philadelphia.*)

Crustaceans replace the statolith with each molt. A tidy experiment has demonstrated the function of the statolith beyond any doubt. A magnet suspended over the head of a crab in which an iron filing had been substituted for the sand grain caused the crab to roll over—it had perceived that up was down, as it were. Similar structures occur in vertebrates, and these will be discussed at the end of this section.

PRESSURE AND MOVEMENT ARE DETECTED WHEN THE MEMBRANE OF A RECEPTOR CELL IS DEFORMED, TRIGGERING A NERVE IMPULSE. DEFORMATION-SENSITIVE NEURONS DETECT SUCH THINGS AS THE MOVEMENT OF HUMAN MUSCLES, THE BENDING OF AN INSECT'S CUTICLE, AND THE FORCE OF GRAVITY ACTING ON A CRAB.

PROPRIOCEPTION Proprioceptors keep the animal informed about the condition and relationships of its various parts. The difference between a well-coordinated individual and one who is not is partly that the former has a better proprioceptive system. Basketball and piano playing are two of the very many activities requiring immediate and accurate input of this type. Many of the receptors discussed above are proprioceptors as are organs of sight and touch receptors when used to provide information about the relative positions of body parts and the relation of the organism as a whole to gravity. The use of proprioceptors can be shown very clearly using insect examples.

In insects, the relative positions of body parts are often indicated by sets of sensory hairs on one part which trigger nerve impulses when they are brushed against other parts. Hair plates in the neck region detect the orientation of the head relative to the thorax (Fig. 8-24). Bending the head to the left, for example, stimulates the sensory cells of the hair plates on the left side while decreasing the sensory input from those on the right. In his studies of prey capture in mantids, A. Mittelstaedt observed that the strike of the legs is too rapid (10 to 30 msec) to be explained by visual guidance. In other words, the muscular contractions that control the entire strike must be programmed in the nervous system before the strike begins. A mantid may stalk its prey or lie in wait for it. When the mantid is close enough to strike, it turns its head so as to face the prey directly. The thorax, to which the striking legs articulate, is often *not* lined up with the prey, which means that the legs must strike *not* straight ahead relative to the thorax but at the point where the eyes are looking (Fig. 8-24). The angle of the strike must be preset on the basis of the angle the head makes with the thorax. Sensory hairs in the neck region provide these data and are used to aim the legs. Cutting the nerves to these hair plates reduces the accuracy of the strike to about 25 percent.

Fig. 8-24 The praying mantis always stares directly at its prey. The angle of strike is equal to the angle the head makes with the thorax. This information is provided by the hair plates at the neck. The width at the arrows is proportional to the impulse frequency in the neurons coming from the hair plates. (*Adapted from Curtis, Helena, BIOLOGY, Worth Publishers, New York, 1968, page 576.*)

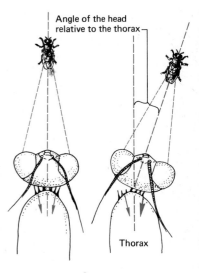

Angle of the head relative to the thorax

Thorax

*PROPRIOCEPTION IS THE PROCESS BY WHICH AN ANIMAL IS MADE
AWARE OF THE POSITION AND FUNCTIONING OF ITS OWN BODY
PARTS. MANY PRESSURE AND STRETCH RECEPTORS ARE PRO-
PRIOCEPTORS.*

VIBRATION RECEPTORS Vibration and sound receptors respond to
compressional waves set up in gases, liquids, and solids. The human ear
is sensitive to vibrations down to about 20 cycles per second (cps).
Vibrations of lower frequencies are "felt" by mechanoreceptors. For
the purpose of description, we will define vibration receptors as those
sensitive to frequencies between 0.1 and 100 cps. In insects, substrate-
borne vibrations are detected by sensory cells in the joints of the appendages. There are many kinds of these, including hair plates.

In fish and some amphibians, the lateral-line system is the primary site
of vibration reception. The system is named for the characteristic canal
running down the side of the body (Fig. 8–25). There are often rami-
fications of the canal system on the head. These canals contain sensory
hair cells and open to the outside through pores. Recently it has been
established that this system gives fish a "sixth sense," something like a
low-frequency sonar. Information about the environment is always impinging on a fish in the form of ripples and pressure waves. These may
be echo waves built up as the fish passes solid objects such as rocks or
waves caused by the movement of other organisms.

SOUND RECEPTION Many structures have evolved to detect the waves
in matter we call "sound." These waves have two characteristics: frequency (distance between peaks) and amplitude (height of the peaks).
The velocity of sound depends on the medium in which it is traveling.

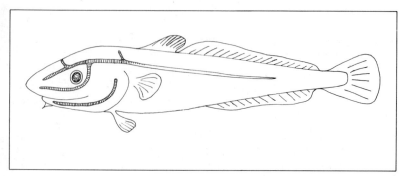

Fig. 8–25 The distribution of lateral
line canals on a fish. Like most sense
organs, the canals are concentrated
in the head region.

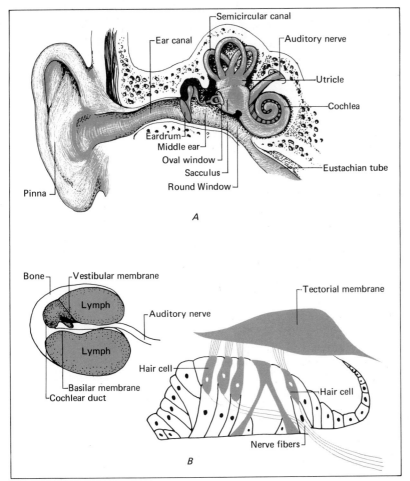

Semicircular canal

Ear canal

Auditory nerve

Utricle

Cochlea

Eardrum
Middle ear
Oval window
Sacculus
Round Window

Eustachian tube

Pinna

A

Bone
Vestibular membrane

Lymph

Auditory nerve

Tectorial membrane

Lymph

Hair cell

Basilar membrane
Cochlear duct

Hair cell

Nerve fibers

B

Fig. 8-26 *(A)* The human ear. The middle ear bones (ossicles) transmit the sound from the eardrum to the oval window of the cochlea. *(B)* The cochlea. Sound-induced vibrations of the basilar membrane rub the sensitive hair cells against the tectorial membrane, thus changing the rate of impulse propagation down the nerve fibers. *(Adapted from BIOLOGICAL SCIENCES, 1/e, by William T. Keeton, illustrated by Paula DiSanto Bensadoun, by permission of W. W. Norton & Company, Inc. Copyright © 1967 by W. W. Norton & Company, Inc.)*

In air at sea level at 20°C, this velocity is 1,125 ft per second; in water, it is about 5,300 ft per second.

The ears of arthropods, called **tympanic organs,** are capable of a great range of sound detection, but we will describe the better-known ear of mammals, more particularly that of man. The outer ear concentrates the sound by reflection to the ear drum (Fig. 8-26). The vibrations registered on the ear drum are passed through the air-filled middle ear, which contains three tiny bones whose shapes have caused them to be given the names "hammer" *(maleus),* "anvil" *(incus),* and "stirrup" *(stapes).* These three pass the vibrations on to the inner ear. In the process, the vibrations are amplified by a factor of 22 because of the ratio of areas between the tympanic membrane and the stapes. The middle-ear chamber connects to the throat through the eustachian tube. Damage to the ear that might be caused by a difference in pressure between the inside and outside (such as may be caused by flying or diving) is prevented by this opening, which acts as a valve, equalizing the pressure inside and outside the tympanic membrane. Ear "popping" is the result of air squeezing in or out through the tube.

ANIMALS HAVE VARIOUS ORGANS FOR THE DETECTION OF VIBRATION (LOW-FREQUENCY COMPRESSIONAL WAVES) AND SOUND (HIGH-FREQUENCY WAVES). IN THE HUMAN EAR, VIBRATIONS ARE TRANSMITTED FROM THE OUTER EAR TO THE EAR DRUM BY REFLECTION AND FROM THE EARDRUM TO THE RECEPTORS BY A SERIES OF THREE TINY BONES.

The stapes is addressed to the oval window of the **cochlea,** the snail-shaped part of the inner ear that changes the mechanical sound energy to the electrochemical energy of nerve impulses. The coiled shape of the cochlea probably has little to do with the biophysics of sound reception. Like a trombone or the cooling coils in a refrigerator, the coiling is a means of achieving length in a small space. The sensory hair cells are located on a membrane known as the basilar membrane (Fig. 8–26). The cochlea works as follows: Vibrations are initiated at the oval window by the action of the stapes. The fluid contents of the cochlea can move a small amount. As the fluid is pushed against the basilar membrane, it bulges into the lower chamber, causing a bulge in the elastic membrane of the round window. The process is reversed when the sound vibration causes the stapes to move away from the oval window. The cochlea is a resonating chamber. Depending on the frequency of the vibration of the stapes, waves are set up in various parts of the cochlea—near the base if the frequencies are high and near the top if the frequencies are low. The shear forces are greatest over the part of the basilar membrane vibrating the most. This initiates impulses in the adjacent nerve fibers. Therefore, the interpretation of pitch (frequency) depends on which part of the basilar membrane is vibrating the most. That is, the brain interprets impulses from the nerves near the base of the cochlea as high-pitched. Amplitude, or loudness, perception is a function of the force of the vibration—the greater the displacement of the membrane, the more the impulses sent to the brain.

THE COCHLEA FUNCTIONS AS A RESONATING CHAMBER IN WHICH A MEMBRANE VIBRATES IN DIFFERENT PLACES ACCORDING TO THE PITCH (FREQUENCY) OF A SOUND. THE FORCE OF THE VIBRATION

*IS PROPORTIONAL TO THE LOUDNESS OF THE SOUND. BOTH THE
AREA OF GREATEST MEMBRANE VIBRATION AND THE FORCE OF
VIBRATION ARE RECORDED BY NERVE ENDINGS ADJACENT TO THE
MEMBRANE AND ARE TRANSLATED BY THE BRAIN INTO "SOUND."*

SENSE OF ACCELERATION   In vertebrates, a second region of the inner
ear, the vestibular apparatus, perceives the orientation and movements
of the head. The lateral-line system and vestibular apparatus are part of
the same system in lower vertebrates; the terrestrial vertebrates have lost
the former. In man, the apparatus consists of three semicircular canals
and two chambers called the **utriculus** and **sacculus.** These are all interconnected and bear an exact geometric relationship to one another (Fig.
8–26). The utriculus and sacculus contain one or more statoliths. In
man and other vertebrates, many small granules embedded in a gelatinous matrix are the inertial mass. These are manufactured by the organism; vertebrates do not depend on accumulating sand grains from the
environment, as do crustaceans. A man tilting his head backward causes
the sensory hairs under the inertial mass to be bent forward. Similarly,
when a person moves forward suddenly, as in a ball game, the inertia of
the mass causes the mass to lag behind. In both cases, there is an automatic response to compensate.

The semicircular canals provide more specific information about the
attitude of the head. Most vertebrates have three tubular canals, which
end in swellings that contain sensory hair cells. Acceleration in any
place stimulates at least one of these groups of cells because of the inertia of the fluid in the canals. Stretch receptors in the neck provide information about the attitude of the head relative to the body. Loss of
inner-ear function drastically affects equilibrium. Humans with such a
condition can still move and maintain balance, but only so long as they
move very slowly. Under this circumstance, pressure receptors and visual
orientation substitute for the inner-ear input, but they do not function nearly
as rapidly as the inner ear.

*THE SENSE OF BALANCE AND INCLINATION IN MAN DEPENDS ON
THE BEHAVIOR OF INERTIAL MASSES IN THE CHAMBERS AND SEMI-
CIRCULAR CANALS OF THE INNER EAR.*

FUNCTIONS OF MECHANORECEPTION AND SOUND RECEPTION
Mechanoreception and sound reception often play important roles in
finding the proper habitat and also in holding onto it once it is staked
out. The former is illustrated by the "settling" behavior of barnacle

larvae. When a barnacle larva has reached the proper stage (it has been a free-drifting larva up until this point), it seeks a place to attach itself. Other factors being equal, it prefers a rough substrate to a smooth one. In other words, where a barnacle spends all its adult life is determined by information received by receptors at the crucial settling stage.

Often animals defend their territories vigorously, even to the point of drawing blood from the trespasser. Usually though, violence is unnecessary because there are ways of advertising occupancy and thus avoiding accidental trespasses. The mockingbird calls day and night from a high perch in his nest area. Such a calling bird is given a wide berth by the other birds, who hear and understand his singing. On a warm spring night the mockingbird's song is an enchanting and peaceful encore to the day, but the meaning to other mockingbirds is less aesthetic. (In terms of natural selection, can you explain why it is advantageous to advertise one's territory?)

Sound serves as an important tool in sexual reproduction for many animals since it produces the signals which bring the sexes together. To appreciate this, one need only step out-of-doors after a spring rain. The calls of crickets, cicadas, birds, and frogs can be deafening at times. Even the calls of fish—for example, croakers and grunts—and lizards (some geckos in the deserts of southern Africa call from their burrows) are thought to have a reproductive function.

Many animals perceive the presence of prey by hearing a sound made by the prey. But insectivorous bats, as well as porpoises and whales, do not depend on chance noises made by their potential prey. As is now common knowledge, these animals hear the echos of their own sounds and can discriminate among these echos with amazing acuity. Bats, navigating solely on the basis of these echo returns, dodge fine wires while flying at full speed and are able to pick flying insects out of the air. A blind-folded porpoise, using the same kind of system, can swim full tilt across a pool and accurately hit a dime-sized object. Presumably this ability also permits them to hunt down fast-moving prey in murky water or at night.

HEARING AND MECHANORECEPTION, AS ONE WOULD EXPECT, SERVE DIVERSE FUNCTIONS IN ANIMALS. THESE RANGE FROM SELECTION OF A "SETTLING SITE" BY BARNACLES TO MATE FINDING IN CRICKETS AND FROGS.

We have only scratched the surface of the vast, and in some ways poorly understood, field of the reception of information from the environment. Yet we know enough to enunciate a rather important principle: the per-

PERCEPTUAL WORLDS

ceptual worlds of organisms are strongly dependent on the nature of their receptors. We live in a world of sight; our chemical senses and hearing are weak and subordinate to our vision in conducting our everyday affairs. Bats and many porpoises, on the other hand, live in a world dominated by sound. They are able to put together an extremely accurate picture of their surroundings solely on the basis of echos of sounds they generate. There is some evidence that eyes are almost superfluous in the adults of some whales. A sperm whale was caught that was found to be blind but perfectly healthy. Hoofed animals usually live in a world of sound and smell, and various kinds of chemoreception control the lives of many invertebrates. Perhaps the strangest of all are the electric fish. These animals are virtually sightless. They generate an electric field and have receptors which permit them to detect deformations in this field. They use this field to orient to their environment; both prey fish and objects such as rocks cause deformations which they are able to detect. These organisms live in a difficult-to-imagine world of deformed electric fields.

Most people tend to think of "reality" as that which they can detect with their receptors. As we will see in the next section, there is nothing even approaching a one-to-one relationship between the information "received" by our receptors and our perceptions. But even if there were, would this be a perception of reality? Does a color-blind person see the real world? Do you see the real world? Does a bee, with its ultraviolet vision, see the real world? Or is the real world made up of smells or deformed electric fields? Perhaps you can see why most scientists are uninterested in such questions as: Is there a sound when a tree falls in a forest and no one is around to hear it? Questions of "what's really out there" are meaningless unless asked in the context of a perceiving system.

ANIMALS LIVE IN VERY DIFFERENT "WORLDS," DETERMINED IN LARGE PART BY THE CAPACITIES OF THEIR PERCEPTUAL SYSTEMS. THE "SOUND" WORLD OF A BAT IS NEITHER MORE NOR LESS REAL THAN THE "SIGHT" WORLD OF A MAN OR THE "ELECTRIC-FIELD" WORLD OF AN ELECTRIC FISH.

SUMMARY

Nervous systems provide more rapid and precise coordination than hormonal systems. The basic elements of the nervous system are neurons, cells specialized for the transmission of electrochemical information (nerve impulses).

A resting neuron has, like other cells, an electrical potential across its membrane. The nerve impulse consists of a reversal in the polarity of the cell membrane involving the rapid passage of sodium and potassium ions across the membrane. A stimulus, in an unknown manner, causes changes in the differential permeability of the membrane to permit this "depolarization," creating an action potential. The formation of an action potential at one point leads to the formation of action potentials at adjacent points, triggering a wavelike nerve impulse.

Nerve impulses spread in all directions from the point of stimulus on a single neuron. This passage through the nervous system is made uni-directional, however, by the one-way synapses which link neurons. Synapses are extremely narrow clefts between neurons; they may be bridged by a transmitter hormone ejected by the presynaptic neuron which de-polarizes the postsynaptic neuron. Less frequently, the synapse is bridged electrically. Synapses can involve inhibitory transmitter substances, which function antagonistically with the excitatory substances.

Animals have evolved receptors which can detect environmental changes in electromagnetic radiation, temperature, chemistry, pressure and movement, vibration, and electric fields. The nature and capability of these receptors are the major determinants of the kind of "world" in which the animal lives.

SUPPLEMENTARY READINGS
Books

Sherrington, C. S. "The Integrative Action of the Nervous System," Yale University Press, New Haven, Conn., 1906.
Van der Kloot, W. G. "Behavior," Holt, New York, 1968.

Articles

Eccles, J. The Synapse, *Sci. Amer.,* **212,** no. 1, Offprint 1001, (1965).
Katz, B. How Cells Communicate, *Sci. Amer.,* **205,** no. 3, Offprint 98 (1961).
Lissmann, H. W. Electric Location by Fishes, *Sci. Amer.,* **208,** no. 3, Offprint 152 (1963).
Lowenstein, W. R. Biological Transducers, *Sci. Amer.,* **203,** no. 2, (1960).
Miller, W. H., F. Ratliff, and H. K. Hartline. How Cells Receive Stimuli, *Sci. Amer.,* **205,** no. 3, Offprint 99 (1961).
Roeder, K. D. Moths and Ultrasound, *Sci. Amer.,* **212,** no. 4, Offprint 1009 (1965).
Rushton, W. A. H. Visual Pigments in Man, *Sci. Amer.,* **207,** no. 5, Offprint 139 (1962).
Todd, J. H. The Chemical Languages of Fish, *Sci. Amer.,* **224,** no. 5, Offprint 1222 (1971).
von Békésy, G. The Ear, *Sci. Amer.,* **116,** no. 4, Offprint 44 (1957).

CHAPTER 9

INTEGRATION III: EFFECTORS AND SYSTEM FUNCTIONS

At this point we have outlined the elements of coordination in hormonal and nervous systems and we have discussed a variety of receptors that comprise the input elements. We will now look at the output elements —the effectors. When we speak of behavior, action, or movement, we are speaking of effectors and their operations.

EFFECTORS

Some effectors are complete in themselves. These *independent* effectors act as a receptor-effector unit. Sometimes they are an organelle within a cell, sometimes a cell, and sometimes a group of cells. Guard cells control the size of stomata without any control such as that exercised over growth movements by auxin. The self-contained harpoons (nematocysts) of sea anenomes and other coelenterates are equipped with their own hair triggers. The oscular muscles that control the flow of water through the pores of sponges are also independent. Figure 9–1 shows the latter two types of independent effectors.

Most effectors receive commands rather than act as independent agents —a generalization that applies, in particular, to higher animals. There are many types of these *dependent* effectors: cilia and flagella (Box 7-3), chromatophores, luminescent organs, and electric organs, as well as the glands secreting their products into the blood (such as the adrenal glands) and glands secreting into body cavities or on the body surface (such as salivary and sweat glands). But the most versatile type of effector tissue is muscle. It would not be possible to enumerate here all the functions performed by muscle, but a few examples will suffice.

Muscles

All movement of body parts, and thus all animal behavior, is due to the coordinated contraction of muscles. In addition to effecting movement, muscle contraction also mediates voice and song in man and other animals, the amount of light entering the eye, the blood pressure, thermo-

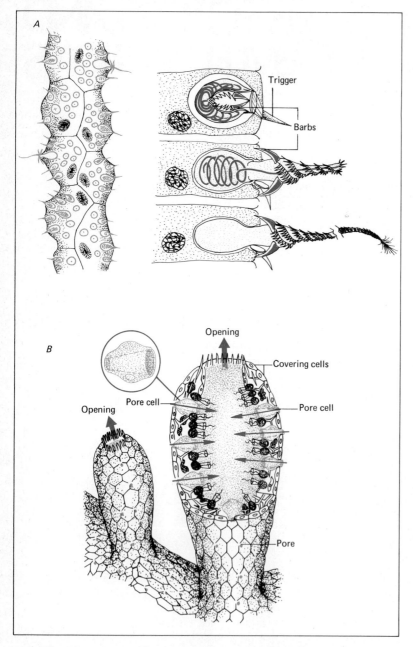

Fig. 9-1 Two kinds of independent effectors found in animals: (*A*) the tentacle of a hydra with the distribution of stinging hairs called nematocysts and the discharge of a nematocyst; (*B*) a simple sponge and the pore cells that regulate the flow of water into the animal by contracting or expanding. (*Adapted from ANIMALS WITHOUT BACKBONES by R. Buschbaum, The University of Chicago Press, 1938; and from Martin Wells, LOWER ANIMALS. Copyright McGraw-Hill Book Company, 1968. Used by permission of McGraw-Hill Book Company and Weidenfeld & Nicolson Publishing Company, Ltd.*)

regulation, movement of food in the gut, and the birth of a child—to name just a few. The arrangement and action of some typical muscles are shown in Fig. 9-2. Notice how the muscles are divided into groups with opposing actions.

EFFECTORS MAY BE DEFINED AS THE OUTPUT ELEMENTS OF INTE-GRATING SYSTEMS, ALTHOUGH SOME ACT AS INDEPENDENT RE-CEPTOR-EFFECTOR UNITS. ANIMALS HAVE MANY DIFFERENT KINDS OF EFFECTOR TISSUES, OF WHICH ONE OF THE MOST IMPORTANT IS MUSCLE.

Muscle is so important that a more detailed examination of its structure is called for. Muscle is not the easiest tissue to understand. Its micro-scopic and molecular architecture is highly variable. Furthermore, its fundamental unit is not always the cell but the **fiber**—a spindle-shaped unit that may be one elongated cell or may be the accretion of many cells, the nuclei of which are scattered about the surface, as shown in Fig. 9–3. The so-called "grain" of meat (except for a few organs such as liver, kidney, and thymus, most meats are muscle) is due to the parallel arrange-ment of these long fibers. Each fiber is usually encased in a double mem-brane, the outer layer of which is continuous with those of adjacent fibers, forming the connective tissue of the muscle. In vertebrate skeletal muscle, this connective tissue is continuous with the shining white **tendons** that bind the muscle to the bone.

Fibrils of many vertebrate and invertebrate muscles are long, threadlike microelements, not to be confused with the larger fibers which are packed

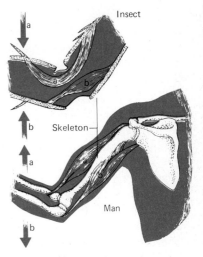

Fig. 9–2 Movement in an arthropod (insect) and a vertebrate (man). In the insect, the muscles attach to an exo-skeleton, in man, to an internal or endoskeleton. The muscles labeled *a* flex (pull in) the appendage, those labeled *b* extend the appendage.

Fig. 9–3 Electron micrograph of a portion of a muscle fiber of a rat. The characteristic banding pattern is shown, as well as sections of mitochondria near the narrow dark bands. (*Courtesy of G. F. Gauthier.*)

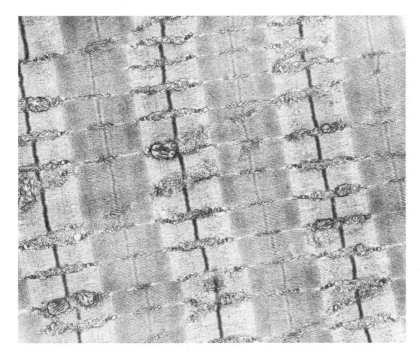

together in analogous fashion to form the muscle itself. Fibrils are of the order of 1 μm in diameter and there may be 1,000 fibrils per fiber. Crammed in among the fibrils are the typical cytoplasmic elements of cells, especially the ATP-producing mitochondria, of which there are many (Fig. 9-3).

Microscopic inspection of the fibrils reveals a highly ordered and repetitive system of banding, or **striations.** These striations are easily visible with the light microscope; they are the optical effect produced by the distribution of protein molecules in the fibrils. When these bands are in register (that is, when similar bands of adjacent fibrils are juxtaposed), the whole fiber appears to be striated. The muscles that operate the skeleton of vertebrates and invertebrates and the heart of vertebrates are of the striated type. Striated muscles are also found in the viscera of insects and crustaceans. The "fast" muscles that snap shut the shells of oysters have spiral striations.

THE FUNDAMENTAL UNIT OF MUSCLE TISSUE IS THE FIBER. IN MANY MUSCLES THE FIBERS ARE STRIATED, THE STRIATIONS BEING PRODUCED BY THE ALIGNMENT OF PROTEIN MOLECULES IN THREADLIKE FIBRILS, WHICH ARE PACKED TOGETHER TO FORM THE FIBER. THE FIBER IS COMPOSED OF ELEMENTS OF MORE THAN ONE CELL, WHICH BECAME FUSED TOGETHER EARLY IN DEVELOPMENT. SKELETAL MUSCLES ARE TYPICAL STRIATED MUSCLES.

The architecture of the muscles of the viscera, uterus, blood vessels, and bladder of vertebrates is based on the cell rather than the fiber and fibril, each cell having one nucleus. The absence of striations in these muscles leads to their being described as **smooth.** It must be pointed out, however, that the classification of muscle types on the basis of striations, the length of fibers, and other systems is always artificial. Nevertheless, a generalization that has fairly broad application is that **tonus muscles** —muscles that maintain a constant partial contraction, such as the body-wall muscles of worms, the abductor muscles of clams, and vertebrate visceral muscles—are unstriated.

The layman makes a distinction between red (dark) meat and white (light) meat in vertebrate skeletal muscle. Histologically, the red muscle has fewer fibrils, has much more of the red, oxygen-storing pigment myoglobin, and sometimes has more fat. The white type is packed with fibrils, has less myoglobin, and may have more glycogen as the ultimate energy source. At this point you might be able to guess the distributions of these muscles.

(Which can be mobilized to produce ATP fastest: fat or glycogen?) Red muscle is not as "fast" as white muscle, but it has more endurance. Marine mammals that may be submerged for long periods have red muscles, as do the slow-moving sloths, which hang upside down for long periods. Can you explain the distribution of the muscle types in a chicken from this information? What would you predict about the leg muscles of ambush hunters such as the leopard or lion?

SMOOTH MUSCLES LACK STRIATIONS. THEIR ARCHITECTURE IS BASED NOT ON THE FIBRIL BUT ON THE CELL. SMOOTH MUSCLES ARE TYPICALLY THOSE WHICH MAINTAIN A STATE OF PARTIAL CONTRACTION, SUCH AS THE MUSCLES OF THE VERTEBRATE GUT. IN VERTEBRATES, RED MUSCLES ARE SPECIALIZED FOR ENDURANCE AND WHITE MUSCLES FOR FAST ACTION.

Muscle Contraction

Does a muscle shorten in a way which is analogous more to an accordian or more to a telescope? It is now thought that the closest analogy is the telescope, at least in vertebrate skeletal muscle. As shown in Fig. 9–4, there are both thick and thin filaments in the fibrils. The thin filaments, apparently composed of a protein called **actin,** are attached to plates at one end and are free at the other. The thick filaments are composed of another protein, **myosin,** and are suspended between the thin actin fil-

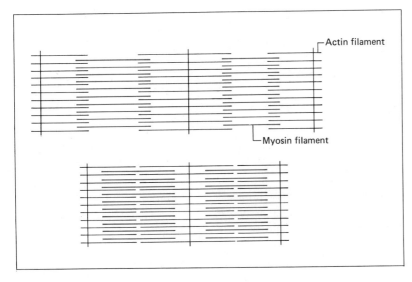

Actin filament

Myosin filament

Fig. 9–4 Contraction is thought to be the result of the telescoping of thin filaments of actin between the thicker filaments of myosin.

aments. Cross bridges between these two kinds of protein have been observed in electron micrographs, and it is hypothesized that the telescopic shortening of muscles might be brought about by a ratchetlike movement of the cross bridges when the thin filaments slide toward the middle of the thick filaments. ATP, the omnipresent fuel, is the energy source, which accounts for the high density of the mitochondria appearing in the electron micrographs. Threads of actin and myosin extracted from muscle contract spontaneously when ATP is added.

MUSCLE CONTRACTION SEEMS TO BE BASED ON THE TELESCOPING OF BUNDLES OF TWO PROTEINS, ACTIN AND MYOSIN. BRIDGES ARE THOUGHT TO CONNECT ADJACENT ACTIN AND MYOSIN FILAMENTS AND TO SLIDE THEM PAST EACH OTHER BY A SORT OF RATCHET ACTION.

NEUROMUSCULAR CONTROL

Now, having looked at the structure and contraction of muscle, it is time to take a close look at the events at the boundary of the nervous and muscular systems.

Synapses occur not only between neuron and neuron but also between motorneurons and muscles. These synapses are contained in the structure shown in Fig. 9-5 called the motor end plate, or neuromuscular junction. This is the site where information is translated into action. In vertebrate skeletal muscle, the transmitter across the synapse is the

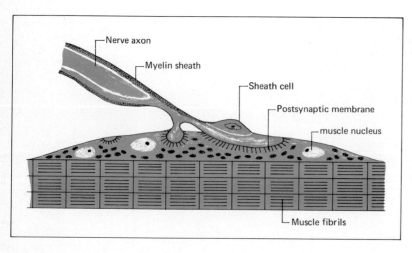

Fig. 9-5 Diagram of a motor end plate in the skeletal muscle of a lizard. *(Adapted from R. Couteaux, Experimental Cell Research, Supplement 5, 1958, Academic Press, Inc.)*

Nerve axon

Myelin sheath

Sheath cell

Postsynaptic membrane

muscle nucleus

Muscle fibrils

chemical acetylcholine. Small amounts of acetylcholine applied to the exposed end plate elicits a muscle twitch. Acetylcholine has no effect when applied elsewhere on the muscle. The enzyme that inactivates acetylcholine can be blocked by the drug prostigmine. In the presence of prostigmine, the arrival of an impulse at an end plate results in a steady sustained contraction, called **tetanus,** rather than a single twitch. When acetylcholine reaches the muscle side of the end plate of vertebrate skeletal muscle, the postsynaptic membrane is depolarized virtually instantaneously and the excitation radiates out over the muscle fiber, followed by a contraction. In the intact animal, there is usually a train of impulses arriving at the end plate, and the separate twitches fuse into a smooth tetanic contraction.

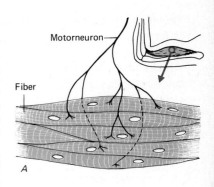

The vertebrate skeletal muscle fiber is basically on all-or-none system, individual fibers being incapable of graded contraction. How, then, is it possible for a vertebrate to exercise the exquisite control needed for a weaverbird to weave a nest or a musician to play a violin sonata? These movements obviously require graded, smooth contraction. For example, your biceps muscle (the one on the upper arm which small boys contract to "show their muscles") doesn't always forcefully snap your hand toward your shoulder when it contracts, even though the individual fibers in the biceps are either contracted or relaxed. The answer depends not on the qualitative nature of fiber contraction but on the quantitative dimension of integration. Each of our limbs is serviced by many thousands of motorneurons, and each nerve innervates a few muscle fibers (Fig. 9–6). The nerve and its slave fibers are a **motor unit.** The control of muscular contraction is achieved by the selective stimulation of the hundreds of motor units in a muscle. These can respond a few at a time, producing a weak contraction; all at once, producing a fast and powerful jerk; or sequentially, producing a gradual shortening (each motor unit maintaining its tetanus as the other units come into play).

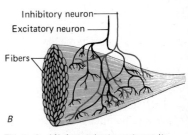

Fig. 9–6 (*A*) A vertebrate motor unit; one motorneuron innervates a few fibers (a muscle consists of hundreds or thousands of such units); (*B*) arthropod skeletal muscle; an entire muscle is sometimes controlled by a few highly branching neurons, some excitatory and some inhibitory.

THE SYNAPSE BETWEEN A MOTORNEURON AND VERTEBRATE MUSCLE IS CONTAINED IN A NEUROMUSCULAR STRUCTURE, THE MOTOR END PLATE. THESE SYNAPSES ARE BRIDGED BY THE TRANSMITTER SUBSTANCE ACETYLCHOLINE, WHICH IS ABLE TO DEPOLARIZE THE POSTSYNAPTIC MUSCLE MEMBRANE. THIS LEADS TO A WAVELIKE DEPOLARIZATION OF THE ENTIRE MEMBRANE. A SINGLE NERVE IMPULSE ELICITS AN ALL-OR-NONE TWITCH FROM THE MUSCLE FIBER. SELECTIVE STIMULATION OF GROUPS OF MUSCLE FIBERS, EACH RESPONDING TOTALLY, PRODUCES GRADED CONTRACTION OF THE ENTIRE MUSCLE.

The muscle action potential of vertebrate skeletal muscle is an exception in animals as a whole. In invertebrates, the muscle-fiber membranes of most muscle fibers or cells do not propagate an impulse although the membrane is locally depolarized. The *strength* of contraction in invertebrate muscles depends on the size of the depolarized area, which in turn depends on the frequency of motorneuron impulses. This is a point worth stressing, because it means that invertebrate muscle is *not* all-or-none but can respond in a graded fashion. To appreciate the implications of this, we need to consider the engineering problems faced by a tiny arthropod such as an ant. The ant has six legs, each with four joints which can move in many directions in a highly coordinated manner. To innervate all its leg muscles would take large cables of axons and masses of coordinating neurons if each muscle fiber was all-or-none and had to be independently controlled, as in vertebrates. Obviously, such a profligate solution would be unsatisfactory for small animals, and another strategy is called for. Vertebrates can produce a graded contraction by firing increasing *numbers* of motor nerves. Arthropods have very few motorneurons, but each one is highly ramified over the fibers that it controls (Fig. 9-6). A graded contraction is achieved by the *rate* of impulses. Hence, with a small number of muscle fibers and just a handful of nerves, insects and other arthropods can make very delicate, quick movements. An example is the clawed appendage of a crab, which can be controlled by a total of 12 motor nerves. A mouse, which may weigh less than the crab's claw, has about 2,600 motor nerves in its front leg (Fig. 9-6).

INVERTEBRATE MUSCLE FIBERS ARE CAPABLE OF GRADED CONTRACTIONS, THE DEGREE CONTROLLED BY THE RATE OF ARRIVAL OF IMPULSES. THIS PERMITS QUICK, DELICATE MOVEMENT WITH A MINIMUM NUMBER OF NERVES AND MUSCLE FIBERS.

Vertebrates have only one kind of motorneuron, but in crustaceans, three kinds of motorneurons may innervate the same muscle: a "slow" type, a "fast" type, and inhibitory neurons. The first works much as described above, in that a slow buildup of tension is achieved by a train of impulses, each of which produces a larger depolarization than its predecessor. Apparently these fibers control slow movements and posture. Fast motor neurons mediate fast movements and probably release a different synaptic transmitter, producing a rapid, twitchlike contraction. Still another synaptic transmitter, GABA (gamma-amino butyric acid), is secreted by the inhibitory nerves. GABA apparently results in hyperpolarization of the muscle membrane by decreasing permeability to sodium. Thus subtle and versatile *peripheral* integration occurs in the crustaceans—in contrast to vertebrates, in which the central nervous system performs these functions far from the location of the effector. This peripheral control is made possible by muscle units that respond in a graded fashion rather than in an all-or-none fashion, so that delicate control of contraction can be mediated by just a few nerves. This integrative strategy must have evolved long ago in the ancestors of annelids because it is characteristic of the entire annelid-arthropod-mollusc line. Muscle units with both fast and slow capabilities are found in all these groups, though peripheral inhibition is not as widespread.

AS MANY AS THREE DIFFERENT KINDS OF MOTORNEURONS MAY BE FOUND IN INVERTEBRATES: SLOW, FAST, AND INHIBITING. THIS PERMITS A DEGREE OF INTEGRATION AT THE MUSCLE SITE THAT IS IMPOSSIBLE IN VERTEBRATES WHICH HAVE ALL-OR-NONE FIBERS AND IN WHICH INTEGRATION IS PERFORCE DONE IN THE CENTRAL NERVOUS SYSTEM.

LEVELS OF INTEGRATION

You will recall that behavior can be diagramed as resulting from the action of a sensory input system, integration, and an effector system. In the previous sections we looked at neurons, the "wires" of the entire mechanism. Next we surveyed the receptors which constitute the sensory input system. After that we examined some effectors, especially muscles. Now it is time to consider the ways in which sensory input is utilized in order to select the appropriate response for the effector system—what we call **integration.** Another definition of integration is that it is the process or processes that result in a meaningful output that is some function of the input. The constraints or limits on the output, and therefore on the function, are determined by natural selection, since it, in turn, defines "appropriate;" or a less formal way of saying this is that an integrative system makes decisions about what kinds of output to produce, based on the patterns of input it receives. These decisions are determined and refined over scores of generations by natural selection, the individuals with appropriate outputs reproducing and those with improper ones not.

The Neuron

The elemental unit of the nervous system is the neuron, and the elemental process is the nervous impulse. At first sight, it is not obvious that a simple neuron can have any integrative capability, especially since the impulse is all-or-nothing. But that alone tells us something because it implies a **threshold** for propagation—the critical amount of membrane depolarization necessary to trigger the impulse. A crayfish "knows" the amount of curvature of its abdomen because of impulses sent to the central nervous system by stretch-receptor neurons which are embedded in muscle fibers that stretch between abdominal segments and connect to the exoskeleton. The rate of firing of these neurons depends on the tension, or stretch, on the dendrites. Even very little stretch depolarizes the dendrite, but the depolarization must reach a certain level or threshold before an impulse is propagated. Hence the threshold setting determines what stimuli can be (are) ignored by the animal.

In some systems thresholds are so low that the neuron responds to the theoretical minimum stimulation, such as in the vertebrate eye, in which a rod apparently responds to a single quantum of light. But when the threshold is above the minimum, the neuron performs an integrative act in that the output, as measured by rate of impulse propagation, is some function of the input—namely, zero below the threshold and something else above it.

Single neurons have other integrative capabilities. Most are subtle and poorly understood as yet. For instance, the all-or-none impulse may be a "law" for the axon, but in the fine branching at the end of the axon the activity may be graded, adding another dimension to output variation.

The final integrative capability of single neurons that will be discussed is **spontaneous activity.** Almost as soon as it became possible to record the activity of single fibers, it was commonplace to find neurons that fired repeatedly and without a stimulus, that is, that fired repeatedly in a constant environment.

The relationship between input and output of spontaneous neurons is highly individual and complex, depending to a greater or lesser extent on the environment of the neuron. Spontaneous activity is the essential feature of **pacemaker** neurons that drive the rhythmic activity of effectors (output organs) such as the heart. The rhythmic contractions of the umbrellas of jellyfish are another such activity. The pacemaker neurons that initiate these swimming pulsations are located in nine ganglia (clusters of neuronal cell bodies) evenly spaced around the margin of the umbrella. An impulse initiated at any one of the pacemakers radiates over the umbrella, inducing a pulsation. When it reaches the other pacemakers, they are reset "back to zero" and the process that results in an impulse begins again. The first one to fire initiates another pulsation and resets the others. All but one of the pacemakers can be removed without eliminating the rhythmic swimming movements, but the rhythm is then neither as regular nor as rapid.

THE INTEGRATIVE SYSTEM OF AN ANIMAL MAY BE THOUGHT OF AS THE SYSTEM WHICH DETERMINES THE RELATIONSHIP BETWEEN

SENSORY INPUTS AND ACTIONS. SINGLE NEURONS HAVE LIMITED BUT IMPORTANT INTEGRATIVE ACTIVITIES. MANY NEURONS HAVE THRESHOLDS OF STIMULATION; SOME, SUCH AS THOSE OF THE HEART PACEMAKER, FIRE SPONTANEOUSLY.

Many receptors show spontaneous activity. They generate impulses at a constant rate in the apparent absence of stimulation. *Excitatory* stimuli are those that increase the frequency of impulses. Those that slow or stop impulse frequency are called *inhibitory*. Often it is difficult to prove spontaneous activity in receptor cells since the receptor cells might be firing in response to small amounts of a ubiquitous material, such as CO_2 or oxygen.

Nevertheless, there do exist spontaneously active receptors in the strict sense. Fish have motion receptors in their lateral-line canals; these are part of a class of receptors known as the accousticolateral system. The sensory cells are innervated from below with both incoming and outgoing neurons. From their tops, little beards of sensory hairs protrude into a gelatinous cap. The cap protrudes into the lateral-line canal, and the shearing force of water currents displaces it and the sensory hairs embedded in it. Water currents in one direction increase the frequency of impulses in the sensory nerve while displacement in the opposite direction slows the propagation rate. Spontaneity, in this case, permits directional sensitivity by establishing a rate which may be increased or decreased. Without spontaneous activity, the receptor could react in only one way: no activity/activity. With spontaneous activity, it can respond in two ways: slowing down or speeding up.

Spontaneity of neuronal activity is thought to be partly responsible for a remarkable but largely unplumbed integrative phenomenon, the **central excitatory state.** Many animals are essentially immobilized if input from auditory, mechanical, and visual receptors is interrupted. The removal of the halteres (sensory organs important in flight) of the fly *Tipula* so weakens it that it is barely able to drag itself along. If all the skin is taken off a frog, it lies still and flaccid, but even a small piece of intact skin is enough to maintain muscle tone and responsiveness. Apparently, the nervous system of many animals is kept "awake" by trains of impulses arriving from spontaneously firing receptor cells.

RECEPTORS OFTEN GENERATE NERVE IMPULSES IN THE ABSENCE OF STIMULATION. THE STIMULI EITHER INCREASE OR DECREASE THE FREQUENCY OF THE IMPULSES. CONTINUOUS INPUTS FROM SENSE RECEPTORS APPEAR TO BE IMPORTANT IN MAINTAINING "TONE" AND CONSCIOUSNESS IN SOME ORGANISMS.

Synaptic Integration

Two people can do things that are impossible for one. Among these are vocal harmonization, marriage, and wrestling. Analogously, two (or more) neurons can integrate in ways that one alone cannot. The locus of such integration is the site of contact or communication, the synapse. For the purpose of this discussion, the synapse is considered a functional unit with unique properties, even though anatomically it is two or more terminals plus the intervening space.

The four principal forms of synaptic integration are spatial summation, temporal summation, facilitation, and inhibition. One way of looking at the nervous system is as a hierachy of filters and transducers (elements which change one energy form to another), with the lowest level being receptors. Receptor neurons synapse with **_interneurons_** (Fig. 9–7), which behave as the primary filters and transducers. The interneuron has

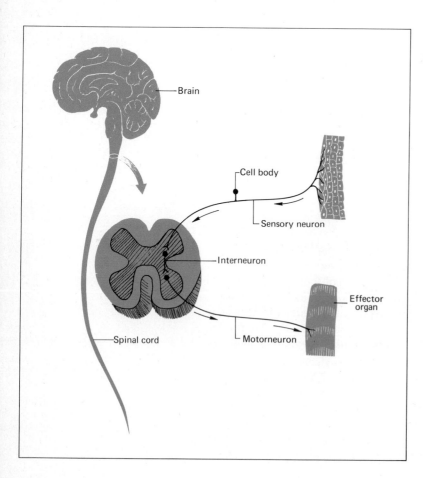

Fig. 9–7 Diagrammatic drawing of a cross section of the spinal cord of a vertebrate. Information in the form of impulses enters the spinal cord dorsally via sensory (afferent) fibers and leaves ventrally via motor (efferent) fibers. Neurons contained entirely within the spinal cord are called interneurons.

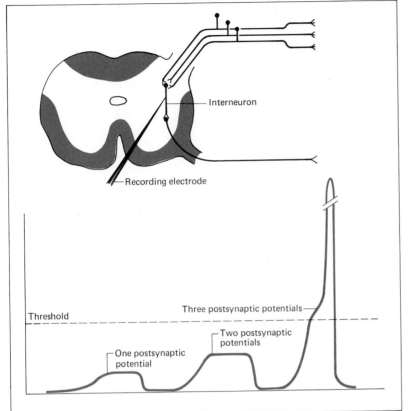

Fig. 9-8 Spatial summation. Intracellular recording of an interneuron: all three sensory neurons must simultaneously depolarize the interneuron before an impulse is propagated.

to "decide" whether to relay the input it received from the receptors. If the threshold for propagation depends on the *number* of simultaneous postsynaptic depolarizations of the interneuron, this is called integration by **spatial summation** (Fig. 9-8). For example, several of the hair receptors on the tail of a crayfish must fire simultaneously if the interneuron with which they synapse is to fire. The word "spatial" refers to the spatial arrangement of the ends of the presynaptic (sensory) neurons, not to the synaptic spaces themselves.

If a single electric shock is given to a sea anemone, it is unperturbed, but another, subsequent shock given anywhere on the body will cause muscular contraction. In some way, the shocks are counted or summed. This is a crude example of **temporal summation.** Motorneurons often exhibit this form of integration (Fig. 9-9). A single impulse arriving at a synapse causes a partial depolarization on the postsynaptic cell body, but this will usually be far below threshold stimulation. But if another partial

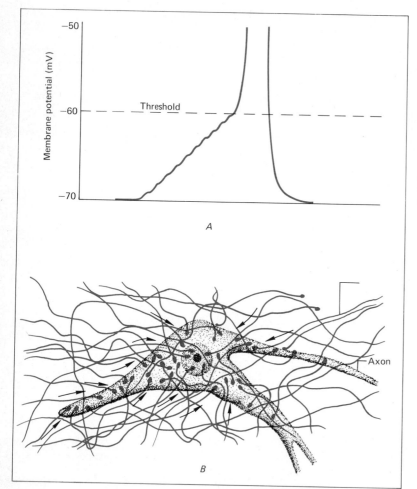

Fig. 9-9 Temporal summation in a motorneuron: (A) graph showing the decrease in resting potential in the membrane of a motorneuron cell body (B) when one or more of the neurons synapsing with the motorneuron is stimulated repeatedly. If the stimulation stops before the threshold of the motorneuron is reached, the membrane potential falls back to the resting potential. (*Adapted from BEHAVIOR by William G. Van der Kloot. Copyright © 1968 by Holt, Rinehart and Winston, Inc. Reprinted by permission of Holt, Rinehart and Winston, Inc. From "How Cells Communicate" by Bernhard Katz. Copyright © 1964 by Scientific American, Inc. All rights reserved.*)

depolarization follows quickly on the heels of the first, before the latter decays, the depolarizations add. In this fashion, the threshold is finally reached. The cell bodies of motorneurons are like little pressure cookers —the valve (impulse) opens when the input of heat (incoming stimulation) is great enough and rapid enough to overcome the loss of heat (decay) by radiation and convection. Temporal summation in motorneurons lasts only milliseconds, but there are cases in which the temporary depolarization, even if it decays back to resting potential, enhances the depolarization of future synaptic potentials. Sometimes these effects remain for hours or days. One theory of learning is that this **facilitation** of synapses following repeated usage engraves a pattern in the nervous system, as running water erodes a channel in the earth. Spatial summation and temporal summation are not mutually exclusive; indeed, they are combined in many synapses.

INTEGRATION AT THE LEVEL OF THE SYNAPSE MAY DEPEND ON THE NUMBER OF RECEPTOR NEURONS (SYNAPSING WITH AN INTER-NEURON) THAT FIRE. THIS IS CALLED INTEGRATION BY SPATIAL SUMMATION. IF IT DEPENDS ON THE FREQUENCY OF FIRING OF THE PRESYNAPTIC NEURON, IT IS CALLED TEMPORAL SUMMATION. TEMPORAL SUMMATION AND SPATIAL SUMMATION OFTEN OCCUR AT THE SAME SYNAPSES. SUMMATION MAY BE INVOLVED IN LEARNING. ONCE A POSTSYNAPTIC CELL HAS BEEN DEPOLARIZED, IT MAY BECOME MORE SUSCEPTIBLE TO DEPOLARIZATION. THIS PHENOMENON IS KNOWN AS FACILITATION.

A synaptic event which integrates in the opposite way to summation and facilitation is **inhibition.** We saw that inhibitory synapses effect an *increase* in potential across the postsynaptic membrane and how the propagation of an impulse depends on something like the sum of the inhibitory and excitory synaptic events. Inhibitory and excitatory synapses are each probably mediated by a different chemical transmitter. This suggests why chemical synapses, in contrast to electrical synapses, are so flexible in their integrative activity. As discussed before, in the course of evolution the primitive electrical communication has been largely replaced by chemical mediation.

Sensory cells, besides transmitting excitatory messages, often transmit inhibitory ones. Thus repeated stimulation of a receptor may lead to an inhibition of a response or perception, as when your ear "tunes out" background noise or when workers in a perfume factory no longer smell the perfume. Such a lowering of sensitivity is called **habituation,** and it permits the organism to "get used to" and ignore repititious and unimportant stimuli that would otherwise distract it. Habituation may well be a result of synaptic events which are more or less diametrically opposed to those causing facilitation. Later we will see how these different integration mechanisms function together in more complex integrative tasks.

HABITUATION, THE INHIBITORY EQUIVALENT OF FACILITATION, IS A LOWERING OF SENSITIVITY OF POSTSYNAPTIC NEURONS FOLLOWING THE REPEATED FIRING OF PRESYNAPTIC CELLS.

Circuits of Neurons

If two neurons and a synapse can perform integrative functions that single neurons can't, it is obvious that circuits of neurons can integrate at even higher levels of complexity. Many kinds of circuits are known to exist, but space allows only the description of one type, the **reverberating circuit.** Figure 9–10 is a diagram of such a circuit. Note that once such a circuit is entered, the action potential continues to cycle. Such reverberating circuits are thought to be responsible for cyclical involuntary functions, such as breathing. There is also evidence that more complex circuits of this type temporarily entrap our perceptions and thoughts. These "short-term memories" are thought to become etched into permanent memories, or *engrams,* if the reverberating circuit is repeatedly initiated. Can you think of a way that a reverberating circuit might be shut off?

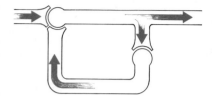

Fig. 9–10 A circuit of neurons in which an impulse could cycle indefinitely.

CIRCUITS OF NEURONS CAN, OF COURSE, PERFORM HIGHER-ORDER INTEGRATION THAN CAN TWO OR MORE NEURONS CONNECTED LINEARLY. REVERBERATING CIRCUITS ARE THOUGHT TO CONTROL CYCLICAL PHENOMENA SUCH AS BREATHING AND MAY BE INVOLVED IN MEMORY.

As you already realize, the neurons of animals do not just randomly associate with each other and with receptors and effectors. Rather, they are organized into **nervous systems.** These vary in complexity from the loosely organized **nerve net,** such as is found in the hydra (Fig. 9–11). to the human nervous system, in which the brain alone has an estimated 10 billion (10,000,000,000) neurons. In the course of animal evolution, there has been a continuous trend toward concentration of neurons into dorsal or ventral **nerve cords,** with the bodies of these nerve cells concentrated in masses, or **ganglia.** There has also been a tendency to concentrate ganglia in the anterior portion of the body region and especially to evolve a head, in which one finds a master ganglion, the **brain.** The reason for this evolutionary trend is clear. Active animals have their sense organs concentrated in the part of the body that confronts novel stimuli first. It is not surprising that the head is also the site of the ganglia serving these organs. More and more functions have become concentrated in this **central nervous system (CNS)** as interconnections within the CNS itself have added new dimensions of flexibility to responses.

In man, the CNS (brain plus spinal cord) coordinates behavioral activities as simple as the reflex jerk in which a finger is withdrawn from a hot stove. It coordinates behavioral activities as complex as painting a mural, flying a jet interceptor, or taking an examination in the calculus. We know a great deal about the control of reflex activities and, unhappily, very little about the performance of the nervous system in painting, flying, or test taking.

The Central Nervous System

Fig. 9–11 A loosely organized nervous system or nerve net in a simple coelenterate, hydra.

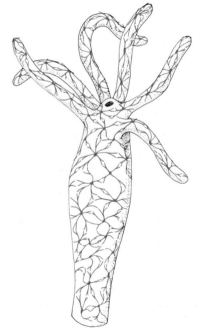

REFLEXES The simplest kind of activity that can be called an integrated behavioral act is a reflex, an act programmed to occur involuntarily when the appropriate situation arises. The elements involved in reflexes are already familiar:

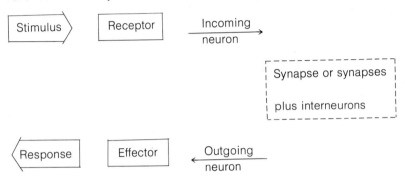

The sucking behavior of the ordinary housefly or blowfly illustrates this. The feet of flies come equipped with an assortment of "taste" receptors, including some for sugar. When a fly steps on something that contains a sufficiently high concentration of sugar to excite these receptors, its proboscis, which is normally folded up against the head, flicks down like the landing gear of an airplane and pumping muscles in the proboscis suck at whatever happens to be there. Even this simple reflex involves many parts of the insect's nervous system. Actually, reflexes grade into more complex types of behavior, so the concept is arbitrary. For instance, a reflex can be conditioned (learned), as Pavlov's famous example showed (Fig. 9–12). Salivation is normally a response (actually, a kind of reflex itself) to the presence of food. In this experiment, the dog was taught to associate the sound of a bell with food; eventually, the bell alone elicited the salivation response. This is called a **conditioned reflex.**

RECEPTORS AND EFFECTORS IN ANIMALS ARE ORGANIZED WITH INTEGRATING ELEMENTS INTO A NERVOUS SYSTEM, WHICH CONTROLS THE BEHAVIOR OF THE ANIMAL. THE SIMPLEST FORM OF BEHAVIOR IS THE REFLEX, WHICH MAY BE GENETICALLY PROGRAMMED (AS IS THE FEEDING BEHAVIOR OF A FLY) OR CONDITIONED (AS WHEN PAVLOV'S DOGS SALIVATED ON SIGNAL).

Vertebrate reflexes are mediated by the CNS: the brain and the spinal cord. Such reflexes as salivation, blinking, and the pupil reflex are mediated by the brain. Others are mediated by the spinal cord. An example familiar to everyone is the response to unexpected pain in the hands or feet. A hand placed on the handle of a hot pan is quickly drawn away. This behavior is not quite as simple as some reflexes because it involves more than two neurons. Furthermore, it is much more complex than some because of nerve connections up and down the spinal cord; these usually bring about compensatory movements in the limbs so as not to throw the person off balance and also often evoke certain culturally dependent and emphatic vocalizations.

Newborn babies come equipped with a set of reflexes, including the sucking reflex elicited by touching the cheek or lips. They also swallow or cough, and in males the penis erects following stroking of the inner thigh. Some other rather anomalous reflexes occur in the infant which are lost after a few months. An infant totally immersed in water will hold its breath and make swimming movements. Another reflex which disappears is the grasping reflex elicited by pressure on the palms and soles of the feet. One recalls how newborn primates cling to the mother's fur.

The simplest kind of reflex is the *stretch reflex* of vertebrates. A physician uses this reflex to determine the health of the nervous system when he strikes the tendon just under the patella (kneecap). As shown in Fig. 9–13, this in turn pulls down the patella, thus slightly stretching the lower leg extensor muscles on the upper thigh (extensors are muscles which straighten a limb; flexors bend it). The receptors of this stretch stimulus are the spindle organs, macroscopic capsules of specialized muscle fibers **(intrafusal fibers)**, the centers of which are elastic but noncontractile. The fibers differ from normal **extrafusal** muscle fibers in that they have sensory nerve endings on them and are capable of *graded* (rather than all-or-none) contraction (Fig. 9–14). The ends of the spindles are attached to extrafusal muscle fibers or to tendons. Any elongation of the muscle compresses the spindle and stretches the central part of the intrafusal fibers and the enveloping ends of a sensory nerve. This depolarizes the receptor and initiates impulses along the axon to a synapse with a motorneuron in the spinal cord. The sensory nerve cell bodies lie in the dorsal root ganglia adjacent to the spinal cord (Fig. 9–14). The motorneurons **(alpha** motorneurons) terminate at motor end plates on extrafusal muscle fibers. A contraction ensues, its degree depending on the number of impulses arriving at the spinal synapse; this, in turn, is approximately equal to the logarithm (base 10) of the force exerted to stretch the muscle. Thus, an order of magnitude increase in impulse frequency will approximately double the force of contraction. The logarithmic relationship is a typical integrative property of sensory neurons. What is the function of the stretch reflex? Imagine how difficult it would be to maintain erect posture if you had to consciously control the dozens of muscles involved. The skeletal receptors are proprioceptors providing the CNS with an account of the movement of skeletal muscles. Without the stretch re-

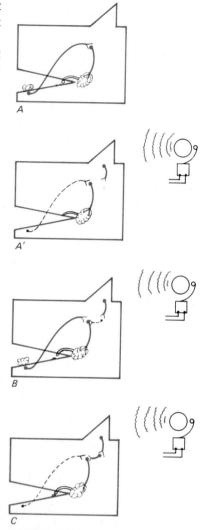

A

A'

B

C

Fig. 9–12 A highly diagrammatic representation of Pavlov's famous experiment on the conditioning of the salivation reflex: (*A*) the salivation reflex is triggered by the stimulus of food on the tongue; (*A¹*) the sound of a bell alone fails to induce salivation; (*B*) the dog gradually associates the sound of the bell with food; (*C*) eventually the new association is established in the form of a neural pathway that induces salivation at the sound of the bell alone.

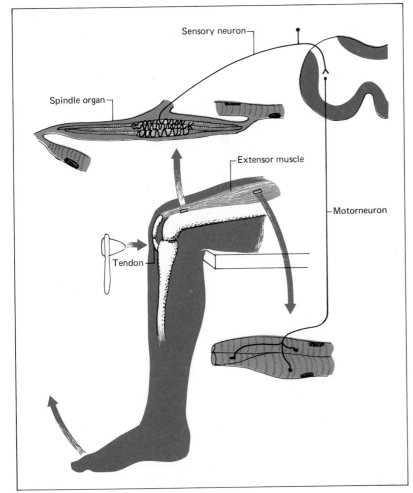

Sensory neuron

Spindle organ

Extensor muscle

Motorneuron

Tendon

Fig. 9-13 Diagram of the knee-jerk or patellar reflex. The hammer blow pulls down the knee cap and this stretches the extensor muscle. The stretch depolarizes the endings of sensory neurons embedded in spindle organs, initiating bursts of impulses down the axons to the spinal cord where a single synapse intervenes between these sensory neurons and the motorneurons. Stimulation of the motorneurons at these synapses results in muscle contraction causing the lower limb to jerk upward.

flex, you would always be falling flat on your face—or just managing to save yourself.

A WIDE VARIETY OF REFLEXES OCCURS IN VERTEBRATES. THE SIMPLEST OF THESE IS THE MUSCLE STRETCH REFLEX, WHICH AUTOMATICALLY CONTRACTS A MUSCLE IF SOME FORCE STRETCHES IT.

The stretch reflex is the single-synapse kernel of a more complex integrative system. Other polysynaptic ("many-synapse") reflexes are woven in and around it. Where the sensory neuron synapses with the motorneu-

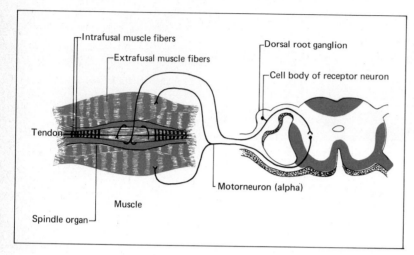

Fig. 9–14 Diagram of the stretch reflex. The spindle organ is embedded in the muscle which is composed of extrafusal fibers. The specialized intrafusal muscle fibers are found only within the spindle organs. Compression or stretching of the elastic centers of the intrafusal fibers triggers the stretch reflex shown in Fig. 9–13.

rons, it gives off other branches that synapse with inhibitory interneurons. Some inhibitory interneurons synapse with motor neurons to the **antagonistic muscles** (flexor leg muscles, in this case). Incoming impulses in the sensory neurons are *always* synaptically transmitted to the inhibitory interneuron, which in turn releases a transmitter that hyperpolarizes the motorneuron, preventing it from eliciting a contraction of the antagonistic muscle at the same time the original muscle is responding to its motorneuron. This circuitry is the basis for a fail-safe system that prevents the simultaneous contraction of a muscle and its antagonist—a mistake that could cause the fracture of one's bones if it took place.

Before passing from the ventral horn of the spinal cord to the ventral root of the spinal nerve, the motorneurons synapse with another type of inhibitory interneuron, called a *Renshaw cell*. These, in turn, synapse with other motorneurons going to the same muscle. Hence every time an impulse is propagated in a motorneuron, it indirectly inhibits other motorneurons in the same "pool." What this means is that unless the motorneurons in a group or pool are firing simultaneously, they inhibit each other. An educated guess about the purpose of this is that the pool of motorneurons serve a common function, which is best performed when they all fire at once.

When a muscle contracts, the tension on the spindle and the sensory nerve is relieved and the spindle therefore becomes much less responsive to stretches. If this condition persisted, an animal would suffer a proprioception information gap and compensatory postural adjustments would be slow and sloppy. The problem could be quickly and efficiently resolved if there were a way of taking up the slack in recently decompressed spindles. Indeed, such a system *is* superimposed on the stretch reflex.

Except for their centers, intrafusal spindle fibers are contractile, and they are innervated by their own **gamma** motorneurons (Fig. 9–15). When the gamma neurons initiate a contraction, the elastic center of the intrafusal fibers are stretched, returning the spindle to its "ready" state. The intrafusal muscle, however, will continue shortening if the gamma neurons continue to fire. In turn, the sensory nerve will increase its firing rate and the reflex system will be activated to again relax the spindle. In this way, impulses flowing through the gamma fibers may be used to adjust muscle tension. In short, the gamma fibers can modulate (control) the amount of shortening of the extrafusal muscle by controlling the stretch of the spindle's elastic center. Hence, the spindle not only responds passively to stretch but also can actively stretch *itself* to an amount controlled by outside sources. These outside sources include the higher centers of integration that initiate most voluntary (nonreflex) contractions, as well as peripheral receptors, such as those that perceive "pain" in the skin. Most gradual movements probably involve activity of both alpha and

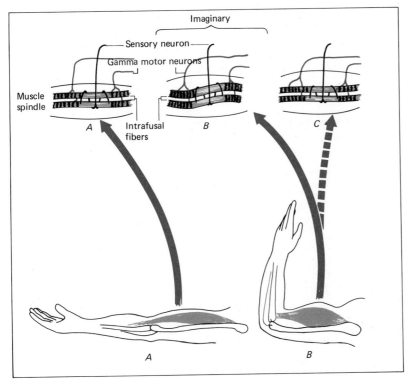

Fig. 9–15 (A) The elastic center of the intrafusal fibers are stretched; any lengthening of the muscle will stretch them further and increase the rate of firing of the sensory neuron. (B) Imaginary condition in the contracted muscle; the flaccid centers of the intrafusal fibers must be straightened out before they begin to stretch and before the firing rate can increase in the nerve; if this condition persisted, the organism would be oblivious to stretching in contracted muscles. (C) In actuality, the slack never develops because the gamma fibers maintain a constant tension by controlling the tone of the contractile regions of the intrafusal fibers.

gamma motorneurons. Rapid movements probably are initiated almost exclusively by the alpha, and continued steady contraction (as in holding something in the arms), by the gamma. Sometimes, however, violent reactions are initiated by the gamma neurons, as in the pain-withdrawal reflex. Sensory input from, say, a burned finger is channeled into the gamma neurons, which contract the spindle and fire the reflex. The gamma neuron more or less plays the role of the physician's hammer in the patellar reflex.

THE STRETCH REFLEX IS INTEGRATED INTO A COMPLEX SYSTEM WHICH, AMONG OTHER THINGS, INHIBITS ANTAGONISTIC MUSCLES AND ADJUSTS THE SENSITIVITY OF THE STRETCH RECEPTORS IN THE CONTRACTING MUSCLE.

Let us turn now from the relatively simple questions of integration at the level of the spinal-cord reflex to the infinitely more difficult problems of integration at the higher centers of the CNS, that is, in the brain.

PERCEPTION AND INTEGRATION AT HIGHER LEVELS The word "higher" has two parallel meanings in connection with man's CNS. Going from the spinal cord up through the brain to the convoluted surface of the cerebral cortex takes one higher physically, but it also ascends a scale of complexity of associations, decisions, and behavior. When biologists speak of "higher" centers, they mean implicitly those brain regions that evolved most recently and control the most complex integrative functions.

VISION AND HEARING The clearest pictures we have of the processes involved in perception are those emerging from studies of vision and

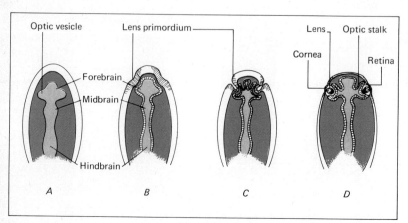

Fig. 9-16 Diagrammatic representation of human eye development: (*A*) outpocketings of the forebrain, known as optic vesicles, grow toward the outer layer of the head called ectoderm; (*B*) the optic vesicles begin to invaginate while the adjacent ectoderm thickens and invaginates toward the retracting optic vesicles; (*C*) the ectodermal invaginations eventually are pinched off, becoming the lenses; (*D*) the overlying ectoderm develops into the transparent cornea.

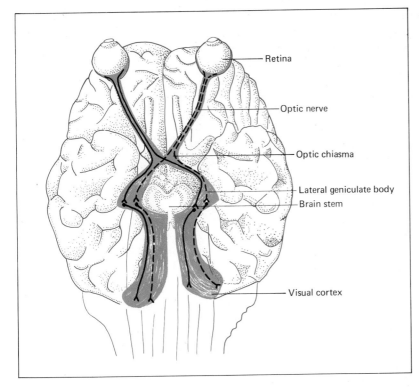

Fig. 9–17 Visual pathways and centers as seen from below. Nerve fibers from the ganglion cells of the retina converge at the optic chiasma where some cross over to the opposite side of the brain. In the lateral geniculate body these fibers synapse with neurons that go to the visual cortex. (*From "The Visual Cortex of the Brain" by David H. Hubel. Copyright © 1963 by Scientific American, Inc. All rights reserved.*)

Retina

Optic nerve

Optic chiasma

Lateral geniculate body

Brain stem

Visual cortex

hearing. We will look at vision first. During the early stages of vertebrate development, an outpocketing pushes out from the brain on each side (Fig. 9–16). When these outpocketings reach the outer skin, they partially contract back into themselves, forming cups—the future retinas. The stalk connecting them to the brain will provide the channel for the nerve fibers leading from the retina to the brain. The retina is highly ordered architecturally. The nerve cells are laid out in regular tiers, facilitating analysis (Fig. 8–13). The retina probably will be the first part of the vertebrate nervous system for which a complete "wiring diagram" of the neuron types and their connections will be drawn.

A first approximation of how vision works is that each light receptor communicates via a direct line to the visual cortex of the brain proper, so that the retinal image is "screened" in the brain without alteration. Anatomical studies show that this is not correct, there being only a few cones in primates (those in the fovea centralis) that are "linelabeled." In the cat, the visual input from the 100 million receptors must be squeezed through more than a million cells of the ganglion. Furthermore, the ganglion cells have overlapping fields of receptors to which they synapse via bipolar cells. Naïve intuition would suggest that this convergence (reduction and overlap) of channels would cause a great loss of visual data, but in fact the opposite is true because of interaction effects. These effects, including spatial and temporal summation, allow for many perceptual abilities that would be impossible with a direct screening system. A $1/4$-in. wire, for instance, that is $1/4$ mile away subtends only 1 sec of arc ($1/3,600°$), or $1/25$ of the width of a cone. How is it possible that we perceive such a wire as a sharp thin line when we consider that the optical system of the eye itself is not good enough to account for this? A piece of film that had the same "grain" as the retina would show a very fuzzy, dim line at least two cones wide—or $1°$ of arc. The key to this sharpening of the image is the interaction of units at higher levels of visual integration. In other words, acuity is vastly increased because a central integrative system permits seeing detail which is much finer than the texture of the cone pattern of the fovea.

VISION PROVIDES AN EXAMPLE OF HIGHER-LEVEL INTEGRATION. IT CAN BE SHOWN THAT THE VERTEBRATE BRAIN IS CAPABLE OF SHARPENING FUZZY IMAGES WHICH ARE TRANSMITTED TO IT BY THE EYES.

How are such perceptual feats possible? With the use of very fine electrodes it is possible to record the impulses in neurons anywhere in the visual pathway. It is possible, for instance, to record from a retinal ganglion cell while the eye is trained on a screen onto which a small spot of light is projected, and the light can be made to fall on any part of the retina by moving the spot. The retinal area that on stimulation produces a change in the firing of the ganglion cell is the **receptive field** of that cell. In the vertebrate eye, the receptive field of a retinal ganglion cell is a circle from $4°$ to $30°$ in diameter containing many rods and cones, and has two zones: the very sensitive **center,** from $0.5°$ to $4.0°$ in diameter, and the annular **surround.** Ganglion cells are spontaneous in that they fire at a steady rate in the absence of a stimulus. For some ganglion cells, light stimulates the center; for others, it inhibits the center. For each type, light has the opposite effect on the surround. Light hitting receptors in the center of the field of one ganglion cell will increase the cell's spontaneous firing rate. Light falling on the surround of the same cell will decrease its firing rate. Other ganglion cells have the opposite arrangement: "off" centers and "on" surrounds. A spot of light precisely covering the "on" center of a receptive field will create a much more vigorous response than a larger spot which also illuminates part of the surround.

THE RECEPTIVE FIELD OF EACH RETINAL GANGLION CELL OF A VERTEBRATE CONTAINS MANY RODS AND CONES AND CONSISTS

*OF A CENTER AND A RING AROUND THE CENTER, THE SURROUND.
FOR SOME GANGLION CELLS, LIGHT HITTING THE CENTER INCREASES
ITS SPONTANEOUS FIRING RATE AND LIGHT ON THE SURROUND
DECREASES IT. FOR OTHER GANGLION CELLS, THE OPPOSITE IS
TRUE. ILLUMINATING THE SURROUND AS WELL AS THE CENTER
TENDS TO INHIBIT THE CENTER RESPONSE.*

Another essential feature of the receptive fields is their *overlap*. Even the
tiniest light spot changes the output of many ganglion cells. Axons in the
optic nerve carry this output to the **lateral geniculate body (LGB)** (Fig.
9–17), the "lower" visual center of the brain. Electrical recording from
the cells of the lateral geniculate ganglia of cats has shown these cells
to have the equivalent of ganglionic receptive fields of center and surround except they are smaller, only 1° to 2° in diameter. There is another
difference between the retinal ganglion fields and LGB fields. In the LGB,
the inhibitory power of the surround over the center of the field is enhanced. This further sharpens the emphasis on contrast in the visual
system—response is to difference rather than degree of illumination.
The overlap of the receptive fields of the ganglion cells can permit this
enhancement of discrimination by the LGB cells. If three different ganglion cells have visual fields with partially overlapping centers, the
LGB cell fires only if all three ganglion cells are simultaneously excited.
This means that the middle field must be illuminated primarily in its center
and the adjacent fields must be illuminated primarily on their surrounds.
This is an example of spatial summation. It would not occur if the centers
of the two outside fields were strongly stimulated, as this would inhibit
their ganglion cells from firing. This exemplifies a process that seems to
be general in sensory integration. At each stage, from the peripheral sense
organs to the "highest" levels of the nervous system, the neurons respond
to more and more specific kinds of input.

*THE RECEPTIVE FIELDS OF THE VERTEBRATE EYE OVERLAP ONE
ANOTHER. IN CATS, THESE FIELDS HAVE EQUIVALENTS TO THE
LOWER VISUAL CENTER OF THE BRAIN (LGB), BUT THE FIELDS IN THE
LGB ARE SMALLER AND THE INHIBITING POWER OF THE SURROUND
OVER THE CENTER IS ENHANCED, SHARPENING CONTRASTS.*

From the LGB, the visual information is passed on to the main center of
the brain, the visual cortex. This rear section of the cerebral cortex
is orders of magnitude more complex than the LGB. A possible model
of connections that would explain how a cell in the visual cortex could
discriminate a line from some other pattern falling on the retina is: the
receptive fields of four LGB cells might be arranged in a straight line.

Each of these receptive fields has an excitatory center. Consequently, the receptive field of the cortical cell will be elongated. If the firing threshold of the cortical cell can be reached by the simultaneous firing of all four LGB cells, then only a correctly oriented line or edge will be perceived by this cell. Another example is given below in the discussion of hearing. Some LGB cells respond to movement, and some to movement in particular directions.

Units which are in the visual cortex can be sensitive to very specific kinds of stimuli, including information which comes from both eyes at once. Some of the so-called "complex" cortical visual cells in a cat's brain respond to moving slits of light anywhere in the visual field. A unit was found that responded to a black rectangle $\frac{1}{3}°$ wide in a 5° by 5° field, especially when the rectangle was moved in one direction. Many other units have been found that respond to very specific kinds of movement and pattern. Most neurophysiologists now believe that the recognition of specific patterns such as letters of the alphabet will turn out to be responses of single cortical cells.

THE VISUAL CORTEX PROVIDES ANOTHER LEVEL OF INTEGRATION BEYOND THE LGB. CERTAIN CELLS RESPOND ONLY TO CERTAIN PATTERNS OF STIMULATION OF THE RETINA OR TO PATTERN COMBINED WITH MOTION. IT MAY BE THAT RECOGNITION OF INDIVIDUAL LETTERS OF THE ALPHABET, FOR INSTANCE, DEPENDS ON THE STIMULATION OF SINGLE CELLS IN THE VISUAL CORTEX.

VISION IN LOWER VERTEBRATES   We see that in the mammals most kinds of sophisticated recognition occur in the brain, especially in the visual cortex. The lower vertebrates, whose behavior rests more on instinct than on learned patterns, do things rather differently, processing more of the visual data peripherally in the retina itself. Fibers in the optic nerve of frogs, unlike those of higher mammals, respond to changes in illumination, moving edges, stationary edges, and convex edges. The fibers in the optic nerve of the pigeon are even more complex than those of the frog and seem to perform analogously to cortical cells in the cat.

The precise nature of perceptual processes in the frog's optic tectum —the visual part of the frog's brain—is still largely a mystery, but two units which have obvious behavioral implications have been discovered. These might be called the "prey-sensing system." The first of these is a cell responsive to jerky movement. It is called a *newness* unit because once it has detected an object, it tends to ignore it thereafter. Presumably the information about the new object is passed on to other centers. The second unit, the *sameness* cell, responds with a burst of impulses when a small object is "noticed" in the visual field. Thereafter it keeps on "muttering"—releasing a background of impulses—so long as the object is visible. Whenever the object moves or changes direction, a further burst of impulses is released against the background mutter. To be lost, a target must remain stationary for about 2 min, at which time the background mutter fades. If the target begins to sneak slowly around at this time, the cell for a time seems oblivious. Then it will suddenly "notice" the target again. Anyone who has watched a frog or a toad notice an insect cannot help but infer the significance of these units.

Some animals give very stereotyped responses to certain classes of patterns. Chickens ignore a moving silhouette of a goose that is passed above them. But the chickens will be alarmed if the same silhouette is moving in the opposite direction, looking then more like a falcon (Fig. 9–18). There is increasing evidence that this sort of response is triggered by single cortical cells whose job it is to recognize specific patterns.

IN LOWER VERTEBRATES, VISUAL INTEGRATION OCCURS TO A GREATER DEGREE IN THE PERIPHERY OF THE SYSTEM THAN IT DOES IN MAMMALS. FOR INSTANCE, IN THE OPTIC NERVES OF FROGS SOME UNITS RESPOND ONLY TO MOVING EDGES AND OTHERS ONLY TO A STATIONARY EDGE. IN THE FROG'S BRAIN THERE ARE SPECIAL PREY-SENSING UNITS; IN A CHICKEN'S BRAIN THERE ARE, IN ALL LIKELIHOOD, PREDATOR-SILHOUETTE DETECTING UNITS. THESE UNITS ARE PRESUMABLY GENETICALLY PROGRAMMED.

HEARING In hearing, as with vision, the best-studied higher vertebrate is the cat. There are fewer receptor cells in the cochlea than in the retina, only 1 million, but the number of units in the auditory cortex, 500×10^6, is the same as in the visual cortex. Auditory pathways are manifold and complex (Fig. 9–19), but only the principal channels will be described here.

Each of the 30,000 incoming nerves of the cochlea synapses with two or three of the neurons from the hair cells, and each of the hair cells is innervated by two or three of the nerve cells. Thus, already there is overlap in the receptive fields and the potential for fine sound discrimination, as in the visual system. Electrical recordings from these neurons have shown that they respond to a narrow range of frequencies at low sound intensities. Besides frequency discrimination, another integrative capacity exists at this low level. Sound of relatively low pitch, up to about 1,500 cps, initiates impulses in the appropriate neurons, and the impulses bear a definite time relation to the pulses of sound energy. Two characteristics of sound provide information about the direction from which the sound is coming: differences in loudness perceived by each ear and differences in time of arrival of pulses.

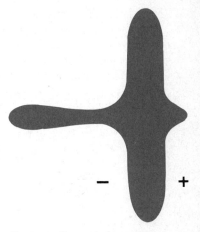

Fig. 9–18 Young ducks and geese run for cover when this model is sailed over them in the + (falcon) direction. When moved in the opposite, − (goose) direction, it elicits no response. (*After Tinbergen*)

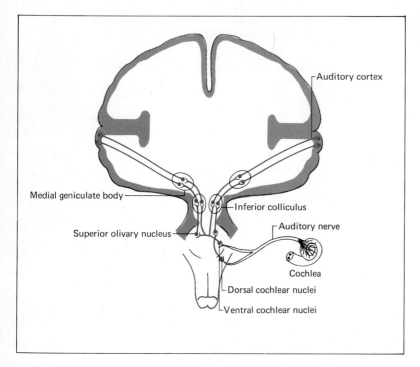

Auditory cortex

Medial geniculate body

Inferior colliculus

Superior olivary nucleus

Auditory nerve

Cochlea

Dorsal cochlear nuclei

Ventral cochlear nuclei

Fig. 9-19 Auditory pathways in man, from the cochlea of the inner ear to the auditory cortex.

In the brain, the incoming nerves synapse with second-order fibers in the cochlear nucleus. Even at this level there is great complexity—the incoming cochlear fibers are mapped out in 13 parallel channels, each of which probably has its own integrative specificity. Recordings from these cells show many kinds of integrative processes. These cells have narrower bands of pitch sensitivity, analogous to the visual system's increase in resolution at higher levels (Fig. 9-20) and probably arising in the same way from the overlap of fields. Some of the cells are inhibited by loud sounds, and some by sounds close to, but not identical with, their peak sensitivity.

The brain region primarily responsible for **binaural localization**—directional sound sensitivity—has many two-horned neurons (Fig. 9-21). One horn of each neuron receives terminals from the right cochlear nucleus; the other horn receives them from the left. About half of these cells are excited by sounds from a particular side, say the right, and inhibited by sounds on the opposite side, say the left. How would you interpret an increase in the firing rate of such a two-horned neuron? The other 50 percent of these neurons are excited by a sound that is loudest on the left side. Furthermore, some of these have fine pitch discrimination; thus sounds of differing frequency can be localized simultaneously.

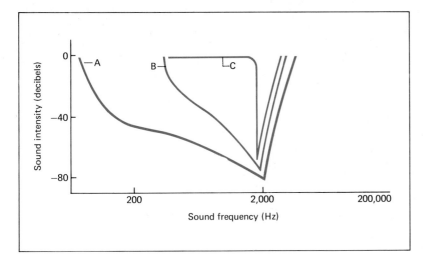

Fig. 9-20 Interaction of neurons brings about the refinement of response to faint sound. In this example, neurons having a maximum sensitivity of about 2,000 Hz respond to a broad range of frequencies in the auditory nerve (*A*); neurons in the midbrain (*B*) are more specialized; in the thalamus (*C*) a neuron only responds to faint sounds in a narrow range around 2,000 Hz. (*Adapted from BEHAVIOR by William G. Van der Kloot. Copyright © 1968 by Holt, Rinehart and Winston, Inc. Reprinted by permission of Holt, Rinehart and Winston, Inc.*)

We can begin to see how an animal could construct a rich three-dimensional sound picture of his surroundings.

As we proceed to the higher auditory centers, the units become more complex and less decipherable. Units in the auditory cortex no longer respond in a one-to-one fashion to pure pitch tones, but they do respond to changing pitch—some to rising, some to falling, and some to both. The cat's auditory cortex has novelty units similar to those in the brain of frogs.

HEARING IN VERTEBRATES, LIKE VISION, INVOLVES OVERLAPPING RECEPTIVE FIELDS FOR FINE DISCRIMINATION. INTEGRATION OF INCOMING SOUND INFORMATION IN THE BRAIN SHOWS PARALLELS WITH THE INTEGRATION OF VISUAL INPUTS. NEURONS RECEIVING INPUTS FROM BOTH EARS PERMIT THE LOCALIZATION OF SOUNDS.

A start, then, has been made at tracing sensory input through various levels of integration in the vertebrate CNS. What has not yet been seriously approached (and what many biologists believe to be outside the realm of empirical science) is the question of how the always discrete, particulate information supplied to the CNS by receptors and incoming neurons is blended into the continuous sense impressions we "see," "hear," "smell," and "feel." The mode of transduction of electrochemical events into perceptions is difficult to imagine, let alone investigate.

THE BRAIN: PRINCIPAL REGIONS   Figure 9-22 shows a diagram of a human brain cut along the median fissure, exposing the cut surface. Any-

Fig. 9-21 Neurons in the olivary nucleus receive input (small, black fibers) from both cochlea via the cochlear nuclei shown in Fig. 9-19. All the fibers synapsing on one of the ends (poles) of the bipolar neuron (*above*) originate in the cochlear nucleus on the same side. This can be shown by destroying the cochlear nucleus on one side, whereupon all of the fibers going to the corresponding pole of the bipolar neuron degenerate (*below*). (*From "Auditory Localization" by Mark R. Rosenzweig. Copyright © 1961 by Scientific American, Inc. All rights reserved.*)

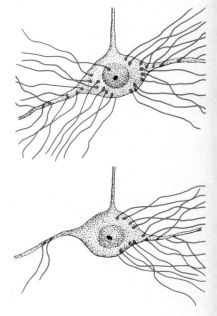

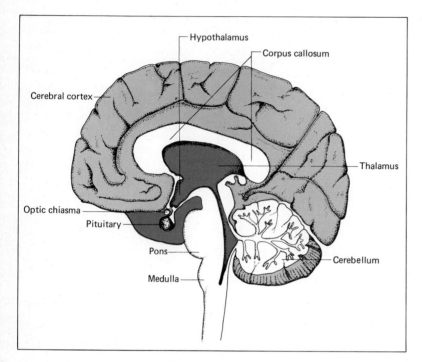

one seeing this for the first time would be struck by the small size of many of the regions compared with the huge overlying cerebral cortex ("skin of the cerebrum") with its fissured surface. This dominance of the cerebral cortex was not always the case. Figure 9-23 shows how the different brain regions have changed through evolution. In fish, the midbrain and hindbrain account for most of the brain mass. In mammals like ourselves, it is the front of the brain, the forebrain (the majority of which is cerebral cortex), that fills most of the skull.

Of the three regions the hindbrain is the first to appear evolutionarily and to appear during embryological development. Its posterior region, the **medulla,** is an enlarged extension of the spinal cord. As such, it is the route taken by sensory neurons passing up the spinal cord to the brain and also the route taken by motor nerves controlling glandular and muscular activities. The medulla is also the site of the lowest centers, those controlling the most primitive, vegetative functions, namely, ventilation and heartbeat. Taste centers occur there, too.

Above and partly anterior to the medulla is the **cerebellum,** the master control center for the coordination of muscular activities. Higher centers give the orders for voluntary actions, but the cerebellum has the circuitry that executes them. Among its roles are the maintenance of equilibrium and posture. Fibers from the inner ear, the eyes, and the stretch receptors all converge on the cerebellum, where their input is analyzed and where signals are continually emanating to the postural muscles of the appendages and trunk.

More subtle tasks are performed by the cerebellum in its role as the monitor of motor functions. A monkey or man missing all or part of its cerebellum is not able to predict the consequences of its movements. Movements can be initiated, but corrective feedback coordination is gone. For instance, when reaching for a glass or when running to answer a phone, the *damping* and *predictive* functions of the cerebellum prevent overshooting the mark by ordering the appropriate braking contractions of antagonistic muscles. When the cerebellum is damaged or removed, the inertia of movements is not opposed. In this condition, monkeys will knock themselves silly by running into walls.

The midbrain in lower vertebrates is a visual and integrating center. The enlarged forebrain has taken over these functions in mammals, leaving the midbrain as a relay center.

THE VERTEBRATE BRAIN IS DIVIDED INTO A FOREBRAIN, MIDBRAIN, AND HINDBRAIN. THE HINDBRAIN CONTROLS BREATHING AND HEARTBEAT AND CONTAINS THE TASTE CENTERS (IN ITS MEDULLA) AND THE COORDINATING CENTER FOR MUSCULAR ACTIVITIES (IN ITS CEREBELLUM).

CONSCIOUS ACTIVITY: THE CEREBRAL CORTEX AND ASSOCIATED CENTERS Many insects and other "lower" animals perform delicate and complex actions, such as nest building, courtship, and web building. Such activities are typically stereotyped, and these animals are unable to learn new, intricate tasks not part of their inherited behaviorial repertory. It is as though the circuits for a finite number of behavior patterns are "prewired" in these organisms. One thing distinguishing the mammals, especially man, is the ability to invent and *learn* intricate skills such as speaking, writing, dancing, and athletic routines. These conscious activities are controlled by the cortex of the forebrain region, known as the **cerebrum** (hence the term "cerebral cortex").

Neurophysiologists now have a rough idea of how conscious activities are controlled. Figure 9-24 illustrates the paths taken by impulses from sensory receptors such as pain or pressure receptors in the skin or visceral organs, stretch receptors in muscles, or receptors in the joints. These sensory nerve fibers branch upon entering the spinal cord, where they synapse directly with motor nerves to elicit spinal-cord reflexes such as the stretch reflex. Other branches are seen to ascend the spinal cord and medulla in specific tracts. Some end in the cerebellum where they provide

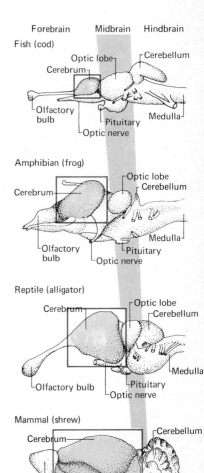

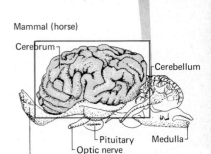

Fig. 9-23 Evolutionary changes in the relative proportion of the cerebrum and midbrain in vertebrates.

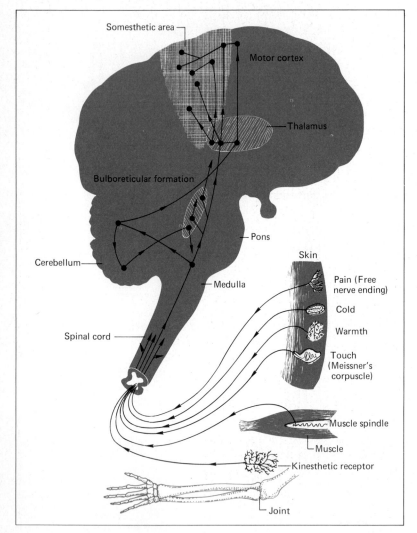

Fig. 9-24 The pathways of sensory information from receptors to the brain of man. (*Adapted from A. C. Guyton, FUNCTION OF THE HUMAN BODY, 3/e, 1969, W. B. Saunders Company, Philadelphia.*)

information for the completely subconscious cerebellar functions already discussed. Other tracts branch to well-defined lower centers controlling other subconscious functions, including blood pressure and gastrointestinal movements. Still other tracts ascend to the ***thalamus,*** the part of the forebrain from which the cerebral cortex evolved. Here at last the impulses cross the threshold into consciousness. With the cortex destroyed but the thalamus intact, it is still possible roughly to localize sensations and determine their type, including pain, pressure, heat, and distension. From the thalamus, fibers radiate to the ***somaesthetic*** (*soma,* "body," and *esthetic,* "feeling") area of the cortex.

By probing nerve trunks of crustaceans with fine electrodes, neuro-physiologists have been able to stimulate the same neuron repeatedly in different individuals. When this kind of stimulation initiates a muscle twitch or a movement, it is inevitably the same twitch or movement, proving that fibers in much of the nervous system are precisely oriented with respect to each other. A predictable spatial orientation of fibers occurs in the lower centers of man's brain and in parts of the cerebral cortex as well. Why is this essential?

We live in a three-dimensional world. When sounds are heard, objects seen, and sensations felt, survival often depends on precise localization of the stimuli. Apparently, the most parsimonious solution to the "problem" has often been a precise mapping of the body, as well as the visual and auditory fields, on the cerebral cortex. This must facilitate localization of stimuli and simplify development of circuits that control appropriate movements.

Grotesque illustrations such as Fig. 9–25 demonstrate this principle as it applies to the somaesthetic cortex. In this figure, the size of the body region represented is proportional to the cortical area actually devoted to it. For instance, a small area on the top of the somaesthetic cortex receives sensory input from the trunk, whereas the sensations from the lips are received by a large area on the side. Inspection of Fig. 9–25 would suggest that the area of cortex allocated to a body part is propor-

Fig. 9–25 Maps of the somaesthetic (sensory) cortex and motor cortex of man illustrating the localization of integration centers. Note the dominance of the hands and mouth. *(Adapted with permission of The Macmillan Company from THE CEREBRAL CORTEX OF MAN by Penfield, W. and T. Rasmussen. Copyright 1950 by The Macmillan Company.)*

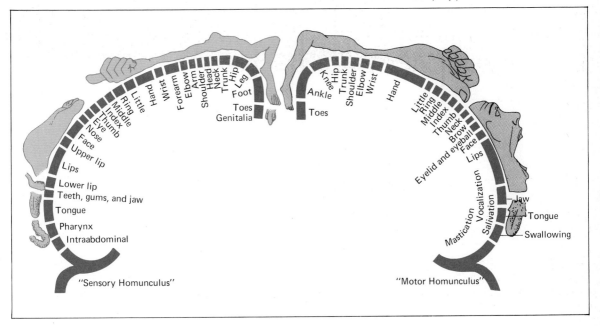

tional to its importance as a source of sensory input from the outer world. The cortex map of a horse would seem even more distorted than that of man since nearly half of the horse's cortex receives input from the nose.

MAN AND OTHER MAMMALS SHOW A MUCH GREATER ABILITY TO LEARN COMPLEX TASKS THAN DO LOWER VERTEBRATES AND IN-VERTEBRATES, WHICH OFTEN HAVE GENETICALLY PROGRAMMED ABILITIES TO PERFORM THEM. THE CEREBRAL CORTEX IS THE CEN-TER OF LEARNING. SENSORY INPUT IS PRECISELY ADDRESSED TO SPECIFIED AREAS OF THE CEREBRAL CORTEX.

Figure 9–26 shows that a band of cortex just behind the somaesthetic cortex is an **association** area for sensory input. The areas of the cortex receiving visual and auditory input also have association areas adjacent

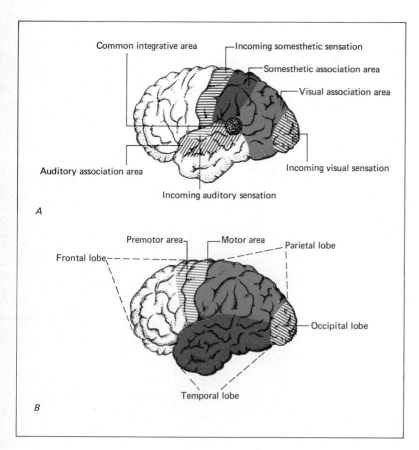

Fig. 9–26 *(A)* Diagram of the surface of the cerebral cortex showing the areas of incoming sensation, their association areas, and the common integrative area; *(B)* the premotor and motor areas of the cerebral cortex and the main lobes of the cerebrum. (*Adapted from A. C. Guyton, FUNCTION OF THE HUMAN BODY, 3/e, 1969, W. B. Saunders Company, Philadelphia.*)

to them. Association areas interpret the information from their respective sources and perceive the more subtle qualities of sensation. The somaesthetic association area, for instance, perceives the shape, weight, and texture of objects, as well as their spatial orientation. Association areas also serve as memory banks of past sensory experiences. During brain surgery under a local anesthetic to the tissues overlying the brain (the brain itself has no pain receptors), electrodes placed on the association areas may elicit in the patient very vivid memories of sensory experiences.

The master association region of the cortex is called the **common integrative area.** To it is fed information from all the other association areas. From it come "orders" to the motor areas of the cortex eliciting appropriate actions via the lower brain centers and the spinal cord. Here, we finally are dealing with the decision and command center of the brain. If it is destroyed, the person becomes an imbecile.

AREAS OF THE CORTEX WHICH RECEIVE INFORMATION HAVE ADJACENT ASSOCIATION AREAS IN WHICH A PRELIMINARY SORTING AND INTEGRATING PROCESS TAKES PLACE AND WHERE MEMORIES ARE STORED. INFORMATION FROM THESE AREAS IS FED INTO THE COMMON INTEGRATIVE AREA, THE MASTER CONTROL CENTER OF THE BRAIN.

The motor area of the cortex has a map similar to that of the somaesthetic cortex. The hand and mouth together are represented by two-thirds of its surface. This is not surprising, since speech and manipulation are the two most complex activities that have evolved, and both require the most highly integrated control.

When an electrode is placed on the motor area, a specific muscle or muscle group contracts on the opposite side of the body. Such contralateral representation is characteristic of the entire brain. For instance, right-handed people typically have a dominant left cerebral hemisphere, and destruction of large areas of the cortex on the right side, including the common integrative area, may not do serious, permanent damage to the intellect.

Complex coordinated movements, such as writing, running a play in football, or dancing, are "learned" and controlled by the **premotor cortex.** The premotor cortex commands the motor cortex and lower centers to activate many muscles in an organized pattern.

You now have a very superficial understanding of how the cerebral cortex integrates behavior. You should not be deluded, however, into believing you understand the brain. We have not even attempted a dis-

Fig. 9–27 A cat with an electrode implanted in the midbrain shows a complex rage behavior when this region is stimulated (*Courtesy of José Delgado.*)

cussion of mental phenomena like motivation, ambition, greed, awe, or transcendent experience. These are challenges for future generations. The more primitive emotions—lust, fear, aggression—will be understood sooner. The electrical stimulation of the midbrain, the thalamus, and the cortex repeatedly elicits these behaviors. A cat with an electrode implanted in its midbrain is illustrated in Fig. 9–27. Stimulating this region produces a complex rage response from the cat.

THE COMMAND AREA OF THE CORTEX HAS A MAP ANALOGOUS TO THAT OF THE RECEPTIVE AREAS. UNDERSTANDING THE WAY IN WHICH THE BRAIN DEALS WITH COMPLEX PHENOMENA SUCH AS AMBITION IS A PROBLEM FOR THE FUTURE (WHICH MAY NEVER BE SOLVED). A GREATER UNDERSTANDING OF CONTROL OF EMOTIONS SUCH AS RAGE SEEMS LIKELY IN THE RELATIVELY NEAR FUTURE.

We said at the beginning of the section on integration that the line between hormonal and nervous integration can be arbitrary. Consider this while you are reading about the autonomic nervous system (ANS).

Many of the neurons that serve the gastrointestinal and other internal organs, the smooth muscle of blood vessels, and the sweat glands, as well as the other organs shown in Fig. 9–28, are unlike the motor and sensory neurons that innervate skeletal muscle. First, the majority are not wrapped by myelin sheaths, so that they have relatively slow rates of impulse transmission. Second, any impulse going to a target organ crosses a synapse between the spinal cord and the target organ. These synapses are located in ganglia outside the spinal cord, some in the target organs themselves and others in ganglia found somewhere between the spinal cord and the target organ. The impulse is carried from the synapse to the target organ via a postganglionic neuron. Recall that no such synapse intervenes in the passage of an impulse from the spinal cord to a skeletal muscle.

Third, there are two divisions of the ANS, and these often have antagonistic effects on the target organs. (In a way, this is analogous to inhibitory and excitatory synapses in the CNS.) These opposite effects are mediated by two transmitter substances, noradrenaline and acetylcholine. In the CNS, only acetylcholine has been found at synapses outside the brain and spinal cord, that is, at neuromuscular junctions.

The two divisions of the ANS are the **sympathetic nervous system** and the **parasympathetic nervous system** (*para* means "next to" or "to the side"). These names are a key to the anatomical arrangement of these two systems, as illustrated in Fig. 9–29. The sympathetic system is "central," in that its neurons enter and leave the spinal cord in the thoracic (chest) and upper lumbar (abdominal) region. The parasympathetic is "lateral," in that its neurons enter and leave the CNS through cranial (brain) nerves and, at the opposite end of the CNS, the sacral region of the spinal cord. Ninety percent of the fibers in the parasympathetic nervous system pass through the *vagus,* or tenth cranial nerve.

THE NERVES OF THE AUTONOMIC NERVOUS SYSTEM ARE SLOW IN COMPARISON WITH THOSE OF THE CNS, AND IN THE ANS, UNLIKE THE CNS, NERVE IMPULSES CROSS A SYNAPSE BETWEEN THE SPINAL CORD AND THE TARGET ORGAN. THE SYMPATHETIC AND PARASYMPATHETIC DIVISIONS OF THE ANS OFTEN HAVE ANTAGONISTIC EFFECTS ON TARGET ORGANS.

Noradrenalin, the hormone usually secreted by sympathetic neurons,

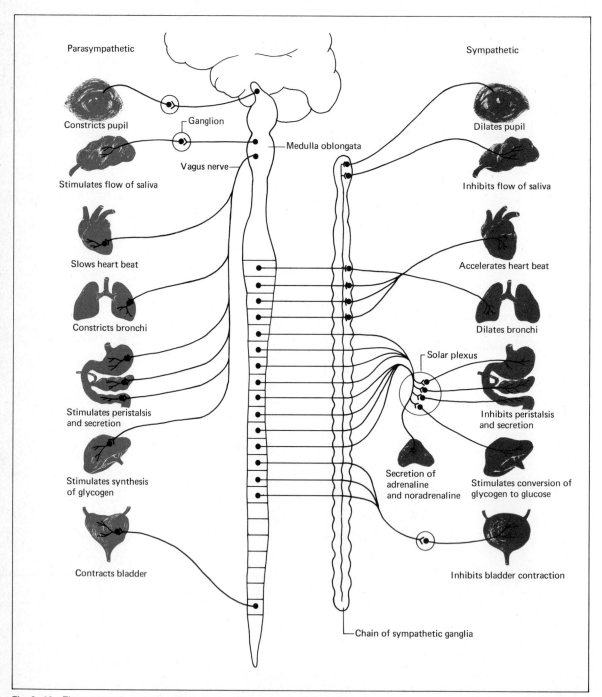

Fig. 9–28 The autonomic nervous system. Some of the synapses of the autonomic nervous system occur in the sympathetic ganglia, including the solar plexus; others occur in the organs themselves. (*Adapted from BIOLOGICAL SCIENCES, 1/e, by William T. Keeton, illustrated by Paula DiSanto Bensadoun, by permission of W. W. Norton & Company, Inc. Copyright © 1967 by W. W. Norton & Company, Inc.*)

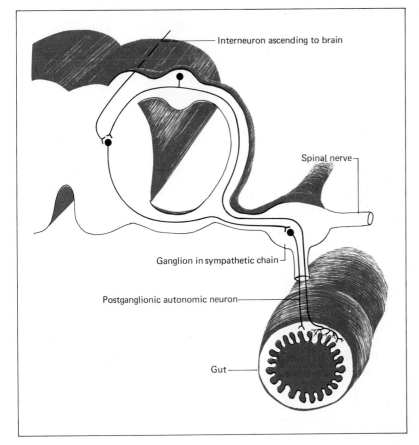

Interneuron ascending to brain

Spinal nerve

Ganglion in sympathetic chain

Postganglionic autonomic neuron

Gut

Fig. 9-29 A typical circuit of the autonomic nervous system; note the synapse within the sympathetic ganglion. (*Adapted from A. C. Guyton, FUNCTION OF THE HUMAN BODY, 3/e, 1969, W. B. Saunders Company, Philadelphia.*)

is the same hormone as that secreted by the medulla of the adrenal glands, that is, the hormone that elicits the "flight-fear-flight" reaction. Noradrenalin reduces the activity of the gastrointestinal tract, increases heart and ventilation activity, and boosts the basal metabolic rate. All these effects prepare the body for strenuous activity. Parasympathetic stimulation has the opposite effects on many of these organs.

One might get the impression that these two ANS systems function alternatively, one being off while the other is on—like the accelerator and brake of a car. Actually, impulses are continuously being propagated in both, and the relative firing rates determine the state of an organ. The consequence of this continuous stimulation is called **tone**—a readiness, in a sense, to respond quickly to stimulation in either direction. For instance, a panic-inducing visual stimulus will cause an increase in sympathetic impulses and a decrease in parasympathetic impulses to the heart, eliciting a greater response than would sympathetic stimulation alone.

The ANS is sometimes thought by students to be independent of the CNS. In fact, the centers that control ANS activity are in the brain, although as you might expect, these centers are in the lower brain areas and hence usually below the threshold of conscious control. The hypothalamus is an especially important control center. In it are areas that regulate heart rate, arterial (blood) pressure, body temperature, and metabolic rate, not to mention the many hormonal functions discussed earlier. Recent scientific studies have confirmed what mystics have known for millenia —that with practice many ANS functions can be brought under conscious control.

NEURONS OF THE SYMPATHETIC SYSTEM SECRETE NORADRENALIN AND TEND TO PREPARE THE BODY FOR STRESS. IN MANY ORGANS THE PARASYMPATHETIC SYSTEM TENDS TO COUNTER THIS EFFECT. THE TONE OF THE BODY IS CONTROLLED IN LARGE PART BY THE TWO-PART ANS, WHICH IS INTEGRATED WITH THE CNS IN THE BRAIN.

NERVOUS AND HORMONAL INTEGRATION

It should be clear at this point that the differences between nervous and hormonal integration in animals are mostly of degree, not of kind. At most synaptic junctures, a transmitter substance, which might be thought of as a "microhormone," carries the message from the presynaptic to the postsynaptic membrane. Furthermore, the secretion of hormones by nerve cells (neurosecretion) is a common and important mechanism. For instance, you will recall that the hypothalamus of the brain is directly involved in the secretion of the hormone vasopressin. The neurosecretory cells almost literally have their "heads in the nervous systems and tails in the endocrine." Similarly, nervous stimulation of the adrenal glands is what causes them to secrete **adrenalin.** A threat or other stimulus, interpreted by the nervous system as calling for readiness for action, activates the secretion of this hormone. Adrenalin increases pulse rate, raises blood pressure, causes the liver to break glycogen down to glucose and release glucose into the bloodstream, increases blood supply to skeletal muscles, and generally makes other preparations for "fight or flight." Although we have considered them separately, nervous and hormonal systems work together at the task of integration.

THERE IS NO CLEAR DISTINCTION BETWEEN HORMONAL AND NERVOUS INTEGRATION. BOTH SYSTEMS INVOLVE CHEMICAL MES-

*SENGERS, AND GLANDS OFTEN ARE STIMULATED TO PRODUCE
HORMONES BY THE RECEIPT OF NERVE IMPULSES.*

SUMMARY

A wide variety of effectors is found in animals, the most important of which is muscle. A muscle contraction is the telescoping of bundles of fibrils, initiated by one or more nerve impulses and powered by ATP from abundant mitochondria. In vertebrate muscles, each unit shows all-or-none contraction and graded contraction is produced by the summation of events in many units. In insect muscles, in contrast, individual units show graded contractions depending on the pattern of nerve impulses received. Events at the nerve-muscle synapse are similar to those at nerve-nerve synapses in vertebrates.

Integration in animals can be performed by elements as simple as a single neuron and as complex as the cerebral cortex of man. In lower animals much integration is peripheral, with little or no involvement of the central nervous system (CNS). In higher animals most of the integration has become central. For example, in the mammalian visual system a complex hierarchical arrangement permits perception of finer distinctions than the lens-retina system is physically capable of resolving.

Little is known about integrative phenomena such as ambition, intelligence, or jealousy, and it is not certain that we will ever fully understand them. We do know, however, that inputs to the cerebral cortex of our brain from various sources are received in spatially segregated areas and that outputs arise from similarly segregated areas. Several levels of integrative centers connect these. Centers in the brain also connect the CNS with the largely "involuntary" autonomic nervous system (ANS), which controls the "tone" of the body.

There is no clear distinction between hormonal and nervous integration in animals. Both systems work together to produce appropriate responses to changes in the external and internal environments.

SUPPLEMENTARY READINGS
Books

Marler, P., and W. J. Hamilton: "Mechanisms of Animal Behavior," Wiley, New York, 1966.

Articles

Di Cara, L. V.: Learning in the Autonomic Nervous System, *Sci. Amer.,* **222**, no. 1, Offprint 525 (1970).
Hayashi, T.: How Cells Move, *Sci. Amer.,* **205**, no. 3, Offprint 97 (1961).
Hubel, D. H.: The Visual Cortex of the Brain, *Sci. Amer.,* **209**, no. 5, Offprint 168.

Huxley, H. E.: The Mechanism of Muscular Contraction, *Sci. Amer.,* **213,** no. 6, Offprint no. 1026 (1965).

Kandel, E. R.: Nerve Cells and Behavior, *Sci. Amer.,* **233,** no. 6, Offprint 1182 (1970).

Luria, A. R.: The Functional Organization of the Brain, *Sci. Amer.,* **222,** no. 3, Offprint 526 (1970).

Smith, D. S.: The Flight Muscles of Insects, *Sci. Amer.,* **212,** no. 6, Offprint 1014 (1965).

3

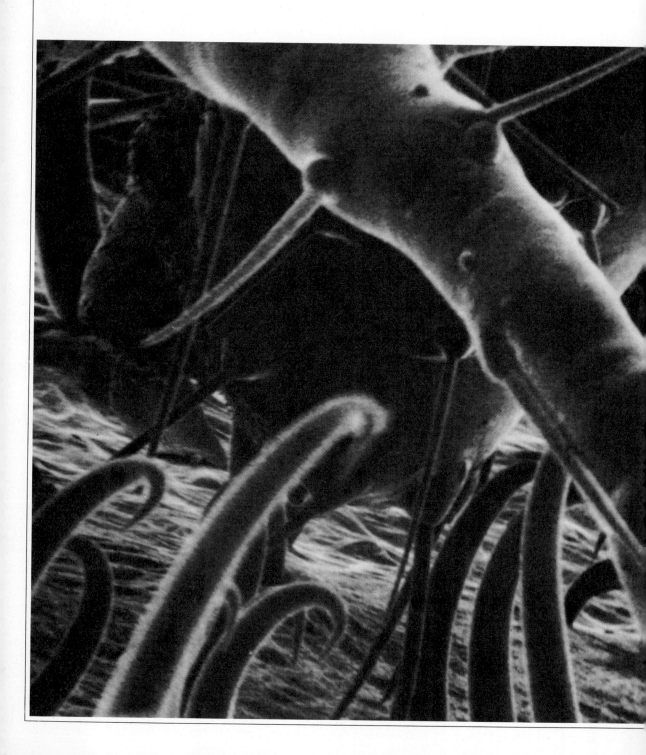

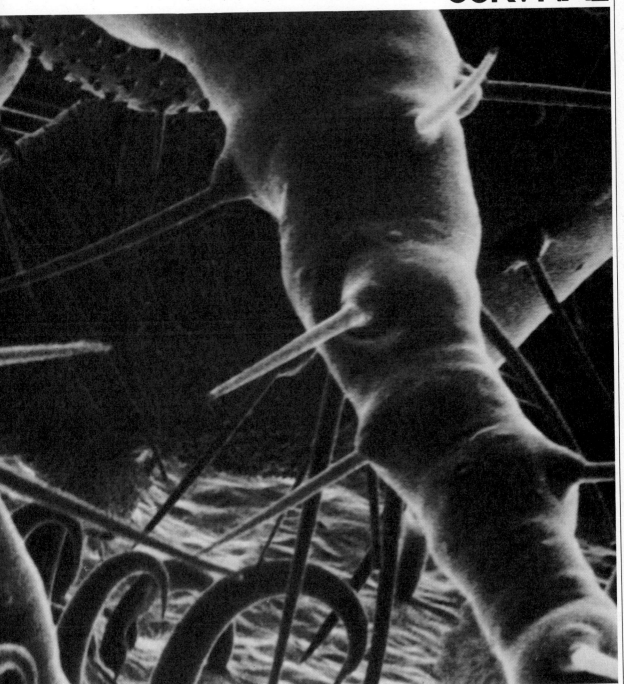

STRATEGIES FOR SURVIVAL

This section, like the last, deals with the individual, but now we will begin to reinsert the individual into the context of ecological relationships and into the flow of time. The theme for Chap. 10 is patterns of individual change through time and how they have developed in relation to both external and internal environments. Life cycles and development in both higher plants and animals are described.

Chapter 11 is an introduction to the varied devices which organisms have evolved to protect themselves from the many kinds of hazards presented by their environment. Defense is discussed at cellular and organismal levels.

[Illustration on preceeding pages shows a caterpillar of Heliconius butterfly caught on a leaf of the passion flower (Passiflora adenopoda). The hooks on the leaves provide a specific, effectively absolute defense against the caterpillars of these butterflies which are important predators on other passion flowers.]
(Courtesy of Lawrence E. Guilbert)

CHAPTER 10

LIFE CYCLES AND DEVELOPMENT

In Chap. 2, we discussed the cytological and genetic features of reproduction common to virtually all organisms. You will remember that in the life of an organism, some of the cells are haploid and some are diploid. Meiosis, or reduction division, is responsible for producing the haploid phase. Syngamy, or fusion of gametes, produces the diploid phase. Diploid and haploid phases thus alternate in what is called the **life cycle** of an organism. The diploid part of the life cycle is called

the **diplophase,** and the haploid part is the **haplophase.** If we look at a variety of different kinds of plants and animals, we find that not all have the same kind of life cycle. Some organisms spend most of their life cycle in the diplophase; others have only a short diplophase.

KINDS OF LIFE CYCLES
Haploid and Diploid Cycles

A UNICELLULAR PLANT: *CHLAMYDOMONAS* In order to examine the range of types of life cycles, we must look at a number of representative types. Let us begin with a unicellular alga, *Chlamydomonas*. *Chlamydomonas* is a free-swimming freshwater alga often found in ponds, where it may be so numerous as to color the water green. As you can see in Fig. 10-1, an individual cell is pear-shaped and about 15μm in length. At one end there is a pair of flagella, which propel the cell through the water. Most of the cell is filled with a large cup-shaped chloroplast with a prominent pyrenoid, the site of starch formation. Below the flagella are two contractile vacuoles and a pigmented, light-sensitive area called the *eyespot*. From time to time, an individual of *Chlamydomonas* may divide by mitosis into four or eight cells, contained within the parental cell wall. These develop flagella and eventually break out of the old cell wall and swim off. Such cells are called **zoospores,** and they develop directly into a typical *Chlamydomonas*. This is an example of **asexual** reproduction.

Periodically, the parental cell divides into 16 or 32 cells which look very much like zoospores. Instead of developing directly, however, these cells fuse in pairs. There are two mating types in *Chlamydomonas*.

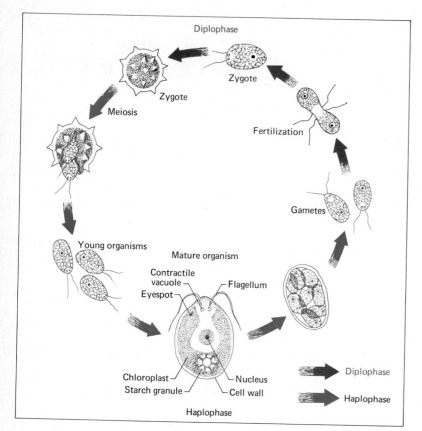

Fig. 10-1 The life cycle of a unicellular alga, *Chlamydomonas. (Adapted from Relis B. Brown, GENERAL BIOLOGY. Copyright McGraw-Hill Book Company, 1970. Used by permission.)*

These are called + and − strains, and they are attracted to one another by substances produced by the flagella. Since these cells fuse in pairs, you know that they must be haploid. They are, in fact, gametes and are identical in appearance. The zygote which is formed develops a thick cell wall and often goes into a resting stage, in which it can survive adverse conditions—for example, the drying up of the pond. Eventually the zygote divides and produces four zoospores. This division is meiotic; therefore, the zoospores are haploid. Half of them will be of the + mating type, and half will be of the − type. The major portion of the life cycle of *Chlamydomonas* is spent in the haplophase. Indeed, the zygote is the only diploid cell in the life cycle.

IN SOME UNICELLULAR ORGANISMS, THE MAJOR PART OF THE LIFE CYCLE IS SPENT IN THE HAPLOID CONDITION AND THE ONLY DIPLOID CELL IS THE ZYGOTE.

A UNICELLULAR ANIMAL: *PARAMECIUM*   Now let us turn to a unicellular animal. *Paramecium* is a much more complex organism than
Chlamydomonas. Some cells may be nearly a millimeter in length, just
visible to the naked eye. *Paramecium* belongs to a group of animals
called ciliate protozoans. The cell is covered with cilia, which are arranged in a complex pattern and which propel the animal through the
water. Within the cell are the usual organelles, such as mitrochondria,
together with more specialized ones which function in nutrition. *Paramecium* reproduces asexually by mitotic division.

Sooner or later in its life cycle, *Paramecium* undergoes sexual reproduction, that is, meiosis and syngamy (Fig. 10–2). This process is rather complex, and we need not be concerned with the details. Individual cells of
Paramecium contain two kinds of nuclei: a large macronucleus and (in the
species we will discuss) a small micronucleus. When sexual reproduction
occurs, cells of different mating types come together in pairs, or **con-
jugate.** Within each of the conjugating cells, the macronucleus breaks
down and the micronucleus divides meiotically to produce four daugh-

Fig. 10–2 Sexual reproduction
(conjugation) in a unicellular animal,
*Paramecium. (Adapted from Relis B.
Brown, GENERAL BIOLOGY,
Copyright McGraw-Hill Book
Company, 1970. Used by permission.)*

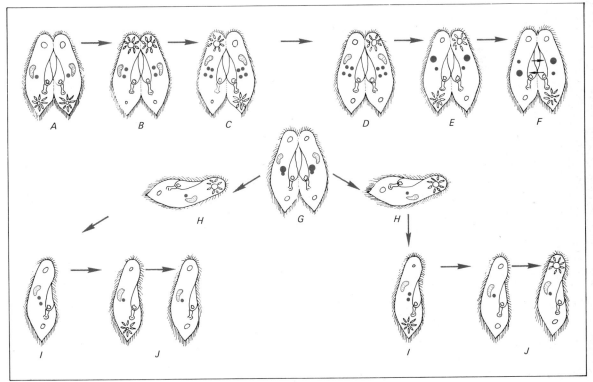

ter nuclei. Three of the daughter nuclei degenerate, leaving one haploid nucleus in each cell. This nucleus then divides by mitosis, forming two gamete nuclei. One of these two nuclei from each cell moves into the cell of the conjugal partner. Nuclear fusion occurs in each cell, and thus each has a new diploid zygotic nucleus.

Thus you can see that *Paramecium* is an organism that spends most of its life cycle in the diplophase. The nuclei produced by meiosis of the micronucleus are the only haploid nuclei in the life cycle.

IN SOME UNICELLULAR ORGANISMS, THE MAJOR PART OF THE LIFE CYCLE IS SPENT IN THE DIPLOID CONDITION AND THE GAMETES ARE THE ONLY HAPLOID CELLS.

MULTICELLULAR ANIMALS Multicellular animals have one basic life cycle. The major portion of the life cycle is spent in the diplophase. Diplophase individuals are of two mating types: male and female. These individuals produce sperm and eggs by meiotic division in organs called **gonads.** Customarily, the eggs are large, rich in stored food, and non-motile. The sperm are small, consisting largely of a nucleus and a fla-gellar tail, and are motile. We will discuss this kind of life cycle in detail below.

In the multicellular plants, a variety of life cycles has evolved, some of them very complex. We will not concern ourselves with these here, except for several types which occur in the vascular land plants, non-vascular land plants, and some aquatic plants.

IN MULTICELLULAR ANIMALS, THE MAJOR PART OF THE LIFE CYCLE IS SPENT IN THE DIPLOID CONDITION AND ONLY THE GAMETES ARE HAPLOID. MULTICELLULAR PLANTS SHOW A NUMBER OF DIFFERENT KINDS OF LIFE CYCLES.

Alternation of Generations in Plants

ULVA In many plants, there is a multicellular body found in both the haplophase and the diplophase. Life cycles which have alternating haploid multicellular bodies and diploid multicellular bodies are said to have **alternation of generations.** A good example is the marine green alga called sea lettuce *(Ulva,* Fig. 10–3). The plant body in *Ulva* is two cells thick and is in the form of a wrinkled sheet, rather like a leaf of lettuce in form and color. Individual cells of *Ulva* have a nucleus and a large chloroplast. Some of the cells at the base of the plant produce fingerlike extensions which attach the alga to its rocky substrate. Collectively, they make up the **holdfast.** This sort of plant body, which is relatively simple and without specialized tissues, is called a **thallus.**

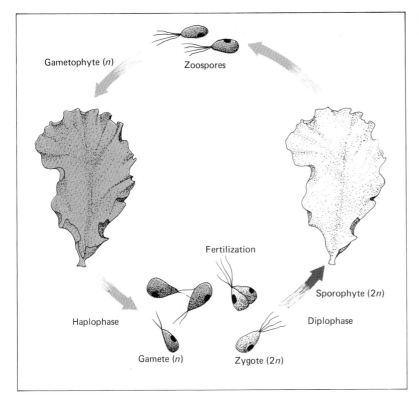

Gametophyte (n)

Zoospores

Fertilization

Haplophase

Gamete (n)

Zygote (2n)

Sporophyte (2n)

Diplophase

Fig. 10-3 The life cycle of the green alga *Ulva*. (Adapted from Paul B. Weisz, THE SCIENCE OF BIOLOGY. Copyright McGraw-Hill Book Company, 1971. Used by permission.)

If you were to examine a number of thalli of *Ulva*, you would be able to recognize two different kinds. These are distinguished not by any difference in overall appearance but rather by the kind of reproductive structures formed. In one type you would find certain cells dividing to produce gametes. Each of these gametes has a pair of flagella, and after swimming about for a time, the gametes will fuse in pairs. The result of such fusion, or syngamy, is the diploid zygote.

The zygote divides by mitosis, forming a multicellular thallus indistinguishable superficially from the gamete-producing thallus. This is the second type of *Ulva* you might find. In this plant, certain cells called **sporangia** divide by meiosis, forming four haploid zoospores. Each of these zoospores has four flagella. Eventually these settle down on the substrate and begin to divide mitotically to produce a haploid thallus.

There are, therefore, two kinds of multicellular plants in the life cycle of *Ulva*. One is a haploid, gamete-producing plant known as a **gametophyte.** The other is a diploid, spore-producing plant called a **sporophyte.** Such a life cycle is said to have alternation of generations. Since the two "generations" do not necessarily succeed one another, a more accurate designation would be "alternation of phases." Historical usage gives precedence to the former, however. Because the two phases are so similar in *Ulva,* that alga can be said to have alternation of undifferentiated generations.

IN SOME MULTICELLULAR PLANTS, THERE IS AN ALTERNATION OF GENERATIONS OF HAPLOID AND DIPLOID INDIVIDUALS IN EVERY LIFE CYCLE. THE DIPLOID PLANT, THE SPOROPHYTE, PRODUCES HAPLOID SPORES BY MEIOSIS. HAPLOID SPORES GERMINATE INTO HAPLOID PLANTS, GAMETOPHYTES, WHICH PRODUCE GAMETES BY MITOSIS. SYNGAMY OF GAMETES PRODUCES SPOROPHYTES.

LAMINARIA In many algae and in land plants, the two phases are very different in size and appearance. In the brown alga *Laminaria,* for example, the sporophyte is a massive plant, often many feet long (Fig. 10–4).

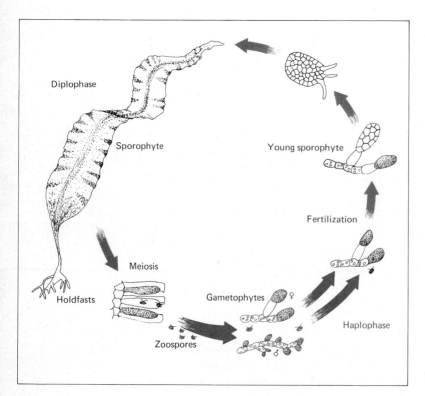

Fig. 10–4 The life cycle of the brown alga *Laminaria.*

Diplophase

Sporophyte

Young sporophyte

Fertilization

Meiosis

Holdfasts

Gametophytes ♀

Haplophase

Zoospores

The thallus is specialized into holdfasts, which grip the rock tightly, a cylindrical stalk, and one or more flattened photosynthetic blades. This plant body is constructed of complexly interwoven filaments. In the stalk, there are phloemlike cells with sieve plates in their walls. In certain cells of the blade which function as sporangia, meiosis occurs, and the haploid daughters then divide mitotically to produce a large number of zoospores.

Zoospores, upon settling down, develop into tiny filamentous gametophytes. There is sexual differentiation among the gametophytes. Half of them are male and produce motile sperm in special structures. The remaining half are female, and they produce large, nonmotile eggs. Notice that in *Laminaria* the gametes are not the same size. The production of gametes of unequal size is common in algae and higher plants. The difference in size may be relatively slight, or it may be great. When it is sufficiently large, the gametes are called egg and sperm. *Laminaria* produces eggs and sperm.

Fertilization of the egg by the sperm produces a diploid zygote, which divides and eventually becomes the large sporophyte of *Laminaria*. In many of the major groups of algae known, there appears to be a trend from unicellularity to multicellularity, from similar gametes to eggs and sperm, and from similar generations to differentiated generations.

ALTERNATING HAPLOID AND DIPLOID GENERATIONS MAY BE SIMILAR IN SIZE AND SHAPE OR VERY DIFFERENT, JUST AS MALE AND FEMALE GAMETES MAY BE SIMILAR OR DIFFERENT. THE MORE SPECIALIZED MULTICELLULAR PLANTS HAVE VERY DIFFERENT GENERATIONS AND VERY DIFFERENT GAMETES.

MOSSES AND LIVERWORTS   Mosses and their relatives, the liverworts, are land plants which do not contain vascular tissue. They do, however, have alternation of differentiated generations. The conspicuous photosynthetic plant of a moss is the gametophyte (Fig. 10–5). At the tips of some of the branches, gamete-producing structures may arise. If one plant produces both sperm and eggs, the plant is said to be *monoecious* (Greek: "one house"). Many mosses are *dioecious* (Greek: "two houses"), however, each plant producing either eggs or sperm, but not both. Within the sperm-producing structure, numerous flagellate sperm are produced by mitosis. The egg-producing structure is shaped like a long slender flask. Within its base is formed a single egg and a row of canal cells. When the egg is ready for fertilization, the canal cells disappear. Attracted by a substance secreted by the egg-producing structure, the sperm swims through a film of water, moves down the canal, and fertilizes the egg.

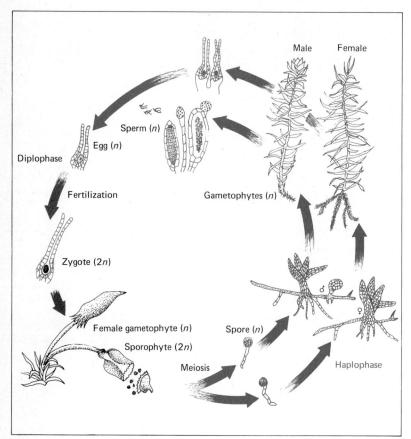

Fig. 10–5 The life cycle of a typical moss.

The zygote is the first cell of the sporophyte which remains attached to the tip of the gametophyte as it grows. The sporophyte in mosses usually has a relatively long stalk, which supports a more or less cylindrical capsule, the sporangium. The sporophyte wall may contain chlorophyll and carry on photosynthesis. Its epidermis has stomata with guard cells. The sporophyte is dependent upon the gametophyte both nutritionally and for support. Within the sporangium, certain cells undergo meiosis and a large number of haploid spores are formed. At maturity, the capsule opens and the spores are shed.

In a moist environment, spores germinate, forming slender filaments called protonemata. This is the horizontal component of the gametophyte, from which vertical leafy branches arise. Thus mosses have an alternation of generations in which the gametophyte is the dominant phase.

*MOSSES AND LIVERWORTS SHOW ALTERNATION OF GENERATIONS
IN WHICH THE HAPLOID GAMETOPHYTE IS THE DOMINANT PHASE AND
THE DIPLOID SPOROPHYTE REMAINS ATTACHED TO AND DEPENDENT
UPON THE GAMETOPHYTE.*

FERNS   There are a great many relatively simple vascular land plants
which also have alternation of differentiated generations.   In all these
forms, however, the sporophyte phase is the dominant one.   The familiar
fern plant, for example, is a sporophyte (Fig. 10–6). On the lower surface
of a fern frond you will frequently see small brown spots or patches. These
are the sites at which sporangia are produced. Spores give rise to a relatively small gametophyte, which produces sperm and eggs. Motile sperm
swim through a film of water to fertilize the egg. The new sporophyte
grows rapidly and soon is independent of the female gametophyte. In

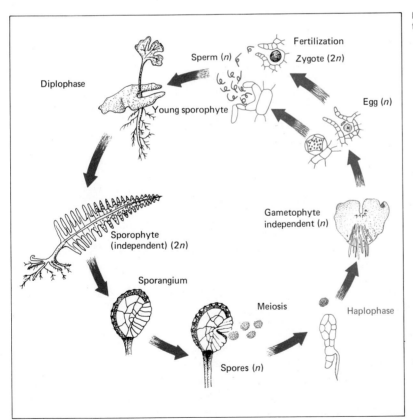

Fig. 10–6 The life cycle of a typical
fern.

the life cycle just described, the female gametophyte produces gametes of both sexes; in other words, the gametophytes are monoecious. In other ferns there are male and female gametophytes. In this case, although the spores produced by the sporophyte are identical in size and shape, they are genetically different, giving rise to either male or female gametophytes.

THE VASCULAR LAND PLANTS HAVE ALTERNATION OF GENERATIONS IN WHICH THE DIPLOID SPOROPHYTE IS PREDOMINANT AND THE HAP-LOID GAMETOPHYTE IS RELATIVELY SMALL AND INCONSPICUOUS.

All the spores produced by the sporophyte of some ferns are identical in size and shape. In other ferns and other lower vascular plants, the spores themselves are of two sizes: tiny microspores and much larger megaspores. These are produced in two different kinds of sporangia: microsporangia and megasporangia. Microspores develop into male gametophytes of relatively few cells. Each megaspore produces a female gametophyte, which is enclosed within the megaspore wall. The latter cracks open, and eggs are produced on the exposed portion of the megagametophyte. The megagametophyte is dependent upon the sporophyte for food stored in the megaspore. Although the exposed portion of the gametophyte may develop chlorophyll, it is doubtful that a significant amount of photosynthesis occurs.

IN VASCULAR LAND PLANTS, MALE AND FEMALE GAMETOPHYTES MAY BE VERY DIFFERENT IN SIZE AND ARE PRODUCED BY SPORES THAT ARE VERY DIFFERENT IN SIZE.

SELAGINELLA   All the common vascular land plants above the ferns produce spores of different sizes. And in these, the evolution from equal-sized spores to megaspores and microspores is carried one step further: to the evolution of the **seed.** In the fern relative *Selaginella* (Fig. 10–7), the developing sporophyte is an embryo within the megagametophyte. Also, the megaspore often germinates within the megasporangium. This is the first step in the evolution of the seed. A true seed exists when the megaspore regularly germinates to form a megagametophyte while enclosed in the megasporangium. When this occurs, some provision must be made for getting the sperm to the egg. *Selaginella*, like ferns and mosses, requires a film of water for the sperm to swim in for fertilization to occur. Only when the megasporangium and megaspore have both cracked open and are both flooded with water can a sperm reach an egg in *Selaginella*. Plants which have evolved seeds have solved this problem in an interesting way. In the seed plants, the entire microgametophyte is brought to

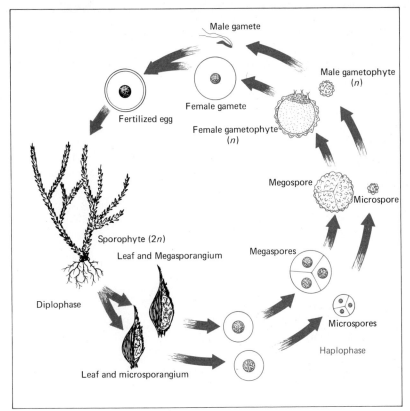

Fig. 10-7 The life cycle of the fern relative *Selaginella*. (*Adapted from BOTANY, fourth edition by Carl L. Wilson and Walter E. Loomis. Copyright 1952, © 1957, 1962, 1967 by Holt, Rinehart and Winston, Inc. Reprinted by permission of Holt, Rinehart and Winston, Inc.*)

the megagametophyte and a special means of getting the male gametes to the egg has evolved.

WHEN THE FEMALE GAMETOPHYTE IS PRODUCED WITHIN THE SPORO-PHYTE AND THE MALE GAMETOPHYTE IS TRANSFERRED TO ITS VICIN-ITY PRIOR TO FERTILIZATION, THE NEW SPOROPHYTE DEVELOPS WITHIN THE FEMALE GAMETOPHYTE, AND THE SEED HABIT IS SAID TO HAVE OCCURRED.

SEED PLANTS   The seed plants include two major groups: the gymnosperms and the angiosperms. Gymnosperms are those plants, mainly trees, whose seeds are produced (usually) on the scales of a cone. Pines, firs, and spruces are examples of gymnosperms. The angiosperms, which are the flowering plants, are a very diverse group, ranging from tiny duckweeds to trees, such as oaks and magnolias. The seed in flowering plants is contained in a structure of the flower called a **carpel.**

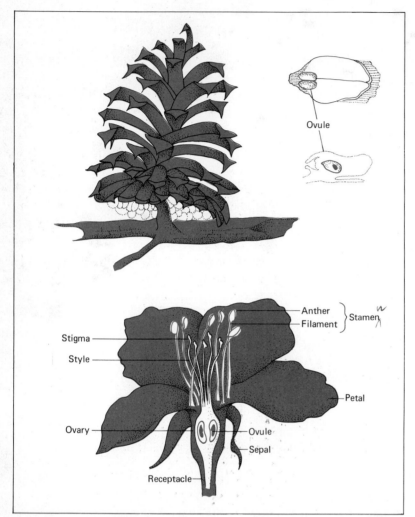

Ovule

Stigma

Style

Anther
Filament } Stamen

Petal

Ovary

Ovule
Sepal

Receptacle

Figure 10-8 shows a pine cone and an idealized flower. In seed plants, the megasporangium is very specialized and is surrounded by several layers of tissue. The specialized megasporangium and its surrounding tissues are called an **ovule.** At one end of the ovule there is a small opening in the surrounding tissues. These layers will eventually develop into the seed coat. You can see in the figure that each cone scale in the pine has two ovules on its upper surface. In the flower, there are several ovules contained within the flask-shaped carpel. The opening in the outer tissues of the ovule points toward the center of the cone in pine. In the flowering plant, it opens into the cavity of the carpel.

Microspores in the pine are produced in another kind of cone. In flowering plants, microspores are produced in appendages of the flower called **stamens.** Since the pine and angiosperm are both sporophytes, meiosis occurs, producing the haploid microspores. Before they are discharged from the microsporangium, the microspores germinate, forming microgametophytes consisting of only a few cells. These microgametophytes are called **pollen grains.**

Megaspores are produced by meiosis in the megasporangium. Each ovule produces four megaspores, and customarily all but one of these disintegrate. The remaining megaspore germinates to form the megagametophyte. In the pine, the megagametophyte consists of many cells, and egg-producing structures are formed at its tip below the opening in the layers surrounding the megasporangium of the ovule. Angiosperms have a simpler female gametophyte, which usually consists of eight cells. Transfer of the pollen grain, or male gametophyte, to the ovule is a process called **pollination,** and it is followed by **fertilization** of the egg by male gametes produced by the microgametophyte. We will discuss pollination and fertilization below.

You can see from the above discussion that all the vascular land plants have alternation of differentiated generations. Even such familiar seed plants as pines and oaks have an alternation of diplophase and haplophase. In the seed plants, the dominant phase is the massive sporophyte. The haplophase is reduced to the few-celled pollen grain (male gametophyte) and tiny female gametophyte contained within an ovule.

IN THE SEED PLANTS, THE DIPLOID SPOROPHYTE IS THE PREDOMINANT GENERATION, PRODUCING HAPLOID FEMALE GAMETOPHYTES INSIDE OVULES AND HAPLOID MALE GAMETOPHYTES CALLED POLLEN.

Alternation of Sexual and Asexual Generations in Animals

In discussing the life cycle of *Paramecium,* we mentioned that this protozoan sometimes reproduces sexually with conjugation and at other times reproduces asexually. This is not an example of alternation of generations, this does not in any sense compare with that found in the vascular land plants.
animals without backbones are able to reproduce both sexually and asexually. Although these are sometimes said to have alternation of generations, this does not in any sense compare to that found in the vascular land plants.

The only group of animals with true alternation of generations is the group of protozoans to which the malarial parasite *(Plasmodium)* belongs.

These protozoans have a very complicated life cycle, carried out in two different hosts—for example, man and mosquito. In parts of their life cycle they are haploid and in other parts diploid.

EXCEPT FOR ONE GROUP OF PARASITIC PROTOZOANS, ANIMALS DO NOT SHOW A REGULAR ALTERNATION OF HAPLOID AND DIPLOID GENERATIONS. ANIMAL LIFE CYCLES USUALLY ARE PREDOMINANTLY DIPLOID, WITH ONLY THE GAMETES BEING HAPLOID.

GAMETES AND FERTILIZATION

Now that you have some impression of the diversity of life cycles, it is time to turn to the subject of gametes in greater detail. Our discussion will be focused largely on the vertebrates and man. Before considering these, however, it might be well briefly to discuss gametes in plants and invertebrates.

Gamete Formation

PLANTS As we have seen above, gametes in plants may be of diverse types. Gametes may be equal in size and shape in the mating types, or they may be different. With rare exceptions, gametes in plants originate by mitosis. In nonvascular plants, they are produced by haploid gametophytes. In vascular plants, the haploid gametophyte may be reduced to only a few cells, but gametes are still produced in them by mitosis. (Remember, meiosis occurs in spore formation in sporophytes.) Male gametes in algae, the mosses and liverworts, and the lower vascular plants are motile. The number and structure of flagella vary from group to group. Figure 10–9 shows several kinds of these male gametes.

In most of the fungi and most of the seed plants, male gametes are nonmotile. Either they come across female gametes fortuitously in the course of being passively moved, or there are special structures which conduct them to the egg. Such special structures include pollen tubes, which will be discussed in detail below.

THE GAMETES OF PLANTS USUALLY ARE PRODUCED BY MITOSIS (MEIOSIS HAVING OCCURRED IN SPORE FORMATION), AND MALE AND FEMALE GAMETES MAY BE SIMILAR OR DIFFERENT IN SIZE AND STRUCTURE. WHEN THEY ARE SIMILAR, THEY USUALLY ARE MOTILE. WHEN THEY ARE DIFFERENT, THE MALE GAMETE IS USUALLY MOTILE AND THE FEMALE GAMETE NONMOTILE. IN SOME PLANTS, BOTH GAMETES ARE NONMOTILE.

INVERTEBRATES   Invertebrate animals are exceedingly diverse. Many unusual and bizarre modes of reproduction have been discovered, and others undoubtedly await discovery. Most animals have just two mating types: male and female. Some protozoans, however, such as *Paramecium,* may have four mating types. Organs of a multicellular animal which produce gametes are given the general name of **gonads.** Gonads which produce sperm are called **testes,** and egg-producing gonads are **ovaries.** Ducts conduct the sperm and ova outside the animal or to specialized organs for internal fertilization or further development.

Many invertebrates are **hermaphroditic;** that is, a single individual may develop both male and female gonads. In some instances, these gonads mature at the same time and eggs produced by one individual may be fertilized by sperm from that same individual. As you will remember, this reduces the amount of genetic recombination which potentially might occur in the fusion of gametes from different parents. The potential variability of the animals is thus reduced. It is not surprising, therefore, that mechanisms which prevent self-fertilization in hermaphrodites have developed. These may take one of three forms. In the first, the animal may not mature testes and ovaries at the same time. The same individual may produce testes at one time and ovaries at another. However, in other species, such as some oysters, the sex may change in the course of life of an individual. An individual may begin life as a male and then, after several months, change into a female. The structure of the gonads changes, so that eggs instead of sperm are produced. Finally, in some worms, such as the common earthworm, and in the garden snail, all individuals have male and female gonads. In sexual reproduction, however, mutual copulation takes place, and self-fertilization is thus avoided.

IN SOME ANIMALS, ONE INDIVIDUAL MAY PRODUCE BOTH MALE AND FEMALE GAMETES. SUCH HERMAPHRODITES MAY BE SELF- OR CROSS-FERTILIZING.

VERTEBRATES   Hermaphroditism is very rare in the vertebrates. Some fishes are known in which change of sex may occur. In higher vertebrates, however, change of sex means acquisition of only the secondary sex characteristics of the opposite sex. We will consider later the origin of the gonads themselves. At this juncture, let us discuss how the sperm and ova are produced.

The testis of a mammal, such as man, is very complex. Figure 10–10 shows a sectional view of a human testis. A tough connective tissue capsule surrounds the organ. Inside the capsule are interstitial cells and seminif-

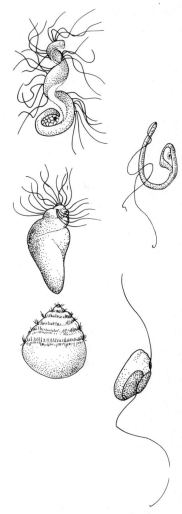

Fig. 10–9   Male gametes of various kinds of plants. From the top, clockwise: a fern, a moss, a brown alga, a cycad, and a horsetail.

erous tubules. The interstitial cells produce the hormone **testosterone,**
which was discussed in Chap. 7. The seminiferous tubules are slender
but exceedingly long tubes which lie closely coiled in the testis. They
consist of germ cells as well as cells whose function is not known precisely. There are also, of course, blood vessels and nerves throughout
the testis.

Within a seminiferous tubule (Fig. 10–10), the germ cells undergo meiosis almost continuously in man. Thus the testis contains mature sperm
as well as all intermediate stages in sperm development. Meiosis of a
germ cell includes the two divisions described in Chap. 2. These result
in genetic recombination and independent assortment of parental centromeres. Four daughter cells, each with the haploid number of chromosomes, are produced. In man, you may remember, the diploid chro-

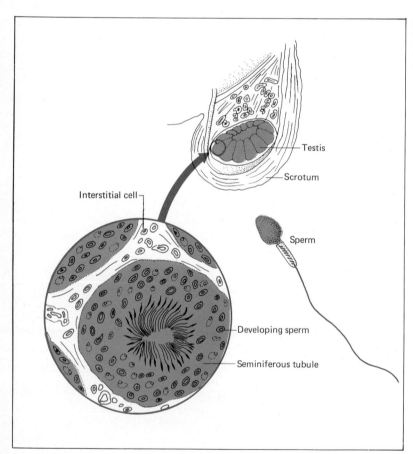

Fig. 10–10 Male gonads in man.
*(Adapted from Relis B. Brown,
GENERAL BIOLOGY. Copyright
McGraw-Hill Book Company, 1970.
Used by permission.)*

mosome number is 46. Therefore, each sperm contains 23 chromosomes. Half of them will have received the X chromosome and half the Y chromosome in each meiotic division.

Each potential sperm cell undergoes a very complex process of differentiation to form a functional sperm. The nucleus becomes very compact and often becomes elongated or strangely shaped. In man it is egg-shaped. One end of the nucleus is covered by a cap called the *acrosome.* This is formed by the dictyosomes (Golgi bodies). The nucleus and acrosome are called the *head* of the sperm. Most of the cytoplasm moves to the opposite end of the nucleus and is lost. The only remaining cytoplasm is organized into the so-called *middle piece* of the sperm and the *tail.* The middle piece includes one or more centrioles, which function as basal bodies for the tail. Around the base of the tail, the mitochondria become arranged in a spiral mass and presumably supply the energy for its movement. The tail is a flagellum with the characteristic arrangement of nine paired fibrils in a circle surrounding two others.

Mature sperm are stored in the testis until needed. When copulation takes place, they are conveyed to the copulatory structure through a duct, which may be associated with glands producing other substances that mix with the sperm.

THE SPERM OF VERTEBRATES ARE PRODUCED IN THE TESTIS BY MEIOTIC DIVISION OF THE GERM CELLS. EACH SPERM CELL IS THE RESULT OF A COMPLEX DIFFERENTIATION PROCESS DURING WHICH MOST OF THE CYTOPLASM IS LOST, THE NUCLEUS IS MUCH CONDENSED, AND THE SPECIALIZED PROPELLANT TAIL IS PRODUCED.

Like the testis, the ovary is enclosed in connective tissue and contains nerves and blood vessels. Relatively few ova are produced, however, compared with the billions of sperm that are formed in the lifetime of an individual. For example, an average human female potentially may produce a few hundred ova, typically 1 every 28 days. A section of a human ovary is shown in Fig. 10–11. You can see that the female germ cells are relatively quite large.

Meiosis begins in a germ cell in the ovary but is interrupted in the prophase of the first division. Large amounts of the food-storage material called *yolk* are produced in the cytoplasm. Yolk consists of glycogen, nucleic acids, lipid material, and proteins. The cell then enlarges tremendously. If you think of an ostrich egg or even an ordinary chicken egg, you will realize how great may be this increase in size. While this growth is going on, the potential egg is enclosed within a mantle of cells called the *follicle.* Follicle cells presumably regulate the passage

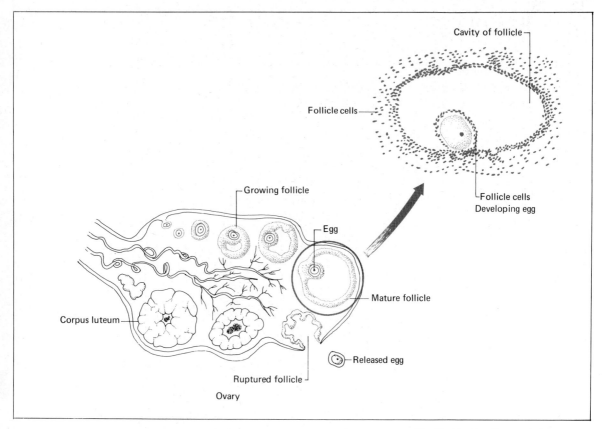

Cavity of follicle

Follicle cells

Follicle cells
Developing egg

Growing follicle

Egg

Mature follicle

Corpus luteum

Released egg

Ruptured follicle

Ovary

Fig. 10–11 Female gonads in man.
*(Adapted from BIOSCIENCE by
Robert B. Platt and George K. Reid
© 1967 by Litton Educational
Publishing, Inc. Reprinted by
permission of Van Nostrand
Reinhold Company. Adapted from
INTERACTING SYSTEMS IN
DEVELOPMENT, second edition, by
James D. Ebert and Ian M. Sussex.
Copyright © 1965, 1970 by Holt,
Rinehart and Winston, Inc. Reprinted
by permission of Holt, Rinehart and
Winston, Inc.)*

of materials from the blood vessels of the ovary to the egg. As you can
see in the figure, the developing egg in man (as is true in all mammals)
does not accumulate as large an amount of yolk as do the eggs of birds
or reptiles. It lies within a large fluid-filled cavity surrounded by follicle
cells.

After cytoplasmic specialization, the developing egg resumes meiosis
(Fig. 10–12). When the end of the first division arrives, however, division
of the cytoplasm is not equal. In fact, it is grossly unequal. One of the
daughters is a tiny cell called the ***first polar body.*** The other daughter
receives most of the cytoplasm with its stored reserves, and this is the
potential egg. Usually in vertebrates, meiosis stops once more until the
egg is fertilized. When the second division takes place, it is again grossly
unequal, and a ***second polar body*** is formed. The first polar body might
be expected also to divide, but this usually does not occur.

You can see that although there are potentially four daughter cells in every meiosis, only one of these functions as an egg in the ovary. In man every egg will receive 23 chromosomes and every egg will contain an X chromosome.

THE EGGS OF VERTEBRATES ARE PRODUCED IN THE OVARIES BY MEIOTIC DIVISION OF THE GERM CELLS. OF THE FOUR HAPLOID CELLS PRODUCED, ONLY ONE DIFFERENTIATES TO FORM THE EGG, WHICH CONTAINS STORED FOOD RESERVES.

Fertilization is the special case of syngamy in which gametes are differentiated into egg and sperm. It involves the penetration of the egg cell by at least the nucleus of the sperm. The two haploid nuclei of the gametes then merge, or each undergoes the usual chromosomal changes and they undergo a joint mitosis. This diploid mitosis is the first division of the zygote. Here we shall discuss fertilization in the angiosperms, or flowering plants, and in certain animals.

Fertilization

ANGIOSPERMS The male gametophyte produces two nonmotile gametes in flowering plants. These are released from the pollen tube (which will be described below) at the apex of the female gametophyte. The female gametophyte in angiosperms consists basically of eight nuclei in seven cells, although there is considerable diversity in the number of cells and mode of formation of this structure. A common type of female gametophyte is shown in Fig. 10–13. The megagametophyte of flowering plants is called an **embryo sac.** At one end is the egg with two associated cells. The significance of these cells is not clear; they may represent all that remains in the flowering plants of the egg-producing structures of the lower plants. At the opposite end of the embryo sac, farthest from the opening of the ovule, are three cells which may be remnants of the sterile cells of the megagametophyte of lower plants. In the center of the embryo sac are two nuclei called *polar nuclei.*

Upon entering the embryo sac, one of the gametes fuses with the egg, effecting fertilization. The zygote is the first cell of the new sporophytic generation, and it lies in the haploid gametophyte, which in turn is enclosed within the old sporophyte. In angiosperms a remarkable event also takes place: the second male gamete fuses with the two polar nuclei to form a triploid nucleus (three haploid sets of chromosomes in each nucleus). This is the first cell of a tissue called the **endosperm.** The endosperm is a nutritive tissue within which the young sporophyte (embryo) grows.

Fig. 10–12 Formation of the egg and polar bodies in meiosis. (*Adapted from Paul B. Weisz, THE SCIENCE OF BIOLOGY. Copyright McGraw-Hill Book Company, 1971. Used by permission.*)

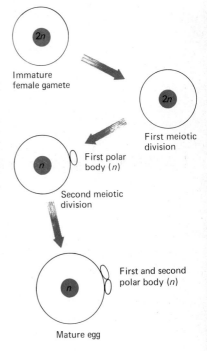

Immature female gamete

First meiotic division

First polar body (n)

Second meiotic division

First and second polar body (n)

Mature egg

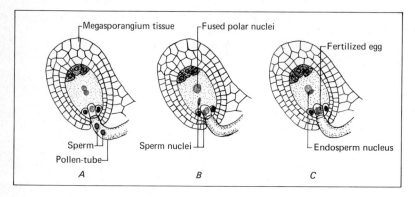

Fig. 10-13 Double fertilization in the embryo sac of an angiosperm. One sperm fuses with the egg, the other, with the polar nuclei. (*Adapted from BOTANY, fourth edition by Carl L. Wilson and Walter E. Loomis. Copyright 1952, © 1957, 1962, 1967 by Holt, Rinehart and Winston, Inc. Reprinted by permission of Holt, Rinehart and Winston, Inc.*)

GYMNOSPERMS   Double fertilization of this sort does not occur in the gymnosperms, such as pine. Nor is the female gametophyte as much reduced. As you can see in Fig. 10-14, the megagametophyte is rather massive and develops egg-producing structures at the end facing the opening in the ovule. When the male gametes are released, more than one egg may be fertilized. Only one embryo develops; the others all degenerate. The survivor, however, grows down into the megagametophyte, which contains the stored nutrients the embryo utilizes in development. When you eat a pine "nut," you first crack open the seed coat, which develops from the outer layers of the ovule and is thus diploid sporophytic tissue. Inside you will find the kernel, which is the haploid female gametophyte; this may have a tiny gray cap, the remains of the sporangium. In the kernel is the diploid embryo sporophyte.

IN THE SEED PLANTS, POLLINATION—THE TRANSFER OF THE MALE GAMETOPHYTE TO THE VICINITY OF THE FEMALE GAMETOPHYTE— MUST OCCUR BEFORE FERTILIZATION. FERTILIZATION OCCURS WHEN THE MALE GAMETES ARE RELEASED AND FUSE WITH THE EGG. IN FLOWERING PLANTS, ONE OF THE MALE GAMETES FUSES WITH OTHER NUCLEI OF THE FEMALE GAMETOPHYTE TO FORM A NUTRITIVE TISSUE CALLED THE ENDOSPERM.

ANIMALS   Fertilization in animals may be **external,** as in many marine invertebrates. The male and female gametes are released into the seawater, where they come together and fertilization occurs (see Fig. 2-1). In other animals, the male introduces the sperm into the female and fertilization takes place **internally.** Where fertilization is external, there are obviously several problems that must be solved. Eggs and sperm must be released at the same time. There is evidence that spawning by one

Fig. 10-14   A longitudinal section of an ovule of a pine.

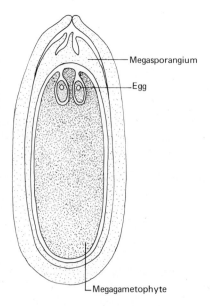

individual stimulates other individuals in the population to do likewise. Sperm and eggs must also come into contact and remain together long enough for fertilization to occur. It has been shown that in the jellylike coating of eggs of marine invertebrates there may be a species-specific substance called **fertilizin.** Fertilizin causes clumping of sperm of the same species, presumably by interacting with a complementary **antifertilizin** in the sperm. Contact between egg and sperm occurs as a result of random motions, and they are held together by the fertilizin-antifertilizin reaction.

Usually only one sperm penetrates the egg, and this is accomplished in complex ways not completely understood (Fig. 10-15). The sperm may contain, presumably in the acrosome, enough of an enzyme to dissolve an opening in the egg membrane. Eggs of some species form a **fertilization**

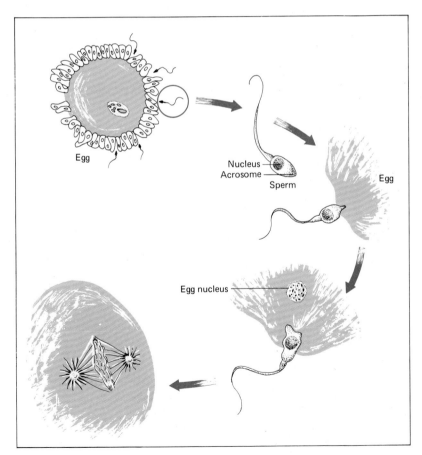

Egg

Nucleus
Acrosome
Sperm

Egg

Egg nucleus

Fig. 10-15 Fertilization in man. From the top, clockwise: the egg surrounded by follicle cells in the fallopian tube; the sperm; the acrosome of sperm releases enzymes that penetrate the cell membrane; fertilization cone of the egg engulfs the sperm head while its tail is left outside; chromosomes of the egg and sperm in the first division of the zygote. *(Adapted with permission of The Macmillan Company from BIOLOGY: OBSERVATION AND CONCEPT by James F. Case and Vernon E. Stiers. Copyright © 1971 by The Macmillan Company.)*

cone from the cytoplasm beneath the sperm. This conical extrusion draws the sperm down through the outer jellylike layer. In some forms, the outer layer, or cortex, of the egg has a layer of granules just under the cell membrane. Entry of a sperm causes rupture of local cortical granules, and this is followed by a wave of rupturing that moves swiftly across the egg surface. The result is release of a substance which lifts the outer layer away from the cell membrane. Presumably this prevents entry of more than one sperm.

Fertilization accomplishes **activation** of the once-dormant egg. Activation includes resumption of meiosis at whatever point it was arrested and the formation of the second polar body. It also begins a process of differentiation of the cytoplasm of the egg. There are changes in the permeability of the cell membrane and in the distribution and arrangement of cytoplasmic constituents. As this occurs, the major axes of cell division of the zygote and of the embryo itself become established. We will go into this and subsequent events below. It should be added here that activation of an egg may be accomplished in ways other than by the sperm. Various chemical treatments, heat shock, injection of protein, etc., may activate an egg, setting in motion the sequential events of early development. The development of an egg without entry of a sperm is called **parthenogenesis.** Parthenogenesis occurs naturally in certain groups of animals. Ordinarily an egg undergoing parthenogenesis becomes diploid by the incorporation of the chromosomes of the second polar body into the egg nucleus. Therefore, if the sex chromosome mechanism is XX-XY, parthenogenetic individuals will be female.

WITH EXTERNAL FERTILIZATION, EGGS AND SPERM MUST BE RELEASED AT THE SAME TIME AND MUST BE HELD TOGETHER LONG ENOUGH FOR SYNGAMY TO OCCUR. SYNGAMY RESTORES THE DIPLOID CHROMOSOME NUMBER AND ACTIVATES THE ZYGOTE FOR FURTHER DEVELOPMENT. RARELY, ACTIVATION CAN OCCUR WITHOUT FERTILIZATION.

Internal fertilization is more complex. Usually, the egg is released from the ovary and carried through a duct to the point at which it will meet the sperm. In man (see Fig. 7-12), one of the fallopian tubes receives the egg, enclosed in its follicle cells, and conducts it toward the uterus. The sperm swim up the fallopian tube, and fertilization occurs in this passage. The follicle cells are held together by hyaluronic acid. The fluid in which the sperm are contained, the semen, contains the enzyme hyaluronidase, which breaks down hyaluronic acid, permitting the sperm to reach the egg. The fertilized egg continues down the tube to the uterus, where it becomes implanted in the blood-vessel-rich lining and continues development.

Since, in man, the zygote and the embryo which develops from it are nourished by the mother, it is not surprising that there is relatively little stored food in the human egg. The same is true for other mammals. In birds and most reptiles, however, the eggs will eventually be laid in a relatively dry environment. After they have been fertilized, they pass through the oviducts, where protective layers, forming the supportive "white" of the egg (albumin), and the drought- and break-resistant shell, are added. In some reptiles, the eggs are retained, shell-less, in the oviducts. A structure analogous to a placenta is formed, and the embryo develops within the body of the female.

AFTER INTERNAL FERTILIZATION, THE ZYGOTES OF MANY VERTE-BRATES ARE USUALLY ENCLOSED IN A SPECIALIZED RESISTANT EGG SHELL, WHICH WILL PROTECT THE DEVELOPING EMBRYO. IN MAM-MALS, THE EMBRYO DEVELOPS WITHIN THE UTERUS OF THE MOTHER.

Synchrony

We have already discussed most of the factors involved in synchronizing the reproductive behavior in plants and animals. Marine and freshwater organisms in which external fertilization occurs are genetically programmed to respond to factors of the environment. When conditions fortuitous for reproduction occur, most of the individuals will have physiologically anticipated this and will be ready to reproduce. Spawning by a few individuals will stimulate others to do so, even more precisely synchronizing their behavior. Such factors as the chemistry of the water surrounding them, its temperature, and very frequently the photoperiod are the critical environmental factors.

Hormonal control of reproductive readiness has also been discussed. You will remember that in flowering plants, flowering is regulated by photoperiod, which affects the pigment phytochrome. There are, as well, responses to temperature. In birds, photoperiod also plays a role in timing reproduction. Changes in photoperiod set in motion long-range interrelated hormonal responses that affect not only the gonads but also the secondary sex characteristics and the behavior of individuals.

COURTSHIP BEHAVIOR In both invertebrate and vertebrate animals, there are often exceedingly elaborate, hormonally controlled sequences of behavior which precede reproduction. These make up what is called **courtship behavior.** Courtship behaviors often involve displays of patterns of colored feathers, skin, or fur. They may include posturing and dancing of various kinds. The details of courtship behavior depend in large part on the dominant perceptual mode of the species. For example, in the active, visually oriented jumping spiders, males court females by waving specialized leglike appendages which look rather like semaphores (Fig. 10-16). In the touch-oriented and web-weaving spiders, the male may pluck or shake the web of a female in courtship, signifying both that he is ready for mating and that he is not a trapped prey. Light is used in courtship in fireflies; different species have different patterns of blinking their tail lights. What do you suppose makes up courtship behavior in man?

Courtship behaviors serve not only to physiologically coordinate males and females. They also serve to advertise reproductive readiness. Finally, they solve the important problem of reducing aggression so that mating can take place. Mating with a large hungry female spider is a dangerous matter. The natural response of a predator is, of course, to attack and consume any smaller organism that approaches it. Similarly, territorial animals and behaviorally dominant animals attack intruders or less dominant individuals. Courtship functions to reduce these aggressive tendencies so that the sex partner, male or female, can be recognized as such and the appropriate behavior can ensue.

Consider the baboon troop we discussed in Chap. 1. As we described, the dominant male is difficult to approach for lesser males and for females. Ordinarily, if they intrude too closely, they will be driven away. A female in estrus, however, advertises this fact and reduces the male's aggressiveness by approaching the male and displaying her buttocks to him. In estrus, the skin of this area is brightly colored and engorged with blood. This so-called **presentation** posture involves bending the body into a position in which the male can easily mate with the female. After perhaps several mildly aggressive encounters, the male accepts the female and copulation takes place.

COURTSHIP BEHAVIOR IS A COMPLEX SET OF ACTIONS, HORMONALLY CONTROLLED, WHICH PRECEDE REPRODUCTION. IT COORDINATES MALES AND FEMALES PHYSIOLOGICALLY, ADVERTISES READINESS TO BREED, AND REDUCES AGGRESSIVE TENDENCIES.

Fig. 10-16 Semaphore communication in courtship behavior of a male spider, which permits mating. (*Adapted from SOCIAL BEHAVIOR FROM FISH TO MAN by W. Etkin, The University of Chicago Press, 1967.*)

Intromittent Organs and Spermatophores

Mating behavior in animals with internal fertilization is distinguished from courtship behavior as being directly involved with insemination. As with courtship behavior, mating behaviors are very diverse. In many invertebrates, such as insects, spiders, and molluscs, the sperm are held together in a package called a **spermatophore.** The male may insert the spermatophore into the body of the female or leave it in a place where she may find it. In octopuses, one of the eight arms of the male is specialized as a spermatophore carrier. After a period of courtship, this arm becomes separated, and the female inserts it into her genital tract. The male regenerates a new arm for subsequent matings.

Mating behavior in most other animals involves the transfer of sperm directly into the female through what is called an ***intromittent organ.*** Legs or other appendages may be modified for the function of sperm transfer. In many invertebrates, modified legs serve this role. Similarly, in those fish with internal fertilization, such as guppies and sharks (Fig. 10-17), modified fins in the male channel the sperm. Many invertebrates, together with most land vertebrates, have intromittent organs which, for simplicity, may here all be called ***penes*** (sing. ***penis***). In some of these groups, the penis is given another name, reflecting its developmental origin. Only the birds (with a few rare exceptions) and amphibians lack intromittent organs.

Amphibians have external fertilization. Males and females of frogs and toads are aggregated into large breeding populations by their characteristic calls. The male clasps the female behind the forelimbs, and when she expels eggs, the male emits sperm. In salamanders, the male drops spermatophores, which the female picks up in her cloaca. Female salamanders emit an attractive chemical in the streams in which breeding takes place. The males locate the females by swimming upstream along the gradient of this chemical messenger. Substances which serve the

Fig. 10-17 A clearnose skate (relative of sharks) showing the paired claspers below the tail. (*Courtesy of Jack Dermid.*)

function of a messenger outside the body are analogous to hormones and are called **pheromones.**

Most birds copulate by pressing the cloaca of the male against that of the female. Sperm are quickly transferred in the brief touching.

Reptiles and mammals have a penis or some modification thereof, which ordinarily is not exposed to view except during copulation. The paired **hemipenes** of male snakes and lizards are kept in sacs in the base of the tail. The penis in mammals is usually protected in some sort of penis sheath until mating takes place.

The penis of a mammal is plentifully provided with blood vessels, which open into large chambers (Fig. 10–18). When the male is ready to mate, these chambers become filled with blood and the entire penis becomes engorged with blood. This increase in the amount of blood serves to expand the penis in size, often remarkably so, and to stiffen it. This increase in size and stiffness of the penis is called **tumescence.** Similar tumescence occurs in the external genital parts of the female, the labia, which also secrete a lubricating solution about the entrance to the vagina. In copulation, the male inserts the penis into the vagina and after a few to many thrusts (depending on the species), **ejaculation** occurs. Ejaculation is the forcible emission of the sperm, together with substances produced by accessory glands such as the prostate. (See Fig. 7–12 and Fig. 7–14.)

The amount of seminal fluid emitted by the male varies with the species. According to Ebert, a single ejaculation ranges from 0.05 ml in a bat to 2 ml in a ram and 500 ml in a boar. The latter quantity contains 6 million cells per microliter (μl) of fluid. In man, the ejaculate is about 3 ml and contains 100,000 sperm cells per microliter. Obviously only a small proportion of the sperm reaches an egg, and only one of these succeeds in fertilization. In man, it takes the sperm about 3 hr to reach the egg in the fallopian tube. There are many factors in the internal environment of the female which are deleterious to sperm.

Wasteful as this amount of sperm production may seem, the strategy of internal fertilization is much less wasteful than is external fertilization. Furthermore, internal fertilization frees reptiles, birds, and mammals, together with the invertebrates which have adopted it, from the need to reproduce in an aquatic environment. Internal fertilization is an important specialization making terrestrial life possible. Of course, acquisition of the shelled egg in reptiles and birds and of viviparity in mammals is also important in the exploitation of the terrestrial environment. It is not surprising that analogous specializations have developed in the higher land plants.

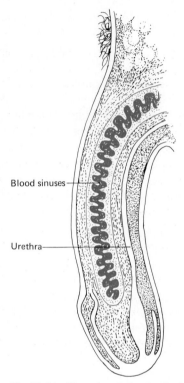

Blood sinuses

Urethra

Fig. 10–18 Diagrammatic longitudinal section of the human penis to show the blood sinuses which function in tumescence. (*Adapted from Kimball, BIOLOGY, 2/e, 1968, Addison-Wesley, Reading, Mass.*)

*INTERNAL FERTILIZATION IS USUALLY ACCOMPLISHED BY A SPE-
CIALIZED INTROMITTENT ORGAN OF THE MALE. THE SPERM MAY
BE STORED IN THE FEMALE OR MAY MOVE DIRECTLY TO THE EGG
OR EGGS. INTERNAL FERTILIZATION MAKES POSSIBLE REPRODUC-
TION IN A TERRESTRIAL ENVIRONMENT.*

Pollination in Seed Plants

Earlier we discussed life cycles in lower plants. In all these forms, fertilization is external. The male gamete, or sperm, swims through a watery
medium or a film of water on the surface of the gametophyte to effect
fertilization. Such plants require, obviously, the presence of free water
in the environment, at least at the time of reproduction. Those mosses and
ferns which live on desert rocks cannot reproduce sexually until rain or
dew provides the necessary water.

With the development of the seed habit and pollination, a process equivalent to egg formation and internal fertilization evolved. As was discussed above, in the seed plants the male gametes alone are not brought
to the egg, the entire male gametophyte is. The male gametophyte in
seed plants is called a pollen grain, and the process of bringing it to the
vicinity of the female gametophyte is known as **pollination.** Referring
back to Fig. 10-8, you will remember that the ovules of gymnosperms
are usually borne on the surface of cone scales. The ovules of flowering
plants are contained within a carpel. Let us examine how pollination
occurs in each of the groups.

At the time of pollination, the female cone of a pine is less than an inch
long. Early in the spring the cone scales separate and each ovule exudes
a drop of sticky fluid through the tiny opening of the ovule. At the same
time, male cones are releasing pollen grains into the air. Each pollen grain
has a pair of air bladders, enabling it to float through the air. Pollen is
produced in tremendous quantities, and it may color the surface of
the ground or of ponds and lakes bright yellow. If a pollen grain falls
onto a cone, it may get caught in the sticky drop at the opening of the
ovule. As this drop of fluid evaporates, it shrinks and draws the pollen
grain down through the opening to the megasporangium.

Subsequent to pollination, the cone scales close. The male gametophyte
then begins to grow a tube, which digests its way slowly through the
megasporangium until it gets to the female gametophyte. There the pollen
tube bursts, and the male gametes fertilize the eggs.

In the flowering plants, or angiosperms, the pollen does not have access
to the ovules since they are enclosed within the carpel. Referring to
Fig. 10-19, you can see that a carpel consists of three basic regions.

Fig. 10-19 Diagrammatic longitudinal
section of the ovary of an angiosperm,
showing pollen tubes and ovules.

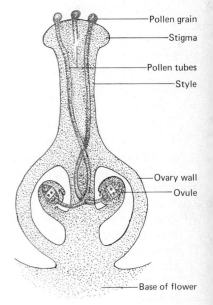

Pollen grain
Stigma
Pollen tubes
Style
Ovary wall
Ovule
Base of flower

The swollen base, or **ovary,** contains the ovules. Usually above the ovary there is a region without ovules called the **style,** which ends in a **stigma.** This stigma is either hairy or sticky, and it is here that the pollen grains land. Each pollen grain then produces a pollen tube, which grows into the stigma, down the style, and into the ovary, where it enters the ovule. The tip of the tube ruptures and releases two male gametes at the end of the female gametophyte, or embryo sac. As we saw above, one gamete fuses with the egg to form a zygote. The other fuses with the two polar nuclei (either the three fuse simultaneously, or the polar nuclei fuse and the gamete is added). The endosperm, within which the embryo grows, is formed by the division of that nucleus, which has three sets of chromosomes.

The pollen tube is analogous to an intromittent organ. Instead of many gametes being produced by a male gametophyte, only two are formed. Male gametes in seed plants (except for a few gymnosperms) are non-motile. They are passively conducted to the egg. Free surface water is not necessary for fertilization, and seed plants can reproduce sexually in any habitat to which they are otherwise physiologically suited. Pollen grains may be wind-dispersed, as described for pine. Angiosperms may be wind-pollinated, but most are pollinated by insects. The interrelationships of flowering plants and their animal pollinators are discussed in Chap. 14.

THE SEED PLANTS HAVE BEEN ABLE TO EXPLOIT THE TERRESTRIAL ENVIRONMENT AS A RESULT OF THE EVOLUTION OF THE POLLEN TUBE. THIS OUTGROWTH OF THE MALE GAMETOPHYTE CONVEYS THE MALE GAMETES TO THE EGG AND IS THUS ANALOGOUS TO AN INTROMITTENT ORGAN.

DEVELOPMENT

The animals and plants we can see without the aid of a microscope commence the diploid part of their life cycle as tiny, relatively simple zygotes. If, as the early preformationists thought they observed, there were a miniature man or homunculus in the egg or sperm as well as in the zygote, then the problem facing the embryo would be a trivial one, at least for man. All that would be needed would be that each part of the homunculus enlarge. That is, growth of the preformed individual would be a sufficient process in **ontogeny,** or **development.** But since the electron microscope reveals no tiny oak in the acorn or tiny chicken on the yolk of a fertile egg, the science called *embryology,* or developmental biology, must account for a much more complicated process—

that of ***differentiation***—tracing the origin of different cell types, tissues, organs, and even types of individuals from a single undifferentiated cell.

Animal development has been divided into a series of stages and processes which succeed one another, as well as partially overlap. For instance, cellular differentiation and growth may both participate in morphogenesis. Figure 10–20 summarizes in a descriptive way the stages of development. You may find it helpful to refer to this figure as you read about the different processes discussed below. The purpose of the following sections is to describe the most typical developmental stages. After that, the "how" questions will be studied.

CLEAVAGE The nucleus may be thought of as both the library and the control center of the cell. Information moves out into the cytoplasm in

Development in Animals

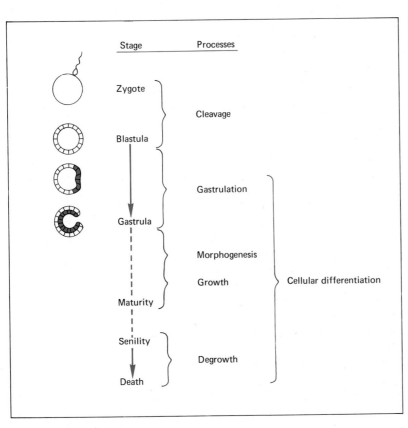

Fig. 10-20 Diagram showing principle stages in the development of an animal.

the form of RNA, and this in turn directs the synthesis of tools (enzymes) and materials. To get on with the business of development, the nucleus of the zygote must direct many of the cytoplasmic processes, but in this regard the zygote suffers from a severe handicap. The zygote is often a very large cell, especially if it contains much yolk, but it has only a single nucleus. Even in the relatively yolk-free sea urchin egg, the ratio of the volume of nucleus to the volume of cytoplasm is 100 times less (about 1:500) in the zygote than it is in cells after a few divisions (about 1:6 in the blastula stage). This means that the first problem to be overcome by the developing organism, or embryo, is the production of more control centers. This problem is solved by the consecutive division of the original nucleus into a large number of nuclei and the distribution of these nuclei throughout the mass of the zygote. Once this is accomplished, every bit of cytoplasm will come under the influence of a nucleus administering a small neighborhood.

Cytoplasmic divisions accompany the nuclear divisions in most animals, and this process of cell division following fertilization is called **cleavage.** The first cell division results in two daughter cells; the next division cleaves each of these, making four cells; and so on, producing 8, 16, 32, 64, ... (Fig. 10–21). At first, the cleavage divisions are synchronous, but eventually divisions in different parts of the embryo are no longer in phase. As shown in Fig. 10–21, insects and other arthropods are an exception to the statement that the cytoplasm divides along with the nuclei. In these organisms, the original zygote nucleus undergoes a series of mitotic divisions and the daughter nuclei migrate from the center of the egg to its surface before the limiting cell membranes are formed.

In yolky eggs, the cleavage process is physically (mechanically) retarded by the particles which make up the yolk. These obstruct the movement of organelles and molecules. As a result, the furrowing of the embryo into distinguishable cells often lags behind the nuclear divisions. In many eggs, including those of amphibians, reptiles, and birds, the yolk is not evenly distributed but is concentrated at one end, the **vegetal** pole. The other end, the **animal** pole, has less yolk and a higher concentration of cytoplasm. Consequently, in the eggs of many animals, the animal pole may have undergone numerous cleavage divisions before the first cleavage furrow divides the vegetal pole into two cells. Another result of this in fish and amphibians is that the cells in the vegetal region, because they undergo fewer divisions, are larger than the cells at the animal pole. As will be described below, these relatively large, yolk-laden cells will become part of the gut in vertebrates and will supply the developing embryo with food until it begins to feed itself.

The yolky egg reaches its extreme in reptiles, birds, and the egg-laying mammals (the duckbill platypus of Australia and the spiny anteaters of New Guinea). These animals have no free-living larval stage and hatch from the egg as functioning individuals, though more or less immature. So much yolk must be stored in their eggs that the cytoplasm is restricted to a small whitish cap at the animal pole. Throughout most of development, and even after hatching in some species, the mass of yolk protrudes from the abdomen as a yolk sac (Fig. 10–22). The yolk is gradually absorbed by blood vessels lining the sac, but the bulk of it is never cleaved into separate cells.

One way or another, the cytoplasm of the egg is divided up among a large number of cells—the results of cleavage. These cells are called **blastomeres.** In most animals, a space called the **blastocoel** forms during cleavage, at which time the embryo is called a **blastula.** Usually the blastula is a hollow sphere with walls one cell thick. As far as can be seen externally, nothing really exciting has happened during cleavage and the formation of the blastocoel. The embryo is still the same size, shape, and color as the zygote. Except for their differences in size, yolk content, and pigment content (in some species), the blastomeres do not

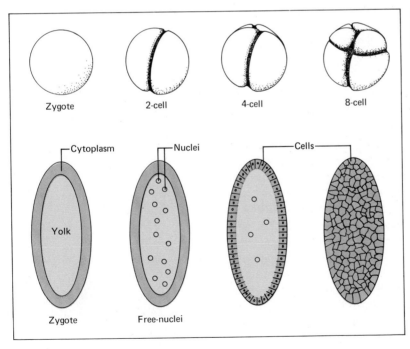

Zygote   2-cell   4-cell   8-cell

Cytoplasm   Nuclei   Cells

Yolk

Zygote   Free-nuclei

Fig. 10–21 Cleavage in a frog (*above*) and an insect (*below*). (*Adapted from Kimball, BIOLOGY, 2/e, 1968, Addison-Wesley, Reading, Mass.*)

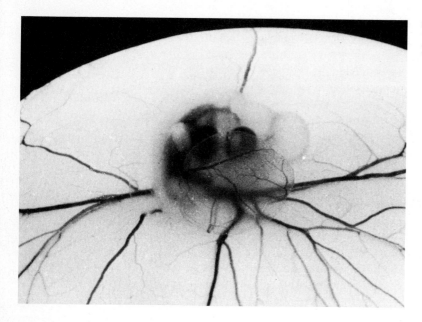

Fig. 10-22 Chicken embryo about 96 hr old, showing yolk sac with extensive blood system. (*Courtesy of V. K. Abbott.*)

reveal any specializations suggesting their developmental fate. But as will be described below, the fate of some of these is already sealed. In some animals, this statement applies to all the cells of the blastula.

DEVELOPMENT IN ANIMALS BEGINS WITH MITOTIC DIVISIONS OF THE ZYGOTE CALLED CLEAVAGE DIVISIONS. CLEAVAGE EVENTUALLY RESULTS IN A LARGE NUMBER OF CELLS, WHICH USUALLY BECOME ARRANGED IN A HOLLOW SPHERE.

GASTRULATION Recall that embryological development results in a spatially heterogeneous organism expressed as many distinct cell types. In most invertebrates and all vertebrates, these cells all stem from three **germ layers** of cells in the developmental stage called the **gastrula** (Fig. 10-23). In vertebrates the cells of the outer cell layer of the gastrula or **ectoderm,** will develop into most of the tissues associated with the outer skin (epidermis) of the adult, including the important sense organs, the nose, hair, feathers, scales, parts of the skeleton in the head and neck, pigment cells, and nervous system. The middle layer of the gastrula, the **mesoderm,** produces tissues of support and movement (bone, cartilage, muscle connective tissue), tissues of transport (heart, blood, and blood vessels), and tissues of excretion and reproduction (kidneys, gonads, and other reproductive tissues). The inner cell layer of the gastrula, the

Fig. 10-23 Section of a gastrula showing ectoderm, mesoderm, and endoderm.

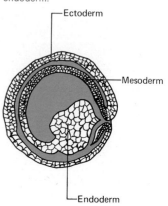

endoderm, produces the organs of internal ventilation and the alimentary canal and associated glands, including the liver and pancreas. With the possible exception of coelenterates, all multicellular animals pass through such a simple three-layered stage. How does the single, undifferentiated layer of cells in the blastula become the triple-layered gastrula?

In this discussion we will limit ourselves to the vertebrates. The simplest way for a single layer to become three concentric layers is ignored by vertebrates. Rather than using a simple series of cell divisions to produce more layers, with the cells on the outside becoming ectoderm, those on the inside becoming endoderm, and those in between becoming mesoderm, vertebrate embryos go through a slow-motion contortionist act during which groups of cells on the outside push their way inward through a slit or pore of their own making. Once on the inside of the embryo, these wandering cells and their progeny develop into mesodermal structures. While this is happening, the future ectodermal cells move about to cover the surface of the embryo, replacing the cells that have gone inside. The future endoderm cells either move inside or are overgrown by ectodermal cells. Which course is followed depends partly on how laden the cells are with yolk. Large, yolky endoderm cells cannot easily move about and tend to be rather passive until much of their burden of stored food is metabolized.

Figure 10-24 shows the stages in the gastrulation of a primitive fishlike vertebrate called a lancelet (amphioxus). Cellular movements in lancelets are not hindered by large amounts of yolk, so gastrulation is a straightforward business of **invagination,** or pushing in of the future endoderm and mesoderm. Think of this as similar to pushing your fist into a balloon. If the gas inside the balloon could slowly escape, the "blastocoel" would disappear and your fist would produce a new cavity. In many animals it is this new cavity that will persist in part to become the gut. This is what happens in the lancelet.

Gastrulation in birds and reptiles is superficially very different from simple invagination. Most of their embryonic material is yolk, and gastrulation involves only a fraction of the surface. The mass of inert yolk rules out invagination as a means of gastrulation. Instead, there forms a crease, or **primitive streak,** through which future mesodermal cells on the surface pass to the inside into the blastocoel (Fig. 10-25). The layer of the blastocoel closest to the yolk becomes endoderm. Mammals such as ourselves have given up egg laying in favor of the placenta and viviparity, but we still retain the same primitive-streak type of gastrulation as our reptilian forebearers.

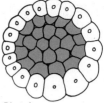
Blastula

Early gastrula

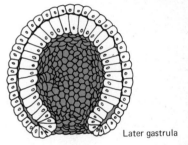
Later gastrula

Fig. 10-24 Formation of the gastrula in the lancelet, *Amphioxus,* by invagination.

Amphibians have an intermediate amount of yolk and an intermediate type of gastrulation. True invagination is impossible because of the yolky endoderm, but a **blastopore** forms on the side of the blastula. Future mesodermal cells slowly move toward it from the surrounding surface of the embryo, passing through it and then radiating out on the inside. By a process of cell movement and division, these cells come to lie in between the ectoderm and endoderm.

FOLLOWING THE HOLLOW-SPHERE STAGE OF DEVELOPMENT, THE YOUNG EMBRYO BECOMES THREE-LAYERED IN A MANNER DEPENDENT UPON THE SPECIES. EACH OF THE THREE LAYERS OF THE EMBRYO PRODUCES CELLS DIFFERENTIATING INTO DIFFERENT GROUPS OF TISSUES.

FROM THREE LAYERS TO AN ORGANISM It is not entirely misleading to think of the gastrula as a three-layered cup, even though the three germ layers are not always distinguishable and the shape of the gastrula is often more platelike than cuplike. The adult animal—a fish, for example —at first would not seem to be cuplike, by any stretch of the imagination, but as you can see in Fig. 10-26, the problem of converting a cup into a fish is topologically quite simple. First, a hole (the future mouth) must be made in the "bottom" of the cup; the result is an elongated doughnut. The inside wall of the tubular organism is the future intestinal tract, and the outside wall is the skin. Glands and organs such as lungs, liver, and pancreas develop from outpocketings of the inner wall into the space between the walls. Limbs, ears, the nose, and so forth develop as outpocketings of the outer wall. Internal organs such as the heart, kidneys, bones, and muscle develop from the middle layer.

IN VERTEBRATES, FURTHER DEVELOPMENT OF THE THREE-LAYERED EMBRYO ESTABLISHES A MOUTH AT ONE END AND AN ANUS AT THE OTHER OF A BASICALLY TUBE-SHAPED ORGANISM. THE OUTER LAYER OF CELLS BECOMES THE SKIN AND NERVOUS TISSUE (AMONG OTHERS), THE MIDDLE LAYER BECOMES THE ORGANS OF SUPPORT AND TRANSPORT, AND THE INNER LAYER BECOMES THE DIGESTIVE TRACT AND ITS ASSOCIATED GLANDS.

We will consider the development of a few organs as examples of the phenomenon as a whole. In reading the following, it should be kept in mind that biologists do not know exactly *how* these changes occur; that is, we know very little about the molecular mechanisms we assume to be the basis of development.

A frog, a bird, and a human embryo just after the completion of gastrula-

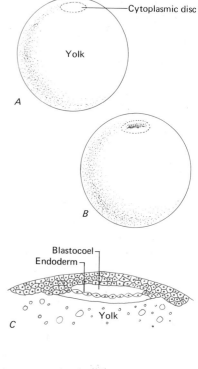

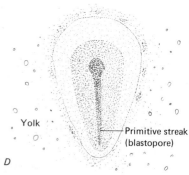

Fig. 10-25 Development of the primitive streak in the chicken. After cleavage of the cytoplasmic disc atop the yolk, a blastula (C) and a gastrula (D) are formed. (*Adapted from BIOLOGICAL SCIENCES, 1/e, by William T. Keeton, illustrated by Paula DiSanto Bensadoun, by permission of W. W. Norton & Company, Inc. Copyright © 1967 by W. W. Norton & Company, Inc.*)

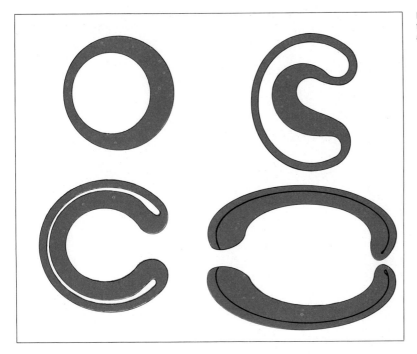

Fig. 10-26 Diagram to show formation of a hollow tube from a cup (gastrula).

tion all show the first **organ rudiments.** One of these is the future **notochord,** a cartilaginous (gristlelike) rod, which in the frog develops from the mesodermal cells invaginating at the top (dorsal lip) of the blastophore. The notochord is the backbone of all vertebrate embryos, and some adult vertebrates, such as sharks, retain it as part of their cartilaginous backbone. Coming from the same source are the two rows of **somites,** which will later participate in the development of back muscles and the vertebrae. The rest of the mesodermal cells are still relatively undifferentiated.

Another organ rudiment already partly differentiated at this stage is the neural plate, a thickened oval layer of ectoderm on the top of the embryo that will develop into the animal's nervous system. It must be rolled into a cylindrical **neural tube.** This is accomplished by **differential cell growth** (some cells growing at different rates and in different directions from others), by **folding** of the plate, and by **fusion** of cell masses. Figure 10-27 shows that the neural-plate cells become elongated and that the edges of the plate fold up to meet at the midline, where they fuse to form a hollow tube. The front end of this tube becomes the brain and the nerves and sense organs that grow out from it. The hind end becomes the spinal cord and the nerves and ganglia that grow out from it.

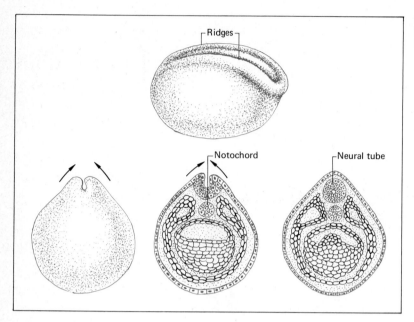

Ridges

Notochord

Neural tube

Fig. 10-27 Formation of the neural tube in a frog embryo by overgrowth and fusion of neural ridges.

As an example of the development of an organ, we will look at the eye. Even before the edges of the neural plate have met and fused along their entire length, paired bulges appear on the front of the future brain (Fig. 10-28). These are the hollow **optic vesicles** which will become the retinas, the pigmented layers, and part of the optic nerves. The optic vesicles continue to push outward until they reach the ectoderm. The front ends of the vesicles flatten out and then invaginate inward, forming a double-layered receptacle—the **optic cup**—attached to the brain by a thin stalk. In time, the outer wall of the optic cup becomes the pigmented layer of the eye, whereas the cells of the thick inner walls assort themselves into the various layers of the retina. Innermost are the rods and cones; outermost are the ganglion cells (see Fig. 8-13). The retina becomes "wired" to the brain by outgrowth of axons from the ganglion cells through the stalk, eventually making proper connections in the lateral geniculate body (see Fig. 9-17).

We have said nothing of the lens. As the optic vesicle contacts and then withdraws from the epidermis, the latter thickens or invaginates (depending on the kind of vertebrate). Eventually, this rudimentary lens is pinched off, after which it develops into a transparent globe. The transparent cornea develops from the epidermis directly in front of the lens.

The eye is almost entirely ectodermal in origin. The optic vesicle was originally part of the neural-plate ectoderm, and the lens and cornea develop from typical ectoderm. Some organs are more complex in their origins. The salivary glands, for instance, develop from interacting endoderm and mesoderm rudiments.

IN VERTEBRATES THE NERVOUS SYSTEM DEVELOPS FROM A TUBE FORMED BY AN ENFOLDING OF SOME OF THE OUTER LAYER OF CELLS. CERTAIN REGIONS OF THIS TUBE BECOME SPECIALIZED AS SENSE ORGANS, SUCH AS THE EYES.

INDUCTION   To this point, we have described some of the developmental processes and events, but nothing has been said of mechanism—how these complex events are controlled and integrated. The original breakthrough came in the 1930s and was brought about by Hans Spemann and his students, who skillfully grafted parts of one embryo onto others, and followed the development of the graft in the new host. In this way, they were able to discover the phenomenon called embryonic induction. One technique they used was to stain the grafts with harmless dyes. An alternative method of marking employed the natural differences in pigmentation between related species. The cells, stemming from a piece of tissue grafted from an unpigmented embryo to a darkly pigmented embryo, were recognizable thereafter because they lacked pigment granules. In one of the most important of these experiments, a piece of the upper lip of the blastopore in an early gastrula of an unpigmented salamander was grafted into the lower part of the blastopore of the early gastrula of a darkly pigmented salamander (Fig. 10-29). The result was essentially two embryos in one, including two notochords and two neural tubes. The graft had somehow caused the host embryo to organize a second system of principal rudimentary organs. When one tissue causes differentiation in another, the process is called **embryonic induction.** Tissue from the upper, or dorsal, lip of the blastopore had such profound and general effects that Spemann called it the *primary organizer.* We mentioned above that this tissue later becomes the notochord and the somites. It is now known that these tissues will induce the formation of the neural tube and the rudiments of the eyes and ears in all vertebrate embryos.

Later experiments showed that much of organogenesis (the origin of organs) can be studied in terms of induction. When the optic vesicle reaches the epithelium, the latter responds by invaginating a sphere of cells that develop into the lens. To test whether this is truly a case of induction, the optic vesicle was transplanted to a site under the skin on the side of the embryo. As would have been predicted, the epithelium

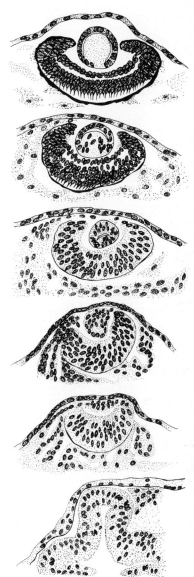

Fig. 10-28   Stages in the formation of the eye. *Below,* outgrowth of brain; *top,* lens induced. (*Adapted from Paul B. Weisz, THE SCIENCE OF BIOLOGY, Copyright McGraw-Hill Book Company, 1971. Used by permission.*)

directly overlaying the isolated optic vesicle was induced to form a lens (Fig. 10–30).

Experiments such as these led to the important conclusion that the cells of an embryo often have a very broad range of *potential* futures but that the local environment determines which of these many possible futures actually develops. Any part of the ectoderm, for instance, can develop into a neural plate if it comes in contact with notochordal tissue. A little later in development, epithelium from anywhere on the embryo has lens-making potential. Development can be regarded as a complex series of inductions, each one along the way eliminating sets of possible futures.

IN THE COURSE OF DEVELOPMENT, DIFFERENT TISSUES INTERACT WITH ONE ANOTHER. SOME TISSUES ARE ABLE TO INDUCE OTHER TISSUES TO FORM SPECIFIC ORGANS. FOR EXAMPLE, THE DEVELOPING OPTIC VESICLE INDUCES THE EPITHELIUM NEAR IT TO DIFFERENTIATE INTO A LENS.

A question that might be asked at this point is the following: Is the range of future potentials of tissues in the young embryo restricted to the characteristics of the species or is the potential actually much greater? For example, can the skin of a mammal such as ourselves produce feathers if the proper inductive stimulus is present? The following experiment is typical of the sort designed to answer this question.

The larvae of the European salamander *Triturus* produce fleshy "balancers" on their heads, whereas the larvae of the North American *Ambystoma* salamanders do not. The absence of balancers in *Ambystoma* could be explained in either of two ways:

1 The proper inducer is absent.
2 The skin does not have the capacity to respond even in the presence of the correct inducer.

To test these alternative hypotheses, the epithelium (epidermis) at the future site of potential balancer formation was removed from embryos of both species and grafted onto the corresponding place in the other species. Each embryo then had the epidermis of the *other* species at the place where balancers can develop. The result was that only the *Ambystoma* larva with *Triturus* epidermis developed a balancer. From this we conclude that the inductive stimulus is present in both species (because the *Ambystoma* environment could induce the *Triturus* skin) but that the *capacity to respond* is peculiar to the species. The *Ambystoma*

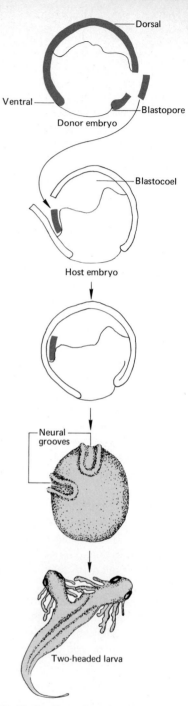

Fig. 10–29 Tissue from the donor embryo (colored) induces formation of a second head in the host embryo after developing into notochord.

skin did not respond to the *Triturus* inducer. Another way of saying this is that hereditary factors in differentiating cells determine the specific nature of responses to induction.

It was once hoped that careful scrutiny of the mechanism of induction would lead to an understanding of development in terms of physics and chemistry. This hope failed to materialize when the identity of the inducer turned out to be much more complex than had been anticipated. Apparently there is no single class of materials—such as nucleic acids, for example—that are inducers. Furthermore, many environmental changes of a purely physical nature (such as changes in pH or salt concentration) can mimic the effects of natural inductive stimuli.

MANY CELLS OF THE EMBRYO HAVE THE POTENTIAL TO DEVELOP INTO ONE OF SEVERAL DIFFERENT TYPES. THE TYPE OF CELL INTO WHICH THEY DEVELOP IS DETERMINED BOTH BY THEIR OWN HEREDITY AND BY THE INDUCERS TO WHICH THEY ARE EXPOSED.

INDUCTION AND RECAPITULATION: A THEORY In the great debate that followed the publication of Darwin's *On The Origin of Species* in 1859, one of the most frequently cited evidences for the common ancestry of species was the remarkable similarity of the embryos in all classes of vertebrates. Among the traits held in common from lamprey and fish to mammals are the notochord (retained as the main part of the skeleton in lampreys), the gill arches, and the tail. In fish, the gills grow out from the gill arches. In reptiles, birds, and mammals, the gill arches as such disappear during development, but parts of them develop into such structures as the ear canal (eustachian tube) and the thyroid and parathyroid glands. The great German biologist E. H. Haeckel (1834–1919) was one of the first to appreciate this tendency of embryos to retain ancestral features. He propounded the principle of recapitulation: "ontogeny repeats phylogeny"; that is, during development an animal retraces its evolutionary history. If this were strictly true, a man would have to pass through a fish stage, an amphibian stage, and a reptile stage, in that order. Obviously Haeckel's principle cannot be taken too literally, but it is curious that we do still retain the notochord and gill arches as part of our embryological baggage. Perhaps the explanation lies in the phenomenon of induction.

We discussed previously how the development of the nervous system depends on induction by the notochordal tissue. It has been shown experimentally that the nervous system will fail to develop if the notochord is absent. This should suggest why primitive features may be retained. They often provide a necessary inductive stimulus in the complex chain

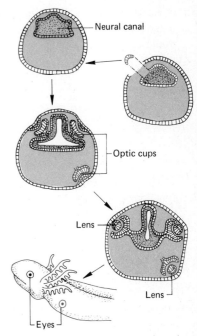

Fig. 10-30 An optic cup transplanted to the body of a host embryo induces lens formation and development of a third eye.

of inductive stimuli. Thus they are highly functional in the developmental processes of the animal. The evolutionary removal of such a key structure would present more difficulties than are involved in retaining it.

THE PHENOMENON OF INDUCTION, IN WHICH THE DEVELOPMENT OF CERTAIN TISSUES OCCURS ONLY AFTER EXPOSURE TO OTHER TISSUES, MAY EXPLAIN WHY EMBRYOS OF VERTEBRATES PRODUCE, IN DEVELOPMENT, PRIMITIVE FEATURES CHARACTERISTIC OF LOWER GROUPS, WHICH ARE LATER LOST OR MODIFIED.

Development in Plants

Development in plants is basically different from that in animals, in large part because of the nature of the plant cell. As we have seen earlier, plant cells are enclosed within a cellulose cell wall. Adjacent cells are held together by intercellular cement and are interconnected by fine cytoplasmic strands passing through tiny perforations in the wall. Plant cells, therefore, cannot move about in the process of morphogenesis, the formation of tissue and organs. Obviously, except in early growth and before the cell walls have stiffened, new cells cannot be added to the interior of a mass of plant cells without causing the outer cells to move. Plant growth and morphogenesis thus involve the addition of cells to the outside of preexisting cells, for the most part. On the other hand, cells may be added at almost any point in the developing animal. In animals, growth often ceases after a specific size and form have been reached. In most plants, growth takes place throughout the life of the plant. As we have seen in Chap. 7, response to environmental stimuli in plants is by changing patterns of growth, controlled by many interacting hormones.

In the nonvascular plants, a variety of different patterns of growth are found. Plants may be single-celled or multicellular. If multicellular, they may be filamentous, flattened sheets of cells, or complex masses of great size. In the fungi and some of the algae, the plant body is formed of filaments or tubes wound about into structures of amazing complexity. In this chapter, we will discuss development in flowering plants, considering first the growth of the embryo and then morphogenesis in the growing plant.

DEVELOPMENT IN PLANTS USUALLY CONTINUES THROUGHOUT THE LIFE OF THE PLANT. MORPHOGENESIS DOES NOT TAKE PLACE BY THE MOVEMENT OF CELLS AND THE ADDITION OF CELLS WITHIN TISSUES. INSTEAD, NEW CELLS ARE CONTINUALLY BEING ADDED TO THE OUTSIDE OF EXISTING TISSUES.

EMBRYO DEVELOPMENT IN FLOWERING PLANTS   As we have seen earlier in this chapter, pollination brings the male gametophyte of a flowering plant to the stigma of the carpel. The male gametophyte, or pollen grain, produces a pollen tube, which grows down the style to an ovule in the ovary. The tip of the pollen tube enters the ovule through a small opening at its tip. The pollen tube ruptures and releases two non-motile male gametes at the end of the female gametophyte. Within the embryo sac, one of the male gametes fuses with the egg nucleus and the other fuses with the polar nuclei. The zygote is a diploid cell which develops into the embryo plant. The product of the fusion of the polar nuclei and a male gamete develops into the nutritive tissue called endosperm, which is polyploid.

Details of the development of the embryo and the endosperm vary from species to species. A common example is that of *Capsella,* the shepherd's purse plant; its stages of development are shown in Fig. 10-31. Before the zygote begins to divide, the endosperm nucleus usually begins to divide rapidly. At first, cell walls are not formed, so that the endosperm is a multinucleate liquid within the embryo sac. Eventually the endosperm develops cell walls, and its cells contain food reserves and essential growth factors. If you have ever opened a fresh coconut, you have seen endosperm in the process of developing. The liquid, or coconut milk, is the noncellular endosperm, and the meat of the coconut is cellular. The developing embryo of a plant may use up all the endosperm. In some species, however, some endosperm remains as a food reserve in the seed before germination.

The first division of the zygote usually is uneven, and the fates of the two cells produced are different. The cell toward the opening of the ovule divides to form a linear series of cells called a suspensor. The other cell divides in all planes, producing a globular embryo. Within this more or less spherical mass of cells, tissue begins to differentiate which will eventually produce the epidermal cells of the plant, the vascular tissue, and the cortex (the cells surrounding the vascular tissue). Further growth causes a flattening of the embryo and the formation of the first leaves of the plant, the cotyledons.

Subsequent growth of the plant is from two regions of the embryo which remain permanently embryonic. These regions are at the end of the embryo that will become stem and at the opposite end, which will become root. They are called **apical meristems.**

DOUBLE FERTILIZATION OCCURS IN THE ANGIOSPERMS, ONE MALE GAMETE FUSING WITH THE EGG TO FORM THE ZYGOTE AND THE

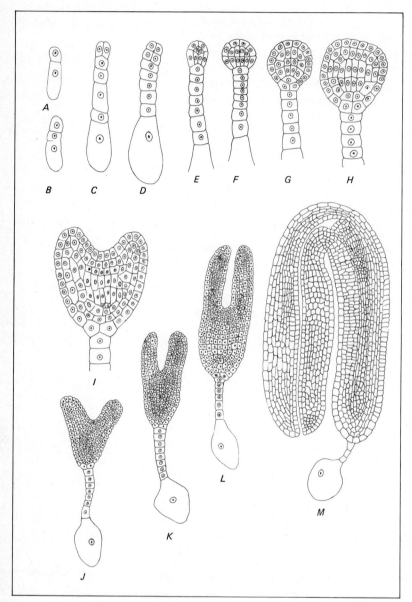

Fig. 10-31 Stages in the development of an embryo of shepherd's grass (*Capsella*). By the stage shown in (*F*), the three primary tissue systems are present: the epidermis, the future cortex, and the future vascular system (center). The first leaves, the cotyledons appear in (*I*). The embryo is curved because the ovule and embryo sac are curved.

OTHER FUSING WITH THE POLAR NUCLEI TO FORM THE ENDOSPERM, A NUTRITIVE TISSUE IN WHICH THE EMBRYO DEVELOPS. THE DEVELOPING SPOROPHYTE CONSISTS OF A SUSPENSOR AND THE EMBRYO PROPER, WHICH DIFFERENTIATES COTYLEDONS AND THE APICAL MERISTEMS OF THE STEM AND ROOT.

APICAL MERISTEMS AND PRIMARY GROWTH The apical meristems are permanently embryonic regions which, by mitotic division, add cells to the tip of the root and stem. Growth resulting from these meristems is called **primary growth.** The root and the shoot (stem plus leaves) have different functions and operate in very different environments. It is not surprising, therefore, to find that the meristems that produce them have different patterns of organization.

The terminal portion of a growing root is shown in Fig. 10–32. As you can see, the end of the root, which is pushed down into the soil as new cells are added, is covered by a **root cap.** The root cap consists of large cells with mucilaginous contents. Presumably it functions to facilitate movement of the root, as well as to protect the apical meristem behind it. The meristem itself consists of small, dense cells which are capable of dividing indefinitely. Some of its products become root-cap cells; the majority form the primary tissues of the root. The most recently produced cells are, of course, closest to the meristem. As one goes up the root from the tip, older and more differentiated cells are found. A longitudinal section of a root, then, shows all stages of differentiation of primary tissues.

Immediately behind the apical meristem, the first sign of differentiation is the gradual enlargement of cells. This occurs by the uptake of water, under the control of hormones, as discussed in Chap. 7. Beyond this region of elongation, root hairs are produced by the epidermal cells, and these serve to increase enormously the absorption surface of the root, as we have seen. Within the central region, the vascular tissues differentiate, with functional cells of the phloem being found first, followed by those of the xylem. Other specialized tissues of the root also differentiate progressively from the tip upward.

THE APICAL MERISTEM OF THE ROOT PRODUCES, TOWARD THE TIP, CELLS THAT FORM A PROTECTIVE ROOT CAP. AWAY FROM THE TIP ARE FORMED CELLS WHICH DIFFERENTIATE INTO THE EPIDERMAL, VASCULAR, AND OTHER TISSUES OF THE ROOT.

The organization of the stem apex is different from that of the root in two aspects. There is no cap over the tip of the stem, which usually is covered by the developing leaves. Also, the apex is divided into nodes, at which leaves are formed, and internodal regions, as shown in Fig. 10–33. The stem tip tends to grow upward, and as with the root, the more highly differentiated cells will be found farthest behind the tip. Leaves originate as tiny bulges, which eventually grow outward and upward. In the angle between the leaf and the stem is an axillary bud, which has the potential to form a branch stem. Vascular tissues and other specialized cells of

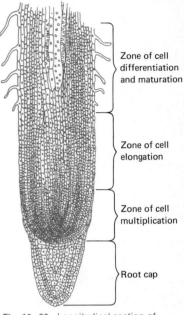

Zone of cell differentiation and maturation

Zone of cell elongation

Zone of cell multiplication

Root cap

Fig. 10–32 Longitudinal section of an angiosperm root tip showing three principle growth zones and the root cap. (*Adapted from COLLEGE BOTANY by Harry J. Fuller and Oswald Tippo. Copyright, 1949, by Holt, Rinehart, and Winston, Inc. Adapted and reprinted by permission of Holt, Rinehart, and Winston, Inc.)*

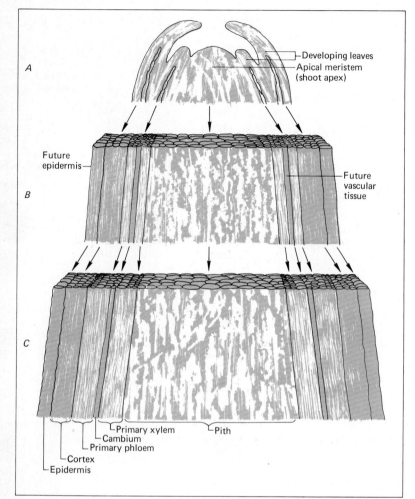

Fig. 10-33 Diagrammatic longitudinal section of a shoot tip showing differentiation of tissues. (*Adapted from BOTANY. fourth edition by Carl L. Wilson and Walter E. Loomis. Copyright 1952, © 1957, 1962, 1967, by Holt, Rinehart and Winston, Inc. Reprinted by permission of Holt, Rinehart and Winston, Inc.*)

the stem differentiate after a period of cell enlargement. Eventually the vascular tissues of the stem become connected with those of the developing leaves, and photosynthetic products can be moved upward from the older functional leaves below.

You can see that primary growth from the apical meristems of root and shoot results primarily in an increase in length of these structures. The girth of primary roots and stems depends upon the size of the apical meristems producing them and the degree of enlargement of cells in the course of their differentiation. Obviously, some means of increasing the diameter of roots and stems must occur to account for the growth of trees and shrubs. Growth in diameter is the result of *lateral* meristems.

*THE APICAL MERISTEM OF THE SHOOT PRODUCES, AWAY FROM THE
TIPS, CELLS WHICH WILL DIFFERENTIATE INTO THE EPIDERMAL,
VASCULAR, AND OTHER TISSUES OF THE STEM. IT ALSO PRODUCES
LEAVES AND BUDS AT NODES. GROWTH FROM APICAL MERISTEMS
IS CALLED PRIMARY GROWTH.*

LATERAL MERISTEMS AND SECONDARY GROWTH Apical meristems
are permanently embryonic areas which add new cells at the tips of stems
and roots. Since differentiation occurs behind these meristems, functional continuity of the xylem and phloem is maintained and the protective epidermis of the plant is not disturbed. Growth in diameter,
called **secondary growth,** presents serious developmental problems.
Transport of water and photosynthetic products must be maintained, and
therefore cells must be added to the xylem and phloem. These vascular
tissues are in the center of roots and stems, however, and if cells are added
there, the outer layers, including the epidermis, will be ruptured. These
problems have been solved by the development of two kinds of lateral
meristems called **cambia** (sing. **cambium**). One of these, the **vascular
cambium,** produces new xylem and phloem. The other, the **cork cambium,** produces layers of cork, a protective tissue taking the function of
the epidermis.

The vascular cambium arises from cells between the xylem and phloem
which remain embryonic instead of becoming vascular cells. Division
of cells of the vascular cambium produces daughter cells either toward the inside of the cambium or toward the outside (Fig. 10-34).
Daughters toward the inside of the stem develop into cells of the xylem,
and those to the outside differentiate into phloem cells. As the mass of
xylem, called **secondary xylem,** grows the cambium periodically extends

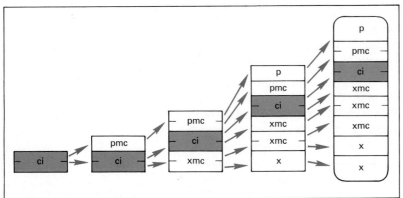

Fig. 10-34 Diagrammatic section to
show operation of the vascular
cambium. The cambium cell (ci)
divides to form phloem and xylem
mother cells which also divide to
produce cells which differentiate into
functional cells of the secondary
xylem and phloem. (*Adapted from
BOTANY, fourth edition by Carl L.
Wilson and Walter E. Loomis.
Copyright 1952, © 1957, 1962, 1967 by
Holt, Rinehart and Winston, Inc.
Reprinted by permission of Holt,
Rinehart and Winston, Inc.*)

Fig. 10-35 Cross section of the trunk of a pine tree showing the annual rings caused by production of smaller, denser cells toward the end of each growing season. (*USDA Photo.*)

itself by divisions in the plane opposite to that producing xylem and phloem.

Cells of the secondary xylem which differentiate in the spring are usually larger than those formed in the summer. In most habitats, no secondary xylem is produced in the winter or dry season. The result is the characteristic banding of the trunk of a tree (Fig. 10-35). Each band is an annual ring representing the secondary xylem laid down during one year. By examining the cells in detail (Fig. 10-36), one can determine which were produced in the spring and which in the summer, as well as such things as the amount of water available and the general health of the tree. The most recently formed cells of the secondary xylem conduct water and dissolved nutrients upward. Toward the center of a trunk, the xylem is no longer functional for conduction but functions only for support. Commonly, this wood becomes impregnated with minerals and other substances and is darker and stronger than the conductive xylem. It is called heartwood, in contrast to the latter, which is called sapwood.

Cells which differentiate toward the outside of the vascular cambium become secondary-phloem cells. The **secondary phloem** and all tissues outside it are continually disrupted by the more rapid growth of secondary xylem. Therefore, there can be no buildup of secondary phloem as there is of secondary xylem. Only the most recently formed rings of secondary phloem are functional in the transport of photosynthetic products.

All the tissues outside the secondary phloem are torn apart in secondary growth, including the epidermis. The role of the epidermis in preventing drying out and gas transport is taken on by the secondary tissue known as **cork,** produced by the cork cambium. As pressure from the secondary growth of the vascular tissue increases, the cork cambium develops from cells inside the epidermis. Divisions of the cork cambium (Fig. 10–37) produce cells mainly toward the outside, which develop heavily waterproofed walls. Interspersed among these are less specialized cells, through which gas exchange may take place. Cork is continuously sloughed off from the outside and replaced by new cork cambia, which develop successively deeper and deeper in the trunk, until they arise in the secondary phloem adjacent to the functional elements of that tissue. Therefore, when you peel the bark (corky tissue) off a twig, the thin-walled cells of the cork cambium, secondary phloem, and vascular cambium are ruptured. Differences among species in the pattern of cork cambia and the manner in which bark is sloughed off lead to the characteristic bark patterns of trees.

In a tree, such as an oak or maple, the vascular cambium may be thought of as a continuous sheath of embryonic cells between the xylem and phloem whenever secondary growth occurs. There are many cork cambia, depending upon the growth pattern of the species. The growth of a tree or shrub is, as you can see, very different from that of an animal. Primary growth results in the addition of cells at the tips of roots and stems. Increase in girth occurs through the activity of lateral meristems or cambia. Cells are nearly always *added onto* existing groups of cells.

INCREASE IN GIRTH, OR SECONDARY GROWTH, IN PLANTS IS THE RESULT OF DIVISIONS OF LATERAL MERISTEMS CALLED CAMBIA. THE VASCULAR CAMBIUM PRODUCES SECONDARY XYLEM (WOOD) AND SECONDARY PHLOEM. THE CORK CAMBIUM PRODUCES A PROTECTIVE LAYER OF BARK.

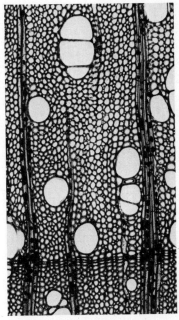

Fig. 10–36 Cross section of the secondary xylem of a silver maple at the boundary between annual rings. Very large vessels are interspersed among smaller tracheids and fibers. The horizontally arranged ray cells are also seen. ×122. *(Courtesy of Philip Feinberg.)*

Fig. 10–37 Cross section of a cork cambium and its derivatives: cork to the outside, thin-walled cells to the inside. *(Adapted from BOTANY, fourth edition by Carl L. Wilson and Walter E. Loomis. Copyright 1952, © 1957, 1962, 1967 by Holt, Rinehart and Winston, Inc. Reprinted by permission of Holt, Rinehart and Winston, Inc.*

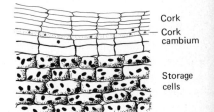

Cork
Cork cambium
Storage cells

Now that we have discussed patterns of development in animals and plants, we may turn to more general aspects of development. Presumably these apply to all organisms. These days, it is common to equate complexity with information. Information can be quantified, and a whole body of mathematical theory called *information theory* has developed around this principle. One of the consequences of this new theory is that it is possible to say how much information is necessary to construct a three-dimensional object such as a crystal.

An organism is a three-dimensional object, and one of the ways to approach the subject of development is through information theory. The problem as we stated it above is basically that of an engineer. How can material that has relatively little structure be made into material that has a high order of integrated complexity? How, in other words, does a building develop out of a pile of cement and sand or a falcon out of the superficially similar cells of the early embryo? The same question can be asked in slightly more precise terms:

1. How many sources of information are there in the development of of an organism?
2. How much information does each of these sources contribute?

We would be misleading you if we were to pretend to answer these questions in quantitative terms. This is not yet possible, though it may be in a few decades. For the time being, though, we can make some qualitative statements about the sources of information.

NUCLEAR INFORMATION If the nucleus is removed from a cell, the cell will soon cease all its activities and die. Obviously, information coming from the nucleus is essential for the maintenance and differentiation of the cell. In general, we know that this information leaves the nucleus in the form of messenger RNA. Can this simple observation be extended to say that changes within the nucleus, or **nuclear differentiation,** are what determine the changes in the cytoplasm that we call **cellular differentiation?** This is really several questions in one, so we will have to look at each individually.

First, there is an elementary possibility that can be quickly disposed of. One of the simplest of the nuclear theories of development is that the genes in the zygote nucleus are parceled out during the cleavage and later cell divisions so that the cells of the adult contain only the few genes that control their special function. According to this theory, the heart-muscle cells, for example, lose all the genes for making brain or digestive enzymes and retain only the genes for making the constituents of heart

The Information for Development

muscle. This idea can easily be shown to be false. For one thing, nearly all the cells of an adult organism contain the complete diploid set of chromosomes and the same amount of DNA as the zygote, so presumably all the genes are still present. Secondly, it is possible to grow a complete organism, such as a carrot plant, from one cell of the adult. One of the cleavage cells of the early blastula may suffice in the case of animals. Even more impressive, though, are recent nuclear-transplant experiments in which the nucleus from the intestine of a tadpole has been substituted for the normal zygote nucleus in the fertilized egg (Fig. 10–38). Such artificial zygotes often develop normally into fully formed adult frogs, thus showing that no genes were lost during the development of the intestinal cells of the donor frogs.

This experiment also disposes of the possibility (at least for this species of frog) that genes are destroyed or permanently "turned off" during development. It would look at though, under certain circumstances, genes can be turned on, even if they previously had been turned off.

Studies of plant cells in tissue culture also bear upon this question. Tissue from plants, such as carrots and tobacco, can be grown in liquid culture (Fig. 10–39). The medium must contain necessary minerals and a source of carbon. It also must contain coconut milk (liquid endosperm), which provides essential growth factors. If the culture is shaken as the cells divide, small clusters of cells or even single cells become separated. These can be isolated from the liquid culture and placed on a solid growth medium. There they develop into cell masses which resemble plant embryos very closely. Under the proper conditions, such embryolike cell groups, even those derived from a single cell, may be induced to form roots, shoots, and eventually entire plants.

NUCLEAR DIFFERENTIATION OFTEN DOES NOT ACCOMPANY CELLULAR DIFFERENTIATION. NUCLEI FROM DIFFERENTIATED CELLS MAY BE ABLE TO CONTROL THE DEVELOPMENT OF AN ENTIRE ORGANISM.

But one cannot, from such experiments, jump to the conclusion that the nuclear information is unimportant. There are many experiments that show that the *specific details of developmental end products depend on the specific nature of the genes.* Figure 10–40 shows the results of a classic experiment by Joachim Hämmerling with two species of *Acetabularia,* green algae that grow to a height of 6 cm or more in warm seas. Each plant consists of a single giant cell. At the base of this cell is a system of anchoring processes, the rhizoid, which contains the nucleus. A green stalk extends from the rhizoid, and in mature individuals, the stalk is capped by an umbrellalike fruiting body. In one species, there are

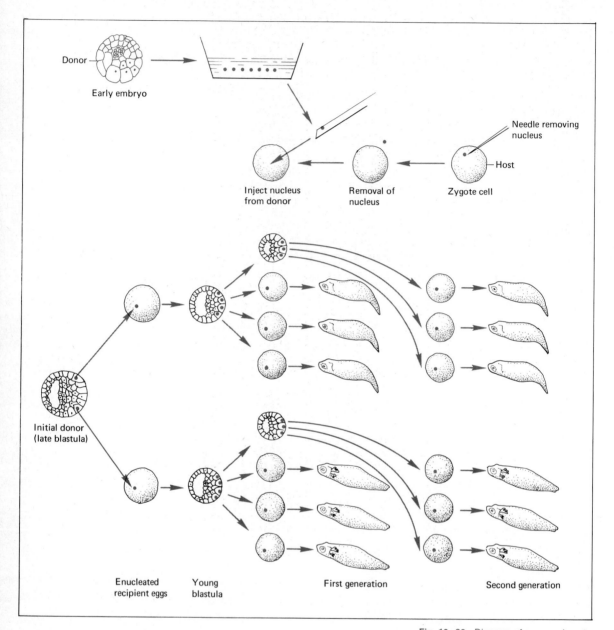

Fig. 10-38 Diagram of an experiment in which enucleated eggs are provided with nuclei from cells of a donor embryo of the blastula stage. Two generations of larvae have been produced. (*Adapted with permission of The Macmillan Company from PRINCIPLES OF DEVELOPMENT AND DIFFERENTIATION by C. H. Waddington. Copyright © 1966 by C. H. Waddington.*)

about 80 rays in the cap; another species has about 30. When the stalk of one species was grafted onto the rhizoid of another, the first cap that formed was intermediate in ray number. When this cap was removed, the second cap to develop was of the type made *by the species contrib-*

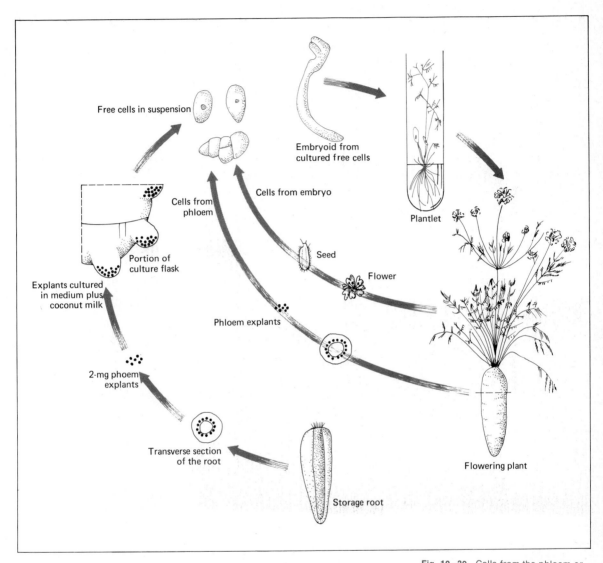

Free cells in suspension

Embryoid from
cultured free cells

Plantlet

Cells from
phloem

Cells from embryo

Portion of
culture flask

Seed

Explants cultured
in medium plus
coconut milk

Flower

Phloem explants

2-mg phoem
explants

Transverse section
of the root

Storage root

Flowering plant

Fig. 10-39 Cells from the phloem or
from an embryo of a carrot can be
grown in tissue culture. Single cells
from such a culture can develop into a
carrot plant. The embryo came from a
plant started from a single cell in
culture. (*Adapted from "Growth and
Development of Cultured Plant Cells,"
Steward, F. C., et al, Science, Vol. 143,
pp. 20–27, Fig. 2, 3 January 1964.
Copyright 1964 by the American
Association for the Advancement of
Science, by courtesy of F. C. Steward
and the American Association for the
Advancement of Science.*)

uting the rhizoid rather than the species from which the stalk had come.
The experiment was interpreted to mean that the information for the first
cap came from both the rhizoid and the stalk but that the information for
the second cap came only from the part with a *nucleus*—the rhizoid. This
is typical of many experiments showing that the nucleus determines
species-specific characteristics.

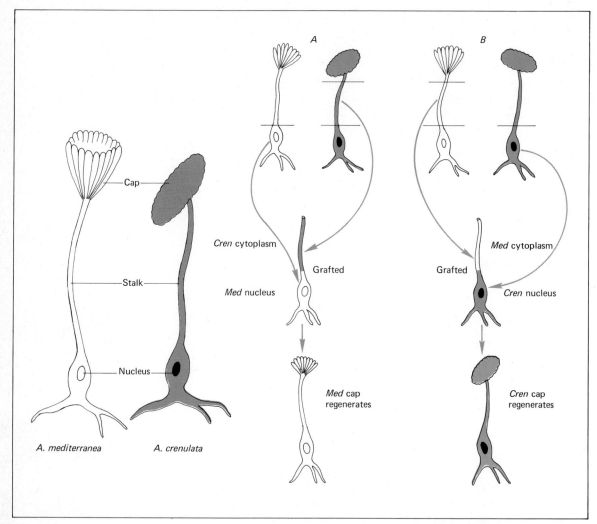

Fig. 10–40 Diagram to show the role of nucleus in regeneration of the cap in the alga *Acetabularia*.

INFORMATION DERIVING FROM THE NUCLEUS OF THE ZYGOTE AND ITS DAUGHTER CELLS IS ESSENTIAL FOR THE DIFFERENTIATION OF CELLS AND CAN BE SHOWN TO BE SPECIES-SPECIFIC.

NUCLEAR-CYTOPLASMIC INTERACTIONS As has been shown by nuclear-transplant experiments, nuclear information is only part of the story of development. The environment of the nucleus is another source of information and one of equal importance. We already have seen, for instance, that cells in the ectoderm can develop into either nerve cells or skin cells, depending on their environment—in this case, the presence or absence of notochordal tissue. Here is a case in which some cue or information in the environment is responsible for a "decision" by the responding cells. Apparently, cytoplasmic information can dictate which parts of the nuclear information are to be used.

Today most biologists think of development as the coordinated switching on and off of genes in the proper sequence so that each gene product is present at just the right time. There is some direct evidence for this sort of control, deriving from experiments with microorganisms, as described in Chap. 4. Very little, however, is yet known about how gene activity is controlled in multicellular organisms.

DIFFERENTIATION IS AFFECTED NOT ONLY BY INFORMATION FROM THE NUCLEUS. INFORMATION FROM THE CYTOPLASM AFFECTS WHAT NUCLEAR INFORMATION IS USED AND HOW. DEVELOPMENT INVOLVES A COMPLEX SET OF INTERACTIONS OF THE NUCLEUS AND THE CYTOPLASM.

ENVIRONMENTAL INFORMATION   Another source of information is the outside world. How important environmental information is depends on how dependent the developing organism is on the outside world. The development of a mammalian embryo in the insulated and regulated incubator of its mother's uterus is hardly affected by humidity and temperature, whereas a free-living insect larva exposed to the physical and chemical extremes of a pond may not survive if it fails to make the proper developmental responses.

The larvae of many flies, including *Drosophila,* have organs of salt balance called *anal papillae.* Larvae grown in a medium containing a high proportion of salt develop larger papillae than do larvae grown in typical medium. It has been shown that the larvae that can develop the largest papillae in a salty medium have the best chance of surviving. This is a case in which nuclear, cytoplasmic, and environmental information all contribute to a developmental process. More particularly, it is an example of the developing organism's ability to correctly interpret environmental data and then use this information to alter its differentiation.

A similar example of the use of environmental information during development is seen in mosquito larvae raised in two media—water poor in oxygen and water rich in oxygen. In the oxygen-poor water, the gills are very prominent structures compared with their relatively weak development in the well-aerated water.

THE ENVIRONMENT OF INDIVIDUAL CELLS, AS WELL AS THAT OF THE ENTIRE ORGANISM, AFFECTS DIFFERENTIATION AND DEVELOPMENT. THE FINAL EXPRESSION OF THE GENETIC MATERIAL (GENOTYPE) OF AN ORGANISM AS ITS PHENOTYPE IS DEPENDENT UPON A COMPLEX SET OF INTERACTIONS WITH THE CYTOPLASM, THE CELLULAR ENVIRONMENTS, AND THE ENVIRONMENT OF THE ORGANISM.

SUMMARY

Life cycles of animals, plants, and microorganisms vary in the portions spent in the diploid and haploid condition. In some unicellular organisms, the major part of the life cycle is spent in the haploid condition and the only diploid cells may be the zygote. In other unicellular forms, the gametes are the only haploid cells and the major part of the life cycle is spent as a diploid cell. Multicellular plants have a variety of different kinds of life cycles, whereas multicellular animals spend the major part of the life cycle in the diploid condition and only the gametes are haploid.

In some multicellular plants, including all the vascular land plants, there is a regular alternation of generations of haploid and diploid individuals in every life cycle. The diploid plant, called the sporophyte, produces haploid spores by meiosis. Haploid spores germinate into gametophytes, which produce gametes by mitosis. In the seed plants, the diploid sporophyte is the predominant generation. Few-celled female gametophytes are retained in the ovules. The few-celled male gametophytes are called pollen grains. After pollination and fertilization, the new sporophyte is retained in the ovule, which develops into a seed.

The sperm of vertebrates are produced in the testis by meiotic divisions of the germ cells. Each sperm cell is the result of a complex process of differentiation, during which most of the cytoplasm is lost, the nucleus is much condensed, and the specialized propellant tail is produced. Vertebrate eggs are produced in the ovaries. Of the four haploid cells produced in each meiotic division, only one differentiates to form the large, nonmotile egg. Internal fertilization, usually accomplished by an intromittent organ of the male, makes possible reproduction in a terrestrial environment.

Development in animals begins with mitotic division of the zygote called cleavage division. Change eventually results in a large number of cells, which usually become arranged in a hollow sphere. In a manner varying from organism to organism, the hollow sphere becomes three-layered. Each of the three layers of the embryo produces cells which differentiate into different groups of tissue. In vertebrates, further development of the three-layered stage establishes a mouth at one end and an anus at the other end of a basically tube-shaped organism.

The outer layer of cells becomes the skin and nervous tissue, the middle layer becomes the organs of support and transport, and the inner layer becomes the digestive tract and its associated glands. In the course of development, different tissues interact with one another, and some tissues induce others to form specific organs. Many cells of the embryo have the potential to develop into one of several cell types. The kind of cells into which they develop is determined both by their own heredity and by the inducers to which they are exposed.

Cells may be added at almost any point in the developing animal, in which growth often ceases after a specific size and form have been reached. In most plants, growth continues throughout the life of the plant. Morphogenesis does not take place by the movement of cells and the addition of cells within tissues. Instead, new cells are continuously being added to the outside of existing tissues.

The developing sporophyte of a flowering plant grows within a nutritive tissue called endosperm. Embryonic leaves called cotyledons and the apical meristems of stem and root are produced. After germination of the seed, these apical meristems begin to divide to form the primary tissues of the plant. Increase in girth, or secondary growth, is the result of division of lateral meristems called cambia. The vascular cambium produces secondary xylem and phloem, and the cork cambium produces a protective layer of bark.

Information deriving from the nucleus of the zygote and its daughter cells is essential for the differentiation of cells. Differentiation is affected not only by information from the nucleus, however. Information from the cytoplasm affects what information is used and how. The final expression of the genetic material (the genotype) of an organism as its phenotype is dependent upon a complex set of interactions of the nucleus, cytoplasm, cellular environments, and environment of the organism.

SUPPLEMENTARY READINGS
Books

Balinsky, B. I.: "An Introduction to Embryology," 2d ed., Saunders, San Francisco, 1965.

Ebert, J. D. and I. M. Sussex: "Interacting Systems in Development," 2d ed., Holt, New York, 1970.

Cutter, E. G.: "Plant Anatomy: Experiment and Observation," Pt. I, "Cells and Tissues," 1969, Pt. II, "Organs," Addison-Wesley, New York, 1971.

Wessels, N. K. and W. J. Rutter: Phases in Cell Differentiation, *Sci. Amer.*, **220,** no. 3, **Articles:**
Offprint 1136 (1969).
Wessels, N. K.: How Living Cells Change Shape, *Sci. Amer.*, **225,** no. 4, Offprint
1233 (1971).

CHAPTER 11

DEFENSE: PREVENTING SELF FROM BECOMING NOT-SELF

Organisms wage a continuing battle to keep themselves intact—to prevent self from becoming not-self. Mechanisms for the defense of self have evolved which operate at all levels of biological organization. In this chapter we are concerned with those at the cellular and organismal levels. Defense at the population level is discussed in Chap. 15.

In addition to the more obvious means of defense, such as speed, armor, unpalatability, and mimicry, all of which serve to protect against direct and overt assaults by large predators, there are more subtle defenses against the equally dangerous attacks of microorganisms. Defense against microorganisms may be nonspecific, operating against a broad spectrum of forms. Other types of defenses are highly specific at the molecular level.

MICROORGANISMS AND TOXINS

One of the nonspecific defenses of vertebrates is a general antiviral agent, a protein called *interferon,* secreted by many tissues. The interferon released by dying cells into the surroundings confers resistance on other, uninfected cells. In fact, recovery from virus diseases such as colds and influenza (e.g., the Asian flu) is probably due to interferon release, at least in the early phases of recovery. Many tissues secrete another protein, the bacteriocide *lysozyme,* that attacks the cell walls of bacteria, exposing the osmotically fragile protoplasts. Recent evidence indicates that other substances, as yet unknown, initiate the attack on the bacteria. Certain amoebalike cells (polymorphs) in the blood and tissues, whose job it is to engulf and destroy bacteria, have high concentrations of lysozyme in their cytoplasm. Bacteria can enter the body most easily through moist, living tissues, including the eyes, nose, mouth, and intestine. Tears, saliva, and the mucus secreted in the nasal passages and intestine contain lysozyme.

Nonspecific Defenses

Lysozyme should not be confused with lysosomes, very small membrane-bound organelles within phagocytic cells (amoebalike scavenger cells) that contain powerful hydrolytic enzymes that digest cellular contents. Death or disease of a cell is accompanied by disintegration of the lysosomal membranes, an efficient "self-destruct" system. Tadpoles digest their tails by liberating lysosomal enzymes in the tail cells.

The skin of mammals is protected from attack by microorganisms through the sterilizing power of secretions of oil from the **sebaceous** glands associated with hair follicles. Saturated fatty acids with from eight to eleven carbon atoms are very toxic to fungi. Their concentration increases in the sebum (oily secretion) after puberty. This is probably why such childhood infections as ringworm disappear spontaneously. Athlete's foot is a fungus disease that occurs in parts of the body (the lower surfaces of the feet and between the toes) devoid of hair follicles and sebaceous glands.

NONSPECIFIC DEFENSES OF ORGANISMS INCLUDE THE ANTIVIRAL AGENT INTERFERON; LYSOZMYE, WHICH ATTACKS BACTERIA; AND SECRETIONS OF THE SKIN.

The Immune Response

The natural mechanical-chemical defenses just described are non-specific; they act against a variety of potential pathogens. An important supplement, for vertebrates at least, is the **immune response,** a potent and adaptable defense that is specific against a *single* pathogen. Human awareness of this response dates far back. In a description of a typhus or plague epidemic 2,500 years ago, Thucydides (460–400 B.C.) stated that fear led to the neglect of ill and dying persons. What attention was provided was the "pitying care of those who had recovered, because they . . . were themselves free of apprehension. For no one was ever attacked a second time, or with a fatal result." A millenium followed before man began to deliberately exploit observations such as these. The momentous discovery was made by the eighteenth-century English physician Edward Jenner (1749–1823), who, like many of his contemporaries, had noticed that individuals who had had the relatively mild disease known as cowpox were somehow protected when a smallpox epidemic swept through the region. Jenner put this observation to the test by inoculating (implanting or injecting) a boy with pus from a dairymaid's cowpox lesion. A later inoculation with the virulent pus from a lesion of a smallpox patient failed to induce any symptoms in the boy. Many repetitions of this famous experiment yielded similar results and led to the adoption of **vaccination** (Latin: *vacca,* "cow") as a means of immunization against smallpox. One hundred years later, Pasteur inoculated chickens with an *old* culture of *Pasteurella aviseptica,* the micro-organism causing chicken cholera, but the disease failed to develop. This alone is not very surprising, but the chickens later failed to become ill when they were inoculated with fresh, virulent cultures. Most of Pasteur's contemporaries would have ignored such an "anomalous" result, but Pasteur saw its implications and went on to establish that as a culture ages, it loses its virulence but not its capacity to induce immunity.

Passing a virulent organism through a host of another species can *attenuate,* or weaken, its virulence to the point where it can safely be in-oculated, producing immunity but not the disease. Human smallpox and bovine cowpox, for instance, are apparently caused by the same organism, a virus, but as the virus "adapts" to one species, it becomes less **pathogenic** (capable of causing disease) for the other. The reason should be obvious. A change of host is a change in environment, so any mutants that reproduce faster (are more fit) in the new host environment will continually replace the less fit genotypes. Usually, when an animal is infected by a microorganism adapted to another species of host, the immune process has a head start. By the time the microorganism has evolved back to the new host type, the latter is already immune. Smallpox, yellow fever, and some polio vaccines (Cox, Koprowski, and Sabin) are made from live viruses that are attenuated by passing them through animal hosts such as monkeys and, more recently, by growing them in tissue cultures. Another safe way to induce immunity to a virulent microorganism is to inoculate with killed or *inactivated* organisms. The Salk polio vaccine is made by exposing the viruses to formalin in a low enough concentration to kill them but not destroy their ability to induce immunity.

Natural selection, the process responsible for attenuation, is not, however, an unmitigated blessing in the field of medicine. It has worked against man, too. The resistance of many bacteria to antibiotics has evolved and continues to evolve almost as fast as new antibiotics are developed and marketed.

VERTEBRATES, INCLUDING MAN, ARE ABLE TO ACQUIRE IMMUNITY TO CERTAIN DISEASES AFTER VACCINATION WITH ATTENUATED, OR WEAKER, STRAINS OF A PATHOGEN OR WITH INACTIVATED MICROORGANISMS. THIS IS CALLED THE IMMUNE RESPONSE.

THE ANTIBODY What is the nature of this mysterious defense system called immunity? During the very exciting decade between 1890 and 1900, the outlines were established. The blood serum (the fluid that remains after removal of blood cells and platelets) of an immunized animal was found to neutralize the toxin produced by the tetanus organism. Soon thereafter, the three fundamental **serological reactions** (reactions between blood serum and particles such as cells or bacteria) were described from studies of mixtures of serums taken from immune animals and the substances and organisms used to elicit the immune response:

1 The serum of an immunized animal can **lyse** (disrupt or break apart) organisms and cells of the same type used to inoculate the animal (Fig. 11–1).
2 The serum of an immunized animal can **precipitate** (cause to sink out of solution) bacteria or other particulate bodies of the same type used to inoculate the animal (Fig. 11–1).
3 The serum of an immunized animal can **agglutinate** (cause to clump and stick together) cells of the same type used to inoculate the animal (Fig. 11–1). Bacteria can be caused to adhere not only to each other but to phagocytic cells.

All these reactions increase the vulnerability of bacteria and other particles to attack by the body's phagocytic cells. The general term **antigen** is used to describe any foreign material, compound, or cell that elicits an immune response. The general term for the agents causing the serological reactions is **antibody.** Understanding the immune response requires a knowledge of the structure and diversity of both antigens and antibodies.

An important clue to the nature of these reactions is the observation that the immunity "power" of serum from an immunized animal is limited and measurable. For example, a certain amount of serum may have the potential to agglutinate about 500 cells. If half this number is added to the serum, these will agglutinate; but if 500 cells were added originally, the agglutination power is depleted. From the nature of the serological reactions and the "depletion effect" just described, you should be able to develop a model of the agency in the serums of immunized animals. Try this before reading on.

Disregarding the existence of some mysterious forces in the serum, there must be a particle of molecular dimensions or larger that attacks cells in order to cause them to adhere to one another. Furthermore, the particle must adhere to the cells since it is lost from the serum. How is this different from the behavior of enzymes? Further still, the agglutination reaction suggests that the particles adhere to two (or more) cells at once (Fig. 11–1). The phenomenon of agglutination also suggests that

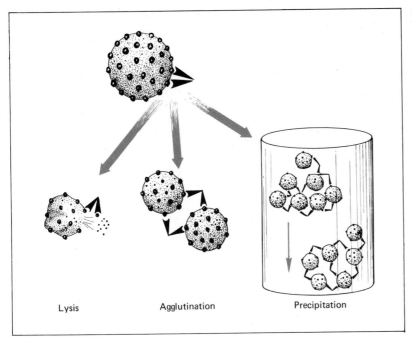

Lysis Agglutination Precipitation

Fig. 11–1 Diagrammatic representation of three types of antigen-antibody reactions.

the cells or objects agglutinated have more than a single site of attachment for particles. So, by deduction alone, it is very likely that antibodies are particles with at least two "sticky" attachment points.

FOREIGN MATERIAL, CALLED AN ANTIGEN, MAY ELICIT ANY OF SEVERAL SEROLOGICAL REACTIONS WHICH PROTECT AN ANIMAL. THE AGENTS RESPONSIBLE FOR THESE REACTIONS ARE CALLED ANTIBODIES.

THE STRUCTURE OF ANTIBODIES   Many kinds of evidence led to the conclusion that antibodies are protein molecules—in particular, gamma globulins—contained in blood serum. When animals are injected with large amounts of antigen, the amount of gamma globulin in the serum increases soon thereafter (the antigen induces the production of gamma globulin). Furthermore, it is possible to obtain relatively pure antibody by centrifuging or precipitating antigen-antibody complexes, then separating off and analyzing the antibodies. They turn out to be gamma globulins. More recently, other plasma globulins have been shown to possess antibody behavior. The arsenal of modern protein chemistry has been brought to bear on these important molecules, and some fascinating facts have come to light. In man, there are several groups of immunoglobulin antibodies: one type appears in small quantities soon after an antigenic stimulation; another appears after a longer delay but in greater amounts. But whatever the type, an antibody is not a single long protein molecule but, in common with many enzymes, is a highly ordered complex of subunits. Antibodies of the globulin type are composed of four chains, one pair of short chains and one of long chains. These are held together by both strong covalent bonds and weaker forces. The assumed arrangement of these four chains is shown in Fig. 11–2.

ANTIBODIES COMMONLY ARE PROTEINS CALLED GLOBULINS, WHICH ARE FOUND IN THE BLOOD. THEY HAVE COMPLEX STRUCTURE, WHICH ACCOUNTS FOR THEIR ACTIVITY AND SPECIFICITY.

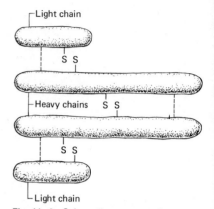

Fig. 11–2 Schematic representation of an antibody molecule: Two light and two heavy chains are joined by disulfide bonds (S—S) and weaker bonds (----).

ANTIBODY SPECIFICITY   Antibodies differ from enzymes in being permanently sticky, whereas enzymes interact with a substrate molecule and then release the products. They are similar to enzymes in that they are very particular about what they react with. Much of our knowledge of **antibody specificity** is due to the skill and creativity of Karl Landsteiner and his associates. They tried using a great variety of chemicals as possible antigens by attaching them to protein carriers. These and later experiments led to the following generalizations:

1 An antigen must be a fairly large molecule—the size of a protein or polysaccharide.
2 The region on the antigen which is recognized by the antibody is quite small—about the size of a few amino acid residues.
3 An antibody is highly specific, reacting only with the antigen that induced it or with one with very similar structure.
4 The specificity resides in the sequence of amino acids at the active (sticky) sites; the sequence of the other parts of the protein chains is relatively constant.

We now have a fairly clear picture of what antibodies are. They are Y-shaped or V-shaped molecules, two ends of which bear active sites for binding antigens. Experiments have shown that both active sites are the same; that is, they bind a particular antigen. But this picture leaves the function of much of the antibody unaccounted for. Half of one short chain and three-quarters of the long chains are relatively constant. It is thought that these large regions participate in *general* antibody reactions, such as the transmission across the membranes of the placenta from the mother to the fetus and the fixation (binding) of **complement,** a group of proteins involved in cell and bacterial lysis. The variable ends of the antibody chains contain the regions of antibody specificity—the active sites.

ANTIBODIES REACT ONLY WITH THE ANTIGEN WHICH LED TO THEIR FORMATION. THIS SPECIFICITY LIES IN THE SEQUENCE OF AMINO ACIDS AT THE TWO ACTIVE SITES FOR BINDING ON THE MOLECULE.

DEVELOPMENT OF THE IMMUNE RESPONSE   All normal adult mammals have levels of antibody sufficient to protect them against the common pathogens.   Exposure to scores of viral and bacterial pathogens is an everyday occurrence, but if these organisms evade the defenses at the body surface, they are "caught" by antibodies circulating in the blood which were elicited at some earlier time. Sometimes the primary infection leads inevitably to disease, as with childhood measles and chicken pox. For other diseases, such as poliomyelitis and tuberculosis, the primary infection may be of negligible proportions but may be large enough to elicit an immune response, with long-lasting protection afforded by a continued production of antibodies against the particular pathogen. Immunization accomplishes the same thing by artificially introducing a noninfecting form of a pathogen.

One of the "miracles" of life is the health of newborn animals, human or otherwise. Babies until the age of 1 or so are often the personification of health, even being free of ordinary colds as well as the more dangerous childhood diseases. If antibodies are so specific and if their evocation requires a particular antigen, how is it that newborns survive? The answer is that the antibodies come from the mother. The first "milk" sucked by a newborn mammal is called **colostrum.** Colostrum is usually rich in the mother's antibodies, secreted by the mammary tissues. This substance is the most important source of antibodies for the newborn of the hoofed mammals. Colostrum contains pepsin and trypsin inhibitors. In an adult pig, sheep, or goat, the intestinal proteolytic enzymes would destroy the antibodies. However, many newborn mammals do not produce gastric hydrochloric acid; the mucosa is permeable, and proteolytic enzymes are not yet being secreted. In primates, including man, the principal source of antibodies is the maternal circulation across the placenta, which is particularly leaky to these molecules. The "constant" region of the antibody is essential for placental transmission. Bottle-fed babies lack what protection there may be in their mother's milk.

NEWLY BORN MAMMALS ARE RELATIVELY DISEASE-FREE, HAVING ACQUIRED IMMUNITY FROM THE MOTHER IN THE FORM OF ANTI-

BODIES IN COLOSTRUM OR FROM THE MATERNAL BLOODSTREAM VIA THE PLACENTA.

ELEMENTS OF THE IMMUNE RESPONSE   The first invasion of tissues by foreign chemicals or microorganisms (or tissue transplants) is called the primary stimulus and evokes a **primary response.**   Usually in 3 to 14 days, but anytime between 1 and 30 days after the initial exposure, antibodies begin to appear in the blood.   Their concentration increases exponentially, reaches a peak, and then declines to a low or even undetectable level.   But even if no antibodies can be found, the body "remembers" the antigen.   A subsequent exposure evokes a **secondary response,** shorter in lag time and greater in amount and duration.   Also, the antibodies produced during a secondary response typically have greater combining power with the antigen.   In addition, the protection afforded by the secondary response is greater and more permanent, sometimes lasting a lifetime.   The long, accurate immunological memory has proved useful to "archeological pathologists" seeking to trace the evolution of influenza viruses.   People tend to respond more strongly to strains of the virus which caused their original attack than to strains to which they were subsequently exposed.   Elderly people therefore, who had the disease in past epidemics, say in the 1918 influenza epidemic, can be tested with various strains to see to which type they respond most strongly.

At the tissue and cellular level, the immune response is paralleled by changes in the **reticuloendothelial system,** expecially tissues of the lymph nodes and spleen.   Following a large injection of antigen, these tissues may proliferate cells and enlarge conspicuously.   The following is thought to be the typical sequence of events, though all the steps are still a matter of continuing research and debate:

1   Antigen is first detected by a hypothesized cell type, perhaps the phagocytic cells called **macrophages.**   These cells are thought to interact with the antigen in some way, but they do not themselves make antibodies.

2   Circulating white blood cells called **lymphocytes** have been seen to come into very intimate contact with macrophages, and intercellular bridges between them are thought to facilitate the transfer of an immunological message.

3   Circulating lymphocytes seem to then send a "message" back to the sites of antibody production: the spleen, lymph nodes, and bone marrow.   At these sites, **plasma cells** produce one or at most two types of antibody specific against the antigens of the invading orga-

nism. The antibodies travel via the bloodstream back to the scene of the battle, there to cause the serological reactions that aid the body's phagocytes to ingest and kill the pathogens. Lymphocytelike ***"memory" cells*** are thought to remain quiescent until a secondary stimulus by a later dosage of antigen, when they become active as part of the secondary response.

The mammalian thymus, at birth a large lymphocyte-containing organ that all but disappears by maturity, is also implicated in the immune response. If it is removed from newborn mice, the animals fail to become immunologically competent (do not produce many antibodies and do not reject grafts), and their lymph nodes remain very underpopulated with lymphocytes. These symptoms disappear if the animal receives a graft of thymus tissue from a mouse of the same inbred strain. But even if this graft is enclosed in a plastic chamber that permits the movements of molecules but not cells, the lymphocyte population of the lymph nodes increases and the animal can respond normally to antigens. This suggests that the thymus secretes a hormone necessary for the development of the lymphoid system, as well as contributing cells to the system.

THE IMMUNE RESPONSE CONSISTS OF A PRIMARY RESPONSE TO THE INITIAL EXPOSURE TO AN ANTIGEN AND A SECONDARY RESPONSE IN WHICH ANTIBODY PRODUCTION IS MORE RAPID AND GREATER IN AMOUNT AND DURATION. THE SECONDARY RESPONSE TO EXPOSURES AFTER THE INITIAL ONE MAY BE EVOCABLE FOR YEARS OR A LIFETIME. SPECIAL CELLS IN THE LYMPH NODES, SPLEEN, AND THYMUS GLAND PLAY ROLES IN THE RECOGNITION OF ANTIGENS AND FORMATION OF THE CORRECT ANTIBODY.

Self and Not-self

One of the most mysterious aspects of the immune system is its capacity to recognize and distinguish foreign material from molecules and cells of the animal's own body—to know what is self and what is not-self. With a few exceptions, antibodies are never made to a vertebrate's own tissues and macromolecules. The exceptions are materials such as eye lens protein, sperm, the thyroid gland, and nerve tissue, components that normally never enter the circulatory system and hence are well isolated from the immune system. This observation has suggested that the immune system must first know—be exposed to—the materials of the body before it can ignore them as potential antigens. This has been experimentally verified by injecting cells of one animal into another newborn animal of the same species before the newborn was able to manufacture antibodies. The recipient individual could thereafter accept tissue from the donor because it could tell its own cells from those of the donor. The same sort of thing, called **tolerance,** can be induced in older animals by relatively massive doses of antigen—as though the immune system were swamped. The tolerance will disappear in the continued absence of the antigen. There are some interesting speculations about the molecular and cellular basis of tolerance, but none has received substantial experimental verification.

THEORIES OF ANTIBODY SYNTHESIS   Antibodies are proteins. Proteins are strings of amino acid residues. And the central dogma of molecular biology is that the order of amino acids in a protein is determined by the sequences of triplets in messenger RNA, which in turn is determined by a complementary template of DNA triplets. A reasonable assumption is that about 10^6 different antibodies can be made by a mammal. Therefore, there ought to be 10^6 different genes that code *just* for antibodies, enough to use up about 14 percent of the genetic material in human cells. This is a tremendous commitment of hereditary material. There are at least two ways that this condition could be fulfilled:

1   Each cell could have 10^6 genes for antibody synthesis. This is the **germ-line theory.**

OR

2   Beginning with a few genes in embryonic lymphoid cells, these genes could mutate at a very high rate during their proliferation and development and could thereby generate the stupendous array of heterogeneous genetic templates needed to code for 10^6 antibodies. Each cell, that is, would have only one or a few kinds of antibody genes, but there would be millions of cells. The thymus gland of a newborn mammal is relatively large. Some immunologists believe that this organ is the source of these individualistic cells, the "generator of diversity." An antigen would then select and activate the small proportion of the entire lymphocyte population that had the matching genetic template; this cell group would then proliferate into a clone of antibody-synthesizing cells. This is the **clonal selection theory.**

The latter theory is now the most popular, but it is much too soon to say that it is established.

ANTIBODIES MAY BE THE PRODUCT OF GENETIC TEMPLATES WHICH DIFFER IN THE VARIOUS CELLS PRODUCING THEM. ANTIGENS COULD

ACTIVATE THOSE CELLS CONTAINING THE NECESSARY TEMPLATE, WHICH WOULD THEN REPRODUCE TO FORM MANY ANTIBODY-PRODUCING CELLS.

IMPERFECTIONS OF THE IMMUNE SYSTEM   Two classes of disease are associated with the immune system: **allergy** and **autoimmune disease.** The basis of allergic reactions is a group of antibodies secreted on the surfaces of the skin and mucous membranes. Sometimes, for unknown reasons, the reaction of the antibody with a certain antigen, such as ragweed pollen, triggers the release of **histamine** from certain cells, and this results in itching, swelling, fluid discharge, and other symptoms of allergy. Allergies to specific substances may be inherited.

Autoimmune disease occurs when the immune system fails to distinguish foreign (not-self) from domestic (self) material. Recall that two conditions must be met for antibody production. First, the antigen must be exposed directly or indirectly to the antibody-producing cells in the spleen or lymph nodes. Second, the antigen must *not* be exposed to the cells of the immune system when tolerance to self is developing. (Antibodies are not formed against the many proteins circulating in the blood and on the exposed surfaces of cells because the body has "learned" to recognize them as self.) Usually, some of an animal's own tissues meet the second condition. The tissues of many glands and organs are so well packaged that their proteins and other potential antigens are never "seen" by the immune system. As a consequence, the immune system is **intolerant** of these antigens should they be accidentally exposed.

Such accidents do occur. When tissue is damaged by radiation, burning, bruising, or attack by disease organisms, some of these antigens may escape into the circulatory system. The resulting formation of circulating antibodies may start a vicious cycle of further destruction of the injured tissue and the production of yet more antibodies. The result can be disastrous. Among diseases of this type are rheumatoid arthritis, Hashimoto's disease of the thyroid, and encephalitis following measles or mumps. The latter is an especially morbid, rapid destruction of the brain following the release into the circulation of antigens from nervous tissue.

THE NATURE OF THE IMMUNE RESPONSE LEADS TO TWO CLASSES OF DISEASE. ALLERGIES OCCUR WHEN THE ANTIGEN-ANTIBODY REACTION LEADS TO RELEASE OF THE IRRITATING SUBSTANCE HISTAMINE. AUTOIMMUNE DISEASE RESULTS WHEN ANTIBODIES ARE FORMED AGAINST AN INDIVIDUAL'S OWN TISSUE.

TRANSFERRING IMMUNITY Immunity has one great advantage over antibiotic therapy in controlling infection: it is highly specific. Antibodies attack a specific molecule or a specific bacterium or virus, for instance, whereas antibiotics are "shotgun" chemicals that can destroy beneficial microorganisms (such as the intestinal flora) along with the destructive. A big disadvantage, though, in the immune response is the time lag of the primary response, a period that often gives the infection or poisonous agent time to establish a foothold or even cause death before the antibody level is high enough to help.

Our species has a cultural trick that partially overcomes the lag. Since antibodies circulate in the blood, the serum of an immune individual (duck, horse, or another human) can be injected into a person in need of immediate protection. Antiserum against snake venom is merely serum from a horse that has been immunized against the venom of one or more snakes by injecting small amounts of the poison into the horse over a period of time. Protection against infectious hepatitis is conferred by injections of serum from people with antibodies against hepatitis virus. Such transferred protection lasts only as long as the transferred antibodies, however—a few weeks or months at most. There is also the danger of evoking an immune response to the foreign serum proteins of the donor species, that is, a kind of allergy. Some people, for instance, are so sensitive to horse serum that snake antiserum from a horse is more dangerous to them than the snake venom.

ANTIBODY FORMATION IN AN INDIVIDUAL REQUIRES TIME AFTER INTRODUCTION OF AN ANTIGEN, AND DURING THIS TIME DISEASE OR DEATH MAY OCCUR. THIS MAY BE OVERCOME BY INJECTING, IMMEDIATELY AFTER EXPOSURE, ANTIBODIES MADE BY ANOTHER ORGANISM. THE FOREIGN SERUM PROTEINS MAY THEMSELVES BE DANGEROUS ANTIGENS, HOWEVER.

Trauma or damage to the vertebrate body marshals into action several homeostatic systems. (Invertebrates do analogous but different things.) Two stages of homeostasis may be thought of as normally acting following a wound: (1) a first-aid system that stops too dangerous a loss of body fluids, especially blood; (2) a repair system that heals damaged tissue and sometimes replaces lost or destroyed parts of organs.

COAGULATION AND WOUND HEALING

There are three recognized mechanisms which prevent spontaneous bleeding and excessive loss of blood following trauma: (1) Immediately after injury to blood vessels, the vessels contract reflexly for about 30

Coagulation

sec. The smaller the caliber of the blood vessel, the more important the part played by **vasoconstriction,** as it is called. (2) The agents of the second mechanism preventing hemorrhage are **platelets,** small anucleate cytoplasmic fragments that circulate in the blood with the red and white blood cells. Where blood-vessel walls are injured, platelets quickly "drop out" of the circulating blood and adhere to the damaged vessel lining. They change shape and clump, forming plugs, which prevent blood loss through small holes in the vessel walls. (3) Clumped platelets release substances which promote further platelet clumping and substances that participate in the third mechanism of stopping bleeding, **plasma coagulation.**

Normal coagulation is the result of a complicated sequence of inter-actions of factors present in blood and tissues. Our knowledge of the **coagulation factors** has come from careful study of patients with bleed-ing disorders (e.g., hemophilia), who usually lack a single clotting factor. There are now 13 factors known, most of which are enzymes that exist in the body as inactive profactors or proenzymes until clotting is ini-tiated. Very generally, clotting can be described as a "cascade" of enzyme activation. Thus, release of a tissue lipoprotein called thrombo-plastin from injured tissue causes the activation of factor XII. Activated XII causes inactive (circulating) factor XI to change to an active enzyme, and activated XI converts its substrate factor IX to an activated form. This, in turn, activates factor VIII, which then activates factor X. Factor X and factor V in the presence of calcium convert the molecule prothrombin to thrombin.

Clots are formed when thrombin splits off two small protein fragments from a protein called **fibrinogen,** converting it to **fibrin.** Spontaneous hydrogen bonding causes polymerization of fibrin molecules into large nets, which trap red and white blood cells and form plugs in damaged blood vessels as well as sheets of "scab" to cover injured areas of the body surface. One of the small peptides split off from fibrinogen stim-ulates vascular contraction, further reducing the possibility of extensive blood loss at sites of trauma.

The relative importance and strength of the platelet plugs and fibrin clots are illustrated by the plight of patients who are missing one of the coagulation factors (as, for example, in hemophilia). Unusual bleeding in these patients may not appear until several hours or days after injury, since their platelets aggregate normally to seal off damaged vessels. Subsequent fibrin formation does not occur, however, so the platelet plugs are never strengthened and bleeding eventually recurs. The most common types of hemophilia are hereditary and sex-linked, as discussed earlier in Chap. 2. Coagulation defects can also be acquired. Vitamin K, for example, is essential to the synthesis of some coagulation factors in the liver, and vitamin K deficiency can cause excessive bleeding.

COAGULATION OF BLOOD IS A COMPLETE DEFENSE MECHANISM PREVENTING LOSS OF BODY FLUIDS. THROUGH THE ACTION OF EN-ZYMES, FIBRIN IS PRODUCED, WHICH FORMS CLOTS TO BLOCK BLOOD VESSELS. CLOTTING IS STIMULATED BY THE BLOOD PLATE-LETS, WHICH CLUMP AND BLOCK OPENINGS IN VESSELS.

Wound Healing

Regardless of whether tissue injury is caused by living agents such as bacteria or by nonliving agents such as heat, cold, electrical energy, or mechanical trauma, an immediate **inflammatory response** occurs which is almost always the same. The inflammatory response serves to destroy, dilute, or seal off the damaging agents and the dead and dying tissue they have produced. Dilation of blood vessels and increased blood flow are the first events in the inflammatory response. These produce redness and heat, cardinal signs of inflammation, at the site of injury.

Damaged cells release chemical mediators, such as histamine, that cause changes in the permeability of vessels and capillaries, allowing fluid and plasma proteins, including antibodies, to pour out into the injured tissues. The characteristic swelling of wounds is the result. Large numbers of phagocytic cells push their way out of the blood vessels to accumulate in the area of injury. There the phagocytes ingest and destroy bacteria, parasites, foreign matter such as dirt, and accumulations of the host's own dead and dying tissue. Pus is composed of live and dead phagocytic cells and tissue which has been partially digested.

As soon as inflammation begins, repair of tissues is also initiated. The fluid which poured out of blood vessels is drained and reabsorbed. Clotting seals off damaged blood vessels. If damage has occurred to tissues whose cells retain the ability to divide during adult life (e.g., epithelial cells of the skin, certain cells of bone, bone marrow, liver, and kidney), injury may be repaired by simple proliferation of these cells, and the original structure and function can be restored. In tissues made of cells which are unable to divide in the adult animal, repair occurs by proliferation of fibrous tissue and results in scarring. Thus, when the connective tissue of an organ is damaged, scarring occurs, and when nerve or muscle cells are injured, scarring occurs. A heart attack means permanent loss of at least some functioning cardiac muscle cells and their replacement by scar tissue.

As skin wound heals, blood clots are rapidly replaced by actively growing connective tissue rich in new small blood vessels. Connective tissue proliferates, pinching off the many small blood vessels, eventually producing a pale, bloodless scar. Epithelial cells grow out to cover the scar with skin.

Many factors influence the speed of repair. Old animals have a decreased capacity for repair. Poor circulation impairs healing, as does poor nutrition—in particular, lack of vitamin C and protein. Inadequate intake of calcium and phosphorus can impede repair of bones. Further stress on injured tissue can delay repair. This is why broken bones are immobilized for a time and why a person with liver damage from a hepatitis infection is warned against consuming alcohol, which damages liver cells.

THE INFLAMMATORY RESPONSE TO WOUNDING INITIATES A SERIES OF REACTIONS, INCLUDING INCREASED BLOOD FLOW, RELEASE OF ANTIBODIES, AND ACCUMULATION OF PHAGOCYTIC CELLS. CELL PROLIFERATION THEN REPLACES DAMAGED OR MISSING TISSUE.

AUTOTOMY AND REGENERATION

A not uncommon tactic in the biological game of survival is to sacrifice a part to save a whole. The nature of plant growth permits plants to lose leaves and branches without permanent damage to the plant. As long as photosynthesis and nutrient uptake can occur, the meristems will replace lost stems, leaves, and roots. In some animals, however, a much more specific defense mechanism has evolved. This permits an individual to lose a noncritical appendage, which is then regenerated.

Autotomy

Some kinds of starfish shed an arm under the slightest provocation. The legs of crabs and some spiders are voluntarily shed when seized by a predator. A violent muscle contraction accomplishes this at preformed sites, analogous to the perforations on a cereal box. Voluntary self-mutilation of this kind is called **autotomy.** Among vertebrates, lizards go unchallenged at this art. Many species have brightly colored blue or red tails that wiggle frantically when lost by the frightened animal. There is a species of sand lizard, the gridiron-tailed lizard (Fig. 11–3) inhabiting the deserts of the Southwest, that invites the loss of its tail by wiggling it temptingly in the air. The color of this lizard matches precisely the sand or soil on which it occurs. When frightened, it runs a short distance, stops suddenly, and seems to disappear except for the wiggling, striped tail. It is probable that a pursuing predator would strike at the tail and miss the protectively colored body.

Appendages may be lost in combat as well. The limbs of larval salaman-

Fig. 11–3 The gridiron-tailed lizard.
(*San Diego Zoo Photo.*)

ders may be bitten off by members of the same species. The males of lizards engage in violent territorial combat. The fights terminate when one lizard (the winner) pulls off the tail of the other.

THE PROCESS OF AUTOTOMY IS THE VOLUNTARY LOSS OF A NON-CRITICAL APPENDAGE, AND IT IS ACCOMPLISHED BY SPECIAL PHYS-IOLOGICAL MECHANISMS.

Regeneration

Of course, not all loss of parts is the result of autotomy. Animals may involuntarily lose noncritical appendages or even large parts of critical ones and still survive. This is because mechanisms exist to replace lost structures. The replacement of lost organs is called **regeneration.** The capacity to regenerate is found in all major groups of organisms, although

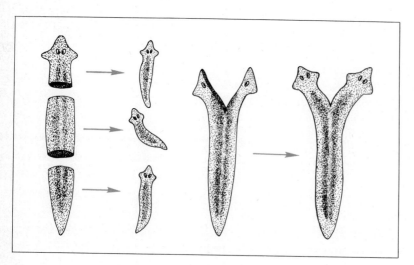

Fig. 11–4 Regeneration in planarian worms.

it tends to be most dramatic in the simpler animals. A small piece cut from the middle of a flatworm will regenerate a miniature flatworm (Fig. 11–4), whereas fishes, mammals, and birds have very limited regenerative abilities.

Regeneration of a foot from the stump of an amputation in a salamander larva is fairly typical of the process (Fig. 11–5). Coagulation of blood on the cut surface is the first thing to happen. Next, epithelial cells of the skin migrate out from the cut edges of the wound and cover the stump. Following the closure of the wound, cells from tissues adjoining the wound dedifferentiate into a homogenous mass of embryoniclike cells called a **blastema,** or regeneration bud. Muscle, bone, and connective tissue all contribute cells to the blastema, though the muscle cells seem to determine the kind of structure the blastema becomes during the later phase of regeneration. Following several generations of mitotic divisions and growth of the blastema cells, a new appendage begins to form in a manner very much like that of normal limb development.

LOSS OF PARTS, EITHER BY AUTOTOMY OR BY NONVOLUNTARY MEANS, MAY BE FOLLOWED BY THE SPECIALIZED PROCESS OF REGENERATION, IN WHICH CERTAIN CELLS DEDIFFERENTIATE AND MULTIPLY.

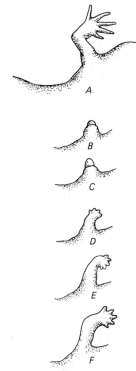

Fig. 11–5 Regeneration of an amputated limb in a salamander. (*Adapted with permission of The Macmillan Company from PRINCIPLES OF DEVELOPMENT AND DIFFERENTIATION by C. H. Waddington. Copyright © 1966 by C. H. Waddington.*)

PROTECTION BY USE OF COLOR AND PATTERN

A quick way to have self turn into not-self is to be devoured by a predator. It is hardly surprising, then, that a few plants and many animals have evolved color patterns which help them to avoid such a demise. After all, once eaten it becomes very difficult to reproduce successfully— and you will recall that reproductive success is the name of the game in evolution. We should not think that the appearance of an animal or plant to us is the same as its appearance to other animals. Most animals have only a limited ability to see colors, which are visualized as tones in these cone-free organisms.

Protective Coloration

Perhaps the simplest way to use color and pattern to avoid being eaten is to use them to produce a resemblance to something inedible. The classic example of this among plants is offered by the so-called "stone plants" of the family Aizoaceae, found in the arid lands of South Africa. Their uncanny resemblance to rocks and pebbles permits them to exist where hungry grazing animals would otherwise destroy them.

Concealing **(cryptic)** coloration in animals takes many forms. In some animals the color pattern is one which simply permits blending in with the inedible (from the predator's point of view) background. The mottled coat of a fawn minimizes its changes of discovery as it lies motionless. The speckled color of one form of the moth *Biston betularia* makes it inconspicuous on a lichen-covered tree trunk; the dark color of the other form produces near-invisibility when it alights on a grimy tree trunk (see Fig. 1–45). The caterpillars of many butterflies are green and very difficult to see on the grass-blades on which they rest. Some animals,

Fig. 11–6 Many fishes, such as the flounders shown here, are able to change color to resemble that of their background. (*Courtesy of R. H. Noailles.*)

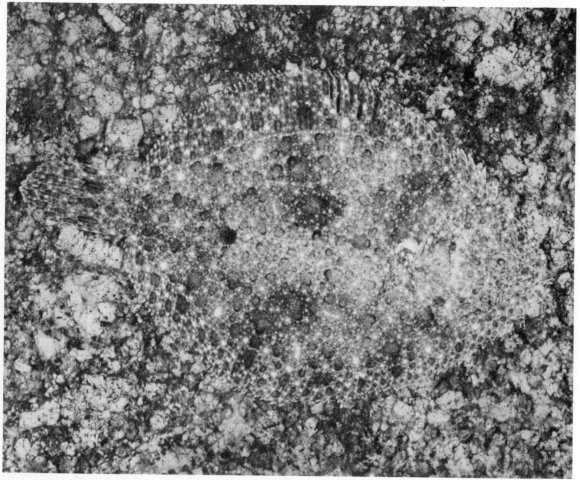

A

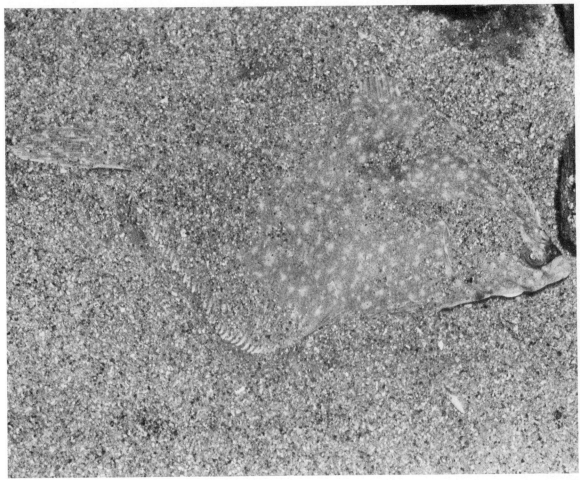

B

such as certain lizards and flounders (Fig. 11–6), are able to modify their color pattern to conform more closely to the backgrounds on which they rest.

The use of color to conceal goes beyond simple matching of the background. Many organisms show **countershading,** which in its simplest and commonest form consists of having the upper part of the body darker than the lower part. When a uniformly colored object is top-lighted, its upper parts are lightened and its lower parts shadowed and darkened. These differences in shading are often critical to the accurate perception of the three-dimensional characteristics of the object. By lightening the underparts and darkening the upper, an organism can compensate for this effect and lose much of its characteristic contours.

Other animals have evolved **disruptive** coloration in order to lose their characteristic contours. Bold patches of color contrast with the background, while others blend with the background. The contrasts distract attention from the form of the organism and often make it nearly impossible to detect—especially if the background contains similar contrasts. One of the commonest uses of disruptive coloration is to conceal the eyes, for eyespots are among the most naturally conspicuous of all patterns. The color pattern around the eye of many animals has evolved into a "mask," which breaks up the outline of the eye and prevents a conspicuous eyespot from spoiling the animal's camouflage.

One of the most effective coloring devices used by animals to escape from predators is **flash-and-dazzle** coloration. This device is employed by many butterflies, especially in the tropics. The upper surfaces of the wings are brightly colored, often metallic. The undersurfaces are cryptically colored, blending with the background on which the insect alights. In flight, the butterfly presents dazzling flashes of color; when it alights, it simply disappears into the background. Anyone who has attempted to follow a butterfly with flash-and-dazzle coloration through a sun-dappled forest will have little doubt about its efficiency—at least as far as human predators are concerned.

Other animals carry concealment one step further: natural selection has resulted in their coming to resemble a specific inedible object. Some plant-eating bugs are shaped and colored like miniature thorns (Fig. 11-7). Certain caterpillars, when disturbed, may become rigid and take on a close resemblance to twigs, as do an entire group of insects known as stick insects (Fig. 11-7). Many different kinds of insects have evolved detailed resemblances to leaves—some reproducing details down to small blemishes, patches of fungus, and so on (Fig. 11-7). The young of one kind of plant bug occurs in both black and white forms. The bugs rest together arranged in a pattern which resembles a bird-dropping. Indeed, bird-droppings are a favorite object to mimic; some small frogs do this, as well as many insects and some spiders.

Of course, not all concealing color and form were evolved to protect prey. In many cases it is the predator that is made inconspicuous as it lies in wait for its prey. Some of the most striking examples are spiders and predacious bugs which blend in with the flowers on which they lie in wait for their insect victims (Fig. 11-8) and the coat patterns of

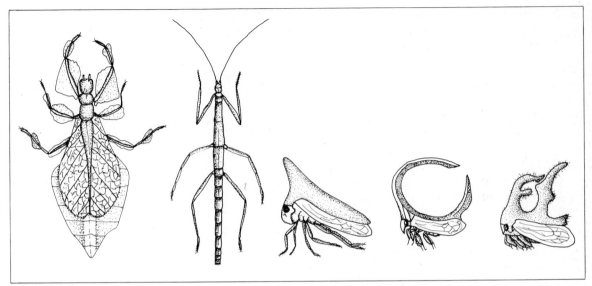

Fig. 11–7 Many insects mimic plant parts, such as leaves, *left;* twigs, *center;* or thorns, *right.*

big cats, which make them less conspicuous when they stalk their prey. The plain tan coat of the lion makes it hard to see on the open, dusty plains it inhabits; the striping of the tiger breaks up its outline in the tall grass where it hunts.

PROTECTIVE COLORATION MAY BE ACCOMPLISHED BY BLENDING WITH THE ENVIRONMENT, USING COLORS AND PATTERNS TO CHANGE THE APPARENT CONTOURS OF THE BODY, OR ADOPTING FLASH-AND-DAZZLE PATTERNS.

Mimicry

Some harmless organisms have evolved a resemblance to harmful organisms, thus sharing, at least in part, the protection from predators enjoyed by the harmful organism. This is known as **mimicry** (also discussed in Chap. 14). It is a very widespread biological phenomenon. For instance, many (perhaps most) species of tropical butterflies are involved in mimicry in one way or another. Some of the most bizarre examples of mimicry involve the resemblance of certain hawkmoth larvae to poisonous snakes and bugs whose backs are formed into "sculptures" which look like predacious ants poised on a leaf. One of the most interesting forms of mimicry is found in insects which have evolved large eyespots accurately portraying vertebrate eyes. These are often on the upper side

Fig. 11-8 A crab spider lying in wait
in a thistle flower has captured a
bumblebee. (*Courtesy of Ross
Hutchins.*)

of the headwings and are normally covered when the insect is at rest.
When the insect is disturbed, however, it suddenly spreads its forewings,
exposing the eyespots (Fig. 11-9). In a series of elegant experiments,
A. D. Blest has shown that naive birds are frightened by this display.
Some butterflies have developed smaller eyespots on the wings. These
may serve to direct the attacks of predators to relatively unimportant parts
of the insect. In many of the hairstreak butterflies, the eyespot is at the

rear of the hindwings and is associated with threadlike "tails" which mimic antennae.

Aggressive mimicry occurs also, in which one organism resembles another in order to be able better to attack its victims. In perhaps the most interesting case, a fish *Aspidontus* closely mimics the cleaner fishes of the Indian and Pacific Oceans. The cleaner fish does a characteristic dance which attracts other fishes. These fishes permit the cleaner fish to approach and search them for parasites, which the cleaner fish eats. The fish being cleaned usually enters a semitrance while it is scoured, opening its mouth and spreading its gill covers so that the cleaner can enter those cavities and do its job (Fig. 11–10).

At stations on the reefs occupied by the cleaner fishes, one can usually find the *Aspidontus,* which belongs to an entirely different family of fishes from that of the cleaner. It bears close resemblance to the cleaner, however, both in appearance and behavior. The "customers" of the cleaning station are fooled and take up their characteristic positions, permitting the *Aspidontus* to approach. The *Aspidontus* then calmly takes a bite out of its victim's fin and calmly eats it. The surprised reaction of the

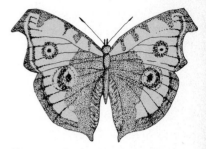

Fig. 11–9 Eye spots on the wings of a butterfly.

Fig. 11–10 A cleaner fish picking parasites off a butterfly fish. (T.F.H.)

victim does not cause *Aspidontus* to lose its cool—it remains at the victim's side, the picture of innocence!

An interesting variety of aggressive mimicry is found in some fishes and reptiles. For example, the angler fish has a special appendage which dangles in front of its mouth. This serves as a lure for other fish, which interpret it as potential food, only to find themselves as food of the angler fish (see Fig. 6–10). Snapping turtles of several kinds use their tongue as a lure. Usually very cryptically colored, they lie in wait with their mouth open, wriggling the tongue, which, to the human observer, resembles a worm. Unsuspecting fishes are lured to their death.

PROTECTION BY USE OF COLOR AND PATTERN OCCURS IN MIMICRY. A PALATABLE SPECIES MAY EVOLVE TO CLOSELY RESEMBLE A NONPALATABLE ONE. ANIMALS MAY DEVELOP PATTERNS AND BEHAVIOR WHICH SIMULATE A DANGEROUS ANIMAL. AGGRESSIVE MIMICRY PERMITS A PREDATOR TO LURE A PREY CLOSE ENOUGH TO BE CAUGHT.

Color, of course, is not the only tool organisms use in their struggle to prevent self from becoming not-self. Some plants and animals, for instance, armor themselves in various ways to discourage those who would devour them. This armor is familiar to all: the spines of cacti, the thorns of roses, the thick bark of trees, the shells of snails and turtles, and the heavy fur of mammals. The weapons used to pierce these armors are well known also: sharp teeth in powerful jaws, piercing mouthparts, and razorlike claws.

MECHANICAL AND CHEMICAL DEFENSES
Mechanical Defenses

Less obvious to the eye, but no less interesting, are the chemical defenses used by both plants and animals. Virtually all plants have evolved compounds whose primary function is to serve as herbivore poisons. Some of these are familiar to us because we employ them for the same purpose. The pyrethrins, which are the active ingredient of "safe" home insecticide bombs, are extracted from daisylike plants of the family Compositae. Nicotine extracted from tobacco is used as an insecticide to protect crop and ornamental plants. Other plant defensive compounds, such as quinine and belladonna, are used as medicines, and many others are familiar as spices. And, of course, some plants (e.g., peyote and marijuana) have evolved compounds which may modify the behavior of herbivores rather than kill them directly. Animals with altered perception are undoubtedly easy prey for predators.

Chemical Defenses

The best-known group of plant defensive chemicals is a heterogeneous collection of compounds known as **alkaloids.** An alkaloid is a plant-derived compound which contains nitrogen. Besides nicotine and quinine, already mentioned, the alkaloids include caffeine, strychnine, cocaine, piperine (the active ingredient in pepper), opium, and mescaline (the active ingredient in peyote). All these presumably disrupt the physiology of various herbivores, but in each case other herbivores have learned to deal with the poisons and feed successfully on the plants. Evolving means of breaking plant defenses can confer numerous advantages on the successful herbivores (as discussed in Chapter 14).

There is growing evidence that some plants have gone even further than a simple poisoning of their insect enemies. A number of plants, such as the American balsam fir, have in their tissues substances similar to insect juvenile hormone. These plants may be able to interfere with the development of some insects which feed on them, perhaps delaying their metamorphosis until cold weather or predators catch up with them. One should not assume, however, that using the insects' own hormones against them represents an unbreakable defense.

Insects undoubtedly can evolve barriers to the uptake of the hormone substitute, ways of differentiating their hormone from the plant substitute (perhaps by modifying their own) or of modifying the concentrations required to produce the proper hormonal balance. Nonetheless, hormonal defense is an attractive strategy for a plant. Not only does it present serious problems to the insect, but it probably represents less of a threat to the plant itself; many of the other poisonous compounds produced by plants are so toxic they cannot be stored in delicate tissues.

Animals also may employ chemical weapons. The biochemical defenses of insects and other arthropods offer a multitude of examples, brilliantly elucidated by Thomas Eisner and his collegues. Because of repellent substances in the blood, many insects taste bad to predators. For example, any insect that is brightly colored in contrasting orange and black or red and black either tastes very bad to predators or is a mimic of a distasteful species. The next logical evolutionary step is to ooze or secrete a repellent liquid when being probed by a predator, so that one does not have to be eaten for the predator to get the message. Many millipedes and insects, including grasshoppers and butterfly larvae, do this. The larvae of one group of beetles accumulate their molted skins and feces, constructing with them a "fecal shield." They can successfully defend themselves against ants by maneuvering the shield. The ant quickly loses interest after biting the distasteful umbrella.

A common defensive refinement is the ability to direct a spray in the direction of the disturbance. Whip scorpions and some beetles effectively aim their sprays. The bombardier beetle *(Brachinus)* has evolved a system which permits it to spray quinones on its enemies at the temperature of boiling water.

Defensive secretions include a wide variety of noxious and highly reactive chemicals, including cyanide (HCN), acids, quinones, and terpenes. A number of devices are employed to protect the animal from its own poisons. Many store and possibly synthesize the chemicals in chitin-lined tubules and "reaction chambers," the inert walls of which act as a barrier between the noxious chemicals and the living tissues.

The use of chemical weapons by animals is not, of course, limited to arthropods. Molluscs, such as squid and octopuses, use "ink" as smoke screens to cover their escape; this secretion also dulls the olfactory senses of potential predators. The chemical defenses of wasps, bees, and skunks are well known, especially to the unwary. Some fishes, such as the lion-fish and stonefish, have poisonous spines which can maim or kill an unlucky attacker. The male duckbill platypus possesses a poison gland which evacuates through a front claw. Of course, chemical weapons are not restricted to defense alone. Many animals, from jellyfish to spiders and rattlesnakes, use venoms to help them capture prey.

PLANTS AND ANIMALS HAVE EVOLVED A WIDE VARIETY OF MECHANICAL DEFENSES. BOTH PLANTS AND ANIMALS ALSO PRODUCE CHEMICALS WHICH REPEL ATTACKERS AND WHICH MAY EVEN KILL OR SERIOUSLY INJURE THEM.

BEHAVIOR IN ATTACK AND DEFENSE

An entire book could easily be written on the techniques by which predators pursue their prey. Many large predators, such as lions, often depend on a long, stealthy stalk followed by a quick charge to catch their prey. Others, such as cheetahs, depend on high-speed chases. Still others, especially wild canines, often utilize teamwork, as they hunt in pairs or packs. Large herbivores, such as antelopes, on the other hand, often live in herds and depend largely on group alertness and flight for survival, with smell being the primary perceptual system used for detecting danger.

Some spiders hunt by leaping on their prey, while others use nets to strain the air for victims. The prey of the former must depend on alertness for survival. While most insects caught in spider webs perish, some moths are able to escape. They are clothed with loose scales which may be shed and left sticking to the web while the moth escapes. However, some insects live in spider webs, and others actually prey upon spiders.

Moths, together with other nocturnal insects, face an unusual peril. They are hunted by bats (Fig. 11–11), which detect and attack them using an ultrasonic "sonar" system. The bat produces high-pitched beeps (inaudible to the human ear) when it is hunting, emitting 10 to 30 beeps per second in its search mode. When a returning echo indicates the presence of prey, the animal switches into attack mode, emitting more than 100 beeps per second. Donald Griffin, who has done much of the work on bat navigation and hunting by sonar, has produced thrilling slow-motion movies of the acrobatics of hunting bats, accompanied by the sonar sounds played at a frequency audible to the human ear. The bat is able to put together a detailed perceptual world on the basis of its sonar returns. The entire pulse-generator, return reception and analysis apparatus is crammed into the bat's small head, one of nature's outstanding feats of microminiaturization.

Many moths have evolved a variety of defenses. They often have a thick clothing of body hairs, which serves to dampen the search pulses, func-

Fig. 11–11 A bat pursuing a moth. (*Courtesy of Ross Hutchins.*)

tioning as a sort of acoustic camouflage. Some of them have ears at the base of the abdomen which serve as receptors for the ultrasonic pulses. On hearing a searching bat, the moth instantly takes evasive measures, stops flying, and drops to the ground. The moth's ears are often infested with colonies of mites, which destroy their ability to detect the pulses. But the mites' behavior is such that only one of two ears on any moth is infested. Can you explain how selection has led to this behavior in the mites? Some moths produce sounds which "jam" the bat's sonar. Finally, there is some evidence that distasteful moths announce their unpalatability to bats by emitting their own characteristic ultrasonic pulses.

Bats, by the way, do not live on insects alone. Many species live on fruit or nectar and are the principal pollinators of many tropical plants. Vampire bats attack sleeping birds and mammals, slashing them with razor-sharp teeth and lapping up the blood. The bite is painless, and the victim is not disturbed. Indeed, birds may die undisturbed while being exsanguinated in their sleep. Finally, some bats have hawklike talons with which they catch fish, detecting the fish ultrasonically when they produce surface disturbances on the water.

Another unusual orienting, hunting, and defense system is that employed by electric fishes. Some fishes, such as electric eels and catfishes, can generate powerful, high-voltage shocks to immobilize prey or blast attackers. But other fish produce low-voltage fields, which may serve both for orientation and for communication in the muddy waters they inhabit. These fish, whose behavior has been studied in detail by H. W. Lissman, have organs which are extraordinarily sensitive to changes in the electrical potential of the water in which they swim. They can detect minor changes in the electric field caused by prey or other objects in the water.

SUMMARY

Organisms have evolved many means of defense at every level of biological organization. At the cellular level there are several nonspecific defenses directed at microorganisms. The protein interferon is an antiviral agent produced by organisms. The substance lysozyme and many skin secretions affect bacteria and protect the skin and mucous membranes.

The immune response of vertebrates, including man, is a much more specific defense against foreign substances, including bacteria and their toxins. The immune response occurs in response to attack from a pathogen. It may also occur following vaccination with attenuated or weaker strains of a pathogen or with inactivated microorganisms. Foreign material, called an antigen, may elicit any of several serological reactions which protect an animal. The agents responsible for these reactions are called antibodies.

Antibodies are proteins called globulins, which are found in the blood. They have a complex structure, which accounts for their activity and specificity. Antibodies react only with the antigen which led to their formation. This specificity lies in the sequence of amino acids at the two active binding sites on the molecule. Newly born mammals are relatively disease-free, having acquired immunity from the mother in the form of antibodies in colostrum or from the maternal bloodstream via the placenta.

The immune response consists of a primary response to the initial exposure to an antigen and a secondary response. In the secondary response, antibody production is more rapid and also greater in amount and duration. Secondary response to exposure after the initial one may be evocable for years or a lifetime. Special cells in the lymph nodes, spleen, and thymus gland play important roles in the recognition of antigens and the formation of the correct antibody from a DNA template.

Coagulation of blood is a complex defense mechanism preventing loss of body fluids. Through the action of enzymes, fibrin is produced, and this forms clots which block damaged blood vessels. Clotting is stimulated by the blood platelets, which clump and also block openings in the vessels. Wounding also results in an inflammatory response. This leads to a series of reactions, including increased blood flow, release of antibodies, and accumulation of phagocytic cells. Proliferation of undamaged cells then replaces damaged or missing tissue.

Organisms may lose vital or noncritical parts in the course of their activities, and many species are able to regenerate these. Regeneration is a rather complex process in which certain differentiated cells lose their specialized nature and resume division. Some animals are able voluntarily to lose a noncritical appendage when attacked. This process of autotomy requires special physiological mechanisms to prevent blood loss, etc.

Protection is achieved in both plants and animals by mechanical and chemical means or by protective coloration. The latter may be achieved by blending with the environment, by using colors and patterns that change the apparent contours of the body, or by adopting flash-and-dazzle patterns. Mimicry is also protective coloration, in which a palatable species comes to closely resemble a nonpalatable one in warning colors and patterns. Aggressive mimicry permits a predator to lure a prey close enough to be caught.

In addition to such mechanical defenses as armor and spines, both plants and animals produce chemicals which repel attackers and which may even seriously injure or kill them.

Cott, H. B.: "Adaptive Coloration in Animals," Oxford University Press, 1940.

Marler, P. R., and W. J. Hamilton: "Mechanism of Animal Behavior," Wiley, 1966.

McGaugh, J. L., N. M. Beinberger, and R. E. Whalen (eds.): "Psychobiology: The Biological Bases of Behavior; Readings from Scientific American," Freeman, 1966.

Portmann, A.: "Animal Camouflage," University of Michigan Press, 1959.

Abramoff, P., and M. F. LaVia: "Biology of the Immune Response," McGraw-Hill, 1970.

Brower, L. P.: Ecological Chemistry, *Sci. Amer.,* **220,** no. 2, Offprint 1133 (1969).

Edelman, G. N.: The Structure and Function of Antibodies, *Sci. Amer.,* **223,** no. 2, Offprint 1185 (1970).

Hilleman, M. R., and A. A. Tytell: The Induction of Interferon, *Sci. Amer.,* **225,** no. 1, Offprint 1226 (1971).

SUPPLEMENTARY READINGS
Books

Articles

4

THE BIOLOGY OF POPULATIONS

In the last two parts of this book, our central focus has been on the individual—how it functions, the strategies which help it survive, and the way it develops through time and passes on genetic information to its descendants. In this section, the focus will change and we will consider the properties of groups of individuals.

In Chap. 12, the properties of populations of organisms belonging to the same species are examined. The evolutionary processes by which populations change their genetic constitution through time and by which single-species populations are split into populations representing two or more species are dealt with in Chap. 13.

Chapter 14 looks at more complex populations, that is, those containing individuals belonging to many species—assemblages known as communities. Chapter 15 deals with the general properties of societies, and Chap. 16 examines the ultimate social animal, man.

(Courtesy of J. Paul Kirouac)

CHAPTER 12

THE ORGANIZATION AND DYNAMICS OF POPULATIONS

In preceding chapters, many kinds of biological organization have been discussed. You have seen how the linear organization of nucleotide units in DNA and RNA molecules functions in the storage, transmission, and translation of biological information. The delicate and intricate organization of cells has been explored, as has been the much more gross organization of a cow's digestive tract. In this chapter, we will begin to look at the kinds of organization which are found in groups of organisms. If you have ever watched a flock of geese heading south for the winter, or the schooling of fishes in an aquarium, or a football game, you have observed some obvious examples of organization in groups of organisms. The individuals in the groups govern their activities at least in part in response to the activities of the other members of the group. There are, as you will seé, many other kinds of organization in groups of organisms, some of which, rather than being obvious, are extremely subtle.

Groups of organisms are commonly called **populations,** a general term which originally referred to the group of people living in a specified place (from the Latin *populus,* "the people"). The word "population" is used broadly in science and mathematics to denote any group of things or creatures under consideration. In biology there are many different kinds of populations, and the kind under consideration must always be specified or made clear from the context in which it is mentioned. All the organisms in an acre of woodland—birds, trees, mushrooms, soil-dwelling insects and bacteria, wildcats, the whole lot—make up a population. All the nesting birds in the woodland are also a population, as are all the nesting robins and, indeed, all the male robins. Furthermore, all the robins in North America constitute another population. All these populations clearly have different biological characteristics. The first two, all the organisms and all the nesting birds in the woodland, are made up of individuals of more than one kind of **species** of organism. They are both multispecies populations, that is, **communities.** The other populations are all made up of a single species, the robin *Turdus migratorius,* and it is this sort of assemblage of organisms of the same kind to which the term "population" is often restricted by biologists. This chapter will concern itself with this latter kind of population.

BIOLOGISTS COMMONLY DIFFERENTIATE TWO KINDS OF POPULA-TIONS: POPULATIONS IN THE STRICT SENSE ARE COMPOSED OF INDIVIDUALS OF A SINGLE SPECIES, AND COMMUNITIES ARE COMPOSED OF INDIVIDUALS OF TWO OR MORE SPECIES.

POPULATION STRUCTURE

Technically, the structure of a population of sexually reproducing organisms has been defined as the resultant of all the factors which govern the pattern in which gametes from various individuals unite with one another. Population density, that is, the number of individuals per unit area, is one element in the structure of a population. For instance, in organisms such as sea urchins, which release their gametes into the surrounding water, any gamete may have an essentially equal chance of uniting with a gamete from any other individual if the population is dense. On the other hand, if the population is sparse, each individual's gametes may have a much higher probability of uniting with gametes from its nearer neighbors than with those from more distant individuals.

Density and Dispersion

Figure 12–1 shows the distribution of individuals in a colony of a checkerspot butterfly in Stanford University's Jasper Ridge Biological Experimental Area. A long series of detailed studies, in which the butterflies were captured, given individual marks, released, and later recaptured, revealed a great deal about movements of individuals in the colony. This provided a good start on understanding the colony's structure. Very few individuals were ever found to transfer among the areas marked C, G, and H, but within these areas they moved about freely. Thus it could be inferred that the chances of a male from area C mating with a female from area H were very small, but his chances of mating with a female from area C were very high. An important question about the population structure remained to be answered, however: What were the chances that those few individuals which were observed to move from one area to another would be able to reproduce successfully in the new area? This question has important implications if we are to understand the evolutionary changes occurring in the colony—it bears on whether it is reasonable to regard the entire colony as an evolutionary unit or whether the populations in the three areas were discrete entities.

BUTTERFLY POPULATIONS   Part of the answer to this question came through a study of the reproductive biology of the butterfly. Copulation

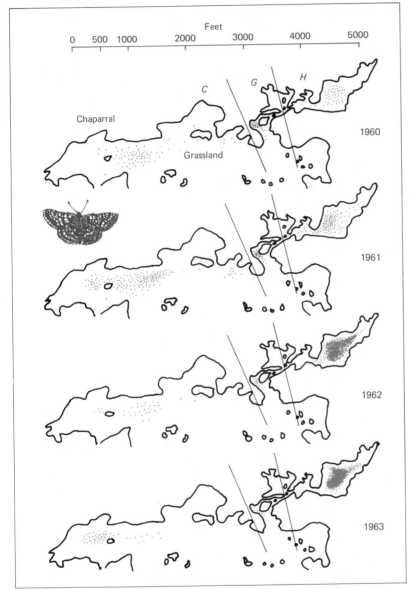

Fig. 12-1 Distribution of adult
Euphydryas editha in years 1960 to
1963. Dashed lines indicate the
borders of areas C, G, and H. Solid
line shows the edge of chaparral and
oak woodland which surround the
grassland in which the butterfly
occurs. Each dot represents the
position of first capture of a butterfly.

in butterflies is protracted, for the male secretes a complex sperm-
bearing structure into the sperm storage sac of the female. In the checkerspot butterfly, when the male has finished depositing this structure, he
blocks the female genital passage with a plug, topped off with a little cap
over the opening. Usually the plug prevents the female from copulating
again, and all her eggs are fertilized by sperm from the same male.
With a practiced eye, one can tell nonvirgin female checkerspots without

a microscopic examination by looking for the tiny cap on the plug; visual surveys as well as a series of dissections have shown virgin females to be a rarity in nature. The reason is clear. The male butterflies normally emerge from the pupa slightly before the females and patrol the colony area tirelessly. As soon as a female emerges, she is normally mated with immediately, often before she has had a chance to fully expand and dry her wings.

Now what can this tell us about the chances of a wandering male having reproductive success away from home? Most individuals that moved from one area to another in the colony did not do so at the very start of the season but rather moved later, when the supply of emerging females was not as large. Therefore, the chance of a successful mating by a wandering male (indeed, the chance of *any* male successfully mating at that time) was presumably much reduced. Furthermore, there is evidence that females who lay their eggs later in the season have less chance of their offspring surviving than do females who lay earlier. The food plant necessary for the larvae is available for only a short time. This means that wandering females who lay their eggs in other areas are less likely to have those immigrant offspring survive. It is also another factor in lessening the reproductive success of migrant males: if they successfully mate, their sperm will fertilize eggs laid late and thus subject to higher mortality.

You can see that study of the reproductive biology of the checkerspot butterfly provides an answer to the question of how successful rare wandering individuals are in carrying genes from one area to another. Because they wander from their own areas late in the season, by the time females reach a new area most of the females there will have mated and been plugged. If they do mate, the resulting late eggs are less likely to produce a successful brood. Immigrant females also are relatively unlikely to transfer genes from population to population because of the small chance of survival of larvae hatching from late eggs.

BABOON TROOPS   Other kinds of animals have different reproductive strategies, and their population structure may be quite different. For comparison, let us turn to primates in Africa. The structure of some baboon populations is rather well known. In Nairobi Park, Kenya, their population density was reported by I. DeVore and S. L. Washburn to be about 10 individuals per square mile. The individuals are not, however, either dispersed randomly or spread evenly over the park (Fig. 12-2). They are organized into highly structured troops. Troops ranging in size from 12 to 87 individuals have been recorded in Nairobi Park, and other troops

Fig. 12-2 Types of dispersion of organisms. Quadrat RA shows individuals randomly dispersed; CL shows clumping; RE shows repulsion. Frame of reference is very important in discussing dispersion. Note that, if the entire quadrangle ABCD is considered, the distribution of individuals is clumped (20 each in CL, RA, and RE, but none in N_1, N_2, or N_3).

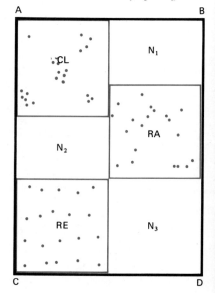

Fig. 12-3 A group of baboons and other animals at a waterhole in the Amboseli Reserve. (*Courtesy of Irven DeVore.*)

of 13 to 185 individuals have been observed in Amboseli Park, 100 miles south of Nairobi (Fig. 12–3). The area normally covered by a troop in the course of its activities (the home range of the troop) is located relative to a "core area." This core area contains a concentration of the requisites for leading a baboon's life: food, water, refuges, and trees for sleeping in. The troops spend most of their time in these core areas, and their presence in other parts of the home range may be very infrequent. Figure 12–4 shows the home ranges and core areas for the baboon troops in Nairobi Park in 1959. In spite of the wide overlaps of home ranges, baboons of different troops were quite isolated from one another by the separation of the core areas. In Amboseli Park, in contrast, members of different troops frequently came together around water holes in the dry season. In neither case was the integrity of the troops compromised—few baboons were observed to change troops, and troops reacted with suspicion and

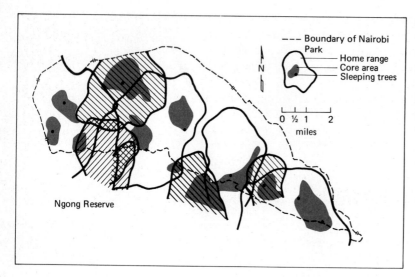

Fig. 12-4 The home ranges and core areas of baboon troops in Nairobi Park. (*Adapted from F. Clark Howell and Francois Bourliere, AFRICAN ECOLOGY AND HUMAN EVOLUTION (Chicago: Aldine Publishing Company, 1963); copyright © by Wenner-Gren Foundation for Anthropological Research, Inc. Adapted by permission of the authors (Irven and Nancy DeVore and Aldine Atherton, Inc.)*

hostility toward strange baboons. The probability of a male in one troop mating with a female in another was obviously very low. This means that there was a high degree of mating among close relatives, since for generation after generation members of each troop mated primarily with each other.

INBREEDING The continued mating together of relatives is known technically as ***inbreeding***—the intensity of the inbreeding being a function of the closeness of the relatives. Brother-sister mating is more intense inbreeding than second-cousin mating. Do you suppose that baboons practice only one specified system, such as continued mating of first cousins? Even human beings do not control pair formation in their populations that closely, although very often some kinds of matings are highly specified. In certain ancient royal lines, brother-sister mating was practiced, presumably to keep the power in the family. And even in recent times, there has been a great deal of inbreeding in royal lines because of the limited supply of "appropriate" eligible mates (Fig. 12-5).

IN ANIMALS, BEHAVIOR OF INDIVIDUALS AND THE WAYS IN WHICH POPULATIONS ARE ORGANIZED EXERT CONTROLS ON THE BREEDING SYSTEM. THEY AFFECT, FOR EXAMPLE, THE AMOUNT OF INBREEDING, THAT IS, CONTINUED MATING OF RELATIVES.

In most human populations, however, formal control of the breeding system (that is, the degree of relatedness among mates) is limited to a series of

Mating Systems

taboos against close inbreeding. Sometimes custom or law prescribes a series of preferred marriages (such as to the wife's sister) if the original partner dies. All known human societies have taboos against incest, that is, against sexual relations or marriage between members of the immediate family (variously defined in different cultures). Outbreeding appears to have been promoted by early man as a method of binding together family groups in a growing tribe, enabling the tribe to grow without fragmenting. With the advantage of numbers, such tribes could subdue or wipe out smaller, inbreeding groups. As anthropologist E. B. Tyler said 80 years ago, "savage tribes must have had plainly before their minds the simple practical alternative between marrying-out and being killed out."

ASSORTATIVE MATING Very widespread in man, in addition, is pairing on the basis of resemblance, or **assortative mating.** Assortative mating is not usually controlled formally by the society. Mating systems are based on phenotypes, not on relatedness. A man marrying his first cousin is contributing to the level of **inbreeding** in the population. A tall man marrying a tall woman is contributing to a pattern of **positive assortative mating.** In human populations, there is often positive assortative mating with respect to height, intelligence, and deaf-mutism. This means, for instance, that the proportion of marriages between two tall people and between two short people is higher than one would expect on the basis of chance alone. In general, both relatives and individuals with similar phenotypes are more alike genetically than nonrelatives and dissimilar individuals. Therefore, both inbreeding and positive assortative mating tend to increase the proportion of homozygous individuals in a population. This, for instance, may lead to the appearance of undesirable phenotypes as the frequency of double-recessive genotypes rises. Of course, it may also lead to the production of desirable phenotypes if these are associated with homozygosity.

MATING ON THE BASIS OF RESEMBLANCE IS CALLED ASSORTATIVE MATING, AND ALTHOUGH NOT FORMALLY CONTROLLED IN HUMAN SOCIETY, IT IS VERY WIDESPREAD IN MAN.

Fig. 12–5 Many members of the Hapsburg dynasty of Europe had a narrow, undershot lower jaw and protruding underlip as shown in this portrait of Charles V (1500-1558). (*The Bettman Archive.*)

Population Structure in Man

Homo sapiens, of course, has populations which have structure in the same sense that butterfly or baboon populations have structure. Indeed, our ancestors of, say, 5 million years ago probably had a pattern of spatial distribution not unlike that described for the baboons. Our forebears doubtless made their livings by hunting and gathering food in a way rather similar to that of baboons, although doubtless with more emphasis by that time on the hunting. It was not until man first invented agriculture, perhaps 10,000 years ago, that his spatial distribution began to become more tightly clumped. The amount of food which could be obtained from a given area increased dramatically with agriculture, sharply reducing the need to be nomadic while simultaneously making a continued presence in one place, to tend and protect crops, a requirement. Man was on the road to becoming an urban beast—a road he still treads. Division of labor became possible when agriculture reached the point where one farmer could grow enough food to supply more than his own family. Other men were then free to pursue different occupations, including the fashioning of transport devices and systems. These, in turn, permitted cities to grow beyond the population size which could be fed on produce of the immediate hinterland, and concentrations of human population became even more dense.

More than 70 percent of the population of the United States today lives in the cities. It was not always thus. In 1800 only some 6 percent of the American population lived in urban areas. In 1850 that had increased to 15 percent, and to 40 percent in 1900—and the latter percentage has now almost doubled. Population densities in some cities now exceed 10,000 per square mile, with Manhattan holding the current record at some 75,000 per square mile. Japan, a crowded country, has an overall density of 700 per square mile, and the United States, a relatively uncrowded country, has 50. The land surface of the entire planet has about 70 per square mile. Contrast that figure with the estimated density of 0.04 per square mile for our ancestors just before they invented agriculture.

Remember that this trend in human population structure toward greater and greater concentrations has not yet reached its zenith. If urbanization were to continue into the future at its present rate, the results would be startling, to say the least. If the urban growth rate which has prevailed since 1950 should continue to 1984, half of the people in the world would become city dwellers. If it should go on until 2023, *everyone* in the world would live in an urban area. By 2044, *all* cities would have a population of at least 1 million, and the largest city would have a projected population of 1.4 billion people (40 percent of the population of the world today!).

As the pattern of human density has changed, many other factors relating to the population structure of *Homo sapiens* have also changed. We will consider the spectacular increase in human numbers in the next section, but note here that it has resulted usually in increased population densities across the face of the earth. In spite of mass movements into cities, population densities in the countryside are in general far above those which existed before the move to the cities began. Human mobility has also increased spectacularly, so that devices such as automobiles and jet aircraft have complicated the problem of determining the pattern of gamete union in the population. Mankind has, however, remained relatively parochial in its mating patterns. In general, one's spouse is still chosen from close neighbors, usually from one's own tribe, clan, caste, religion, or social group. Jet-setters are a very small fraction of the human population and are themselves a rather clannish group. Undoubtedly the greatest factor changing the structure of the human population has been urbanization, for besides bringing diverse peoples into close proximity it also tends to break down ancient cultural patterns and encourages crossbreeding among previously inbreeding groups.

THE DRAMATIC CHANGE IN THE SIZE OF THE HUMAN POPULATION, TOGETHER WITH CHANGES IN MOBILITY AND DENSITY, HAS GREATLY AFFECTED THE POPULATION OF HOMO SAPIENS.

We have discussed population structure in butterflies, baboons, and man to give you an impression of the many factors affecting the pattern in which gametes come together. You undoubtedly noticed that the word "population" was used quite loosely, its meaning made clear primarily by the context. For instance, we spoke of the Jasper Ridge *colony* of butterflies and the three populations in areas C, G, and H. It would have been equally proper to speak of the Jasper Ridge *population* as consisting of three *subpopulations*. For convenience, we speak of the population of baboons of Kenya as being divided into a series of *troops*. If we concentrate on a particular troop, it becomes, of course, the *population* under study. What we call a population is merely a convenient or appropriate group of individuals to be studied.

"POPULATION STRUCTURE" IS A CATCH-ALL TERM FOR THOSE FEATURES OF A POPULATION THAT CONTROL THE PATTERN IN WHICH GAMETES FROM THE INDIVIDUALS IN THE POPULATION UNITE. THE STRUCTURE OF A POPULATION INCLUDES SUCH THINGS AS THE DENSITY OF INDIVIDUALS, THEIR SPATIAL ARRANGEMENT, AND THE SYSTEM OF BREEDING OR MATING.

POPULATION DYNAMICS

Virtually all biological populations are constantly changing in size. In general, the number of individuals in one generation will not be identical with the number of individuals in the preceding or succeeding generations. If we are investigating a closed population, that is, one with no individuals moving in (immigration) and no individuals moving out (emigration), then it will remain constant in size only so long as births and deaths exactly balance one another.

THE POPULATION OF MAN   Let's take a look at a closed population
—the population of *Homo sapiens* on the planet earth (Fig. 12–6). We
know much more about the dynamics of this population than we do about
the dynamics of most other populations, but the general approach to its
investigation is not unique. If one wants to investigate changes in the
size of a population, the first step (after defining and delimiting the
population) is to count the numbers of individuals at different points
in time, that is, to take censuses. This is almost always more easily
said than done. Even when the subject is modern man, there are serious
problems in getting accurate census data. The 1960 United States Census
is thought by some to have missed several million people, and population estimates for many underdeveloped countries may be off 10 percent or more. The United Nations mid-1968 estimate for the population
size of mainland China was 728 million people, but other estimates range

Changes in Population Size

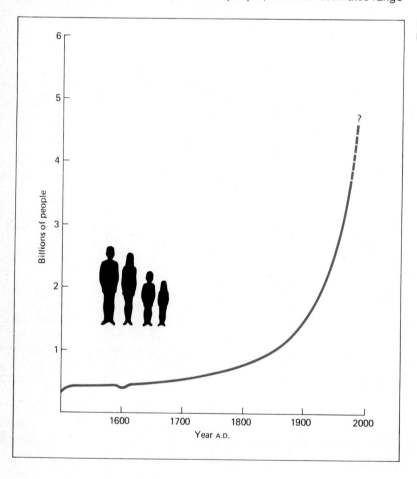

Fig. 12–6 Growth of the human
population.

from 700 to 950 million. But difficult as it is to estimate human population sizes, it is simple compared with the problems encountered in censusing populations of deep-sea fish, butterflies, small mammals, tropical rain forest trees, and so forth. A great deal of the time of those who study human population dynamics (demographers) and those of other organisms (population ecologists) is spent devising sampling methods and statistical techniques for determining what changes in numbers have taken place. Only when the "what?" has been determined is it feasible to go on to the more interesting "why?"

CONSTANT CHANGE IN SIZE IS A FEATURE OF VIRTUALLY ALL BIOLOGICAL POPULATIONS. SIZE STABILITY WILL OCCUR ONLY WHEN BIRTHS BALANCE DEATHS IN A CLOSED POPULATION OR WHEN BIRTHS PLUS IMMIGRANTS BALANCE DEATHS PLUS EMIGRANTS IN AN OPEN POPULATION. THE FIRST AND OFTEN MOST DIFFICULT TASK OF THOSE STUDYING POPULATION DYNAMICS IS THAT OF CENSUSING—DETERMINING THE SIZE OF THE POPULATION AT VARIOUS POINTS IN TIME.

Assuming that the first "man" lived 600,000 years ago, about 77 billion *Homo sapiens* have since been born. If we date the start of mankind at 1½ million years ago, then roughly 96 billion men have lived. Since the present population of the planet is some 3.8 billion people, roughly 4 to 5 percent of all the people who have ever lived are alive today. Of course, these figures are all based on educated guesses since we have no census data worth talking about until the middle of the seventeenth century. But various ingenious ways of estimating earlier populations have been found, ranging from counting rooms in the ruins of archeological sites to extrapolating from densities of modern tribes living a hunting and food-gathering existence. Using these methods, it is generally agreed that about 5 million people were living at the time of the agricultural revolution—say 6000 B.C. It took about 8,000 years for that number to increase to 500 million, around 1650, which means that, on the average, the population almost doubled every thousand years. It took only 200 years for the next doubling, as the population reached 1 billion about 1850. It took only about 80 years for the next doubling, reaching 2 billion around 1930. The next doubling has not yet been completed, but at current growth rates there will be 4 billion people in the mid-1970s. This increase in size and growth rate is shown in Fig. 12–6. How did it come about?

BIRTH RATES   In order to answer that question, one must look at births and deaths in our closed population and, in particular, at births and deaths per unit time—birthrates and death rates. This is precisely what one would do to answer this sort of question about the population dynamics of any organism which does not reproduce vegetatively. Of course, some biologists feel silly about discussing the birthrate in a population of, say, oak trees, and so they have invented the term "natality" (to be balanced by "mortality"), but we won't worry about that.

The birthrate in human populations is normally expressed as the number of live births per thousand people per year. For the world population, this is calculated by dividing the estimated number of births during a year by the estimated world population on July 1 of that year. This gives the number of births per person, which can then be multiplied by 1,000 to get the conventional **birthrate.** The use of a rate per thousand is convenient as it permits sufficient accuracy for most purposes while often doing away with the need for a decimal point. For instance, around 1967–1968 the birthrate in the United States fluctuated between 15.7 and 19.0, and in most cases a simple 17 was sufficient for comparisons. (Many underdeveloped countries in the same period had birthrates around 40!)

In 1968, the world birthrate was estimated to be 34 and the world death rate 14. Since they are not equal, the population is clearly not constant in size; indeed, since input is way above output, we know that the population is growing rapidly. How rapidly is determined by examining the difference between the birth and death rates—in this case, 34 minus 14, or 20. That is, for each thousand people in the world population, there was an excess of 20 births over deaths. Conventionally, this growth rate of 20 per 1,000 is expressed as a rate per 100 (percent). Thus the world growth rate was 2 percent per annum in 1968. Since the population was some 3.5 billion, we can easily calculate the approximate number of people added to the population as $0.02 \times 3,500,000,000 = 70,000,000$. Think of it—70 million people added to the world population in 1968, an increase of roughly one-third the population of the United States in a single year!

Populations grow by laws similar to those governing compound interest. Just as in compound interest the interest earned itself draws interest, so in population growth the organisms added to the population themselves reproduce. The time it would take a population growing at a rate of 2 percent per year to double would be not 50 years, as one might expect at first glance, but only 35 years. Should the human population continue to grow at that rate, by the year 2003 there would be 7 billion people jammed onto this planet. In many ways, the growth of the human population is reminiscent of the outbreak phase of an "outbreak-crash" sequence known in other animals. Fruit flies, for instance, often rapidly build huge populations in rotting fruit. The first generations of flies find food in abundance, and the population grows rapidly. Eventually, however, the food resource is exhausted, and the population "crashes" (Fig. 12–7). Compare the shape of the human curve with that of the fruit fly. Since the resources of the earth are clearly finite, one can easily imagine *Homo sapiens* imitating the dynamics of *Drosophila* populations.

THE LAWS DESCRIBING POPULATION GROWTH ARE SIMILAR TO THOSE GOVERNING COMPOUND INTEREST. EXPERIMENTS WITH VARIOUS ANIMALS HAVE SHOWN A RAPID INCREASE IN POPULATION SIZE UNTIL FOOD RESOURCES ARE EXHAUSTED. THE POPULATION THEN CRASHES, OR DRAMATICALLY DECREASES IN SIZE.

BIRTHRATES VERSUS DEATH RATES   Growth or shrinkage of a closed population is simply a function of the difference between the birthrate *(b)* and the death rate *(d)*. This difference, $b-d$, is conventionally labeled *r*. When *r* is positive, it is the rate of increase; when *r* is negative, it is the rate of decrease. In most populations of organisms, not only is the population size varying continuously but *r* is varying continuously also. For instance, look at Fig. 12–8, which shows the estimated total size of the Jasper Ridge checkerspot colony and estimates of the sizes of

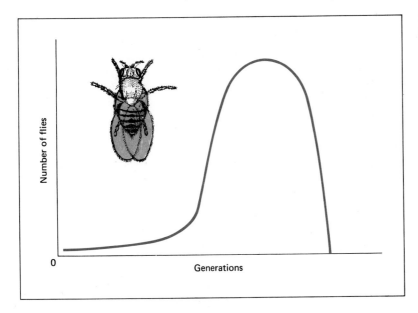

Fig. 12–7 Outbreak-crash type of population curve as seen in a population of the fruitfly *Drosophila*.

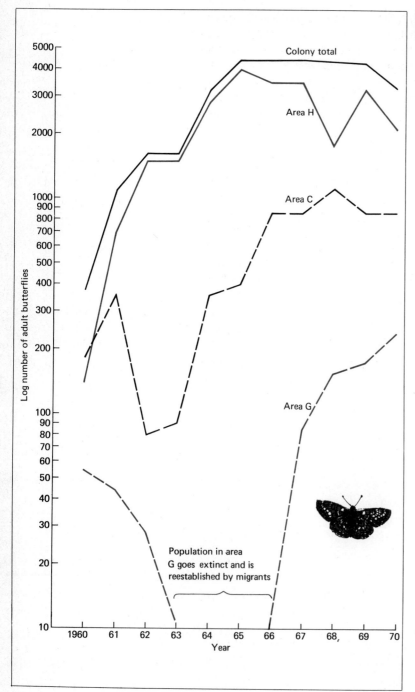

Fig. 12–8 Changes in the size of populations of *Euphydryas editha* during a 10-yr period.

the three populations in areas C, G, and H. In contrast to the pattern shown by the human population of the world, r shows frequent changes of sign. Notice that the fluctuations are more pronounced in the individual populations than in the colony as a whole. Of course, the colony estimate is simply the sum of the estimates of the individual populations, and since the population in C may decrease in a year in which it increases in H, changes may help cancel one another out in the total, smoothing the total line.

There is, of course, a similar smoothing effect in the world-population chart shown in Fig. 12–6. In contrast, look at Fig. 12–9, where the pattern of population size changes in Europe between 1000 and 1700 shows striking reversals in the sign of r, an effect of two visitations of the bubonic plague, called the Black Death. Famines, plagues, war, and natural disasters have often caused localized reversals in the pattern of constant growth of the human population, and so some occasions these have been serious enough to cause slight decreases in the world population. These are indicated by the minor irregularities in the curve of world-population growth.

In both the human and the checkerspot populations, most of the variation in size seems to be caused by variation in the death rate. Before the agricultural revolution, when the human population was growing very slowly, both birth and death rates were probably in the vicinity of 50. Then, at first slowly following the agricultural revolution and then more rapidly following the industrial and medical revolutions, the human death rate dropped. But the birthrate remained close to its primitive level, and as a result r (remember, that's $b-d$) increased, as did the population size.

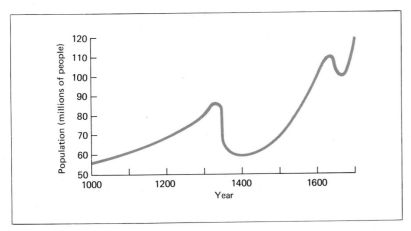

Fig. 12–9 Population size changes in Europe, showing the effects of bubonic plague epidemics. (*From "The Black Death" by William L. Langer. Copyright © 1964 by Scientific American, Inc. All rights reserved.*)

Following World War II, Western countries exported their technology for "death control" to the underdeveloped regions of the world. Penicillin, DDT, and their relatives wrought public health miracles, and death rates plunged to new lows, often to under 10 per thousand. With birthrates over 40 and death rates under 10, some countries, especially in Latin America, now have growth rates of over 3 percent per year; at that rate, these countries will double their populations in about 20 years.

VARIATION IN POPULATION SIZE IS MOST OFTEN CAUSED BY CHANGES IN THE DEATH RATE. IF THE DEATH RATE GOES DOWN, POPULATION SIZE INCREASES, UNLESS OTHER FACTORS INTERVENE. THE HUMAN POPULATION EXPLOSION IS LARGELY THE RESULT OF DECLINING DEATH RATES.

The consequences of such a population growth rate will be discussed in the final chapter of this book. In parts of the world, however, birthrates have also declined. This is primarily true of the so-called "developed" nations. As industry replaced farming for much of the population, attitudes toward children changed. Instead of being viewed primarily as farm labor and old-age insurance, children were viewed as liabilities, expensive to feed and expensive to educate. Under those circumstances, the birthrate dropped, since people desired to have fewer children. In most industrialized countries today, birthrates are well under 20, and growth rates characteristically yield "doubling times" of 60 to 150 years, as opposed to the 20 to 30 years of underdeveloped countries.

Man, then, is capable of rapidly altering his birthrate in response to changing conditions. Not so the checkerspot butterfly. Females are genetically programmed to lay an average of about 700 eggs, and only natural selection over a number of generations will alter that significantly. As a general rule in animals, each female (or pair, if males participate in rearing the young) will have the number of offspring that guarantees the maximum number of descendants in the next generation. That, you will recall, is the name of the game in natural selection—the best reproducers are the ones whose kind of genetic information is passed on to posterity. This does not mean that it would be advantageous for our checkerspot butterfly female to double the number of eggs she lays. Natural selection has presumably already maximized her reproductive contribution. More eggs might mean smaller eggs since each female has only a finite amount of energy she can put into reproduction. Smaller eggs would mean smaller larvae, with an inferior start in life. Perhaps the cost of doubling egg number would be the reduction of larval chance of survival to only 25 percent of its former level and a great reduction in the number of offspring surviving to maturity. David Lack and his colleagues have done detailed studies of the clutch size (number of eggs laid) of British great tits, a species of songbird in England. They have shown that the average clutch size is, in general, carefully adjusted so that a maximum number of young are fledged. Pairs that produce either fewer or more eggs than the average tend to have fewer young leave the nest.

Jasper Ridge checkerspots, then, are unlikely to change their birthrate rapidly. As noted above, their population fluctuations, like those in man, are accounted for primarily by changes in the death rate. We have not yet been able clearly to determine what factors are the primary determinants of the death rate in this colony of butterflies. One possibility is parasites, which themselves vary a great deal in abundance. A more likely factor is weather, which may lead to drowned larvae, frozen larvae, desiccated larvae, growth of fungi dangerous to the larvae, good conditions for enemies of the larvae, and above all, destruction of food resources by desiccation. If weather plays a large part, it will have to be through its influence on the "microclimate" in which the larvae exist, since the weather in general is essentially identical over the entire colony. Exposures and drainage vary from area to area, and you will recall, size changes are often in opposite directions in different areas. In this example and in many similar cases, especially those involving insects, the suspicion is that a great many factors enter into the determination of population size and that they may change in importance with time. At one point, population growth may be halted by a food shortage, at another by a cold snap, and at another by a population boom of parasites. At still another time, a rapid rise in population may result from weather particularly favorable to the food plant or particularly disadvantageous for a predator. In one interesting study, it has been possible to trace population booms in insects called psyllids, which feed on eucalyptus trees in Australia, to flooding. The trees are physiologically stressed by the flooding, and reduction in their natural defenses seems to make conditions ideal for the insects.

CHANGES IN POPULATION SIZE ARE USUALLY THE RESULT OF THE INTERACTION OF A NUMBER OF DIFFERENT FACTORS WHICH AFFECT, AMONG OTHER THINGS, THE DEATH RATE AND BIRTHRATE.

Many insect populations show large fluctuations in size. For instance, in a careful study of an isolated population of leaf beetles, the maximum number of adult beetles in the spring generation was found to fluctuate from 1,000 to 33,500 over a 12-yr period. Similarly, a colony of the moth *Panaxia* showed sixfold to eightfold changes in population size over a 6-yr study. Many other insect populations have shown similar patterns of fluctuation. Much more extreme fluctuations have been observed in populations of forest insects, such as the spruce budworm (the larva of a moth). This insect could be described as "rare" in many years. When conditions are favorable, however, huge outbreaks occur, with the population density increasing many thousands of times. Similar patterns are often found in other forest insects (Fig. 12–10). Other invertebrates which have been investigated tend to show fivefold to fiftyfold changes in population size.

In contrast, many vertebrate populations show much less fluctuation, especially those which display the behavior pattern known as **territoriality.** Territoriality is the defense of some segment of the home range against other individuals of the same species and, in some cases, of many species. The beautiful music of songbirds is actually an announcement by the males which says, in essence, "This is my territory; stay out or I will chase you out." The display of the male Siamese fighting fish carries a similar message. One can easily map the territory of a ground-nesting bird such as the meadowlark. Simply get a stuffed male and set it up at a point roughly equidistant between two nests. Then watch which of the two males attacks it. By moving the model about, it is possible to make an accurate map of areas which have reality only in the minds of the birds. But this ethereal quality does not make the boundaries any less sharp.

Territoriality

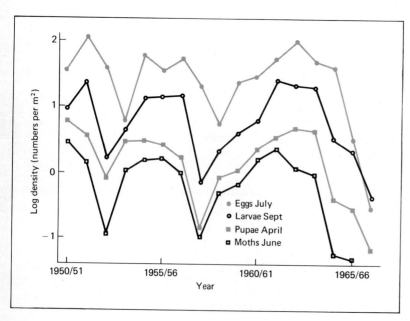

Fig. 12–10 Changes in population size during 17 yr of various stages of the pine looper moth. Note that density is plotted logarithmically. (*Adapted from T. R. E. Southwood, ed., INSECT ABUNDANCE, 1968, Blackwell Scientific Publications.*)

	TERRITORY		FLOCK		TOTAL IN POPULATION	
YEAR	NO.	PERCENT	NO.	PERCENT	NO.	PERCENT
1957	92	36	161	64	253	100
1958	103	35	189	65	292	100
1960	112	39	178	61	290	100
1961	111	40	168	60	279	100

TABLE 12-1. Census Data for a Population of Female Australian Magpies

After all, where they are unmarked, the boundaries between nations, countries, etc., possess little more reality unless you accept the conventions inherent in maps. Indeed, people surround themselves with a kind of mental territory. If you don't believe it, try sticking your nose within 1 in. of your teacher's face the next time you have a discussion with him.

At any rate, territories generally are flexible in size but tend to lose that flexibility as they reach certain minimum dimensions. If a species of bird nests in oak woods and if its minimum territory size is about an acre, then a 20-acre wood is going to have a maximum nesting population of 20 pairs. Surplus birds will be excluded, forced to move on and live in less desirable and more dangerous habitats or form nonbreeding flocks. This results in a rather neat "control" of population size. The maximum is set by the available space for territories, and the population is kept near the maximum by the replacement of territorial birds which die by individuals from the dispossessed flock. A fine example of this has been worked out by Robert Carrick, who studied a population of Australian magpies. Table 12-1 shows 4 years of census data for adult females, with the birds divided into territory (all hens that have had an opportunity to breed) and flock (nonterritorial individuals). The constancy shown here is extreme, even for birds, and probably represents about the maximum level of "control" of population size through territoriality.

Sometimes groups of animals defend a common territory. The well-known behaviorist Jane Van Lawick-Goodall has described the violent territorial clashes of clans of hyenas living in the Ngorongoro crater in Africa. For instance, she was able to observe one clan systematically increase its territory at the expense of another. Howler monkeys also reputedly defend group territories.

TERRITORIALITY IS THE DEFENSE OF SOME PART OF THE HOME RANGE OF AN ANIMAL AGAINST OTHERS OF THE SAME SPECIES. IT OFTEN EXERTS A CONTROL ON POPULATION SIZE WHICH PREVENTS WIDE FLUCTUATIONS.

TERRITORIALITY AND POPULATION SIZE There is considerable disagreement among biologists about both the general nature of population-size changes and the reasons behind them. Unfortunately, properly gathered census data are available for few organisms, and it is also apparent that in many cases the "whys" of size change are exceedingly complex. A major source of controversy centers on the amount of feedback there is from the density of a population to its growth rate. Obviously, in examples such as those of the territorial birds, there is a great deal. When the population exceeds the number for which territory space is available, birds are chased out and the population is reduced. Should it fall below the number necessary to saturate the habitat with territories, immigrants will be able to establish themselves and the population size will increase. This happens, although it is the general feeling that territoriality did not evolve as a population-control device.

The idea that populations would evolve such devices to ensure that there would never be too many individuals for an area to support properly has proved a very seductive one. After all, if you have 60 pairs of birds trying to raise their young in a 20-acre wood, *all* the young might die of starvation. But birds play the game of life by the rules of natural selection. Presumably, territoriality evolved because it confers some advantage on individuals. Space means food supply, and the energy expended in defense of a territory may be more than compensated by not having to share the resources of that territory with other males of the same species. More space means more food, but if one male tries to expand his territory to, say, twice its former diameter (assume, for the sake of argument, circular territories), he ends up with more than three times the perimeter to defend. If he lets his territory shrink, he reduces the amount of resources at his command. Presumably, natural selection sets the minimum territory size for any given habitat—it is to the individual's reproductive advantage to defend a certain-size chunk of that habitat. If a mutant or recombinant bird appeared which could raise more than the average number of young on a smaller or larger territory or could steal from its neighbors' territories, that genotype would be favored by selection and would take over the population. With some possible exceptions which need not concern us here, there is no way in which a nonhuman population can "decide" to limit its size, because individuals which do the maximum reproducing are, by definition, those which will prevail evolutionarily. Territoriality may often function to regulate population size, but it did not evolve to serve that function.

ALTHOUGH TERRITORIALITY MAY DETERMINE POPULATION SIZE AND PREVENT MARKED FLUCTUATIONS IN SIZE, THERE IS NO MECHANISM KNOWN BY WHICH IT COULD HAVE EVOLVED AS A REGULATOR OF POPULATIONS.

Control of Population Size

A population whose size is kept relatively constant by feedback mechanisms, such as that described above, is often said to be showing **density-dependent regulation.** In other words, the more crowded the population, the greater will be the impact of the regulatory factors. Other populations, such as those of the checkerspot butterfly, which fail to show such clear-cut feedback control are thought to show **density-independent regulation.** In particular, factors such as mass kills from frost or drought are supposed to limit the population before it reaches a level where density effects can function because of shortage of some resource or another. Of course, should checks such as weather fail to function, density will sooner or later enter into the picture, if only when the organism becomes so abundant that it runs out of food or space. It is, however, difficult to imagine a factor operating on a population which isn't density-dependent. In a dense population of caterpillars, more individuals proportionately may be forced to spend the night in relatively exposed and undesirable positions and a frost may kill a higher percentage than it would have in a sparse population. A swarm of bees can survive much lower temperatures than can a single bee, because the swarm forms a complex ball in which metabolic heat is produced by the wing muscles and individuals move from the cold outside to the warm center of the ball. Predators often will form a "search image" and feed on the commonest species of prey available, taking a much higher proportion of individuals from a dense population than from a sparse one. Many other examples could be mentioned, but the principle is clear: there are different degrees of density-dependence, but few if any factors operate in a truly density-independent manner.

REGULATION OF POPULATION SIZE MAY BE DESCRIBED AS DENSITY-DEPENDENT OR DENSITY-INDEPENDENT. DENSITY-DEPENDENT REGULATION IS THE RESULT OF FEEDBACK MECHANISMS, WHICH FUNCTION SO THAT A DENSITY INCREASE BEYOND A CERTAIN POINT TENDS TO CAUSE A POPULATION REDUCTION. DENSITY-INDEPENDENT REGULATION IS THOUGHT TO BE THE RESULT OF FACTORS, SUCH AS WEATHER, NOT DIRECTLY AFFECTED BY DENSITY.

It is important for you to remember that when we discuss "control" or "regulation" of population size, we are not speaking of phenomena similar to, say, the regulation of our blood temperature. In the latter, a feedback system has evolved to maintain the temperature at a level which maximizes physiological efficiency. We might say that the optimum temperature is determined without reference to the system which maintains it. There is no such extrinsically determined optimum which a population-regulation system has evolved to maintain. At any given moment, a population is as large as conditions permit, although there may be a lag in growth. Sometimes the most important condition is something such as territory space available, in which case the population may remain relatively constant. In other situations, the maximum may be determined by a rapidly rotating spectrum of factors or by one or a few factors which themselves change radically in magnitude; the population will then show wide amplitude fluctuations.

You have doubtless noticed that most of our discussion of population dynamics has been largely oriented toward animals. The reason is simple: most of the research done in population dynamics has been done on animals. Some work, however, has been done with algae. Very large fluctuations in the size of populations of colonial and single-celled algae have been observed. Some of these, such as the famous "red tides" in the Gulf of Mexico and huge algal blooms in various lakes and in Long Island Sound, have been traced in part to changes in the availability of nutrients. Organic runoff from duck farms, sewage, and nitrate-rich waters draining from farmlands treated with inorganic fertilizers are among the factors causing these changes. In general, changes in the size of populations of higher plants have not been closely studied, except for such things as recording the disappearance of chestnut populations before the ravages of chestnut blight or the extinction of local populations under the blade of a developer's bulldozer. Most numerical changes in plant populations have been considered in a community context, under the general heading of succession. And that's where we'll consider them, in Chap. 14.

IN SOME ORGANISMS, FEEDBACK FROM THE DENSITY OF THE POPULATION HAS A PROFOUND INFLUENCE ON CHANGES IN POPULATION SIZE; IN OTHERS, THE DENSITY-DEPENDENT ACTION OF VARIOUS FACTORS IS LESS PRONOUNCED AND OBVIOUS. POPULATIONS ALWAYS APPROACH A MAXIMUM SIZE FOR THE CONDITIONS PREVAILING; THEY ARE NOT "REGULATED" TO CONFORM TO SOME EXTRINSICALLY DETERMINED MAXIMUM.

Biologists have made impressive progress toward an understanding of the population dynamics of many economically important organisms, especially those of exploited animals such as whales and food fishes. In a classic case, Paulik and his coworkers, using the techniques of systems analysis, were able to come up with a program that saved the Canadian-American West Coast salmon fishery from extinction. They had to feed into their computer masses of data, not just on the dynamics of the salmon populations of various spawning streams but also on net size and efficiency of the various fishing organizations. Their solution to the problem was accepted by the concerned parties, and a continued yield of salmon was ensured.

In a contrasting case, biologists hired by the International Whaling Commission were, in 1963, able to make extraordinarily accurate predictions about the catch. For instance, they stated that if unrestricted Antarctic whaling were carried on in the 1963–1964 season, not more than 8,500 blue whale units would be taken. A blue whale unit is a scoring device for apportioning the catch among whaling nations; it consists of one blue whale or two of the smaller fin whales or six of the even smaller sei whales, etc. When the catch was recorded, it was found that 8,429 blue whale units had been taken, an error in the prediction of less than 1 percent! Unfortunately, the advice of these biologists on the conservation of whales in order that a steady harvest could be maintained has been ignored by the whaling industry. The result is that the industry is now about to go out of business, having virtually exterminated the most economically valuable whale species (Fig. 12–11). Nevertheless, whale meat is still an important component of organic fertilizers and pet foods. In spite of these predictive successes with fisheries, biologists have a long way to go before their knowledge of the dynamics of most populations will approach anything that might be described as satisfactory. In few areas of population biology is there a greater lack of basic data, and especially of viable generalities, than in population dynamics.

AS HIS OWN POPULATION SIZE INCREASES, IT IS BECOMING MORE AND MORE IMPORTANT THAT MAN THOROUGHLY UNDERSTAND THE DYNAMICS OF OTHER ORGANISMS—BOTH THOSE USEFUL OR POTENTIALLY USEFUL TO HIM AND THOSE WHICH COMPETE WITH HIM FOR RESOURCES OR DO HIM DIRECT INJURY.

SUMMARY

In the broadest sense, a biological population is any specified group of individuals. Most biologists recognize two general kinds of populations: those made up of individuals of a single species (populations in the strict sense) and those consisting of more than one species (communities). In this chapter, we have dealt only with the former kind. One set of population characteristics is grouped under the term "population structure." Technically, the structure of a population is defined as the sum of all factors governing the pattern in which gametes from various individuals unite. Population density, systems of mating, and level of inbreeding are all parts of the population structure.

A characteristic of all biological populations is that they change in size.

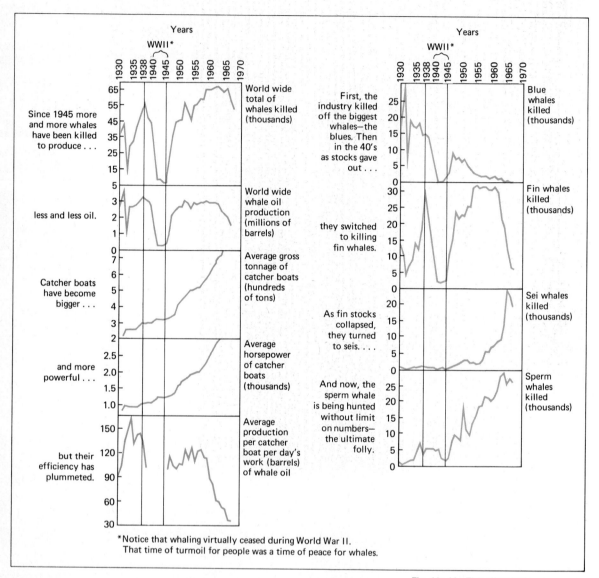

Fig. 12-11 The effects of over-exploitation on whole populations. *(Adapted from Roger Payne, Among Wild Whales, The N.Y. Zoological Society Newsletter, November, 1968.)*

The study of patterns of change in population size and the causation of such change is the study of population dynamics. If immigration and emigration are not involved, the study of population dynamics boils down to the study of birthrates (input) and death rates (output). Rate of population growth (or shrinkage) is a function of the difference between these two. Many different patterns of population-size changes have been observed. The population of the entire species *Homo sapiens* has been growing almost continuously for about 10,000 years. This is not true for any other vertebrate except man's domestic animals and pests. In many ways, the pattern of population growth in man is reminiscent of the outbreak part of an outbreak-crash sequence. Invertebrates, especially insects, often show fivefold to fiftyfold changes in population size when observed over a dozen generations or so. In some insect populations, especially those of certain forest insects, huge outbreaks may occur when conditions are ideal. These outbreaks may produce populations many thousands of times larger than those of nonoutbreak periods. Most invertebrate populations do not seem to be closely regulated by density-dependent factors. In contrast, many vertebrate populations, especially those of territorial animals, show relatively small fluctuations. In these, behavioral factors tend to place rather narrow limits on density.

MacArthur, R. H., and J. H. Connell: "The Biology of Populations," Wiley, 1966.
Kormondy, E. J.: "Concepts of Ecology," Prentice-Hall, 1969.
Andrewartha, H. G.: "Introduction to the Study of Animal Populations," 2d ed., Methuen, 1970.
Wilson, E. O., and W. H. Bossert: "A Primer of Population Biology," Sinauer Associates, 1971.

SUPPLEMENTARY READINGS
Books

CHAPTER 13

EVOLUTION IN POPULATIONS

The last chapter started with the structure of populations. The factors which contribute to the structure of a population—density, mating system, etc.—generally may be understood when studied over a rather short time span, say a generation or two. The chapter ended with a consideration of population dynamics, that is, changes in population size through time. In order to understand population dynamics, it is necessary to consider events in a population over a series of generations. The emphasis in studies of population structure is on relatively static, short-term features: position in space, rank in a dominance hierarchy, choice of a mate. In contrast, the study of population dynamics focuses, as we will also see in this chapter, on change through time. Here we will be especially concerned with change through time in the genetic properties of populations, that is, with evolution. As we hope is already obvious, evolution cannot be divorced from population dynamics—except when it is necessary in order to force a nonlinear subject such as biology into the linear format of a textbook.

THE PRESERVATION OF VARIATION

No two individuals are identical. Indeed, in sexually reproducing organisms, no two individuals have identical sets of genetic information, although identical twins will differ ordinarily by only a few mutations. This nearly ubiquitous genetic variation is the raw material of evolution. Variation, you will recall, is generated by mutation and greatly enhanced by recombination. In interbreeding populations, this variation tends to be maintained as the population reproduces itself. To see how this operates, we must consider those genetic properties of populations that are not properties of the individuals which make up the populations.

Gene Pool, Genotype Frequency, and Gene Frequency

The total genetic information possessed by a population is called the *gene pool* of that population. If the gene pool could be described completely, one would know not only what kinds of information were present but also the proportions (frequencies) of the different kinds and the way in which these kinds were distributed among individuals. This means that if, say, the only genetic information in the gene pool of a pea population concerned the smoothness or wrinkling of the peas, we would know exactly the proportions of the "smooth" alleles and the "wrinkled" alleles and also how these alleles were distributed among individuals—that is, the proportions of homozygous smooth, heterozygous smooth, and homozygous wrinkled pea plants. Since we are considering only the genetic information possessed by the population that is concerned with the smoothness or wrinkling of the peas, the proportions of alleles and of genotypes should each add up to 1. If one-third of the alleles are "smooth," then two-thirds must be "wrinkled." (We are putting "smooth" and "wrinkled" in quotes here to remind ourselves that these are, of course, not properties of the genes but of the phenotypes produced by the interaction of these genes with the environment.)

We know that there is much more genetic information in a real population of peas than that concerned only with the smoothness or wrinkling of the peas. But any attempt to deal with all this information at once leads to unmanageable complexity. Therefore, in our discussion of population genetics, we will generally restrict ourselves to discussions focused on one locus at a time. Indeed, population geneticists rarely consider more than a very few loci simultaneously. In so doing, they must always keep in mind that they are artificially simplifying the situation to make it easier to study, just as biologists use artificial polynucleotides to study genetics at the molecular level.

THE GENE POOL OF A POPULATION IS THE TOTAL GENETIC INFORMATION POSSESSED BY A POPULATION. IN PRACTICE, THE POPULATION GENETICIST STUDIES JUST ONE LOCUS—THE FREQUENCIES OF ITS VARIOUS ALLELES IN THE POPULATION AND THEIR DISTRIBUTION AMONG THE INDIVIDUALS.

The mathematical treatment of the distribution of gene and genotype frequencies was developed in the 1920s and 1930s, principally by R. A. Fisher (the father of modern statistics) and J. B. Haldane in England and by Sewall Wright in the United States. The most fundamental idea in population genetics, however, dates to a simple observation made simultaneously in 1908 by an Englishman, G. H. Hardy, and a German, W. Weinberg. It is called the **Hardy-Weinberg law.** Before examining the law itself, you must know the exact meaning of the terms "genotype frequency" and "gene frequency."

A **genotype frequency** is the frequency of a kind of individual making up a population—that is, a kind with respect to the locus under consideration.

Let's construct a hypothetical population. Assume that only two alleles are known at a locus being considered: *A* and *a*, with *A* fully dominant. Three kinds of individuals may then exist in the population: homozygous dominants (genotype *AA*), heterozygotes (genotype *Aa*), and homozygous recessives (genotype *aa*). If there are *N* diploid individuals in the population, *D* homozygous dominants, *H* heterozygotes, and *R* homozygous recessives *(D + H + R = N)*, then the genotype frequency of *AA* is *D/N*, that of *Aa* is *H/N*, and that of *aa* is *R/N*. Genotype frequencies, then, are calculated by simply dividing the number of individuals with the genotype by the total number of individuals in the population. The dominance relations of the alleles have no bearing on the calculation of genotype frequencies (unless, of course, one cannot distinguish between two genotypes), and a notation of dominant and recessive is used here and below merely for convenience in referring to alleles and genotypes.

THE FREQUENCIES OF INDIVIDUALS WITH EACH OF THE POSSIBLE COMBINATIONS OF ALLELES AT A PARTICULAR LOCUS MAKE UP THE GENOTYPE FREQUENCIES OF THE POPULATION AT THE LOCUS.

A **gene frequency** refers to the proportion of an allele in the gene pool in relation to the other alleles at its locus, without regard for their distribution among individuals. In the population above, there are *N* diploid individuals, which may be thought of as carrying 2*N* genes at the *A-a* locus. Each *AA* individual has two *A* alleles, and each *Aa* individual has one *A* allele. Since there are *D AA* individuals and *H Aa* individuals, the total number of *A* alleles in the population is 2*D* + *H*. The proportion *p* of *A* alleles in the population is

$$p = \frac{2D + H}{2N} = \frac{D + \frac{1}{2}H}{N}$$

The quantity *p*, the proportion of *A* alleles in the population, is called the **gene frequency** of *A*. You will note that population geneticists usually use the word "gene" where, technically, the word "allele" is correct. By convention, the gene frequency of the other allele *(a)* is *q*. Since, in our hypothetical organism, these are the only two alleles at the locus, *p* + *q* = 1 and *q* = 1 − *p*. As an exercise, write down the formula for *q* which is analogous to the one for *p* displayed above. Two populations can have the same gene frequencies and different genotype frequencies, as shown in Table 13–2. If, however, two populations have the same genotype frequencies, they cannot have different gene frequencies.

IN CONTRAST TO GENOTYPE FREQUENCY, GENE FREQUENCY IS

*THE PROPORTION OF EACH ALLELE IN THE POPULATION AS COM-
PARED WITH OTHER ALLELES AT THE SAME LOCUS.*

The Hardy-Weinberg law describes a theoretical situation in which a population is undergoing no evolutionary change. It describes how, in the absence of evolutionary factors, the reproductive process will maintain a genetic *status quo* and preserve whatever variability is present in the population.

Hardy-Weinberg Law

If there is random mating in our hypothetical population and if the gametes produced by the mates combine at random, then there is complete random union of gametes with respect to the *A-a* locus in the population. As each gamete is haploid, it can contain only one allele. The frequency of gametes containing the *A* allele will be the same as the gene frequency of *A*—that is, p. This should be obvious because *AA* individuals will produce two *A*-carrying gametes to every one *A*-carrying gamete produced by *Aa* individuals. If each individual produced 100 gametes, the formula for p would be $\dfrac{100D + 50H}{100N}$, which is the same as the formula for p given above. Similarly, the frequency of gametes containing the *a* allele would be q.

It is convenient to look at reproduction in such a population as if the gametes from the adults were literally released into a large pool, within which they combine at random. A population in which gametes combine at random is said to be **panmictic.** Such a random procedure is actually approached in some marine organisms which simply release their gametes into the sea and depend on chance for fertilization. Each fertilization is analogous to the random selection of one male and one female gamete from the gamete pool. As we saw earlier (Box 1–1), the probability of two independent events happening together is simply the product of the probabilities *of each happening separately*. The probability of selecting an *A* sperm from the gametic gene pool is p. The probability of selecting an *a* egg from the pool is q. The probability of selecting an *a* sperm is q and of selecting an *A* egg is p. The probability of selecting an *A* sperm and an *a* egg is $p \times q$. Therefore, $p \times q$ is the probability of an *A* sperm fertilizing an *a* egg, producing a heterozygous individual, *Aa*.

Looking at all the fertilizations in our pool simultaneously, we can see that random union of the gametes gives us

$$[p \ (A \ sperm) + q \ (a \ sperm)] \times [p \ (A \ ova) + q \ (a \ ova)] = (p + q)^2$$
$$= p^2 \ (AA \ individuals) + 2pq \ (Aa \ individuals) + q^2 \ (aa \ individuals)$$

TYPE OF MATING*	FREQUENCY OF MATING	PROPORTIONS OF OFFSPRING		
		AA	*Aa*	*aa*
$AA \times AA$ ($p^2 \times p^2$)	p^4	p^4		
$AA \times Aa$ ($p^2 \times 2pq$)	$2p^3q$	p^3q	p^3q	
$Aa \times AA$ ($2pq \times p^2$)	$2p^3q$	p^3q	p^3q	
$Aa \times Aa$ ($2pq \times 2pq$)	$4p^2q^2$	p^2q^2	$2p^2q^2$	p^2q^2
$AA \times aa$ ($p^2 \times q^2$)	p^2q^2		p^2q^2	
$aa \times AA$ ($p^2 \times q^2$)	p^2q^2		p^2q^2	
$Aa \times aa$ ($2pq \times q^2$)	$2pq^3$		pq^3	pq^3
$aa \times Aa$ ($2pq \times q^2$)	$2pq^3$		pq^3	pq^3
$aa \times aa$ ($q^2 \times q^2$)	q^4			q^4
Totals	1.00 †	p^2 ‡	$2pq$ §	q^2 ¶

TABLE 13–1. Matings and Offspring in a Population in Hardy-Weinberg Equilibrium (*Adapted from Paul R. Ehrlich and Richard W. Holm, THE PROCESS OF EVOLUTION. Copyright McGraw-Hill Book Company, 1963. Used by permission.*)

*The frequency of types of both males and females is given by the terms of the expression $p^2 + 2pq + q^2$; therefore, with random mating the frequencies of the different matings are $(p^2 + 2pq + q^2)(p^2 + 2pq + q^2) = p^4 + 4p^3q + 6p^2q^2 + 4pq^3 + q^4$.

†The sum of this column is $p^4 + 4p^3q + 6p^2q^2 + 4pq^3 + q^4 = (p^2 + 2pq + q^2)^2 = [(p + q)^2]^2 = [(1)^2]^2 = 1.00$

‡The sum of this column is $p^4 + 2p^3q + p^2q^2 = p^2(p^2 + 2pq + q^2) = p^2(1) = p^2$.

§The sum of this column is $2p^3q + 4p^2q^2 + 2pq^3 = 2pq(p^2 + 2pq + q^2) = 2pq(1) = 2pq$.

¶The sum of this column is $p^2q^2 + 2pq^3 + q^4 = q^2(p^2 + 2pq + q^2) = q^2(1) = q^2$.

Populations in which the genotype frequencies are given by this simple binomial expansion are in an equilibrium condition. Of course, we are not considering genes located on the sex chromosomes. We have already seen that such genes show a characteristic sex-linked inheritance. The equilibrium we have been discussing applies only to genes on autosomal (nonsex) chromosomes. This equilibrium is described by the Hardy-Weinberg law, which may be stated briefly as follows:

If alternate forms of an autosomal gene are present in a large panmictic population, then, in the absence of mutation, differential migration, and selection, the original gene frequencies of these alleles (p_1, p_2, p_3, . . . , p_n) will be retained from generation to generation and, after one generation, the proportion of genotypes will also reach an equilibrium. The genotype equilibrium frequencies are given by the terms of the expansion ($p_1 + p_2 + p_3 + \cdots + p_n)^n$.

Notice that we have generalized the formula to include situations in which there are more than two alleles at a locus. An algebraic demonstration of the maintenance of Hardy-Weinberg equilibrium is given in Table 13–1. If an arbitrary initial population is chosen with, say, gene frequencies of $p = 0.2$ and $q = 0.8$ and genotype frequencies $AA = 0.10$, $Aa = 0.20$, and $aa = 0.70$, the population reaches equilibrium in one generation and then remains there (Table 13–2). Note that there is no change whatsoever, *after the first generation*, in the *gene frequency*, which can then be determined simply by taking the square root of the frequency of the proper homozygote. Under the conditions described, there is a genetic inertia in interbreeding populations. To a very large degree, overcoming the genetic inertia of populations (especially the changing of gene frequency) is what is described as "evolution."

Consider, for example, the observed distribution of the blood-group alleles *M* and *N* in a sample of 1,279 people from England (Table 13–3A). The genotype frequencies are obtained simply by testing each individual's

GENERATION	GENOTYPE FREQUENCY			GENE FREQUENCY	
	AA	Aa	aa		
0	0.10	0.20	0.70	$p = 0.2$	$q = 0.8$
1	0.04	0.32	0.64	$p = 0.2$	$q = 0.8$
2	0.04	0.32	0.64	$p = 0.2$	$q = 0.8$
3	0.04	0.32	0.64	$p = 0.2$	$q = 0.8$
. . .	. . .	. . .	. . .	. . .	. . .
N	p^2	$2pq$	q^2	p	q

TABLE 13-2
(Adapted from Paul R. Ehrlich and Richard W. Holm, THE PROCESS OF EVOLUTION. Copyright McGraw-Hill Book Company, 1963. Used by permission.)

blood type. The gene frequencies p and q are then calculated from the genotype frequencies. The genotype frequencies expected if the population were roughly in Hardy-Weinberg equilibrium can then be calculated as p^2, $2pq$, and q^2, given in the second group of columns of the table. As you can see, the observed and calculated frequencies are very close, the difference being easily explained by chance. This means that, as far as can be detected from this sample, the population is in Hardy-Weinberg equilibrium *relative to the M-N locus*. Mating is presumably random with respect to that locus. The usual interpretation is that such a system has no forces acting on it that are tending to favor certain genotypes or otherwise change their frequencies. Unless the sample of organisms is quite large, however, it is difficult to detect the effects of relatively weak forces.

Consider, in contrast, data from another sample of 1,073 individuals (Table 13-3B). In this case, the calculated genotype frequencies are very different from the observed, and a statistical test indicates that this is no accident. The reason is clear: we produced the data by combining genotype frequencies from a sample of 569 Greenland Eskimos and 504 Ainu (aborigines from Japan). This immediately violated the random-mating assumption of Hardy-Weinberg equilibrium, and that violation was manifest in the departure of observed genotype frequencies from those calculated.

THE HARDY-WEINBERG LAW DESCRIBES THE EQUILIBRIUM OF GENE AND GENOTYPE FREQUENCIES WHICH A POPULATION WOULD ACQUIRE IN THE ABSENCE OF ALL EVOLUTIONARY FORCES. DE-

	GENOTYPE FREQUENCIES (%)			GENE FREQUENCIES (%)	
	MM	MN	NN	p (= freq. M)	q (= freq. N)
A Observed in sample	28.38	49.57	22.05	53.165	46.835
Calculated from gene freq. as expected assuming H-W equilibrium	28.26	49.80	21.94		
B Observed in sample	52.66	31.87	15.47	68.60	31.40
Calculated from gene freq. as expected assuming H-W equilibrium	47.06	43.08	9.86		

TABLE 13-3

VIATIONS FROM THE EXPECTED EQUILIBRIUM SUGGEST THE OPER-
ATION OF SUCH FORCES IN A POPULATION.

THE EVOLUTIONARY FORCES

The Hardy-Weinberg law is important primarily because it describes the situation in which there is *no* evolution, and it thus provides a theoretical baseline for measuring evolutionary change. Now we will turn to some of the factors which cause populations to deviate from Hardy-Weinberg equilibrium, that is, factors responsible for evolutionary change.

Nonrandom mating, small population size, mutation, differential migration, and natural selection all may cause changes in the gene and genotype frequencies in a population. That is, they may cause deviation from a theoretical Hardy-Weinberg equilibrium condition. Since changes are the essence of evolution, each of the above may be thought of as an evolutionary force. In this section, we will discuss briefly all but natural selection, which is by far the most important of these forces and will be examined closely in the next section.

Nonrandom Mating

Various kinds of nonrandom mating were mentioned in the last chapter; it is easy to see how they would lead to changes in the genotype frequencies in a population if the mating pattern were based on genetic differences. Suppose that, in a population, *AA* individuals mate only with other *AA* individuals, *aa* individuals only with other *aa* individuals, and *Aa* only with other heterozygotes. With such rigorous genetic positive assortment, you can see that the genotype frequency of the heterozygotes in the population will rapidly decline. Both kinds of matings within the homozygous types will produce only more homozygotes, but half of the offspring from the heterozygous matings will be homozygotes. Remember that in *Aa* × *Aa,* you expect one-quarter of the offspring to be *AA* homozygotes and one-quarter to be *aa* homozygotes. Therefore, each generation, the heterozygote frequency will dwindle as roughly half of the offspring from heterozygous crosses move into the homozygous lines and there is no compensating production of heterozygotes from the homozygous crosses. On the other hand, if mating were negatively assorted, the genotype frequency of the heterozygotes would be higher than would be expected if the population were in Hardy-Weinberg equilibrium.

THE HARDY-WEINBERG LAW ASSUMES MATING TO BE COMPLETELY
AT RANDOM IN A POPULATION. IN FACT, MATING IS RARELY AT
RANDOM, AND THE NATURE OF THE MATING SYSTEM IS AN IMPOR-
TANT EVOLUTIONARY FORCE.

Remember that, as we found in our earlier discussion of genetics, the expected values of the Mendelian ratios were not achieved exactly because of **sampling error** (see Box 1–1). Sampling error occurs throughout the reproduction of a population, reproduction which may be thought of as a succession of samplings. The gametes are a *sample* of the genetic information possessed by the adults. The zygotes are formed by the fusion of a *sample* of the total population of gametes. And the next generation of adults is a *sample* of the zygotes originally formed. Even in the absence of all other evolutionary forces, this sampling error will cause variation in gene frequencies. The magnitude of this variation in frequencies is a function of population size. If a population is very large, fluctuations in gene frequency will be small. If a population is very small, the fluctuations will be large.

Consider an analogous situation in tossing an honest coin. You always expect the frequency of heads to be 0.50. If you flip the coin four times and get one head and three tails, you will have a head frequency of only 0.25. You know intuitively, however, that one head and three tails is not an unusual result of flipping a coin four times. Indeed, by an extension of the methods used in Box 1–1, you can calculate that such a result would be expected one-quarter of the times that you ran the experiment with four tosses. You can also calculate that once every $(1/2)^4 = 16$ times you would get no heads at all—a head frequency of 0.00.

What if you ran a different experiment, flipping the coin 1,000 times? A result of 250 heads and 750 tails, giving a head frequency of 0.25, would be very unusual, to say the least. With that large a sample of flips, one would not expect so great a deviation from the expected head frequency of 0.50. In fact, if the coin were truly honest, you would never expect this large a deviation, any more than you would expect to get all heads, which would occur about once in every $2^{1,000}$ times you ran the experiment. Since $2^{1,000}$ is a number vastly greater than the estimated number of seconds since the universe began, you can see that you could expect to be flipping coins a long time before you would get a run of 1,000 straight heads. A much more likely result of 1,000 flips would be, say, 527 heads and 473 tails, giving a head frequency of 0.527—quite close to the expected frequency. The general rule then is: the larger the population of flips, the closer the head frequency will tend to be to the expected frequency of 0.50.

Picture now two populations of gametes, each with gene frequencies of $p = q = 0.50$. That is, each population consists of two kinds of gametes, *A* and *a*, in equal numbers. Suppose a sample of *only one zygote* is to be produced by the union of *two* of these gametes drawn at random. The only way for the gene frequency to remain unchanged is for a heterozygote

Small Population Size

to be formed, and this will happen only half the time ($2pq = 0.50$). The chance of totally losing the genetic variability in the population is also $1/2$, since one-quarter of the time an AA individual will be formed ($p^2 = 0.25$) and one-quarter of the time an aa individual will be formed ($q^2 = 0.25$). In either case, the population (one individual) will become genetically homogeneous at that locus. Either allele A or allele a will be lost.

Suppose, however, that in the other population, four zygotes are to be formed by a random union of the gametes. The gene frequency is *less* likely to remain exactly $p = q = 0.50$. In fact, it will do so less than 40 percent of the time ($6p^2q^2 = 0.375$). But deviations in frequency will be considerably smaller, and the chance of losing either allele is reduced from $1/2$ to $1/8$ (p^4 = chance all four gametes $A = 0.0625$; q^4 = chance all four gametes $a = 0.0625$; chance of either event $= p^4 + q^4 = 0.125$). Similarly, if five zygotes are formed from 10 gametes, the chance of losing one or the other gene is reduced to less than 0.002 ($0.5^{10} + 0.5^{10}$).

IN SUMMARY, AS THE SIZE OF THE POPULATION INCREASES, THE CHANCE OF THE GENE FREQUENCY REMAINING PRECISELY THE SAME GETS SMALLER, BUT AT THE SAME TIME THE CHANCES OF LARGE VARIATIONS IN GENE FREQUENCY FROM GENERATION TO GENERATION ALSO RAPIDLY DECREASE, AS DO THE CHANCES OF LOSS OF AN ALLELE AT A LOCUS.

In actual populations, it is generally necessary to consider the possible effects of sampling error only when the number of breeding individuals gets below 50 or so. As size drops toward zero, these effects will become pronounced and may alter or even override those of other evolutionary forces. These other forces may be thought of as directional in the sense that, once they are understood, one can predict the direction in which gene frequencies will change. Sometimes one can predict that the forces will hold the gene frequency at some equilibrium value. Sampling error, in contrast, has no direction. For instance, although you may predict that the chances of getting precisely 500 heads and 500 tails in 1,000 flips of a coin are virtually nil, you have no way of predicting which direction —more heads or fewer heads—the deviation will take. Similarly, although statements may often be made about the expected **magnitude** of changes in gene frequency induced by sampling error, it is not possible to predict the **direction** of change. In a population not subject to other evolutionary forces, sampling error will cause gene frequencies to drift unpredictably from value to value. For this reason, the effects of sampling error on gene frequencies in a population are commonly described as **genetic drift** (Fig. 13-1).

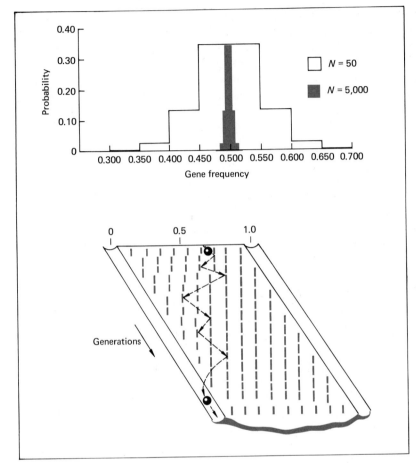

Fig. 13-1 The effects of genetic drift. *Above,* the probability distribution of p in 50 and 5,000 offspring from a parental population in which the gene frequency was $p = 0.50$. The chances for large fluctuations in gene frequency, due to sampling error are much greater for the smaller group of offspring. *Below,* the possible consequences of drift in a small population. The gene frequency is analogous to a pinball; moving down the slope (through time) it ricochets from value to value as long as it stays on the table. In the absence of mutation and migration, the values of 0 (loss) and 1.0 (fixation) are dead ends; that is, gene A is either fixed or lost. (*Adapted from Paul R. Ehrlich and Richard W. Holm, THE PROCESS OF EVOLUTION. Copyright McGraw-Hill Book Company, 1963. Used by permission.*)

THE HARDY-WEINBERG LAW ASSUMES POPULATIONS WHICH ARE VERY LARGE. AS POPULATIONS BECOME SMALLER AND SMALLER, THE EFFECTS OF SAMPLING ERROR IN DETERMINING GENOTYPE FREQUENCIES BECOME PROPORTIONATELY LARGER AND LARGER. IN VERY SMALL POPULATIONS, GENE FREQUENCIES MAY CHANGE UNPREDICTABLY.

These two forces are discussed together because they operate basically in the same manner, by adding or subtracting alleles from the population gene pool. Suppose that a population is made up entirely of *AA* individuals. The gene frequency of *A* could be moved from 1.00 either by

Mutation and Differential Migration

an *A* allele mutating into an *a* allele *(A → a)* or by the arrival of an *Aa* migrant from another population. Mutation is a reversible process. Therefore, one would expect that, in a population where mutation *A → a* was going on at a certain rate, gene-frequency change would be opposed by the force of back mutation *(a → A)*. In the absence of other forces, mutation and back mutation will cause gene frequencies to change until an equilibrium value is reached. That value will be determined by the relative magnitudes of the mutation rate and the back-mutation rate. If the rates are equal, then the equilibrium value will be at $p = q = 0.50$. If the rate of *A → a* is higher than the back-mutation rate, then the equilibrium value of *a* (gene frequency of *A*) will be less than 0.50.

Migration, in order to be an evolutionary force, must be **differential.** In order to change gene frequencies, either immigrants or emigrants must have a genetic constitution different from that of the population as a whole. Adding 10 *AA* individuals to a population of 1,000 *AA* individuals will not change its gene frequency one bit. If both differential immigration and differential emigration are occurring simultaneously, their joint effects may be calculated in a manner similar to that employed for mutation. If they are having opposing effects, as do mutation and back mutation, gene frequencies will move toward an equilibrium point. Differential migration is often, for obvious reasons, referred to as **gene flow.**

THE HARDY-WEINBERG LAW ASSUMES THAT PARTICULAR ALLELES WILL BE NEITHER DIFFERENTIALLY ADDED TO NOR DIFFERENTIALLY SUBTRACTED FROM A POPULATION. MUTATION AND DIFFERENTIAL MIGRATION (GENE FLOW) DO JUST THIS AND THEREFORE MAY BE IMPORTANT EVOLUTIONARY FORCES.

Selection

Other evolutionary forces are relatively insignificant when compared with the overriding importance of **selection.** It is the process which resulted in dinosaurs, roses, whales, mosquitoes, pine trees, viruses, and man. It is the link between changes in the environment and changes in the genetic material of the species attempting to survive and reproduce in that environment. It makes evolution possible. And yet, for all its importance, it is not really proper to think of selection as a discrete thing or force. You will recall from Chap. 2 that it is really a simple concomitant of reproduction and heredity, an inevitable result of like imperfectly begetting like. Selection is the **differential reproduction of genotypes.** Sometimes we distinguish between natural selection and artificial selection, which is imposed by man.

THE HARDY-WEINBERG LAW ASSUMES THAT ALL GENOTYPES IN A POPULATION REPRODUCE EQUALLY SUCCESSFULLY. IF THERE IS

GENOTYPE	AA	Aa	aa	**TABLE 13-4**
FREQUENCY	0.25	0.50	0.25	
FITNESS	1	1	0	
SELECTION COEFFICIENT	0	0	1	

DIFFERENTIAL REPRODUCTION OF GENOTYPES, THEN EVOLUTIONARY CHANGE MAY OCCUR AND SELECTION HAS OCCURRED.

DIRECTIONAL SELECTION   Let's take a look at a specific selective situation, an example of directional selection.   Suppose we had a hypothetical population in Hardy-Weinberg equilibrium for the alleles at the A-a locus, with $p = q = 0.50$. The genotype frequencies would then be as shown in Table 13-4. We are now going to impose selection on the population by sterilizing or shooting all aa individuals before they can reproduce. Assuming that AA and Aa individuals are equally good reproducers in the environment of this population, **reproductive values** may now be assigned to all three genotypes. These reproductive values, known as **fitnesses,** conventionally run from 0 to 1. The value 1 is assigned to the genotype of genotypes which, on the average, reproduce the most, and values less than 1 are assigned, proportionately, down to zero (no reproduction).

The fitnesses for our three genotypes are also shown in Table 13-4, as are the **selection coefficients** (the difference between the assigned fitness and the maximum fitness of 1). Notice that in this example the fitnesses and selection coefficients are all 0 to 1. In other cases, intermediate values would occur, perhaps fitnesses of 0.98 (selection coefficient of 0.02) or fitnesses of 0.30 (selection coefficient of 0.70). At least one genotype is always assigned the fitness of 1, but often no genotype in an array will have a fitness of 0.

The results of imposing such a set of fitnesses generation after generation is easily worked out algebraically (see Box 13-1). As you would expect, since the selective regime consists of preventing reproduction by aa individuals, the overall result is a steady reduction in the frequency of gene a. The amount of reduction per generation is given by the formula

$$\Delta q = \frac{-q^2}{1 + q}$$

where Δq = change in gene frequency per generation. That is, the rate of reduction of the gene frequency *is itself a function of the gene frequency.* The higher the gene frequency of a, the more rapidly a alleles will be removed from the population. Table 13-5 shows the rate of drop of q in a single generation of selection for various values of a. Notice

GENE FREQUENCY (q)	DECREASE PER GENERATION
0.9	0.426
0.5	0.167
0.1	0.009
0.05	0.0024
0.01	0.000099

TABLE 13–5
(Adapted from Paul R. Ehrlich and Richard W. Holm, THE PROCESS OF EVOLUTION. Copyright McGraw-Hill Book Company, 1963. Used by permission.)

that as q gets smaller, it becomes more and more difficult to reduce it further. This is an important and not immediately self-evident result, which, as you can see (by consulting Box 13–1), was achieved by some rather simple algebra. It, for instance, places many eugenic schemes (programs for the genetic improvement of *Homo sapiens* by selective breeding) into immediate perspective. Suppose a particular undesirable allele *a* had a gene frequency of $q = 0.01$ in the human population, so that q^2 *(aa)* individuals made up 0.0001 of the individuals (1 defect per 10,000 "normals"). It would take 100 generations (roughly 2,500 years) of a program of sterilization of defective individuals to halve the gene frequency and reduce the number of defective individuals to 1 in 40,000. The social problems of carrying out such a program for such a protracted period make it highly unlikely that such meager results would justify the effort and probable disruption which would be involved.

AS SELECTION AGAINST A RECESSIVE ALLELE OCCURS, THE FREQUENCY OF THAT ALLELE DECREASES AND THE PROPORTION OF THAT ALLELE IN HETEROZYGOUS RATHER THAN HOMOZYGOUS INDIVIDUALS INCREASES. THEREFORE, THE RATE OF REMOVAL OF THE RECESSIVE ALLELE DECREASES IN TIME.

DIRECTIONAL ARTIFICIAL SELECTION   The selective regime just described is an example of directional selection. The regime moved the frequency of allele *a* toward a lower and lower value. Artificial directional selection is often applied by plant or animal breeders to quantitative characteristics whose genetic background is not fully understood. It consists basically of using organisms at one end of the distribution of a character as the parents in the next generation (Fig. 13–2). A farmer carries out directional selection when he uses only the best egg producers among his chickens or only the fattest of his swine as breeding stock. Usually considerable improvement in domestic animals can be achieved by directional selection. Although a really fat swine would be unable to survive in nature because it could not escape predators, the farmer creates an environment in which this is no longer a problem.

Of course, this doesn't mean that a farmer can go on producing fatter and fatter swine ad infinitum. Sooner or later, something in the system

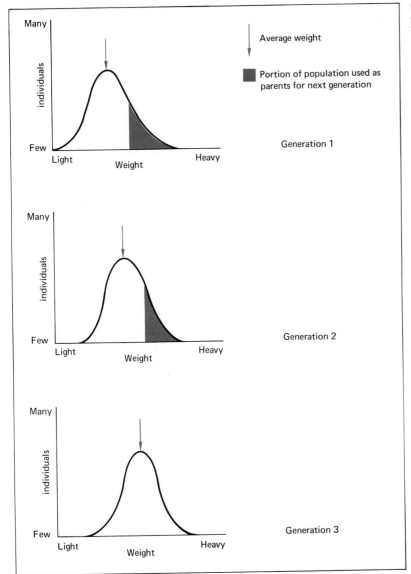

Fig. 13-2 Three generations of directional artificial selection in which the mean weight is shifted.

will give. Perhaps sows will begin to have trouble giving birth to increasingly large piglets, and mortality among mothers and offspring will increase. Or perhaps heart efficiency will be unable to keep up with the need for increased vascularization. Or perhaps the swine will start getting fatal infections from sores produced by perpetually dragging their bellies across the barnyard. Whatever happens, sooner or later natural selection will counter the artificial selection practiced by the farmer and "progress" will come to a halt. Eventually, if the farmer keeps trying, mutant or recombinant swine may appear with broader pelvises, more efficient hearts, or tougher bellies, in which case further gains in weight may be possible.

A general pattern for experiments in artificial directional selection is illustrated in Fig. 13-3. Progress is rapid at first, with the selected line or lines rapidly diverging from the control line in which parental individuals are selected at random. Then progress slows or comes to a halt —a "plateau" is reached. If selection is continued, progress may resume after a period of no change, or it may never resume. Selection, on the other hand, may not be continued—it may be "relaxed." In the example shown in Fig. 13-3, selection consisted of breeding, in each generation, from the *Drosophila* with the greatest (high line) or smallest (low line) number of bristles. When selection was relaxed, parents were selected at random from the line in each generation. As you can see, some of the relaxed lines moved back toward the control level. In other cases of relaxed selection, regression toward the control line was not observed.

ALTHOUGH ARTIFICIAL DIRECTIONAL SELECTION MAY BE APPLIED TO A POPULATION, NATURAL SELECTION OCCURS AS WELL, AND THE FINAL EFFECTS ARE THE RESULT OF THE INTERACTION OF ALL ASPECTS OF SELECTION.

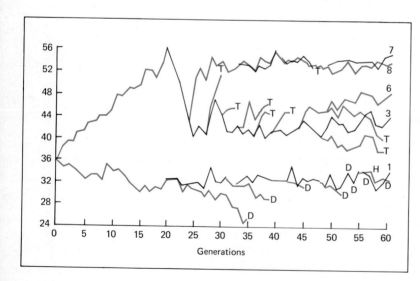

Fig. 13-3 Artificial selection for abdominal bristle number in *Drosophila melanogaster*. Mean number of bristles is plotted against number of generations; lines under selection are black; lines not under selection are colored. T indicates deliberate termination of a line, and D, that it died out through sterility. The numbers indicate different lines; from line 1, all the selections were for a low number of bristles except that marked H, which was for a high number of bristles. (*After Mather and Harrison.*)

These results may be explained in a general way as follows. As the first rapid progress under artificial selection is made, changes occur in the individuals with the most extreme expression of the character. These changes reduce the reproductive capacity of the individuals. The changes may be relatively gross and obvious, such as those in our hypothetical swine or a simple loss of fertility in the eggs in a *Drosophila* experiment. In each batch of individuals chosen as the parents of the next generation, the less extreme outreproduce the more extreme, and eventually no more progress in artificial selection is possible. Natural selection has countered artificial selection. Or a plateau may be reached because the genetic variability of the population (with regard to this character) has been exhausted. In essence, this means that all the individuals now have the genotype that, within the constraints imposed by linkage, have the kind and arrangement of genetic information which produces the most extreme phenotype possible.

Eventually either mutation or a rare crossover or two, permitting further reorganization of the genotype, may provide the genetic basis for further selective advance. But if this does not occur, the end of the line has been reached. What if selection has been relaxed? Further events will depend on how effective at reproducing are the organisms with, say, high bristle number in comparison with those of lower bristle number. If the move to high number has been achieved at the cost of the exhaustion of genetic variability but with no reduction of mating ability, egg production, egg fertility, larval vigor, and all the other elements included in "reproductive ability," then the relaxed line will continue to have many bristles. If, on the other hand, the genetic changes necessary to produce those extra bristles partially fouled up the egg-production apparatus, so that bristly females lay fewer eggs, then natural selection will cause the line to regress toward the control.

FOR CONVENIENCE, SELECTION IS NORMALLY STUDIED ONE LOCUS OR ONE CHARACTER AT A TIME; NEVERTHELESS, IT IS THE ENTIRE ORGANISM THAT IS THE BASIC UNIT OF SELECTION. PROGRESS IN CHANGING ONE CHARACTER CANNOT BE BOUGHT AT THE PRICE OF DESTROYING THE REPRODUCTIVE VALUE OF THE ENTIRE GENOTYPE. SUCCESS IN A GIVEN ENVIRONMENT IS THE RESULT OF THE ORGANISM BEING A MORE OR LESS WELL-INTEGRATED SET OF COMPROMISES.

DIRECTIONAL NATURAL SELECTION   Directional selection has been observed in some natural populations, a classic example being that of industrial melanism in moths, discussed in Chap. 1. Another well-known

Fig. 13–4 Banded and unbanded forms of the land snail *Cepaea hortensis. (Courtesy of W. H. Dowdeswell.)*

case involves the highly variable land snail *Cepaea nemoralis* (Fig. 13–4). These snails vary, among other things, in the amount of banding on their shells (from none to many) and in the color of their shells (yellow or brown). Every small, localized population of the snails is quite different, and the pattern of variation is complex and, at first sight, seems little related to variation in the environment. Some investigators concluded that the variation is the result of the peculiar genetic constitution of the individuals that founded each population and the subsequent occurrence of genetic drift. A closer investigation revealed, however, that selection played a major role in determining the observed patterns.

Careful investigation of the details of the background revealed, for instance, that the rougher and more complex and broken up the substrate, the higher the frequency of many-banded snails. Snails in hedgerows tend to be much more banded than snails in grassland. A fortunate aspect of the situation was that thrushes, chief predators of the snails, have the convenient (for biologists) habit of cracking them on a favorite rock, a so-called "thrush anvil." By comparing the kinds of shells found around thrush anvils with those of living snails in the population, it was possible

TABLE 13-6

	BANDED	UNBANDED	TOTAL	% UNBANDED
Random sample	264	296	560	52.8
Thrush-eaten	486	377	863	43.7

to determine whether or not the thrushes were eating a ***random sample*** of the snails. As you may have guessed by now, they were not. For instance, Table 13-6 gives the percentages of banded and unbanded shells in a random sample of 560 individuals taken from a bog by biologists and 863 shells taken from around thrush anvils adjacent to the bog. Remember that these latter individuals represent the snails eaten by the thrushes.

The difference between the two samples is statistically significant, indicating that unbanded individuals in the bog are less likely to be eaten than banded individuals. Presumably, the relatively uniform background of the bog made it more difficult for the birds to spot the unbanded individuals.

The thrushes were clearly functioning as a ***selective agent,*** causing differential reproduction among the snails by a most direct method. Further investigations have shown that *Cepaea* are also under selective pressure from other factors in their environments, including temperature and humidity. Snails of some shell types, for instance, are also more resistant to desiccation than snails of other shell types. But the last word on the evolution of populations of these interesting creatures is certainly not in. For instance, certain of their characteristics remain unchanged over wide areas and then change very rapidly in a short distance to another widespread form. Similar phenomena have been observed in the distribution of eyespot characters in *Maniola* butterflies (Fig. 13-5), where a change from one stable widespread pattern to another occurs along a narrow boundary which shifts in position from year to year. Presumably the explanation lies in a balance of selective forces, but the phenomenon is far from understood.

Sometimes directional selection may be countered by another evolutionary force to produce a stable situation. In the 1950s, biologists noticed that although most adult water snakes *(Natrix sipedon)* on the islands of Lake Erie were bandless or only weakly banded, most of the young were strongly banded (Fig. 13-6). In nearly all of eastern North America, adult water snakes are strongly banded. On the flat limestone rocks of the Lake Erie islands, however, the bands had almost disappeared, presumably because the bandless snakes were more protectively colored against the background of uniformly colored rock. In the more usual marshy or reedy water's-edge habitat of the snake, banding presumably made the snakes less conspicuous. By comparing samples of newly born, half-grown, and adult snakes, it was possible to show that

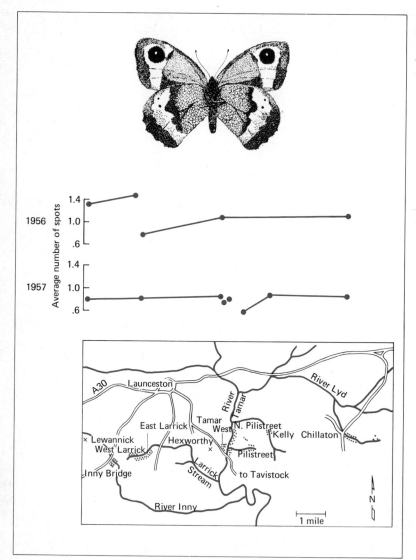

there was differential mortality—those young snakes that were unbanded or lightly banded had a much better chance of survival than the heavily banded young. This immediately raised a question: Why, if selection were favoring unbanded snakes so strongly, had not the island populations become totally unbanded? Why was there still such an abundance of genes for banding in the island gene pool?

At least a partial answer was suggested. On the mainland surround-

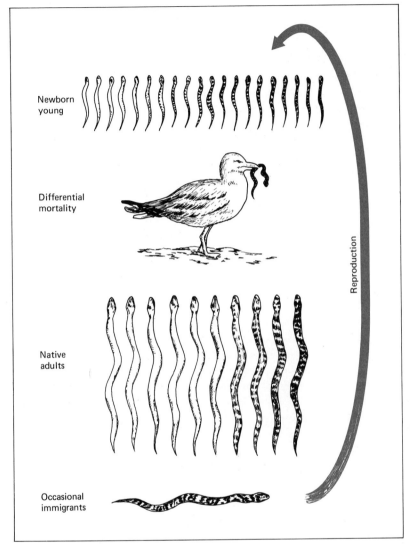

Fig. 13-6 Differential mortality in water snakes of the genus *Natrix*. It is thought that the visual predators (sea gulls) consume more banded young and therefore the proportion of unbanded adults increases. Occasional immigrants keep the ''banded genes'' in the population. (*Adapted from Paul R. Ehrlich and Richard W. Holm, PROCESS OF EVOLUTION. Copyright McGraw-Hill Book Company, 1963. Used by permission.*)

Newborn young

Differential mortality

Reproduction

Native adults

Occasional immigrants

ing Lake Erie, water-snake populations were typically banded. Water snakes are strong swimmers and were often seen far out in the lake. Perhaps banded snakes from the shore provided a continuous migration of ''banded genes'' into the island populations. It would have been nice if further investigations could have provided a firm answer, especially since the significance of migration as an evolutionary force in natural populations is a matter of some debate at present. But the solution to the problem is forever lost to us. As you will learn in the last chapter, Lake Erie has been killed by pollution, and the once huge water snake populations have been decimated.

Plants, of course, like animals, are subject to directional selection. Evolution in plants has not, however, generally been studied in the same way as it has in animals, and so examples precisely parallel to those given above are difficult to find. In one study, two halves of a pasture were planted with a variety of plant species and then one half was grazed and the other half protected from grazing. When plants of the same species from the two halves were both placed in the uniform conditions of an experimental garden, after 3 years it was found that those plants from the protected half grew into upright individuals, whereas those from the grazed half tended to creep along the ground. Clearly, selection had favored genotypes with low growth on the grazed side—taller individuals were eaten. On the ungrazed side, height was necessary. Short individuals would be shaded out by taller ones and would die or set less seed.

In similar transplant experiments, it has been shown that many plant forms characteristic of various habitats are genetically produced. For instance, J. Clausen, D. Keck, and W. Hiesey showed that alpine populations of yarrow *(Achillea)* are short not merely because low temperatures or a brief growing season keep them short. They are genetically dwarfed; even when planted in a warm experimental garden near sea level they did not grow tall. In contrast, dandelions *(Taraxacum)* with identical genotypes produce a wide variety of growth forms in response to warm, cold, wet, and dry environments. One great advantage of perennial plants for this sort of investigation is that they can be **cloned.** Single genetic individuals can be divided and reproduced vegetatively, so that identical genotypes can be grown in different environments. This provides a powerful tool for investigating the "nature versus nurture" problem—that is, for determining the relative contributions of genetic information and environment to the ultimate phenotype. Plant studies of this sort have been extended from such obvious characters as height and leaf form to more subtle physiological characters, such as rates of photosynthesis and respiration.

ALTHOUGH IT IS DIFFICULT TO ISOLATE THE VARIOUS EVOLUTIONARY FORCES AND STUDY THEM SEPARATELY, THERE ARE WELL-DOCUMENTED STUDIES OF EVOLUTION IN NATURAL POPULATIONS. THESE INVOLVE NOT ONLY ASPECTS OF THE PHENOTYPE THAT CONFER PROTECTION FROM PREDATORS BUT ALSO BASIC PHYSIOLOGICAL PROCESSES OF PLANTS AND ANIMALS.

MAN AS AN AGENT OF SELECTION   Man has, in the last thousand years or so, managed to alter every environment on the face of the earth. Even in areas which do not show obvious signs of his presence, higher radiation levels (from fallout), the presence of synthetic molecules of his manufacture (especially chlorinated hydrocarbons such as DDT), and reduced input of solar energy (because of the smog veil which is now planetwide) attest to the activities of *Homo sapiens*. As a great modifier of environments, man has, of course, also acted in many different ways as a selective agent. The case of industrial melanism is an obvious example. Others involve man's attempts to poison organisms which attack him directly, destroy his artifacts, or compete with him for food and fiber. The attacking organisms, mostly insects and bacteria, show genetic variability among individuals in their susceptibility to poisoning. There-fore, applying an insecticide or bacteriocide (antibiotic) to a population amounts to *directional selection for resistance* to the poison. Most of the organisms will probably be killed in the first application, but the relatively resistant survivors will produce a relatively resistant gen-eration of offspring. Repetition of this process eventually may produce a resistant strain of the target organism. Indeed, myriad insect populations are now resistant to various pesticides, and bacterial resistance to antibiotics has greatly reduced the efficacy of penicillin and other "won-der drugs."

Man himself is, of course, subject to continuous natural selection. It is a common error to assume, as some have, that modern science has removed from the shoulders of man the "burden of natural selection." As you now realize, selection is not a "thing" which can be removed; it will be with man so long as he is genetically variable and reproductive differentials exist among individuals. True, today's selection pressures may be quite different from those of 1,000 years ago, but they are prob-ably not as different as many people believe. Agility and coordination may rarely be needed today for escape from wild animals, but they are extremely useful for preserving the life of both driver and pedestrian. People in our society who are diabetic now normally live to reproduce, with the aid of insulin. Therefore, selection pressure against the genes which predispose to diabetes has probably been relaxed, as has been that against genes for myopia, genes predisposing individuals to tuber-culosis, and those associated with numerous other disorders. But pres-sures on genes which predispose to asthma, bronchitis, lung cancer, cholesterol clogging of arteries, hypertension, various mental diseases, and so forth, have been greatly intensified. Man is probably being se-lected for resistance to chlorinated hydrocarbons, increased dosages of radiation, and nitrite poisoning (see Chap. 16).

MAN HAS BECOME AN IMPORTANT AGENT OF SELECTION FOR HUMAN AS WELL AS OTHER POPULATIONS. THE CHANGES HE HAS MADE HAVE ALTERED SOME SELECTIVE AGENTS AND INTRODUCED MANY NEW ONES INTO THE ENVIRONMENT OF MAN AS WELL AS THAT OF OTHER ANIMALS AND PLANTS.

POLYMORPHISM Man, in addition to being faced by directional selection pressures, is also subject to selection which, rather than changing gene frequencies, holds them stable. For instance, in parts of Africa some people die before puberty of a painful anemia, one which characteristically makes a percentage of the red blood cells lose their normal round shape and become sickle-shaped (Fig. 13–7). It was discovered that people with this sickle-cell anemia are homozygous for an allele Hb^S. People homozygous for another allele at the locus, Hb^A, do not suffer from anemia, nor do heterozygous individuals, $Hb^A Hb^S$. It is possible, however, to detect heterozygotes because, if their blood is exposed to low oxygen tension, many of the red blood cells will "sickle." Heterozygotes are said to have sickle-cell trait. The homozygous $Hb^A Hb^A$ adults have, in their red blood cells, only normal adult hemoglobin, hemoglobin *A* (thus the allelic label Hb^A). The sickle-cell anemics have only a different hemoglobin, hemoglobin *S*. Individuals with sickle-cell trait (heterozygotes) have both kinds of hemoglobin, with *A* predominating over *S* about 3:1.

In some populations, the gene frequency of Hb^S is in the vicinity of 0.20, much higher than one would expect if the array of fitnesses were similar to those discussed in our first example in which the homozygous recessives were lethal (Box 13–1). It turns out that, in fact, the *heterozygous* individuals are reproductively favored over both homozygous types. The sickle-cell anemics, on the one hand, usually die of their anemia before they can reproduce, so they face a selection coefficient of very nearly 1. The $Hb^A Hb^A$ homozygotes, on the other hand, are relatively susceptible to a nasty form of malaria, which is found in the same places as sickle-cell anemia. The heterozygotes, however, are relatively immune to the malaria and do not get the anemia. The array of fitnesses is then

GENOTYPE: $Hb^A Hb^A$ $Hb^A Hb^S$ $Hb^S Hb^S$
FITNESS: $1 - s_A$ 1 $1 - s_S$

It can be shown that such an array of fitnesses, with the heterozygote favored over both homozygotes, leads to a stable gene-frequency equilibrium. The equilibrium point depends solely on the values of the selection coefficients against the two homozygotes. If p is the gene frequency

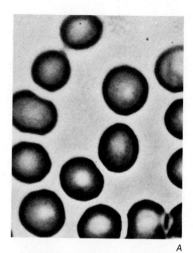

A

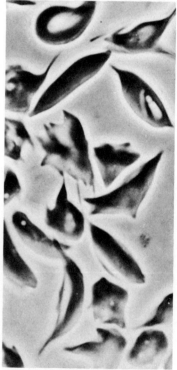

B

Fig. 13–7 *(A)* Normal human red blood cells; *(B)* sickled red blood cells. About × 1000. *(A, courtesy of Eric V. Gravé; B, Jeroboam, N.Y.)*

Hb^A, then the equilibrium value of p is $\hat{p}$ (called "p hat")—given by the simple formula

$$\hat{p} = \frac{\hat{p}_S}{\hat{p}_A + \hat{p}_S}$$

Should the selection coefficients be equal, then $\hat{p} = \hat{q} = 0.50$.

The sickle-cell anemia story is a favorite of ours, as it illustrates the essential unity of the biological sciences. Molecular biologists have been able to elucidate the exact nature of the difference between the two hemoglobins. It consists of a single amino acid residue substitution out of some 300 residues in the molecule. A glutamic acid in S replaces a valine in A. Quite a spectacular phenotypic change results from this seemingly insignificant molecular substitution! Cell biologists are now beginning to understand the mechanism by which people with sickle-cell trait are protected from malaria. The malarial organism *(Plasmodium)* is an intracellular parasite in man, living inside of the red blood cells and feeding on the hemoglobin. Hemoglobin S is more viscous than hemoglobin A. In the process of consuming hemoglobin S, the *Plasmodium* changes the surface of the red blood cell in a manner which permits the liver to identify it as damaged and destroy it. Organism biologists (physicians) first described the problem by diagnosing the anemia. And, finally, population biologists discovered the basic reason that the Hb^S gene persists in populations. Sickle-cell anemia remains an important health problem for Black Americans, but recent public education programs should aid in its control. Lack of understanding of its nature has caused some of its victims to be ashamed of having sickle-cell anemia, an attitude which is not in any way justified.

Under a selective regime favoring heterozygotes, all three genetic forms, both homozygotes and the heterozygote, remain in the population. Biologists refer to such a situation as a **balanced polymorphism** (Greek: "many forms"). Whenever two or more quite different forms of a species occur together in a population and the rarest of the forms is common enough so that its presence cannot be accounted for by mutation rate alone, a polymorphism is said to exist. Man, for instance, is polymorphic for eye color in many of his populations. He is never polymorphic for phenylketonuria, a serious genetic defect which occurs in a frequency which can be explained by mutation alone.

A balanced polymorphism is one maintained by selection favoring the heterozygote. There are other ways of maintaining polymorphisms. For instance, in the industrial-melanism case cited in Chap. 1, populations contained mixtures of dark and cryptically colored forms and the former were in the process of replacing the latter. This is a case of **transient polymorphism,** and it is no longer called a polymorphism when the frequency of cryptic moths drops to the point at which their presence could be explained by invoking mutation alone.

THE EXISTENCE IN A POPULATION OF TWO OR MORE DIFFERENT FORMS OF A SPECIES IS CALLED A POLYMORPHISM WHEN THE RAREST FORM IS TOO COMMON TO BE ACCOUNTED FOR BY MUTATION ALONE.

The equilibrium established in balanced polymorphism is a **stable equilibrium,** meaning that if the gene frequency is moved away from the equilibrium point (perhaps, say, by the arrival of immigrants), it will tend to *return* to the equilibrium point. Hardy-Weinberg equilibrium, on the other hand, is a **neutral equilibrium.** If the gene frequency is changed by an outside force, it will *remain* at the new value. An **unstable equilibrium** may be established in population genetics when selection is against the heterozygote, favoring both homozygous types over it. In such a case, there will be an equilibrium point at $p = q = 0.50$, but any deviation from that point will lead to *loss of one allele or another.* Since such a precarious balance cannot be maintained in real life, it is not surprising that a simple case of polymorphism maintained by selection against heterozygotes has not been uncovered in nature. There is selection against heterozygotes at the Rh (Rhesus factor) blood-type locus in human populations, but this situation is complicated by other factors. If you wish to picture physical analogs of stable, neutral, and unstable equilibriums, think of a ball bearing in the bottom of a hollow cone, a ball bearing on a perfectly flat sheet of glass, and a ball bearing balanced on the point of a cone, respectively.

STABILIZING SELECTION AND DISRUPTIVE SELECTION Two other types of selection will be mentioned briefly (Fig. 13-8). One is *stabilizing selection,* in which individuals near the mean are more likely to produce offspring than those at the extremes of the population. This sort of selection presumably goes on constantly in most populations. For instance, extremely short and extremely tall human beings are less likely to reproduce than those nearer the mean. In direct opposition to this is *disruptive selection,* in which extreme individuals are at an advantage over those nearer the mean. Laboratory experiments using rigorous disruptive selection have led to the establishment of two quite different populations from a single parental population. It is not known how important disruptive selection is in nature, but for its occurrence quite special conditions are required. One case in which it may be important

Portion of population which are parents for next generation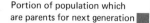

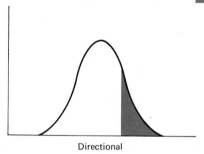

Directional

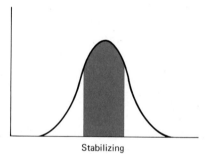

Stabilizing

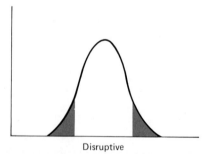

Disruptive

Fig. 13-8 Diagram to show three basic types of selection. The portions of the population which are parents for the next generation are colored.

is in the establishment of mimetic polymorphisms, a subject which is discussed below.

SELECTION IS THE DIFFERENTIAL REPRODUCTION OF GENOTYPES. IT MAY LEAD TO DIRECTIONAL CHANGES IN THE CHARACTERISTICS OF A POPULATION, TO THE MAINTENANCE OF POLYMORPHISM IN A POPULATION, TO THE ELIMINATION OF EXTREME GENOTYPES, AND TO THE FRAGMENTATION OF A POPULATION.

DIFFERENTIATION OF POPULATIONS

As we pointed out at the beginning of this text, one of the most prominent characteristics of life is its diversity: instead of a single living film coating the surface of the planet, life comes packaged in a diversity of different kinds of units. These kinds, you will recall, we label **species.** Darwin, the father of modern evolutionary thought, was so impressed with diversity that he called his monumental work *On The Origin of Species.* Let us now look at how diversity was and is generated, or how populations become differentiated from one another, or in Darwin's terms, how species originate.

Geographic Variation

No two places on the face of the earth are identical; the physical substrate for life is heterogeneous. The amount of solar radiation varies geographically. Climate varies geographically. Soils vary geographically. Water salinity varies geographically. The heterogeneity of the physical environment means that populations of organisms in different areas are inevitably exposed to different selection pressures. This means that organisms are going to vary geographically. In sexually reproducing organisms, no two populations will be identical genetically; natural selection sees to that. As you will see at the start of the next chapter, all organisms are restricted in their distributions. No species has managed to persist in anything like all the possible habitats available to it. This means that no two areas will have exactly the same biological communities. Now, of course, one species of organism often exerts selection pressures on another (remember those malarial organisms and sickle-cell anemia). Therefore, there is a living component to the heterogeneity of the environment, which may be even more important than the physical component—a further reason that geographical variation is ubiquitous.

You have already been introduced to examples of geographic variation. Melanism in moths varies geographically in response to selection pressures. So do color and banding in *Cepaea* shells and banding in water snakes. Yarrow plants growing at high altitudes are genetically different from those growing lower. Variation may be easily continuous across wide areas, or it may be discontinuous, the population being divided into subpopulations of discrete types. An example of the analysis of geographic variation in response to selection pressures is given in Box 13–2.

Perhaps the best-known example of geographic variation, and certainly the one that has been studied most thoroughly, is geographic variation in *Homo sapiens* (Fig. 13–9). Man varies in such diverse characters as the

A

Fig. 13–9 Two different solutions to the problem of losing excess heat are being tall and slender or being small: (*A*) The young girls of the Nuba tribe of Southern Sudan illustrate the first means; (*B*) the group of pygmies illustrate the second. (*A, Courtesy of George Rodger, Magnum Photos; B, American Museum of Natural History.*)

B

shape of his head, the frequencies of many kinds of blood-group genes, skin color, hair type, the distribution of genes controlling the ability to taste certain chemicals, and tooth size. Sometimes it is possible to suggest what environmental variables are responsible for the variation and thus identify the selection pressure. Skin color and hair type tend to be related to environmental differences in solar radiation, tooth size to diet, some blood groups with parasitic disease, and so forth.

VIRTUALLY ALL ORGANISMS VARY IN SEVERAL TO MANY TRAITS ACROSS THE EXTENT OF THEIR GEOGRAPHIC RANGE. SUCH GEOGRAPHIC VARIATION MAY BE CONTINUOUS OR DISCONTINUOUS, MARKED OR SLIGHT.

Speciation

How does one get, then, from geographic variation within a species to the formation of new species? The answer is that there is a continuum between levels of differentiation; when geographic variation has gone to the point at which discontinuities in the pattern of variation are readily observed, then taxonomists (those biologists who specialize in classification) will regard each of the geographic isolates as different species instead of parts of the same species. The usual standard applied is one

Fig. 13-10 Two color types of a California salamander (*Ensatina*) which occur within several yards of one another. (*After Stebbins.*)

of discontinuity, not of *degree of difference*. Look at the two salamanders in Fig. 13-10. They would seem to be "different kinds" by anybody's standards. And yet they are usually considered part of the same species by taxonomists, even though they live in the same area and rarely interbreed. The reason can be seen by looking at Fig. 13-11. They are connected along a geographically circuitous route by a continuum of forms.

Continua perpetually make problems for those attempting to put organisms in pigeonholes. Just as there is no logical place to divide a spatial continuum, as in the salamander case, so there is often no place for a taxonomist logically to divide a continuum in time. Suppose that a single population does not fragment into a number of species as it evolves through time but continually changes in response to changing environments. A paleontologist tracing its fossil remains may find that over a period of millions of years the population has changed dramatically.

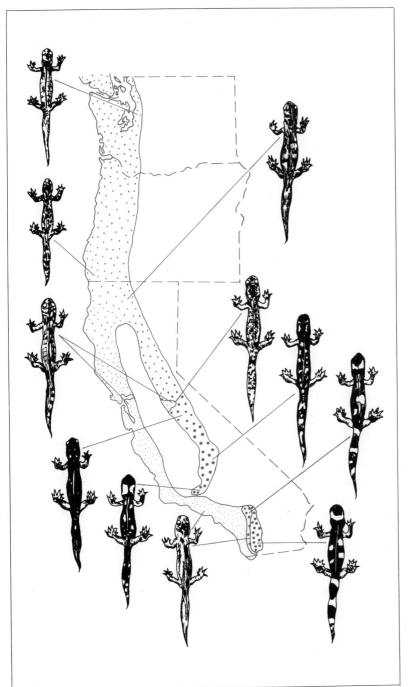

Fig. 13–11 The distribution of color types of *Ensatina* in California. The two forms shown in Fig. 13–10 are at the bottom. (*Adapted from Paul R. Ehrlich and Richard W. Holm, PROCESS OF EVOLUTION. Copyright McGraw-Hill Book Company, 1963. Used by permission.*)

The individuals at the end of the sequence are most certainly a "different species" from those at the start of the sequence. But at what point in time did the population stop being the early species and become the late species? In biological terms there is no answer to the question; it is asked only because man must impose a structure on nature if he is to communicate about it. In practice, the paleontologist is often blessed by gaps in the fossil record which make convenient places for "stopping" one species and "starting" another.

POPULATIONS ARE USUALLY RECOGNIZED AS DISTINCT SPECIES IF THEY ARE READILY DISTINGUISHABLE FROM OTHER POPULATIONS AND IF POPULATIONS WITH INTERMEDIATE CHARACTERISTICS ARE NOT KNOWN (THAT IS, IF THE VARIATION IS DISCONTINUOUS).

Biologists now realize that there are many kinds of relationships among populations. A same-species–different-species division is adequate for taxonomic purposes but inadequate to an understanding of the process of the differentiation of populations. This process seems to be controlled by a balance of selective forces. In a simplified scheme, one might think of these forces as being of two types. The first are the forces of genetic change; they tend to fit the organism to the environment, to tailor each population to its location and time. The second are forces of conservatism; they tend to maintain a "good thing," a well-integrated organism that is a functional whole. In most organisms, these forces produce a kind of compromise.

As a result of selection pressure from certain kinds of malaria parasites, human populations tend to switch to hemoglobin S—that is, in a sense, to "track" the environmental change. But more conservative selection pressures oppose the move; hemoglobin A is necessary for survival, and a compromise is found. If the pressure for change had escalated sharply, say through attack by a new strain of *Plasmodium* which wiped out $Hb^A Hb^A$ individuals as well as a good portion of the heterozygotes, two things might have happened. First, the population might have "found" a new tactic to counter the threat—perhaps the production of a hemoglobin sharply different from either A or S. The population of humans could then have differentiated further in response to environmental pressure. Failing to do this, it might have become extinct.

Again, in a simplified scheme, we might think of two general strategies as being open to populations. One is to track environmental changes genetically, that is, to permit selection continuously to alter the genetic properties of the population so the organisms are "in tune" with the environment. This is an especially attractive strategy for organisms which have short generation times and are numerous. Short generation times permit keeping up with environmental change by producing new genotypes as required. In contrast, a Temperate Zone organism with a 2-year life cycle cannot first produce individuals adapted genetically to heat during the summer and then those suited for cold during the winter. It must follow a strategy of evolving one kind of genotype which permits viable interaction with a variety of environments, that is, which can meet many environmental changes with appropriate phenotypic responses. One response might be a virtual suspension of metabolic activity when conditions are bad (a strategy employed by many insects in winter and by pond microorganisms when the pond dries up). Another might be doing increased physiological work to maintain homeostasis (human beings shivering in the cold). Of course, no species uses one strategy to the complete exclusion of the other. Bacteria, which are primarily environment trackers, commonly show altered metabolism in response to nutrient changes in their substrate. Human beings, who have added cultural homeostatic mechanisms to their physiological ones (we don't just shiver, we put on a sweater!), also change genetically in response to environmental changes.

POPULATIONS MAY RESPOND TO ENVIRONMENTAL CHANGE EITHER BY PRODUCING GENETICALLY DIFFERENT OFFSPRING MORE SUITED TO THE CHANGED CONDITIONS OR BY EVOLVING A GENOTYPE WHICH IS SUCCESSFUL UNDER MANY ENVIRONMENTAL CONDITIONS.

The checkerspot butterflies on Jasper Ridge seem to have a more or less intermediate strategy. Their species, *Euphydryas editha,* occurs in isolated colonies throughout the San Francisco Bay area, living mostly in places where there are outcrops of serpentine rock. Climate and exposure vary considerably from colony to colony, and yet samples from various colonies are at least superficially quite similar. Taxonomists have always considered them to represent not just a single species but a single geographic unit within a species, that is, a subspecies. These butterflies, with a 1-year life cycle and great similarity from population to population, would seem to be at the homeostatic end of the scale. They represent a successful strategy for dealing with the kinds of environmental variation found in serpentine outcrops in the bay area. Presumably, selection for environment tracking would tend to break up the "winning combination" which has evolved and would thus be countered by selection for maintaining that combination. To a large degree, *Euphydryas editha* seems to have found a good way to play the game of life successfully, and individuals that try to change the rules end up at a reproductive disadvantage. To this extent, they are like us and unlike bacteria, which can meet most environmental changes *only* by "changing the rules" (that is, by changing genetically).

But that is not the whole story. We know that, at least in some of its characters, *Euphydryas editha* tracks the environment. A careful study of inconspicuous wing-marking differences has shown these to be changing through time in the Jasper Ridge population. These changes are genetic, caused by selection by an unknown agent, and are sometimes synchronized in isolated populations. Apparently, at least part of the *E. editha* genotype may respond continuously to selection without causing serious upset to the overall "winning combination." As you must realize by now, there is still a great deal to learn about evolution in natural populations, especially with regard to the kinds of strategy matters we have been discussing above. We know, from the types of laboratory selection experiments discussed earlier, that populations are not always free to change in response to selection pressure. Our understanding of the mechanisms which tend to limit the degree of genetic change (**genetic homeostasis,** as I. Michael Lerner called such limitation) is unfortunately very poor.

It now seems likely to many biologists that the major factor controlling the differentiation of populations is the balance between selection for environment tracking and selection for "homeostasis." Suppose an organism has evolved a successful strategy for survival on low coral atolls. If it were able to disperse from atoll to atoll, one might expect its populations to vary little among atolls; they would be living in similar environments and would have a standard successful strategy for coping with them. This is precisely what happens to a large number of atoll plants. In contrast, however, when these plant groups move to high islands, the inland populations are subjected to different selection pressures and often differentiate. Similarly, the North American tiger swallowtail butterflies, *Papilio glaucus* (eastern) (Fig. 13-12) and *Papilio rutulus, P. multicaudatus,* and *P. eurymedon* (western), are strong fliers and show little differentiation throughout their range.

In a certain part of its range, however, *P. glaucus* occurs with another swallowtail butterfly, *Battus philenor* (Fig. 13-13). The latter butterfly does not have the yellow and black "tiger" pattern of *P. glaucus;* it is largely black. It is also poisonous to predators, which *P. glaucus* is not. Wherever *P. glaucus* occurs with *B. philenor,* selection produces a dark female form of *P. glaucus* which mimics *B. philenor* (Fig. 13-13). Mimicry will be discussed in detail in the next chapter; here the point is that strong tracking selection has been able to break up the winning tiger pattern—one which has been adopted by the four widespread North American butterflies mentioned, as well as by a great many butterflies

Papilio glaucus

Papilio rutulus

Papilio eurymedon

Papilio multicaudatus

Fig. 13-12 Four North American swallowtail butterflies.

of diverse groups in other parts of the world. Often two or more forms are found in the same population, as in many *P. glaucus* populations where both mimetic and nonmimetic forms occur. In populations of an African swallowtail butterfly, there are female forms which mimick different models, the mimetic forms varying geographically to match geographic variation in the models. Such **mimetic polymorphism** is common in butterflies. Apparently it is made possible by a rearrangement of the genetic material and the establishment of a single "switch" gene. This gene permits a dramatic switch in pattern without seriously affecting the rest of the genotype. For reasons which are poorly understood, the polymorphisms are restricted to the females.

There are, of course, many other examples of species which show relatively little obvious differentiation over much of their range, but as we indicated earlier, geographic variation is the rule. The house sparrow, *Passer domesticus,* has evolved differentiated populations in the continental United States and Hawaii (where it was introduced from Europe) in as little as 50 years. *Cepaea* populations tend to be different everywhere and to change rapidly in color and pattern (they even genetically track seasonal changes in foliage). A very large number of moth populations in the United States, England, and continental Europe have rapidly differentiated in areas subject to industrial pollution. The length of appendages in mammals varies in relation to temperature. Short appendages mean less body area per unit volume and greater ease of conserving heat; longer appendages mean easier radiation of excess heat.

In plants, genetically based geographic variation may occur in very small areas. This is found in plants which are wind-pollinated, extreme outcrossers, and unable to fertilize themselves, such as the grasses *Festuca rubra* and *Agrostis tenuis.* These grasses form genetically differentiated populations which may occupy areas in nature only a few feet in diameter. And these populations are surrounded by plants of a different population. The differentiation of such localized populations is dependent on the interplay of natural selection and gene flow, as well as on the mating system and genetics of the organisms. The advantage of particular genotypes reproducing under a particular, often extremely local, set of conditions may be so pronounced that other genotypes are systematically eliminated.

Plants which grow in soil polluted by wastes from lead and copper mines have, in several studies, been shown to be special genotypes which show resistance to the poisons. These genotypes are not normally successful when transplanted to nonpolluted soil. Reciprocally, plants growing immediately adjacent to the pollution do not have these geno-

Battus philenor (model)

Papilio glaucus (mimic form of female)

Papilio glaucus (tiger form of female)

Fig. 13-13 Mimetic and nonmimetic forms of the swallowtail butterfly *Papilia glaucus* and the model *Battus philenor.* (*After Brower.*)

types and die if transplanted onto the polluted soil. Selection keeps the populations strongly differentiated.

The basic process of speciation, then, involves the differentiation of populations under different selection pressures. Eventually two populations will become so distinct they will be considered "different species." Two major questions, only partly independent, remain to be answered. The first has to do with the importance of gene flow between differentiating populations. The second has to do with the possibility of speciation occurring within a single interbreeding population.

A BALANCE BETWEEN HOMEOSTASIS AND THE ABILITY TO RESPOND GENETICALLY TO CHANGING ENVIRONMENTAL CONDITIONS MUST BE MAINTAINED IN ALL POPULATIONS. GEOGRAPHICALLY DIFFERENT POPULATIONS OF THE SAME SPECIES ARE EXPOSED TO DIFFERENT SELECTION PRESSURES. IT IS TO BE EXPECTED THAT MANY SUCH POPULATIONS WILL EVENTUALLY DIFFERENTIATE TO THE DEGREE THAT THEY ARE RECOGNIZED AS DISTINCT SPECIES.

Gene Flow

At one time it was the more or less unanimous opinion of evolutionists that geographic isolation was a necessary prerequisite for speciation to occur. The basic idea was that if any substantial amount of migration (gene flow) occurred between populations, differentiation would be prevented. It now seems likely that in many (if not most) cases, gene flow is not the cause of observed lack of differentiation and that differentiation can occur in the presence of strong gene flow. As an extreme example, many populations of asexual organisms show relatively little differentiation. Single asexual species have wide distributions and show relatively little geographic variation. This is so even though, by definition, no gene flow may occur between the populations. In sexually reproducing organisms, often relatively little differentiation is observed between populations which are known to be completely isolated or to rarely exchange genes. This is true for those populations of the butterfly *Euphydryas editha* discussed above, and there are many similar examples in butterflies. Many species of plants are found in both the Northern and Southern hemispheres, and their populations may show little differentiation although they are separated by gaps of many thousands of miles. Warmwater marine organisms often show little differentiation on opposite sides of the Isthmus of Panama, although they have been isolated for 2 million generations or more. Presumably it is selection which keeps these organisms relatively uniform over wide areas. It could not be gene flow.

On the other hand, in the face of strong selection, gene flow would be unable to prevent differentiation since the genetic contribution of immigrant individuals would be selectively eliminated. Before industrial melanism developed, evolutionists would have been tempted to think that *Bison betularia* showed little geographic variation in England because its populations were linked by gene flow. Now it is clear that no amount of flow of genes for light, variegated pattern could have prevented populations in the Manchester area from turning black. The immigrants themselves, or their offspring, would simply have been eaten by birds!

Much of the confusion about the role of isolation in the differentiation of populations seems to be traceable to the existence of reproductive isolating mechanisms between species. It is a commonplace to observe some level of reproductive failure when individuals from two differentiated populations are crossed, and exhaustive analyses of the kinds of isolating mechanisms responsible have been done. These range from physical incompatibility of genitalia to death of sperm or zygote to failure of embryogenesis to sterility of hybrid offspring. It is, of course, also a commonplace to observe, among animals, that when two closely similar species occur in the same place (are **sympatric**), hybridization between them is rare. Indeed, males of one species rarely court or recognize females of the other, and vice versa. In plants, hybridization is somewhat more commonly observed, but most species do remain distinct when sympatric.

The reasons for these patterns may be found in the selective regime which often exists when two differentiated but closely related populations come together and occupy the same area. There will be some hybridization, but hybrid individuals will not be as successful as the offspring of crosses between parents belonging to the same population. In the course of differentiation, the genetic information of the two populations will have been reorganized and there will have been alterations in developmental pathways. Therefore, it is less likely that the hybrid individuals will have as well-integrated genetic-developmental systems as nonhybrid individuals. This imbalance may be so serious as to prevent development altogether, or it may only cause the hybrids to be slightly less successful reproductively. Often differentiated populations which become sympatric may have become differentiated in habitat choice as well as in other characters. They may become sympatric in places where two habitats meet, and yet there may be no habitat suitable for the hybrids.

In plants, developmental difficulties in hybrids are not as serious a problem as they are in animals, for as you will recall, developmental systems in plants are usually not as complex and precisely regulated as those in animals. Hybrid failure in plants is more likely to be due to absence of a suitable hybrid habitat, but many other mechanisms of incompatibility between species are known. Plants, of course, are much less likely to show **prezygotic** isolating mechanisms than are animals. Prezygotic mechanisms are those, such as behavioral mechanisms, which function before a zygote is formed. For instance, differences in color, odor, or courtship calls often prevent animals from "recognizing" or mating with individuals of other species.

Whatever the mechanism of failure or partial failure of hybridization, the selective regime established is similar. Maximum fitness goes to those individuals which do not hybridize; hybridizing individuals make a smaller reproductive contribution to the next generation. This selection against individuals that tend to hybridize quickly reinforces barriers to hybridization.

Further evidence for this reinforcement of barriers to hybridization is seen in the phenomenon of **character displacement,** which is widespread in animals. Character displacement is said to occur when two species are less alike where they are sympatric than where they are **allopatric** (geographically separate). That is, when two species have partially overlapping ranges, they tend to be *most different* in the zone of overlap. For instance, the very closely related Asiatic nuthatches, *Sitta tephronota* and *S. neumayer,* are clearly differentiated by plumage pattern as well as bill length, but only where they are sympatric (Fig. 13–14). Where their ranges are separate, they are almost indistinguishable.

Examples of character displacement are exceedingly rare in plants, which generally do not have elaborate prezygotic isolating mechanisms. It is interesting that an example has been found in the pollination system of certain Mexican fuchsias. Where the taxa (taxonomic groups; sing. *taxon*) are allopatric, their flowers are very similar and a wide variety of pollinators is found. On the other hand, when they occur together, the taxa are very different in floral morphology and in color and each uses different pollinators, one bumblebees and the other hummingbirds.

It is quite possible for isolating mechanisms to evolve fortuitously. If evolving populations are allopatric long enough or if differentiation has occurred rapidly enough, the populations may simply "not recognize" each other when they become sympatric. But, especially in animals, isolating mechanisms are often a product of contact between differentiated populations. They evolved in response to a wastage of gametes by individuals which tended to hybridize.

CLOSELY RELATED SPECIES OF PLANTS AND ANIMALS USUALLY ARE REPRODUCTIVELY ISOLATED FROM ONE ANOTHER BY MECHANISMS WHICH PREVENT HYBRIDIZATION, REDUCE FERTILITY OF OFFSPRING, ETC., AND SELECTION WILL REINFORCE SUCH MECHANISMS.

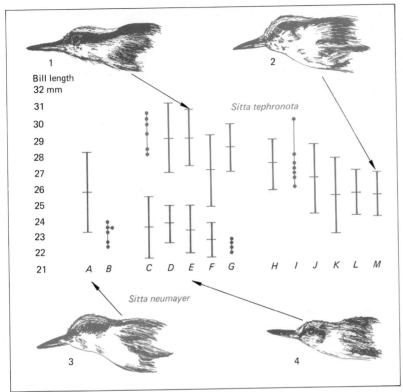

Fig. 13–14 Character displacement
in Asiatic nuthatches. Bill length and
facial strips in the two species are
very different in the areas where they
occur together (1 and 4) but are quite
similar where they occur alone (2 and
3). Areas *A* and *B* are west of the zone
of overlap; *C*, *D*, *E*, *F*, and *G* are in
the zone of overlap; and *H*, *I*, *J*, *K*, *L*,
and *M* are east of the zone of overlap.
(*Adapted from Paul R. Ehrlich and
Richard W. Holm, PROCESS OF
EVOLUTION. Copyright McGraw-Hill
Book Company, 1963. Used by
permission.*)

Sympatric Speciation

Finally, let's turn for a moment to the second question posed above.
Is speciation possible within a single interbreeding population? On
the basis of present knowledge, the answer is: Yes, it is possible in
animals under special laboratory conditions, but it is unlikely to occur
often in nature. As a result of applying strong disruptive selection to
Drosophila populations in the laboratory, the production of two reproductively isolated daughter populations has been reported in one case and
considerable differentiation in others. But whether such powerful disruptive selection occurs frequently in nature, if at all, is not known. Our
suspicion is that it is rare at best.

In plants, however, sympatric speciation may occur in the special case
in which the "single interbreeding population" consists of several parental
types and hybrids among them. You will remember from Chap. 2 that in
the reduction division necessary for the formation of gametes, homologous chromosomes must synapse, or pair. In a diploid hybrid between
two different species, it usually is found that the chromosomes of these
species are different enough that pairing cannot occur properly. Meiosis
thus fails, viable gametes cannot be formed, and the hybrid is sterile. In

many plants, however, doubling of the chromosome number may take place in the diploid hybrid. The resulting organism would then be a **tetraploid** plant instead of a diploid plant. As you can see in Fig. 13–15, each chromosome would then have a homologue and fertility would be restored. This sort of doubling of chromosome number may occur in various ways and may be repeated. Higher and higher levels of polyploidy may occur. Such fertile polyploids are usually thought of as distinct species since they cannot be successfully crossed with either parent.

Polyploidy has been very important in the evolution of plants. It has been estimated that at least half the species of flowering plants are of polyploid origin. Many important cultivated plants are polyploids. A good example of a polyploid series in plants is provided by the cultivated wheats of the genus *Triticum*. The most ancient species have seven pairs of chromosomes. There are also 14-paired and 21-paired species, which were cultivated later. A number of species of the genus *Triticum* are involved in this complex. Of course, man played a crucial role in identifying the various improved forms of wheat and cultivating them. Modern-day bread wheat has 21 pairs of chromosomes and is therefore a hexaploid.

Fig. 13–15 Formation of a tetraploid in the fern *Asplenium*. Each of the parental species has 72 chromosomes which pair to form 36 bivalents at meiosis. The diploid hybrid with 36 chromosomes from each parent and 72 unpaired chromosomes (univalents) are found at meiosis. When the chromosome number is doubled (to 144) in the hybrid, making it a tetraploid, each chromosome has a homologue, 72 bivalents occur, and fertility is restored. (*Adapted from Paul R. Ehrlich and Richard W. Holm, PROCESS OF EVOLUTION. Copyright McGraw-Hill Book Company, 1963. Used by permission.*)

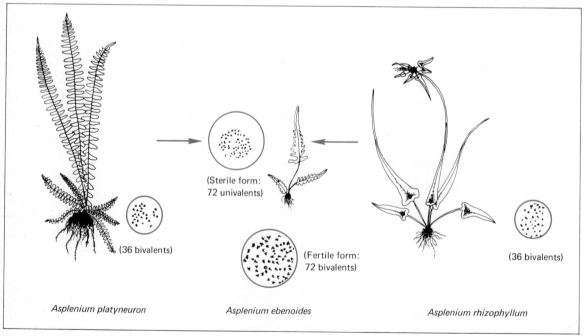

(36 bivalents)

(Sterile form: 72 univalents)

(Fertile form: 72 bivalents)

(36 bivalents)

Asplenium platyneuron

Asplenium ebenoides

Asplenium rhizophyllum

*NEW SPECIES MAY ARISE IN PLANTS AS A RESULT OF POLYPLOIDY,
AN INCREASE IN CHROMOSOME NUMBER. POLYPLOID SPECIES MAY
ARISE SYMPATRICALLY OR FOLLOWING HYBRIDIZATION OF ALLO-
PATRIC SPECIES.*

It is generally accepted procedure to consider two entities which are
separated by reproductive isolating mechanisms to be different species.
Thus two sympatric differentiated populations which did not extensively
hybridize will be given different Latin names unless they are connected
by a ring of intermediates, as in the salamander case discussed above.
Whether or not differentiated allopatric populations are placed in separate
species depends, in the vast majority of cases, on a subjective evaluation
of the degree of difference necessary for considering one a separate
kind from another. Standards differ among taxonomists studying different groups; unfortunately, there is today no way to equate degree
of difference between, for example, two populations of oak trees with
degree of difference between two populations of toads.

It is important to emphasize here that the process of evolution within
populations and the process of speciation described here are quite
adequate to explain the total scope of biological evolution. No further
basic mechanism need be adduced to explain the origin of birds from
reptiles or the differentiation of plants from animals. As these processes
have produced more and more kinds of organisms, the growing complexity
of the biological world has itself created new selection pressures and
modified old ones. Each new herbivore changed the selection pressures
on its food plant, provided a way of life for a series of potential parasites, and created a food source for potential predators. Given the billions of years that evolution has had to operate, very small selection
coefficients ($s < 0.01$) would account for all the observed changes from
primitive single-celled (or noncelled) life to dinosaurs, redwoods, and
man.

*THE CYTOGENETIC MECHANISMS OF ORGANISMS AND THE EVOLU-
TIONARY FORCES OF MUTATION, DRIFT, DIFFERENTIAL MIGRATION,
AND NATURAL SELECTION, COMBINED WITH A HETEROGENEOUS
AND CHANGING ENVIRONMENT, CAN ACCOUNT FOR ALL OF BIO-
LOGICAL EVOLUTION.*

The total amount of genetic information possessed by a population is **SUMMARY**
called its gene pool. For a particular locus it is possible to calculate
gene frequencies (allelic frequencies) and genotypic frequencies. The

frequencies of individuals with each of the possible combinations of alleles at a locus make up the genotype frequencies. Gene frequencies are the proportions of each allele in the population, without regard for the way in which they are distributed in individuals.

The Hardy-Weinberg law describes the equilibrium of gene and genotype frequencies which would occur in a population in the absence of all evolutionary forces. Deviations from the expected equilibrium indicate the operation of such evolutionary forces in the population. The major evolutionary forces are nonrandom mating, small population size, mutation, differential migration, and selection. Selection is the most important of these and is defined as the differential reproduction of genotypes. It is difficult to isolate the various forces and study them separately. Nevertheless, there are excellent studies of evolution occurring in natural populations and involving not only the aspects of the phenotype which afford protection from predators but also basic physiological processes, such as rates of respiration or photosynthesis and reaction to toxic materials. Man has become an important selective agent for all other organisms, as well as for himself. In addition to changing many natural agents of selection in various ways, he has introduced many new ones.

The environment is geographically heterogeneous. Virtually all organisms show geographic variation in one or more of the traits across their range. Such variation may be slight or conspicuous, continuous or discontinuous. Populations are usually called distinct species if they are readily distinguishable from neighboring populations and if there is a clear discontinuity in variation from population to population. The differentiation of species occurs as geographically different (allopatric) populations are exposed to different selection pressures, which accentuate their differences in time. Closely related species commonly are reproductively isolated from one another by mechanisms which prevent hybridization, reduce fertility of offspring, etc. In many plants, new species may arise as a result of polyploidy, an increase in chromosome number.

The consequences of complete removal of the recessives from the population (homozygous recessives lethal, $s = 1$) are shown in Table Box 13-1.

BOX 13-1. HOMOZYGOUS RECESSIVES COMPLETELY UNSUCCESSFUL

The relationship between the gene frequencies of any two consecutive generations is

$$q_n + 1 = \frac{q_n}{1 + q_n}$$

where the subscripts indicate the generation numbers. Thus

$$q_0 = q_0$$

$$q_0 + 1 = q_1 = \frac{q_0}{1 + q_0}$$

$$q_1 + 1 = q_2 = \frac{q_0/(1 + q_0)}{1 + q_0/(1 + q_0)}$$

$$= \frac{q_0/(1 + q_0)}{(1 + q_0)/(1 + q_0) + q_0/(1 + q_0)}$$

$$= \frac{q_0/(1 + q_0)}{(1 + 2q_0)/(1 + q_0)}$$

$$= \frac{q_0}{1 + 2q_0}$$

$$q_2 + 1 = q_3 = \frac{q_0}{1 + 3q_0}$$

TABLE BOX 13-1
(Adapted from Paul R. Ehrlich and Richard W. Holm, THE PROCESS OF EVOLUTION. Copyright McGraw-Hill Book Company, 1963. Used by permission.)

GENER- ATION	BEFORE OR AFTER SELECTION	GENOTYPE FREQUENCIES AA p^2	Aa $2pq$	aa q^2	GENE FRE- QUENCY OF a q
0	Before	p^2	$2pq$	q^2	q
	After*	$\dfrac{p^2}{p^2 + 2pq}$	$\dfrac{2pq}{p^2 + 2pq}$	0	$\dfrac{q}{1 + q}$
1	Before†	$\dfrac{1}{(1 + q)^2}$	$\dfrac{2q}{(1 + q)^2}$	$\dfrac{q^2}{(1 + q)^2}$	$\dfrac{q}{1 + q}$
	After‡	$\dfrac{1}{1 + 2q}$	$\dfrac{2q}{1 + 2q}$	0	$\dfrac{q}{1 + 2q}$

*$p^2 + 2pq$ represents the total after the aa (q^2) genotypes are removed. To find the frequencies of the two remaining genotypes, they must be expressed as proportions of the total. These two frequencies are obtained simply as follows:

$$\frac{p^2}{p^2 + 2pq} = \frac{p(p)}{p(p + 2q)} = \frac{1 - q}{1 - q + 2q} = \frac{1 - q}{1 + q}$$

and similarly, $\dfrac{2pq}{p + 2pq} = \dfrac{2q}{1 + q}$ $\dfrac{1 - q}{1 + q} + \dfrac{2q}{1 + q} = 1$

q = gene frequency of a = $\dfrac{1}{2}\dfrac{2q}{1 + q} = \dfrac{q}{1 + q}$ $p = 1 - q = 1 - \dfrac{q}{1 + q} = \dfrac{1}{1 + q}$

†The genotype frequencies before selection are obtained by using the new p and q values and expanding the following:

$$\left(\frac{1}{1 + q} + \frac{q}{1 + q}\right)^2$$

For example, the zygotic frequency of AA = (new p)^2 = $\left(\dfrac{1}{1 + q}\right)^2 = \dfrac{1}{(1 + q)^2}$

‡Frequencies after selection are calculated in the following manner:

$$\frac{1/(1 + q)^2}{1/(1 \pm q)^2 + 2q/(1 + q)^2} = \frac{1}{1 + 2q}$$

These successive q's (gene-frequency values) fall into a harmonic series, that is, a series whose terms are the reciprocals of those in an arithmetic series. When the initial gene frequency is known, the gene frequency for any succeeding generation may be found by substituting in the equation

$$q_n = \frac{q_0}{1 + nq_0}$$

The change in gene frequency per generation is again symbolized by Δq and is given by the following equation:

$$\Delta q = \frac{q}{1 + q} - q = \frac{-q^2}{1 + q}$$

Note that the **rate of change** of gene frequency is itself a *function of the gene frequency*. When the gene frequency is high, the gene is removed from the population rapidly.

BOX 13-2. GEOGRAPHIC VARIATION IN BODY SIZE

Often, one of the first observations a biologist makes when he has two or more collections of an organism before him is that they differ in average size. Naturally, he wants to know why this is. When there is a consistent trend in size, say from north to south or from lowlands to highlands, we say that there is a **cline** in size. Repeated observations of similar clines in many different species have led to certain generalizations. One of these, Bergmann's rule, is that mammals increase in size toward cool regions, generally the North Pole in the temperate and subarctic regions. Rarely, however, is nature so regular. In man, for instance, both the tallest groups (the Nilotics) and the shortest groups (the Pygmies) occur within a few hundred miles of each other in equatorial Africa (see Fig. 13-9). Even when a pattern exists and is predictable, it is often difficult to establish its cause or causes.

Take, for example, insular reptiles. Given enough time, tortoises, lizards, and occasionally snakes become gigantic on islands. The largest tortoises occurred on many islands in the Indian Ocean (before being exterminated by man) and on the Galapagos Archipelago in the Pacific. The largest living lizard, the 10 foot Komodo dragon, occurs only on the Lesser Sunda Islands. There are many other examples. Two general "explanations" exist for this insular gigantism in reptiles: (1) Islands tend to have fewer species of predators. If predators of reptiles tend to prefer larger individuals for prey or if smaller prey are more agile, then smaller prey would have a reproductive advantage on the mainland.

On islands, this advantage of small size is lost, and size increases as a consequence of physiological selection pressures. (2) Islands have fewer species, in general, and fewer large mammals in particular, so any given island species has relatively fewer competitors. Hence, island reptiles have more food and a wider range in size and kind of food items available to them. In such a reptilian Garden of Eden, the larger individuals have an advantage because they can attack, reach, or swallow large food items as well as small ones. And since larger amounts of food are available to them, they will have a reproductive advantage over smaller individuals.

These possible causes are not mutually exclusive. That is, they could both be operating at once. Also, there is always the possibility that we have missed the real cause altogether. The truth is all the more elusive because gigantism has slowly evolved over thousands or even millions of years and there are no experiments we can perform to test our hypotheses.

Fortunately, the appropriate experiment is sometimes performed by the chance distribution of islands and their faunas. Such an instance is the occurrence of a certain stock of lizards on the islands within the confines of the Gulf of California, Mexico. The lizard is the side-blotched lizard. It is a small relative of the iguana and the common swift or fence lizard and occurs widely in the more arid parts of western North America. Only on some small islands, though, does this rather ordinary lizard achieve a semblance of grandeur. There it has evolved into races that are nearly twice the size of any of its mainland relatives. Fortunately, there are enough well-isolated islands in the Gulf of California inhabited by these lizards to permit the biogeographer to employ modern statistical tests to assist in determining which of the possible causes of gigantism is the most likely. It will suffice for our purposes to discuss the alternative causes in rather qualitative terms.

The distribution of predators seems to be an unlikely explanation of the size differences. Lizard-eating birds and snakes occupy even the smallest of the islands, and there is no obvious reason why the lizard predators on small islands would tend to prey on the smaller lizards while the predators on the larger islands would prefer larger lizards.

What other factor that is correlated with island size would explain gigantism? The variable that seems to best explain the differences in body size of the island lizards is the number of lizard species of the same family (Iguanidae) coexisting with it. The side-blotched lizard is nearly the smallest member of the family to begin with. It is as if it were kept small by competition from larger species.

There are two kinds of evidence that something like this has really led to gigantism in these lizards. First, the body size of these lizards is even more highly correlated with the number of sympatric species of lizards than it is with the size of the island. Second, the largest of the island forms do, indeed, eat new and different foods. The big side-blotched lizards on the island of San Pedro Martír seem to behave like crabs. Unlike any of their relatives, they feed on intertidal invertebrates at low tide. Even more astounding is the way they tear at pieces of dead fish dropped by the nesting blue-footed boobies. Other side-blotched lizards rarely, if ever, eat inanimate food.

In this example, the best explanation of gigantism seems to be the absence of other lizards, though there probably are other factors involved. Gigantism in most other reptiles has not been approached statistically, so it would be foolish to cast the competition hypothesis very widely. The important point is that circumstances sometimes permit the biologist to test hypotheses in the field with statistical tools. Sometimes this is the only kind of test possible when the concern is geographic variation in size, shape, color, or pattern.

SUPPLEMENTARY READINGS

Books

Ehrlich, P. R., R. W. Holm, and D. Parnell: "The Process of Evolution," McGraw-Hill, 1973.

Mayr, E.: "Populations, Species, and Evolution," Belknap Press, Harvard University, 1970.

Wilson, E. O., and W. H. Bossert: "A Primer of Population Biology," Sinauer Associates, 1971.

Articles

Cavalli-Sforza, L. L.: "Genetic Drift" in an Italian Population, *Sci. Amer.,* **221,** no. 2, Offprint 1154 (1969).

Curtis, B. C., and D. R. Johnston: Hybrid Wheat, *Sci. Amer.,* **220,** no. 5, Offprint 1140 (1969).

ORGANIZATION OF COMMUNITIES

As you saw in the last chapter, evolution in the course of the vast reaches of geological time has resulted in the diversity of organisms we see today. Among the plants there are about 275,000 species of angiosperms, 650 of gymnosperms. Mosses and liverworts number about 15,000, and it is estimated that there are 20,000 species of algae and 75,000 of fungi. It is thought that there are at least 1.5 million animal species; some 800,000 species of insects have been described. The vertebrates number some 40,000 species. Some people think that the total number of living species of animals may be as high as 10 million.

THE STRUCTURE OF COMMUNITIES

The present diversity of species has resulted largely from the pressures of selective agents in the environment. We may think of these selective agents as all the factors of the environment including other organisms. All the organisms existing together at a particular place and time may be called a **community.** The word has no special biological meaning; it is merely a useful descriptive term. All the organisms in a community— plants, animals, and microorganisms—are interrelated in complex ways, as we will see below. They also exchange energy and materials with the physical environment. The community plus that physical environment is known as an **ecosystem.** Very likely any evolutionary change in one species affects nearly all others in a community to a greater or lesser degree.

Nevertheless, the distribution of any species is a result of the interaction of individuals of a species with those factors of the environment which impinge upon its functioning. Each metabolic or behavioral process of an organism has a characteristic range of tolerance for each factor of the environment. If an extreme of the intensity span of an environmental factor, such as temperature or moisture, exceeds one of these tolerance ranges, the organism cannot function. It must move to a new environment, die, or produce new genotypes with a different range of tolerance. Since the environment is always changing, it may be thought of as continuously presenting a series of physiological challenges to organisms. Similarly, organisms expanding their range find the environment to be a complex mosaic of factors arranged along smooth or stepped gradients. To expand its range, a species must produce genotypes able to exist in novel environmental situations.

It is quite true that the different species in a community have in common the ability to operate efficiently in the particular regime of environmental factors present there. They also have in common, however, a longer or shorter period of evolving together—of coevolution—in this particular place. As members of each other's environment, they have mutually adjusted not only to temperature and moisture. They have also adjusted to each other as producers of shade, as producers of nitrogenous wastes, as sources of food, as agents of pollination or seed dispersal, or as predators. Many generations of coevolution have interwoven the members of a community into the complex nexus mentioned in Chap. 1.

A COMMUNITY IS ALL THE ORGANISMS WHICH LIVE TOGETHER IN A PARTICULAR PLACE AT A PARTICULAR TIME. THEY SHARE THE ABILITY TO LIVE AND REPRODUCE UNDER THE LOCAL ENVIRON- MENTAL CONDITIONS. THEY USUALLY HAVE COEVOLVED IN RELA- TIONSHIP TO ONE ANOTHER AS FACTORS OF THE ENVIRONMENT.

Arrangement of Individuals by Function

Within a community, organisms become arranged in response to environmental gradients. In the more or less stable, self-sustaining community that we call an ecosystem, however, there are certain ecological roles which are played by different kinds of organisms. We have already mentioned these ecological roles, or functions, in Chap. 1. They are called trophic levels (Fig. 14–1). For all living things, the ultimate source of energy is the energy of the sun. Thus an ecosystem must have **primary producers,** or green plants, that are able photosynthetically to trap a portion of this solar energy. Ecosystems also have **primary consumers,** the **herbivores.**

Primary consumers may be thought of as harvesting a portion of the crop of producers. They convert a portion of the energy and matter contained in the green plants into herbivore tissue. At the same time, they utilize some of this energy in their own metabolic, homeostatic, and reproductive activities. Some of this energy is lost as heat, some as a result of inefficient utilization, and some as feces and other "waste" materials. Thus there is a loss of usable energy in the transfer between the producer level and the primary consumer level.

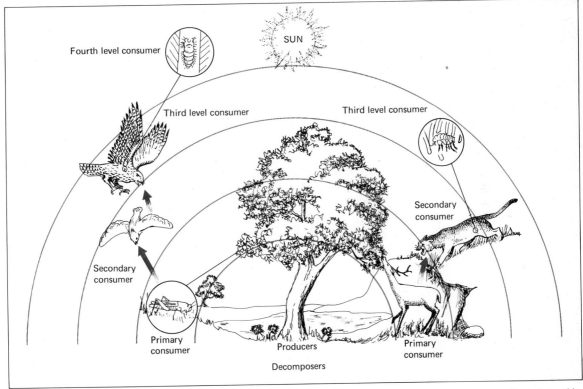

Fourth level consumer

SUN

Third level consumer

Third level consumer

Secondary consumer

Secondary consumer

Primary consumer

Producers

Primary consumer

Decomposers

Fig. 14–1 Diagram to illustrate trophic levels in a community. The sun is the source of energy which is trapped by the producers (green plants) and transferred to primary, secondary, and higher-level consumers (parasites).

Also found in most ecosystems is another nutritional, or **trophic,** level. This is the level of **secondary consumers,** the **carnivores.** We may think of these as harvesting a portion of the crop of herbivores. Again, only a portion of the energy and matter of the herbivores is converted into carnivore tissue; some is used in carnivore activities, some lost as waste, some lost as a result of inefficiencies, and some lost as heat. You can see that there is, once again, a net loss of energy in the transfer between the second and third trophic levels. Higher-order consumers may be found in some communities: carnivores which eat carnivores, parasites, etc.

The Flow of Energy

It should be clear that as one goes up the trophic levels from producers to consumers there is less and less energy available for use at each level. You are probably familiar with the calculation of the amount of energy in food by determining the amount of heat, expressed as calories, released when it is oxidized. Similar calculations can be made of the

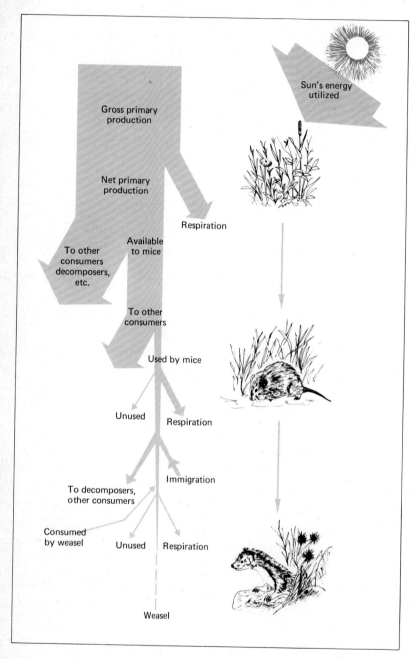

Fig. 14–2 The flow of energy in a field community. At each stage of energy transfer, there is a loss of usable energy. (*Adapted from Figure 3–6 (p. 42) in ECOLOGY AND FIELD BIOLOGY by Robert L. Smith. Copyright © 1966 by Robert Leo Smith. By permission of Harper & Row, Publishers, Inc.*)

amount of energy, expressed as calories, in the organisms at any trophic level of an ecosystem. It is possible to calculate the amount of energy received by producers and to see how much is lost at each successive transfer. Such studies have shown that, on the average, the amount of energy of the producers is reduced 90 percent in the first transfer to the herbivores. At each succeeding transfer, there is a further reduction by 90 percent.

In some communities, there is not enough energy to maintain more than one carnivore level. For instance, it has been shown in a study of a community in an old field that the plants support a population of herbivores, in the form of meadow mice (Fig. 14–2). Only 1 percent of the sun's energy is converted into plant tissue. The mice eat only about 2 percent of the plant material, and the energy they receive must be used in part for their own maintenance. The energy required by the mice amounts to a 68 percent loss to higher trophic levels. The carnivore in this food chain is the least weasel (a kind of small weasel), who must survive on what energy it obtains from the mice. Weasels use 93 percent of the energy they assimilate in maintenance. This means that virtually all the energy originally trapped by the green plants is used by the time the mice and weasels have utilized what they require for their own activities. Therefore, another carnivore preying on the weasel could not permanently live in the community. The weasel population would not grow fast enough to balance its losses from predation.

For most communities one can see that the amount of energy, the total biomass (the total mass of living material), and the number of individuals all decrease as one goes up the trophic levels. These relationships are often expressed in the form of pyramids. Figure 14–3 shows a pyramid of energy, a pyramid of biomass, and a pyramid of numbers. In these diagrams you can see the number of producers required to support higher levels, and from this you can gain an understanding of how little energy is transferred from one level to the next.

THE NUMBER OF INDIVIDUALS, THE TOTAL MASS OF LIVING MATERIAL, AND THE AMOUNT OF ENERGY CONTAINED THEREIN GENERALLY ALL DECREASE FROM ONE TROPHIC LEVEL TO ANOTHER AS ONE GOES UP THE FOOD WEB. ENERGY, IN ITS MOVEMENT UP THE TROPHIC LEVELS, IS EVENTUALLY TRANSFORMED INTO STATES NO LONGER USABLE BY BIOLOGICAL SYSTEMS.

In many communities, the situation is not as clear-cut. Although it is possible to link a series of organisms, such as grass (producer), mouse (herbivore), and weasel (carnivore), in what is called a **food chain,** such simple chains are rare in nature. In most ecosystems, there are many different kinds of producers and many kinds of herbivores with overlapping food preferences. There may be several carnivores also with

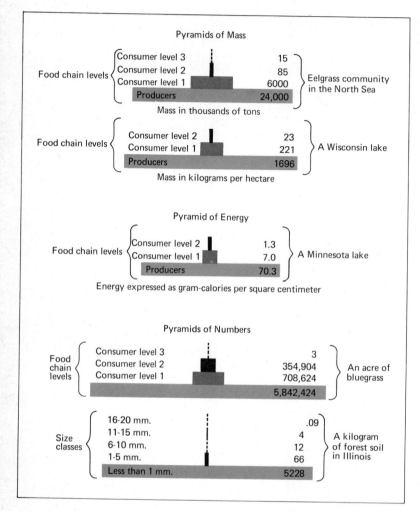

Fig. 14–3 Pyramids of mass (*top*), energy (*middle*), and numbers (*lower*) in various communities. (*Adapted from Eugene P. Odum, FUNDAMENTALS OF ECOLOGY, 2/e, 1953, W. B. Saunders Company, Philadelphia.*)

overlapping prey types. In addition, many communities have **omnivores,** such as raccoons, bears, or men, who eat both plant and animal material. A more realistic model is to think of a **food web** in the community in which the complex feeding interrelationships of the organisms may be displayed (Fig. 14–4).

So far, we have emphasized the loss of energy as one goes up the trophic levels. From this one may deduce that the number of trophic levels and the number of individuals of top carnivores may be limited by the biomass of primary producers. If the number of top carnivores increases, the herbivores may become extinct and the carnivores will have to utilize a lower energy level or perish also. The ecologist LaMont Cole has calculated

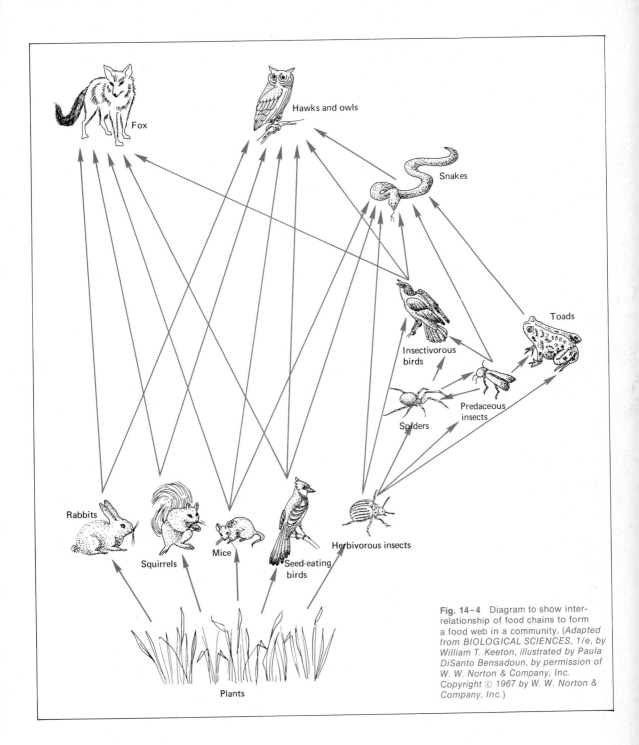

Fox

Hawks and owls

Snakes

Toads

Insectivorous birds

Spiders

Predaceous insects

Rabbits

Squirrels

Mice

Seed-eating birds

Herbivorous insects

Plants

Fig. 14-4 Diagram to show interrelationship of food chains to form a food web in a community. (*Adapted from BIOLOGICAL SCIENCES, 1/e, by William T. Keeton, illustrated by Paula DiSanto Bensadoun, by permission of W. W. Norton & Company, Inc. Copyright © 1967 by W. W. Norton & Company, Inc.*)

that for Cayuga Lake in New York, algae called plankton are the main producers (Fig. 14–5). Small aquatic animals feed on these and can convert 1000 cal or energy stored as plankton into animal tissue amounting to 150 cal. The fish called smelt, which eat these small animals, can make 30 cal of fish tissue from each 150 cal of animals. A man eating the smelt can manufacture 6 cal of human tissue from 30 cal of fish. However, if he prefers to eat trout, which feed on smelt, he gets only 1.2 cal from each 1000 cal that start the chain in the form of algae. Obviously, if a tribe of Indians lived on food from the lake and could not emigrate as they grew in numbers, they would have to first eliminate trout and feed on smelt. Eventually, if their population size increased, they would have to eliminate the smelt and other animals and live on algae in order to obtain the necessary number of calories to support their population.

Matter, in contrast to useful energy, is not used up in ecosystems. Matter is transferred along with energy at each trophic level, and as with energy, there is a loss as a result of inefficient utilization and from "waste" prod-

The Cycling of Matter

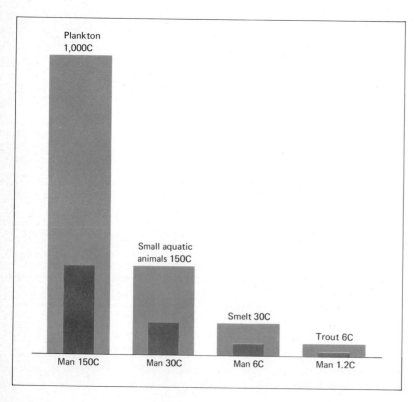

Fig. 14–5 Diagram showing energetic relationships of trophic levels in Cayuga Lake from producers (*left*) to third level consumers (*right*). Insert at each level shows the amount of energy a man would obtain at that level. (*Data from Cole.*)

Plankton
1,000C

Small aquatic
animals 150C

Smelt 30C

Trout 6C

Man 150C   Man 30C   Man 6C   Man 1.2C

ucts. Neither of these represents a *net* loss to the community, as is the case with energy. Another class of organisms in the community operates to keep matter cycling. These are the **decomposers.** Decomposer organisms are primarily microorganisms—bacteria and fungi. Other kinds of plants and animals may also be considered decomposers.

You will remember from Chap. 4 that producer organisms use simple inorganic compounds to make more complex organic molecules. These are used by consumer organisms. The "waste" products and dead bodies of higher trophic levels are made up, of course, of these large organic molecules. It is the decomposer organisms that carry out the function of breaking down complex molecules to simpler ones that cycle in the system. Matter cycles not only between organisms in an ecosystem but also between the living organisms and dead organisms and between organisms and their nonliving environment. This is why we have put the word "waste" in quotation marks. Feces, urine, and dead bodies are not waste. They are the food of decomposers and are necessary for the ecosystem to maintain itself. As we will see later, man is the only organism producing genuine waste (matter not usable by other organisms). Some elements cycle more rapidly than others. In fact, some cycle so slowly that from the human perspective, rather than geological time, their movement is one-way—out of the system.

Let us look at three cycles of great importance in the functioning of ecosystems. These are cycles of the elements carbon, nitrogen, and phosphorus. But do not forget that all other elements—sulfur, iron, magnesium, etc.—are involved in some sort of cycle. Also keep in mind that all the cycles of matter are interrelated in various ways.

CHEMICAL ELEMENTS ARE NOT USED UP BUT, RATHER, CYCLE IN ECOSYSTEMS. DECOMPOSER ORGANISMS, BY BREAKING DOWN COMPLEX MOLECULES INTO SIMPLER ONES, PLAY A CRUCIAL ROLE IN THE CYCLING OF MATTER.

CARBON CYCLE As you already know, carbon is the basic constituent of all the macromolecules characteristic of living systems. Life on earth is possible only because of the properties of this element. The major reservoir of carbon is the gaseous carbon dioxide free in the atmosphere and dissolved in the oceans and fresh water (Fig. 14-6). The process of photosynthesis provides the major pathway by which carbon is withdrawn from this reservoir. Oxygen released into the atmosphere is one of the important products of photosynthesis, in addition to carbohydrates. An essential part of the carbon cycle is the movement of carbon molecules

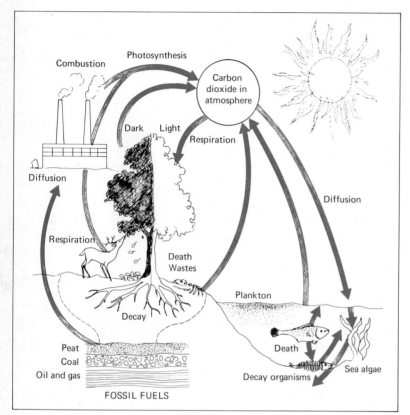

Fig. 14-6 Diagram of the carbon cycle.

from the CO_2 pool to green plants and up the higher levels of the food web. From plants and from animals at every position in the food web, including the decomposer organisms, respiration returns CO_2 and, therefore, carbon to the reservoir.

Not all the carbon built into molecules of organisms by the process of photosynthesis is returned rapidly to the CO_2 reservoir. Some of it leaves the carbon cycle for long periods and enters the crust of the earth. This happens when decomposition is not complete and leads, in ways not fully understood, to the buildup of deposits of coal, oil, and natural gas. One of the greatest periods of the laying down of these **fossil fuels** was the Carboniferous period, some 300 million years ago. Our modern technology makes it possible for us to use up this only partially oxidized organic material by burning it. Of course, we consume it infinitely more rapidly than it can be laid down. Carbon is also temporarily withdrawn from the cycle by the deposition of limestone, often through the activities of organisms. For example, marine algae and coral animals are responsible for extensive coral reefs in tropical waters. These reefs are made up of carbonate rock laid down mainly by algae. Carbon eventually is returned to the CO_2 pool by the burning of fossil fuels and the weathering of limestone rocks, such as reefs.

CARBON CYCLES PRIMARILY FROM A GASEOUS RESERVOIR IN THE ATMOSPHERE OR WATER TO ORGANIC MOLECULES OF LIVING OR FOSSIL ORGANISMS. OXIDATION OF THESE COMPOUNDS RELEASES CARBON DIOXIDE TO THE RESERVOIR.

NITROGEN CYCLE   The atmosphere is almost 80 percent nitrogen, an essential constituent of both proteins and nucleic acids. Nitrogen cycles in ecosystems in a series of complex pathways, partially diagramed in Fig. 14–7. Unlike oxygen and carbon dioxide, the nitrogen in the atmosphere cannot be used directly by most organisms. Some bacteria and

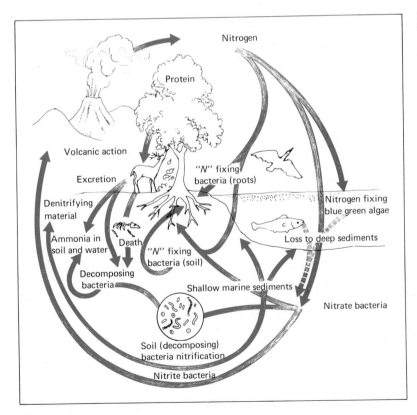

Fig. 14–7   Diagram of the nitrogen cycle.

blue-green algae, however, can utilize gaseous nitrogen to form compounds which can be used by other organisms. Some of these nitrogen-fixing microorganisms live free in the soil. Others are found inside nodules on the roots of certain plants of several different kinds, most notably the legumes, plants of the pea family (Fig. 14–8). Nitrogen is made available to the higher plants by these microorganisms in the form of amino acids. Amino acids may even be released into the soil. Nitrogen-containing compounds made by the plants become available to the animals which eat them. Eventually plants and animals die, and the decomposers begin their role in the cycle.

Decomposition of plants and animals produces ammonia. Many forms of animal excretion also contain ammonia or produce it on decomposition. One group of bacteria, called the **nitrite bacteria,** convert ammonia to **nitrites.** Another group of bacteria, the **nitrate bacteria,** change nitrites to nitrates, which are the commonest form in which higher plants obtain nitrogen from the soil. Therefore, a loop of the nitrogen cycle may be completed without the formation of gaseous nitrogen. Bacteria of another group, known as the **denitrifying bacteria,** break down nitrates, nitrites, and ammonia, liberating nitrogen, which returns to the gaseous reservoir in the atmosphere.

In most communities, there is a considerable quantity of nitrogen in the soil in the form of **humus,** partially decomposed organic material. This is derived from fibrous or woody tissues of plants, insect and other animal remains, and animal waste products. Humus increases the capacity of soil to retain water and results in a porous structure in which oxygen and other gases can move. Cycles of leaf fall, the death of short-lived plants and animals, and the death of widely dispersed root systems or fungal growths maintain humus in the soil. Good agricultural practice also maintains the humus supply necessary for good growth. Of course, adding humus adds carbon to the soil as well as nitrogen.

Unfortunately, when inorganic nitrate fertilizers are used exclusively, the humus level may not be sustained. In experiments with such fertilizers, it has been found that although crop yields were high, as compared with untreated control plots, the humus and organic nitrogen content of the soil dropped. These changes in the humus content of the soil reduced the efficiency of nitrate uptake by the plants. Further reduction in uptake of nutrients occurred as the soil became more compacted, rendering oxygen penetration more difficult. The nitrates not utilized were leached out by water moving through the soil or were converted to forms of nitrogen not available to the plants, some of which entered the atmosphere.

Fig. 14-8 The root system of a soybean plant showing the nodules in which nitrogen-fixing bacteria live. (*USDA photo.*)

You can see that the productivity of the soil in the long run is not restored simply by adding more inorganic nitrate fertilizer. Additional applications result in large amounts of nitrates being flushed from the soil in surface waters or lost into the atmosphere. The use of such fertilizers in the United States has increased some twelvefold in the past 25 years. One result has been a rise in the nitrogen content of our surface water, atmosphere, and rain. Increasing the nitrogen content of ponds and lakes often encourages the rapid growth of certain algae. These are species that cause algal blooms, which cover huge areas and then die in a relatively short time. As the algae decay, the decomposer organisms consume oxygen in large amounts, and this may result in the death of fish and other animals. This process (essentially overfertilization) has occurred in many lakes of the world. It is called **eutrophication.**

The nitrogen cycle in the oceans is essentially similar to the terrestrial cycle described above. One major physical difference is that the bodies of dead and moribund organisms sink inevitably to the depths, where the nitrogen-containing compounds in their bodies are cycled by decomposer organisms, mostly bacteria. The result is that the surface waters, where all photosynthesis occurs, are constantly being depleted of nitrogen and the bottom waters enriched. Here lies the explanation for the desertlike (very low) productivity of the open seas, nearly 90 percent of the ocean. Recent oceanographic studies have confirmed that nitrogen compounds are usually the factor which limits photosynthesis and population growth of phytoplankton.

NITROGEN MOVES FROM A GASEOUS RESERVOIR IN THE ATMOSPHERE TO THE COMPLEX ORGANIC MOLECULES OF ORGANISMS. IN THE COURSE OF BREAKDOWN OF PLANT AND ANIMAL TISSUES, NITRATES ARE RELEASED, AND NITROGEN MAY BE RETURNED TO THE ATMOSPHERE. INTERMEDIATE BREAKDOWN PRODUCTS (HUMUS) ARE CRUCIAL FOR MAINTENANCE OF SOIL STRUCTURE AND PRODUCTIVITY.

PHOSPHORUS CYCLE   Phosphorus is an essential element in adenosine triphosphate, ATP, and other important compounds. Therefore, the phosphorus cycle is crucial (Fig. 14–9). In contrast to carbon and nitrogen, phosphorus is stored primarily in a sedimentary-rock reservoir. Release from these reservoirs comes through erosion and leaching and through mining by man. Some of this becomes available for use by plants and thus enters the living part of the ecosystem. It may pass along the trophic levels and eventually be returned to the soil by decomposers. Among the vast quantities of materials washed to the sea annually are 3.5 million tons of phosphorus. Only 3 percent of this is returned biologically in the form of bird guano, the feces of fish-eating birds. Eroded phosphorus may be utilized in marine ecosystems or deposited as shallow or deep marine sediments. Relatively small amounts may be returned by upwelling currents. Most, however, can be restored only by geological processes leading to uplifting of sediments. At the rate we use phosphates in agriculture, there is not enough time to wait for these slow processes.

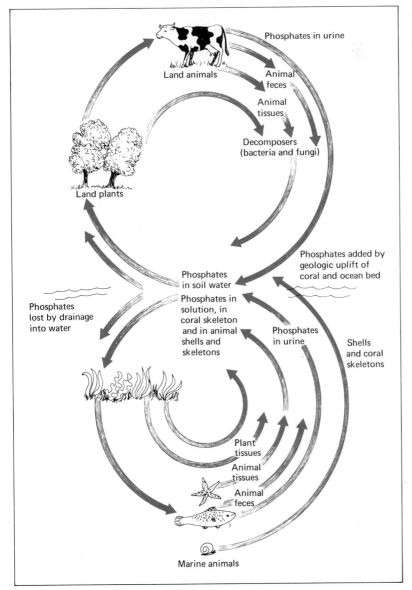

Fig. 14-9 Diagram of the phosphorus cycle. (*Adapted from Raven, Peter H. and Helena Curtis, BIOLOGY OF PLANTS, Worth Publishers, New York, 1970, p. 277.*)

Man must mine phosphate deposits or produce phosphates artificially. As with nitrogen, phosphorus is a common waste product in soil water systems. It participates with nitrogen in causing eutrophication, the over-fertilization of ponds and lakes, which is discussed in Chap. 16. Recent studies by the Institute of Ecology indicate that supplies of phosphorus will probably run out during the next century unless patterns of population growth and phosphate use are changed. They state that: "Thereafter, . . . yields are limited to the rate at which phosphorus is released from its insoluble salts. It is estimated that these natural rates of mobilization would support a world population between one and two billion people." That population size is about one-half of today's population and perhaps one-thirtieth of the projected population around the end of the next century.

PHOSPHORUS CYCLES FROM A SEDIMENTARY-ROCK RESERVOIR TO THE COMPLEX MOLECULES OF ORGANISMS. PHOSPHORUS IS LOST TO THE RESERVOIR FOR VERY LONG PERIODS OF TIME BECAUSE IT IS RETURNED TO LAND ECOSYSTEMS ONLY BY GEO-LOGICAL PROCESSES.

CONCENTRATION OF SUBSTANCES IN FOOD WEBS In considering the cycling of matter in ecosystems, we must not think of some specific quantity of an element or compound flowing among the constituent species. Some organisms concentrate certain elements in their tissues; other species concentrate different ones. An interesting example of such concentration involves substances created by man. In contrast to naturally occurring compounds, these are substances with which organisms have had ecological contact for relatively few generations. We refer to such synthetic substances as biocides (pesticides and herbicides), widely used by man in his attempts to control nature.

It has been remarked that the pollution-absorbing capacity of soils, rivers, oceans, and the air is a great "natural resource." One might think that if 1 gal of poison is added to 1 billion billion gal of water, the highest concentration to which anything will be exposed is on the order of 1 part per billion billion. Given the necessary mixing and diffusion, this might be approximately true in a purely physical system. It clearly is not the case for biological systems.

For example, filter-feeding animals may concentrate poisons to levels far beyond those found in the surrounding medium. Oysters have been known to accumulate a chlorinated hydrocarbon insecticide to 70,000 times its concentration in surrounding water. The two principal reasons for this extraordinary concentration in organisms are (1) the very great solubility of chlorinated hydrocarbons in the lipid components of tissues, in contrast to their near insolubility in water, and (2) the resistance of these compounds to enzymatic breakdown. A number of examples of this accumulation of toxic substances have now been studied. A particularly clear one is that in the Long Island estuary studied by Woodwell, Wurster, and Isaacson. In this estuary the filter-feeding clam contains over 50 percent more DDT than the mud snail at the same trophic level.

From a reconsideration of the changes that take place up the trophic levels, you should be able to understand how concentration occurs. At each step in a food chain, the biomass is reduced. But losses of DDT along a food chain are small compared with the amount that is transferred upward through the chain. As a result, a relatively constant amount of DDT is packaged in a steadily *decreasing* biomass. The parts-per-million concentration increases dramatically, as you can see by looking at Fig. 14–10. The concentrations in birds are tens to many hundreds of times higher than at the producer level. In predatory birds, the concentration of DDT may be a million times that in estuarine waters. We

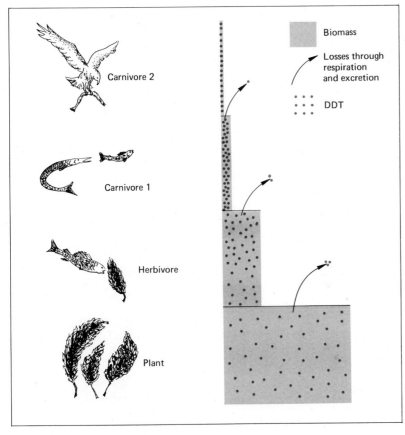

Fig. 14–10 Concentration of DDT along a food chain. The colored area shows the total biomass at each trophic level. The blank dots show the amount of DDT at each trophic level. Arrows show loss of DDT at each transfer from level to level. (*From "Toxic Substances and Ecological Cycles" by George M. Woodwell. Copyright © 1967 by Scientific American, Inc. All rights reserved.*)

will discuss the effects of this sort of concentration in Chap. 16, but it should be obvious to you now that they are grim indeed.

SOME SUBSTANCES BECOME CONCENTRATED IN ORGANISMS OF HIGHER TROPHIC LEVELS AS A RESULT OF BEING VERY STABLE AND NOT EXCRETED BY PLANTS OR ANIMALS. AS THEY ARE CARRIED UP THE FOOD WEB, VIRTUALLY THE SAME AMOUNT OF SUBSTANCE IS CONTAINED WITHIN A SMALLER AND SMALLER TOTAL BIOMASS.

Arrangement of Individuals in Space

The many factors of the environment that are important in the functioning of organisms are arranged in complex gradients. Organisms arrange themselves in space along these gradients. The larger organisms themselves modify factors of the environment, creating new gradients. Such organisms, trees in particular, which have a pronounced effect on the local climate, are called **ecological dominants.** In a redwood forest, for example, *Sequoia sempervirens* is an ecological dominant. It may or may not be associated with Douglas fir, *Pseudotsuga menziesii.* These trees shade the lower levels of the community and change the spectral quality of the light reaching there. They interrupt the rain and change its pattern of distribution. Their leaves accumulate water particles from fog blowing through them, and they may be responsible for considerable precipitation during the rainless summer.

Other organisms more or less stratify themselves in the forest. Some live on the forest floor; some live a subterranean life beneath it. Some climb about the trees. Others occur as **epiphytes** (plants growing upon other plants) at particular levels in the foliage. Thus, the redwood forest comes to have a structure, or morphology.

The same is true for both freshwater and marine communities, in which organisms may arrange themselves in characteristic patterns on the substrate or, for example, along an intertidal zone. Vertical stratification may also occur in floating or swimming organisms. Plankton, which are floating organisms, may float at different levels. Some fish occur only in surface waters, and others are found only at great depths. In response to changing environmental factors or to internal rhythms, marine animals such as the euphausid shrimp may change their level daily. Echos from their massed bodies are responsible for the "deep-sounding layer" detected by sonar depth-determining devices.

Probably the greatest degree of stratification occurs in the tropical rain forest (Fig. 14–11). Here, beneath the canopy of very tall trees, there may be several lower levels of tree and shrub canopies. Very little light reaches

Fig. 14–11 At the margin of a cleared area the tropical forest can be photographed and vegetation grows to the forest floor. Note the abundance of vines and epiphytes in this Amazonian forest. (*Courtesy of Zani Vendal, Jacana.*)

the forest floor, which is remarkably constant in temperature and moisture. There is a profusion of epiphytes in the tropical forest and of lianas, woody vines which climb high into the trees. It is not uncommon to find some species restricted to a particular level in the forest which has a characteristic temperature and humidity. Two species of mosquitoes, for example, may live at different levels and their populations never come into contact.

No matter where a particular organism finds itself in the community, it is there because the spectrum of environmental factors at that point permits it to carry out most effectively the way of life it has become fitted for in coevolution. This leads us to an important concept in ecology: the concept of ***ecological niche.*** This concept has never been defined to everyone's satisfaction, although most biologists believe it to be useful. For our purposes we may define an ecological niche as the specific habitat in which an organism carries out its functioning, together with *its way of life* in that place. This definition may make it hard to come to grips with the concept, but it is useful for descriptive purposes. The specific habitat of an earthworm may be the humus-rich soil of an open forest. This is a part of the definition of the worm's ecological niche. Also to be included are the ways in which worms live, grow, and reproduce in this habitat. Burrowing in the soil, ingesting soil particles, withdrawing nutrients from them, copulating with other hermaphroditic members of the same species—all these are included in the definition of the ecological niche of an earthworm.

The concept of ecological niche always includes the way in which organisms function in a particular environment. Definitions of ecological niches may be quite narrow, as was that for the earthworm. On the other hand, they may be very broad. For example, we might define "flight in air" as an ecological niche. The environment, of course, is the atmosphere. The way of life in this environment includes muscle-driven wings, a lightweight body, ability to see in three dimensions, and many other things. This broadly defined ecological niche is occupied by, among others, butterflies, birds, bats, and moths. We can subdivide this broad niche into smaller ones. Most birds and butterflies fly during daylight hours. Most bats and moths fly at night. "Aerial flight by day" and "aerial flight by night" are two possible subdivisions. Each of these may be subdivided as well, until we get down to individual species, such as the earthworm.

THE HABITAT OF AN ORGANISM AND THE WAYS IN WHICH IT CARRIES ON ITS LIFE IN THAT HABITAT ARE TOGETHER KNOWN AS THE ECOLOGICAL NICHE OF THE ORGANISM.

Like all other things on earth, communities are always in a state of change. Some of these changes are relatively easy to detect and study; others are more difficult to deal with. We will first consider short-range changes that are easily perceived. These are called **successional changes.** We will also briefly discuss the kinds of changes thought to have occurred in long-range geological time.

Any ice-free part of the earth's surface that is devoid of living things (such as a newly emerged island) gradually becomes populated with plants, animals, and microorganisms. Similarly, areas from which the natural community is removed are repopulated if left undisturbed. The changes that go on in this process are known as **primary** and **secondary succession,** respectively. In primary succession, the first organisms to appear are called **pioneer** organisms. Pioneer organisms are able to resist the unameliorated factors of the physical environment. Such plants as lichens and mosses are usually among the first in a terrestrial situation, and they begin a process of environmental modification. On the surface of bare rock, for example, their respiration produces weak carbonic acid when there is moisture, and this begins to etch the substrate. Their dead bodies accumulate in cracks caused by heating and cooling or freezing and thawing. This creates humus, in which animals and larger plants can grow. Wind-dispersed seeds arrive, and their growing roots further break up the rock. Of course, microorganisms, though less conspicuous, play important roles as decomposers and as nitrogen fixers. Pioneer communities are succeeded by a series of intermediate successional stages. In the course of time, the area is able to support larger plants, such as shrubs and small trees, as well as larger animals. The total number of species gradually increases. Some species leave the community, for they are specialized for early successional stages. Others, specialized for later stages, move in. At first, autotrophic organisms predominate and therefore production through photosynthesis *exceeds* metabolic use as measured by respiration. More and more energy and matter are stored in the form of plant bodies. Organic matter as humus is incorporated into the substrate, and this creates soil. As larger and larger plants grow, their deeply ranging root systems act as chemical pumps. Substances are withdrawn from the deeper soil layers, translocated to the leaves, and then returned to the upper layers when the leaves fall and decompose.

As the proportion of heterotrophic organisms increases, respiration comes to more or less equal photosynthetic production. The various interrelationships of organisms increase in number and complexity. Eventually a relatively stable, long-lived community that is characteristic of a particular climatic area arises. This is called a **climax community.** Climax communities intergrade with one another in various ways, depending upon the nature of the environmental gradients. In some places, succession may not proceed to the expected climax. This may be the result of a special soil, or **edaphic,** situation. For example, the pine forests of the southeastern United States are thought to occur there principally because of the sandy soil. The latter is unsuitable for the deciduous forest to be expected on the basis of climate. Disturbance by man is often a factor in preventing succession from reaching a characteristic climax. Grasslands in many parts of the world are thought to be subclimax stages maintained by fire.

Similar successional stages occur in a freshwater pond or lake. The various stages are easy to visualize because succession involves the gradual filling in of the pond from the margin. Figure 14–12 shows a section through a pond with a marginal vegetation in which soil gradually builds up. Eventually the pond is completely filled and may remain a bog or a swamp until the characteristic climax community develops.

ECOLOGICAL SUCCESSION IS THE SERIES OF STAGES IN THE DEVELOPMENT OF A RELATIVELY STABLE, LONG-LIVED CLIMAX COMMUNITY ON LAND OR IN WATER. IN THE COURSE OF SUCCESSION, THE NUMBERS AND KINDS OF ORGANISMS INCREASE AND THE HABITAT IS GREATLY MODIFIED.

There are fewer and fewer places in the world today where primary succession goes to completion. Man as a consumer harvests from communities to obtain matter and energy. Energy is found in communities either in the form of stored energy, as in wood, or in the more immediate products of photosynthesis. One of man's first effects on climax communities is to use up the stored energy, usually by burning it for fuel or simply to get rid of it. Without this reserve of energy, the community is immediately destabilized. And, of course, with the destruction of the trees, many habitats are destroyed either directly or indirectly.

Man, however, is generally not interested in harvesting the primary production of a climax community. In such a community, as we saw above, photosynthesis equals respiration because of the large number of heterotrophic organisms. Man must convert a climax community to an earlier successional stage in which production exceeds respiration and in which he is virtually the only heterotroph allowed. In man's early evolution, such communities were gardens of many kinds of useful plants. Today they are fields of corn, wheat, rice, sugar cane, pineapples, etc., often extending for many square miles. Such artificial communities do not have

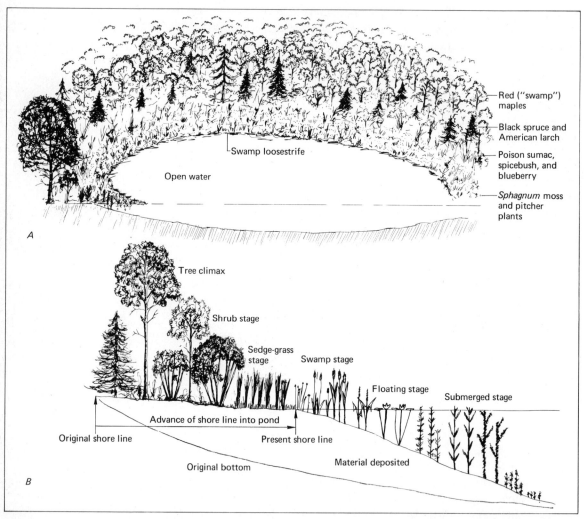

Red ("swamp") maples

Black spruce and American larch

Poison sumac, spicebush, and blueberry

Sphagnum moss and pitcher plants

Swamp loosestrife

Open water

A

Tree climax

Shrub stage

Sedge-grass stage

Swamp stage

Floating stage

Submerged stage

Advance of shore line into pond

Original shore line

Present shore line

Original bottom

Material deposited

B

the built-in stability that natural communities have achieved through the coevolution of their varied constituents. Man attempts to achieve stability artificially by injecting matter and energy through the use of fertilizers, pesticides, herbicides, tractors, etc. We will discuss this further in Chap. 16.

Except in the very broadest terms, we have little information about evolutionary changes in communities. The fossil record would indicate that in the course of evolution, there has been a trend from simpler organisms

Fig. 14–12 Plant succession in (*A*) a bog and (*B*) a pond. (*Adapted from Kimball, BIOLOGY, 1/e, 1965, Addison-Wesley, Reading, Mass. Adapted from BOTANY, fourth edition by Carl L. Wilson and Walter E. Loomis. Copyright 1952, © 1957, 1962, 1967 by Holt, Rinehart and Winston, Inc. Reprinted by permission of Holt, Rinehart and Winston, Inc.*)

Evolutionary Changes in Communities

to more complex ones. There has also been an overall increase in the number of species. These species have evolved, of course, in the context of communities. Thus there must have been an increase in complexity of communities, which are today associations of the simplest and the most complex organisms functioning together.

We may speculate about some of the forces involved in coevolution. For example, following the reasoning of G. E. Hutchinson, let us consider a predator at the $n + 1$ trophic level. Selection will presumably tend to maximize the effectiveness of the predator in dealing with its prey at the nth level. If the predator becomes too efficient, the prey organism may become extinct and the predator will have to either move to feeding on the $n - 1$ level or become extinct itself. Thus there will be an evolutionary trend toward *shortening of food chains.*

At the same time that the predator is becoming more specialized for catching its food, the prey organism is also "experimenting" by producing variable offspring. Any genotype which modifies prey behavior and facilitates escape will be favored. Thus there will be a tendency toward *diversification* at the nth (prey) trophic level. Concomitantly, however, a beneficial strategy will be *diversification at the predator level.* As we will see below, there will be a sort of stepwise coevolution occurring at all trophic levels. Going along with this will be a tendency for finer and finer subdivisions of ecological niches as the environment, including organisms, is exploited in more and more specific ways.

IT IS GENERALLY ASSUMED THAT IN THE COURSE OF EVOLUTION, COMMUNITIES HAVE BECOME MORE DIVERSIFIED AND COMPLEX AND THAT ECOLOGICAL NICHES HAVE BECOME MORE AND MORE FINELY SUBDIVIDED.

RELATIONSHIPS OF SPECIES IN COMMUNITIES

The organisms making up a community act upon one another in a great diversity of ways. Some of these actions do not play any significant role in the structure of the community—they are more or less accidental. An elephant may crush a small plant, or a man may accidentally step on an ant. In neither case is a "community relationship" involved. Indeed, an inanimate object such as a rolling rock could easily substitute for either elephant or man in these examples. For the sake of convenience, we will divide our consideration of interspecific relationships into three classes: those in which the action is largely one-way, those that are primarily bidirectional interactions, and those in which a nexus of interactions must be examined to understand the properties of the system.

Many organisms reap benefits from association with other organisms when the second organism is not affected in any significant manner. For instance, certain flies hover over the raiding columns of army ants, waiting for the cockroaches which are flushed from cover by the ants. They then swoop down and lay their eggs on the cockroaches, which they parasitize. The army ants are not affected by the flies, which benefit greatly from the relationship. Gophers, churning up the soil on Stanford's Jasper Ridge, provide relatively moist, favorable soil for the growth of a small annual plantain. This plantain is a larval food plant for the checker-spot butterfly *Euphydryas editha,* and some of the most suitable larval habitats are on the gopher diggings, where the plantain lasts in fresh condition longer than in most places. The gophers are not influenced by the butterfly (they do not feed on the plantain), but the butterfly benefits greatly from the activities of the gopher. Rodent burrows, in general, provide suitable habitats for insects and mites which are not thought to interfere with the rodent in any significant way.

Similarly, virtually all sessile marine organisms shelter "guests" which do not either benefit or harm the host. Even man is involved in a series of one-way relationships. Most Americans, for instance, provide shelter for a tiny mite in the pores of their skin (Fig. 14–13). This follicle mite does no harm, and people are generally unaware of its presence.

IN MOST COMMUNITIES, SOME SPECIES BENEFIT FROM THE PRES-ENCE OF ANOTHER WHICH IS NEITHER HARMED NOR BENEFITED IN THE RELATIONSHIP. USUALLY THE ENVIRONMENT IS MODIFIED OR SHELTER IS PROVIDED.

Unidirectional relationships tend to blend imperceptibly into two-way relationships. For instance, human activities (such as in developing parks) in cities have greatly increased the distribution of birds such as the robin and the English sparrow. Man has clearly had a much stronger influence on the birds than vice versa, but through their dietary activities (eating insects, seeds, earthworms, etc.) and because of their esthetic value, they certainly do influence man. The microorganisms which live in the intestines of mammals derive great benefit from their hosts. On the other hand, they represent a vast spectrum of degrees of effect on the host; some have essentially no effect at all, others manufacture essential vitamins, and others cause fatal diseases.

One of the most widely recognized two-way relationships is that of host and parasite. The parasite gains; the host loses. Many birds and mammals, for instance, are attacked by parasitic mites, which live in their nests and suck the blood of the host whenever it is in residence.

One-way Relationships

Fig. 14–13 A human hair follicle mite *Demodex folliculorum.* ×300. *(Courtesy of William Nutting.)*

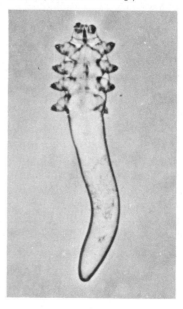

Two-way Relationships

There are specialized ticks which wait patiently in the sandy depressions in which kangaroos relax, waiting for their blood meals until the occupant returns. Man himself is attacked by many parasites. The protozoan *Endoamoeba histolytica* sometimes attacks as many as 50 percent of the human beings in some tropical populations, in which it causes a variety of problems by feeding on human cells. For example, by eating into the walls of the intestine, it causes amoebic dysentery, a disease characterized by crampy pain and bloody diarrhea.

Some of the parasites which attack man have other organisms as intermediate hosts. A good example of this is malaria, which you will remember is caused by small protozoans of the genus *Plasmodium*. The life cycle of one of these malaria parasites is shown in Fig. 14–14. It will give you an idea of how complex a parasitic cycle can be.

There is no sharp line between parasite-host relations and predator-prey relations. In general, parasites are thought of as tiny organisms that attack their hosts to a degree which does not usually cause the death of the host. Predators, on the other hand, are those large animals which pounce on and devour their prey. In what class, then, does one

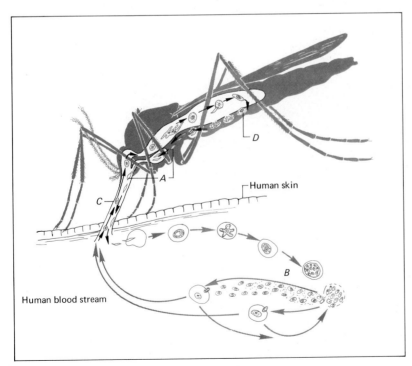

Human skin

Human blood stream

Fig. 14–14 Life cycle of a malarial parasite, *Plasmodium*. (A) Elongate cells are injected into the blood stream during the bite of a mosquito; there they enter blood cells and reproduce asexually, (B) producing large numbers of daughter cells every 48 hr (resulting in the periods of fever as toxins are liberated); eventually cells that will produce gametes are formed (C), and these are sucked up by the mosquito along with its meal of blood. In the mosquito's stomach, fertilization takes place and the zygote enters the wall of the stomach (D). The zygote divides, forming the elongate cells (A) which may be injected by the mosquito into man in a subsequent bite. (*Redrawn from LIFE: AN INTRODUCTION TO BIOLOGY by Simpson, Pittendrigh and Tiffany, copyright © 1957 by Harcourt Brace Jovanovich, Inc. and reproduced with their permission.*)

place the fly which lays its egg on the food plant of a caterpillar and whose larvae hatch out of the eggs inside the caterpillar and gradually devour it? Some biologists refer to insects, mostly flies and wasps, whose larvae slowly consume other insects (always resulting in the death of the host) as *parasitoids*. This partially solves the problem, but you should remember that parasite → parasitoid → predator is a continuum. Obviously, also, the relationships of herbivores to their food plants have almost exact parallels along that continuum. Indeed, it has become more and more common for biologists to refer to herbivores simply as predators on plants.

A special kind of two-way relationship has resulted in the "organisms" that are called **lichens.** Lichens occur in a variety of forms: tightly appressed to rocks, loosely attached to bark or soil, or as mosslike growths several inches high (Fig. 14–15). They are quite variable in color and in their biochemistry. Lichens are often found as pioneer organisms in succession, for they can survive in very severe habitats.

In fact, a lichen thallus is a combination of an alga and a fungus growing together (Fig. 14–16). The alga, of course, photosynthesizes, and the fungus obtains nutrients from the alga, which it parasitizes. The alga is protected and kept relatively moist by the fungus, so it also benefits from the association. The algae and fungi can be separated and grown in pure

A

Fig. 14–15 Two different lichens, both growing on tree limbs. (*A, courtesy of Jack Dermid, B, courtesy of Ross Hutchins.*)

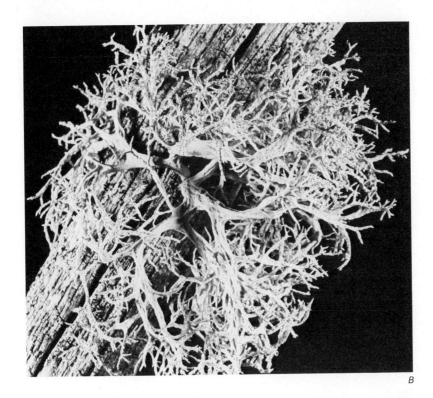

B

culture. Their form and biochemistry are quite different under these circumstances. Some 17,000 kinds of lichens are known.

An excellent example of a very important two-way relationship is that between most seed plants and fungi. The fungi are called **mycorrhizas,** and they are found forming layers over the surface of the roots of the plants or within the cells of the roots. Although it has been known for a long time that many plants will not grow without the associated fungus, it has only recently been discovered what role the fungi play. Many plants are unable to utilize some minerals—for example, phosphorus—even though the minerals are present in the soil. The fungus has the ability to convert the minerals into forms the plant is able to assimilate. In association with other plants, the fungi break down complex organic molecules to simpler ones the seed plant is able to use. It is possible that mycorrhizal fungi may provide vitamins, hormones, or other substances as well.

Mycorrhizas were first recognized as necessary for the normal growth of orchids. It was later found that the great majority, if not all, of perennial plants require mycorrhizas for proper growth. The commonest forest

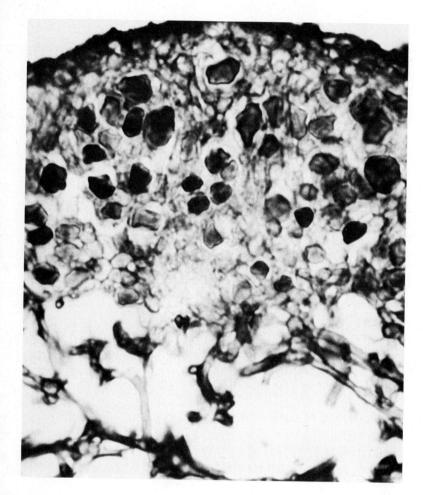

Fig. 14-16 Cross section of a lichen, (*Physcia*) showing the fungal hyphae and the dark, rounded algal cells. × 1680. (*Jeroboam, N.Y.*)

trees, for example, do not grow beyond an early stage if they are not provided with the necessary fungi.

Whenever there is a two-way interaction in which one member of a pair of species is benefited and the other harmed, a selectional "race" is set up. For instance, when a plant species is subjected to the attack of an herbivorous insect, any individual plant which has genetically determined characteristics that help it to escape attack is immediately placed at a selective advantage. In other words, the attacking insect population places selective pressure on the plant population. As a result the defenses of the plant population will be enhanced. It will tend, for example, to develop spines, deposit poisonous alkaloids in its leaves, or complete its flowering before the insect emerges. An example of this sort of selection pressure was recently described in which a small lycaenid butterfly was shown to destroy up to 70 percent of the flowers of a population of lupine plants (Fig. 14–17). There was reason to believe that the plant had been placed under selection pressure to flower early so that it would have a chance to produce seeds before the butterfly could destroy them. The early flowering resulted in some flowers being killed by frost, a selection pressure countering that for early flowering produced by the butterflies.

Of course, the mechanical and chemical defenses of the plants in turn place selective pressure on the herbivores. Speaking anthropomorphically, they must develop countermeasures which will permit them to continue feeding. In the case of the lycaenid butterfly, that might be a speeding up of development or eventually a shift in dietary habits to lupine leaves (where it would face competition from another lycaenid) or to some other plant. In other insects, such a countermeasure might consist of the development of stronger jaws for piercing mechanical defenses or detoxifying systems for handling poisons. In any case, each move by the plant "requires" a countermove by the insect, and vice versa. Failure to keep playing the game has extinction as its consequence—a price which has undoubtedly been paid many times. The sort of stepwise reciprocal evolutionary interaction which is characteristic of plant-herbivore and predator-prey systems is called **coevolution.** A special instance of coevolution has resulted in the very important relationship between pollinators and flowering plants.

Pollination, as was mentioned in Chap. 10, is the transfer of the male gametophyte, or pollen grains, to the stigma of a flower. This can be accomplished by wind, as occurs in grasses and many trees such as oaks and elms. The great majority of flowering plants, however, show a range of simple to astonishingly complex specializations for pollination by insects or other animals (Fig. 14–18). At the same time, insects involved in pollination have also become specialized for the gathering of pollen or nectar. Insects are attracted to flowers by their distinctive fragrance, color, or shape. They may collect pollen or nectar for food, or they may be tricked into visiting the flower for some other reason, as we will see. Whatever the case, in their visits pollination is accomplished.

The more specialized flowers usually have their pollen and nectar protected in such a way that only a particular kind of insect can obtain it, and this has, of course, resulted in selection for specialized pollinators. Conversely, indiscriminate visiting of flowers and wastage of pollen and nectar undoubtedly led to the evolution of discriminatory devices of flowers. In this coevolutionary system, the insects benefit from the pollen and nectar produced by the plants.

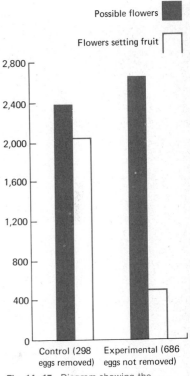

Possible flowers ■
Flowers setting fruit □

Control (298 eggs removed) Experimental (686 eggs not removed)

Fig. 14–17 Diagram showing the effects of predation by a small lycaenid butterfuly on a Colorado lupine. In the control, eggs of the butterfly were removed from inflorescences of the lupine. In the experimental population, eggs were counted but not removed.

A

B

Fig. 14–18 Four examples of animal pollination: (*A*) a bumblebee visiting the open flower of a day lily; (*B*) a bumblebee visiting the closed flower of a snapdragon; (*C*) a nectar-feeding bat visiting flowers of an agave; (*D*) a whitewing dove pollinating a saguaro cactus. (*A and B, courtesy of Ross Hutchins; C and D, USDA photo.*)

There are two interesting instances of specialized modes of pollination in which it is difficult to say that benefit occurs for the insect. In certain milkweed plants of South African deserts, the flowers are colored purple and red and have a strong odor of rotting meat. Carrion insects, such as certain flies, are attracted to these flowers and accomplish pollination in their visits to lay eggs. The fly larvae are unable to live on plant tissue, however, and soon die. Male insects are also tricked by several species of orchids whose flowers closely resemble female insects. Certain bees, wasps, or beetles are attracted to the flowers by their shape, color, and odor and attempt to copulate with them. Pollination in this case is accomplished by what is known as *pseudocopulation.*

Other animals besides insects have become associated with plants in pollination relationships, and the same kind of coevolutionary processes have occurred. Groups of birds (for example, hummingbirds) and of bats have become specialized for pollination, while groups of plants have developed shapes, colors, odors, and structures to promote this relationship.

OFTEN IN COMMUNITIES, TWO DIFFERENT KINDS OF ORGANISMS ARE INVOLVED IN A RELATIONSHIP RESULTING IN MUTUAL BENEFIT OR IN HARM FOR ONE AND BENEFIT FOR THE OTHER. IN EITHER CASE,

THE RELATIONSHIP MAY BECOME VERY SPECIALIZED IN THE COURSE OF COEVOLUTION.

Although many coevolutionary systems may be sensibly viewed as simply involving two species interacting, other species of a community are inevitably affected when *any* species undergoes evolutionary change. In this section, we will look at some of the coevolutionary interactions in which we can clearly see a network (reticulum) of relationships—systems which developed through **reticulate** coevolution.

Reticulate Coevolution

Perhaps the most unusual coevolutionary system related to herbivory is that composed of swollen-thorn acacias and the acacia ants, which are necessary for the survival of the acacia. Acacias are thorny tropical shrubs or trees. The ant acacias lack the bitter-tasting chemicals which defend other acacias against most herbivorous insects. Instead, *ants* serve as the defensive system of the ant acacias, attacking herbivorous insects and killing vines and other plant competitors of the acacias. Colonies of ants live in the swollen thorns which line the stems. If the ant colony is removed, the ant acacias are killed by herbivorous insects. The ants feed on specially modified glands of the acacia which produce protein-containing granules. This system, then, involves ants and acacias in a mutually advantageous relationship, interacting with insect and plant "enemies" of the acacia.

Fig. 14-19 Batesian mimicry in butterflies; (*A*) The Monarch butterfly, *Danais plexippus*, is mimicked by (*B*) the Viceroy butterfly, *Limenitis archippus*. (*American Museum of Natural History.*)

Some of the most thoroughly studied reticulate interactions center around the phenomenon of mimicry. In one kind of mimicry, **Batesian mimicry,** a distasteful, poisonous, or otherwise dangerous "model" organism is imitated by a tasty or harmless "mimic" organism. Ordinarily, the model is strikingly colored or conspicuous in some other way. It "announces" that it is nasty. The mimic imitates the announcement but is unable to deliver the punishment. Bad-tasting red and black monarch butterflies are mimicked by tasty viceroys (Fig. 14-19); stinging striped bumblebees are mimicked by stingless flies. The mimic can be demonstrated to gain protection through the confusion of predators, and a strong selective advantage accrues to those individuals which most closely resemble the model.

The situation is somewhat analogous to a host-parasite system, since the mimic "parasitizes" the protective image of the model and thus reduces the amount of protection enjoyed by the model. For instance, in an area where a model butterfly occurs alone, a bird predator may well have nothing but bad experiences to associate with the model's color-pattern "announcement." However, if one or more mimics occur in the same area, some of the bird's experiences associated with that color pattern will be pleasant. Therefore, a selective advantage will accrue to those variants of the model species which least resemble the mimic, and the model will evolve away from the mimic, with the mimic in hot pursuit. This is an oversimplification, since the nature of a model-Batesian mimic coevolutionary race depends on numerous other factors, including the relative population sizes of the two species and the possibilities of the mimic itself becoming distasteful.

The coevolutionary web involves not just the evolutionary adjustments of model and mimic. Presumably, predator populations are evolving with respect to their ability to discriminate between model and mimic and in respect to their capabilities of dealing with the defenses of the model. Furthermore, plant-herbivore coevolution is normally also involved, at least when the models and mimics are butterflies. There is considerable reason to believe that many distasteful butterflies have simply found ways of biochemically dealing with the poisons of certain well-protected plants and putting these poisons to their own use. For instance, the monarch butterfly contains heart poisons which render them indigestible and distasteful to most vertebrate predators. These are derived directly from their milkweed larval food plants. The plants are also generally avoided by herbivores, which gives the monarchs an added advantage— absence of competition for larval food and a reduced chance of being accidentally devoured by a cow. Clearly, great benefits may accrue to the "winner" of a coevolutionary race.

A second type of mimicry is also widely recognized. This is **Müllerian mimicry,** in which two or more unrelated distasteful or dangerous organisms come to resemble each other. Two butterfly species involved in a Müllerian complex are shown in Fig. 14–20. The advantage of Müllerian mimicry lies in the reduction in the number of warning patterns which a predator must learn. Avoidance of all species in the complex is reinforced by experience with any one. There is, of course, no sharp dividing line between Batesian and Müllerian mimicry. For instance, all butterflies are probably distasteful to one degree or another, at least in comparison with moths. It is quite possible that many situations now thought of as Batesian are partially Müllerian and that Batesian mimics often evolve into fully Müllerian mimics by becoming more distasteful.

By focusing attention on a butterfly mimicry complex, one may trace a webwork of coevolutionary interactions throughout a community. Butterflies coevolving with plants and predators, plants coevolving with other herbivores and the microflora and microfauna of the soil, predators coevolving with their prey and parasites, and so forth. Looking at co-

Hirsutis megara

Lycorea ceres

Fig. 14–20 Müllerian mimicry in butterflies: Two distasteful butterflies of different families are closely similar. (*From "Ecological Chemistry" by Lincoln Pierson Brower. Copyright ©* 1969 by Scientific American, Inc. All rights reserved.)

evolutionary interactions is just one way to approach the questions of community relationships and diversity—a way which has the advantage of stressing their dynamic nature.

COEVOLUTION IN COMMUNITIES MAY LEAD TO A COMPLEX SERIES OF RELATIONSHIPS AMONG MANY DIFFERENT ORGANISMS. AN EXAMPLE IS MIMICRY, IN WHICH PALATABLE HERBIVOROUS ORGANISMS MIMIC NOXIOUS ONES OR SEVERAL NOXIOUS SPECIES COME TO RESEMBLE EACH OTHER. THESE MIMETIC COMPLEXES COEVOLVE NOT ONLY WITH EACH OTHER BUT WITH PREDATOR AND PLANT POPULATIONS.

Now that we have discussed coevolution and the functioning of ecosystems, let us turn to a consideration of how the latter are distributed on the surface of the earth. In order to discuss the major features of the distribution of communities, we must first look at our planet in broad ecological terms.

COMMUNITIES IN SPACE

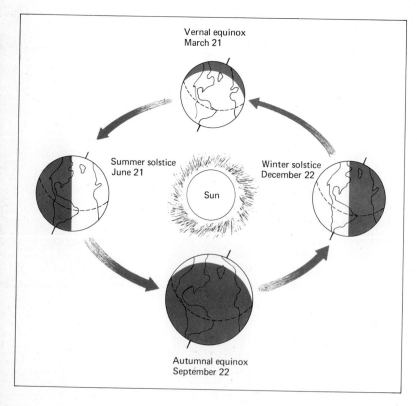

Vernal equinox
March 21

Summer solstice
June 21

Sun

Winter solstice
December 22

Autumnal equinox
September 22

Fig. 14–21 Revolution of the earth around the sun and the tilting of the earth on its axis are responsible for the seasons. (*Adapted from MacArthur & Connell, THE BIOLOGY OF POPULATIONS, 1966, John Wiley & Sons, Inc.*)

Third planet from the sun, the earth revolves about it every $365\frac{1}{4}$ days. Every 24 hr it rotates about its axis, which is tilted 23° from vertical. As you can see in Fig. 14–21, this means that once a year the North Pole is pointed toward the sun and the South Pole is pointed away from it. For the Northern Hemisphere, this is the longest day of the summer, while it is the shortest day of the winter in the Southern Hemisphere. At the opposite point of its orbit, the converse obtains. Thus the revolution and rotation of the earth produce the annual, seasonal, and diurnal rhythms to which all organisms must adjust.

The heating effect of the sun's rays is greatest when they are most nearly perpendicular to the surface. This means that the equatorial region will be heated most, and warm air being lighter than cold air, a rising current of air will be formed, which sinks again at about 30° north and south latitudes. Such a mass of moving air is called a *cell*. At 30°, another cell of atmospheric circulation is established which has down-moving air at 30° and up-moving air at 60° north and south, as you can see in Fig. 14–22. Upward-moving air is cooled, and its water-holding capacity is reduced. Water is then released as precipitation. This accounts for the moist tropics on either side of the equator between the Tropic of Cancer and the Tropic of Capricorn. Downward-moving air is dry, and as it falls and becomes warmer, its water-holding capacity increases. It is not

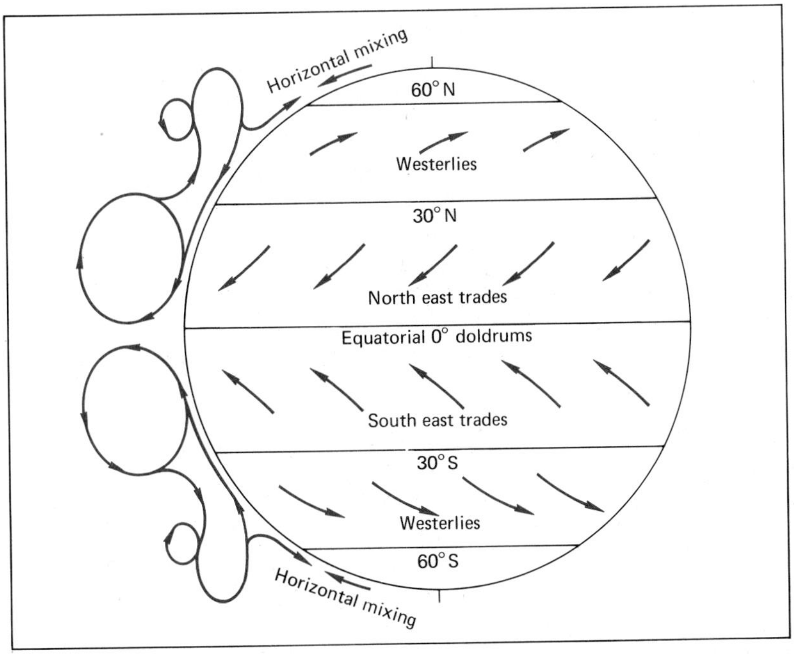

The Physical Substrate of Life

Fig. 14-22 Unequal heating and rotation of the earth set up the major atmospheric cells and the characteristic wind patterns. (*Adapted from THE EARTH AS A PLANET by Kuiper, 1954, The University of Chicago Press, as redrawn in MacArthur and Connell, THE BIOLOGY OF POPULATIONS, 1966, John Wiley & Sons, Inc.*)

surprising that the world's great deserts are found at about 30° north and south of the equator.

Unequal heating thus creates the major patterns of air currents and precipitation belts. The rotation of the earth introduces another factor, which is responsible for the characteristic major wind patterns, for example, the easterly trade winds of the tropics and the prevailing westerlies of the temperate zones. Similar major patterns are set up in the oceans, whose chief currents are shown in Fig. 14–23. All these patterns are modified by local topographic features, such as mountain ranges, changes in coastline, and broad level expanses of land. Air blowing against a mountain, for example, is forced to rise. It thereby becomes cooled, losing its water as rain on the windward side. Descending on the leeward side, it is dry and "water-hungry," and this results in a so-called "rain shadow." North-south mountain ranges in temperate regions typically have less luxuriant vegetation on the eastern slopes than on the west.

As you are undoubtedly aware, there is a temperature gradient north and south to the poles from the equator. This amounts on the average to 0.5°C for each degree of latitude. Across this temperature gradient, as

Fig. 14–23 The major oceanic currents.

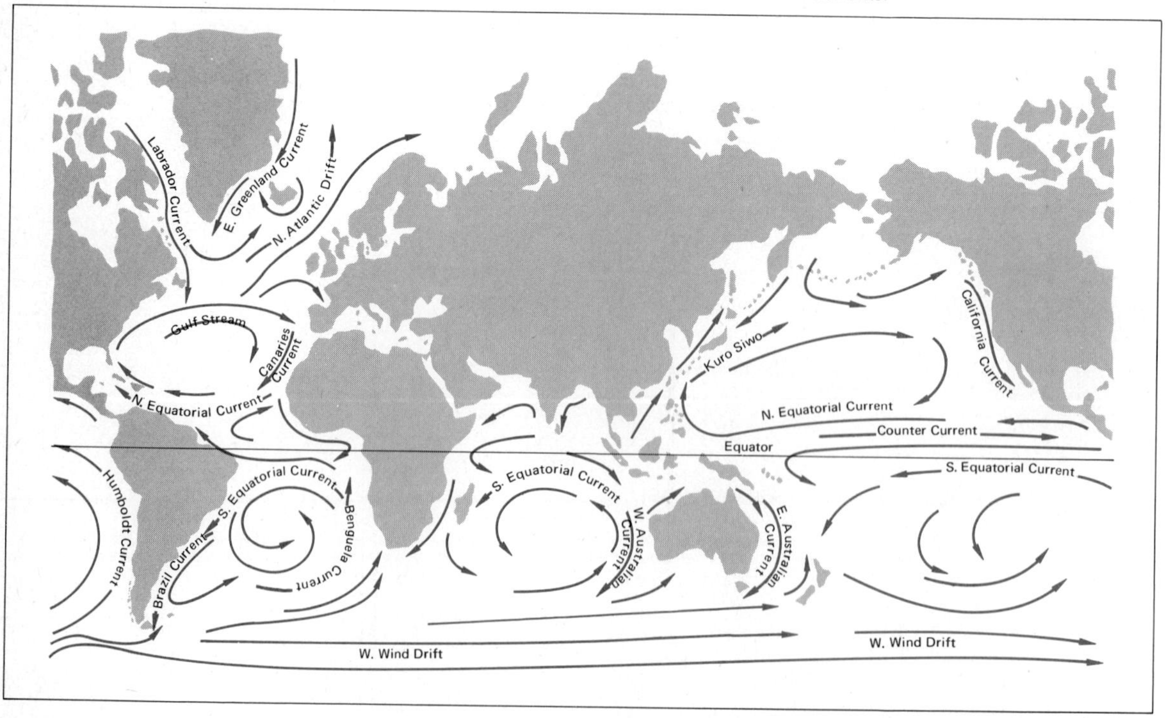

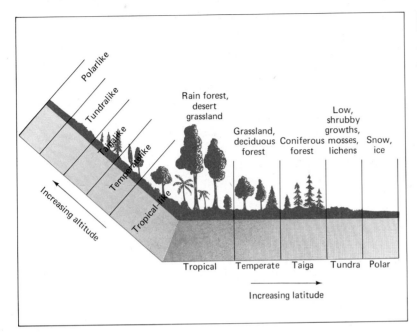

Fig. 14-24 The effects of latitude and longitude on the distribution of plants. (*Adapted from Paul B. Weisz, THE SCIENCE OF BIOLOGY. Copyright McGraw-Hill Book Company, 1971. Used by permission.*)

modified by local conditions such as mountains or rivers, are distributed the major communities (Fig. 14-24). We will return to them shortly. It should be noted at this point that a temperature gradient also exists as one goes up a mountain. This gradient is of the order of 0.5°C for each 100 m of elevation. This means that there can be mountain peaks with a permanent snowcap in the vicinity of the equator. It also means that organisms occurring in temperate or polar regions can move toward the equator by moving up the slope of a mountain until their usual temperature regime is found. Plants and animals may thus spread north and south along such mountain ranges. On the other hand, east-west mountain ranges can act as barriers to the spread of organisms to and from the equator. In order to cross an east-west mountain range, an organism must ascend through an increasingly colder set of habitats and then descend through them on the other side.

THE ORIENTATION OF THE EARTH WITH RESPECT TO THE SUN, ITS ROTATION, AND ITS REVOLUTION RESULT IN UNEQUAL HEATING AND COOLING, WHICH SETS UP PATTERNS OF CIRCULATION OF THE ATMOSPHERE AND THE MOISTURE IT CARRIES. THUS ARE SET UP THE MAJOR CLIMATIC REGIONS OF THE GLOBE, WHICH ARE MODIFIED BY LOCAL FEATURES SUCH AS MOUNTAIN RANGES, RIVERS, AND OCEANS.

Organisms are found in virtually every spot on the surface of the planet. They have become specialized for life in hot springs where the temperature may reach 77°C. They are also found on and in the ice of glaciers and the polar caps. Four-fifths of the surface of the earth is salt water. The oceans form the largest habitat available. It is generally believed that life originated in the sea or along its margins. Before the advent of terrestrial plants, the barren earth must have presented a severe habitat indeed. Except for occasional sallies from the sea, terrestrial life probably was not possible for animals. Not until the green algae became pioneer organisms invading the land did the succession resulting in the world as we know it today begin.

Life in the Sea

The sea is perhaps the most stable of all habitats. Its major zones are shown in Fig. 14–25. Osmotically, the ocean is more favorable than any other. The ocean's great expanses are ideally suited for the growth of unicellular organisms, and here we find many groups of marine algae as constituents of the free-floating **plankton.** Something of the order of three-fourths of the planet's photosynthesis is carried out by phytoplankton, which are thus the world's greatest producers of oxygen. Also part of the plankton are unicellular animals and the larval and adult stages of many invertebrates and vertebrates. Larger swimming forms of both invertebrates and vertebrates may range widely through the waters—as,

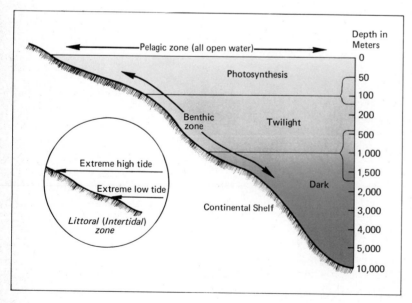

Fig. 14–25 Diagram showing the major zones of the ocean. (*Adapted from BIOSCIENCE by Robert B. Platt and George K. Reid © 1967 by Litton Educational Publishing, Inc. Reprinted by permission of Van Nostrand Reinhold Company.*)

for example, whales, which swim from tropical waters to polar seas. Others have more restricted ranges.

Photosynthesis can occur only to the depths that sunlight penetrates. The water filters the light, and its spectral composition changes with depth. Different groups of algae have peak photosynthesis at different wavelengths and intensities of light. They are appropriately distributed along this gradient. Below the depths that light penetrates is a cold heterotrophic world of darkness, dependent on the upper layers for nourishment. This is the region of the **bathyal** zone. Bathyal organisms show many specializations for life in the dark and at great pressure. Many of them have light-producing organs. Often arranged in patterns and varying in color, these may serve as recognition symbols. Deep-sea squid have a luminescent "ink" instead of the dark-brown sepia of their surface relatives. Only a few kinds of organisms are able to survive in the greatest depths of the oceans, the **abyss.** These abyssal organisms include bacteria, which decay the dead organisms falling from above.

The margins of the sea provide a special way of life for many kinds of plants and animals. Rocky, sandy, and muddy margins, or **littoral** habitats, as they are called, furnish a wide variety of different niches. Coral reefs, produced by coral animals and their associated marine algae, whose photosynthesis results in the precipitation of insoluble salts from the sea, are a fascinating ecological niche; they are inhabited by a great variety of animals, often of brilliant colors and bizarre forms.

There are relatively few species of marine life. Often these are the only surviving members of once widespread and important phyla of strictly marine plants and animals—for example, echinoderms (crinoids, sea urchins, starfish, brittle stars, and sea cucumbers). In comparison with terrestrial faunas and floras, those of the sea are remarkably sparse. The reasons for this are not entirely clear. The rate of species formation perhaps is slow in the oceans as compared with on the land because barriers to dispersal are few. The relatively stable habitat may not pose as many environmental challenges as does the land. Also, from the viewpoint of the terrestrial observer, there seem to be fewer but broader ecological niches. This is, of course, merely another way of saying the same thing and that we do not know enough about the seas around us.

Life in Fresh Water

The freshwater habitat is much more varied and generally less stable than the marine one. Fresh water may exist as huge lakes or small ponds, or it may occur as slow-moving rivers or swift streams. Furthermore, fresh water is also an osmotically unfriendly habitat. As we have seen before, freshwater organisms tend to take up water since their cells are considerably more concentrated than the water of a pond or stream. Hence the many means for solving this problem, which range from contractile vacuoles of protozoans to the kidneys of freshwater fish.

Life in fresh water is a specialized ecological niche. You can see that each of the subdivisions—pond, lake, stream, river—requires further specializations on the part of the organisms that function there. Two points about the freshwater habitat deserve special mention. First of all, fresh water originates from precipitation from clouds. Although unequally distributed, it plays a major role in the normal erosive processes at work on the land surfaces. Continuously dissolving away the land, they carry, in solution or suspended, enormous amounts of material into the sea. Remember the phosphorus cycle discussed above.

LaMont Cole has calculated that each year water is evaporated from the oceans equivalent to a depth of 1 m. Only one-sixth of this amount is evaporated from land and fresh water, a sizable amount being from transpiring plants, of course. The total amount of evaporated water approximately equals the annual precipitation—an amount of water measuring about 100,000 cu miles! The proportion of the annual precipitation which falls on land exceeds the evaporation by about 9,000 cu miles of water, which flows into freshwater ponds and lakes and eventually into the oceans. The great and small rivers of the world annually wash to the sea 4 billion tons of dissolved inorganic material, 400 million tons of dissolved organic material, and 2,000 million tons of materials which are not dissolved.

Interference with the normal movement of water by man may have disastrous ecological consequences not easily foreseen. Improper irrigation can largely ruin the soil for plant growth and has led to the destruction of formerly great civilizations. For instance, irrigation water has a higher concentration of salts than rain water. Unless proper techniques are used, evaporation often will lead to a buildup of salts in irrigated soils. Such salinization was one of the factors which caused the decline of civilization in the Tigris and Euphrates valleys, and the conversion of that once-lush area into desert.

The second point about the freshwater habitat concerns temperate lakes and ponds. Such a lake in the Temperate Zone experiences an interesting annual cycle (Fig. 14–26). During the winter, much of the water is frozen. Ice (water at 0°C) is less dense than liquid water—it floats. This unique property of water protects freshwater organisms from being frozen. Once a blanket of ice develops, it effectively retards further heat loss from the body of water. Therefore, it is rare for a pond or lake to freeze solid. When the spring thaw begins and the water reaches 4°C, it becomes more dense and therefore sinks downward. As a result,

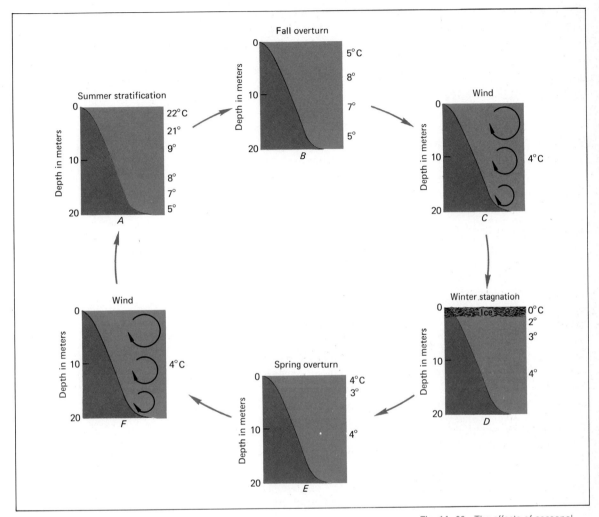

Fig. 14–26 The effects of seasonal temperature change in a temperate zone pond. (*Adapted from Raven, Peter H. and Helena Curtis, BIOLOGY OF PLANTS, Worth Publishers, New York, 1970, p. 555.*)

the water of the lake becomes thoroughly mixed—oxygen, nutrients, and small organisms alike. During the summer, the sun heats the water and a stratification of temperature zones occurs, from warm water at the top to cold at the bottom. With the onset of winter, as the water cools before freezing, another *overturn* may occur, again mixing the water. Most lakes are also stratified in the winter.

Life on Land

The land provides the richest variety of habitats. Some organisms are subterranean and play an important role in the mixing of soil constituents and in the circulation of air and water in the soil. Some organisms are aerial, as we have seen. The existence of trees, often of great height, creates the arboreal ecological niche. This is an extremely important niche in the tropical forest, but unfortunately one which is very difficult to study.

OF THE THREE MAJOR HABITAT AREAS OF THE EARTH, THE SEAS WERE PRESUMABLY THE FIRST TO BE INHABITED; FROM THERE, FIRST PLANTS AND THEN ANIMALS OCCUPIED LAND AND FRESH-WATER HABITATS. THE LATTER TWO, IN COMPARISON WITH THE OCEAN, HAVE MANY MORE SPECIES AND A WEALTH OF SPECIALIZED ECOLOGICAL NICHES.

PATTERNS OF DISTRIBUTION OF ORGANISMS IN SPACE AND TIME

If one looks at the distribution of animals and plants over the surface of the earth with a very broad view, certain patterns can be seen. There are vast areas, of particular climatic types, occupied by characteristic types of vegetation. These are usually associated with characteristic types of animals. Such groupings of plants and animals are known as **biomes.** A biome is a climatically determined association of plants and animals playing the ecological roles we have discussed earlier in this chapter.

One can look at major distribution patterns of organisms from a different point of view, however. In different parts of the world, but in the same biome, ecological roles are played by different actors. Instead of mapping the biome, one can study the kinds of organisms present. One can identify the major constellations of taxonomic groupings. This would enable us to identify what are called **biogeographic realms.**

Biomes

Depending upon the scale chosen, one can recognize a few biomes of great extent or many biomes of smaller area (Fig. 14–27). In this chapter, we will discuss seven major terrestrial biomes. These are the tropical forest, the savanna, the desert, the grassland, the temperate deciduous forest, the taiga, and the tundra.

The most obvious defining feature of a biome is the growth form of the conspicuous plants, and this is determined in large part by the climate. Less easily seen characteristics of the plants, together with the kinds of animals and their structural and physiological specializations, are also important. In the next few paragraphs we will discuss the seven biomes in turn, beginning with the equatorial region of the earth and moving poleward north and south.

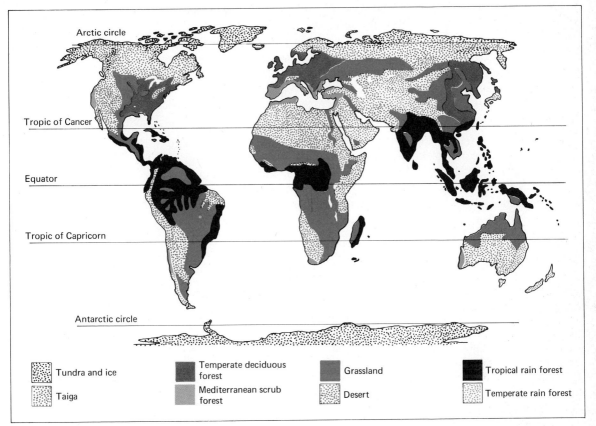

Fig. 14-27 The major biomes of the world. (*Redrawn from LIFE: AN INTRODUCTION TO BIOLOGY by Simpson, Pittendrigh and Tiffany, copyright © 1957 Harcourt Brace Jovanovich, Inc. and reproduced with their permission.*)

The **tropical forest** is found in a broad belt surrounding the planet, mostly between the Tropic of Cancer and the Tropic of Capricorn. There are several subtypes of tropical forest that may be recognized. We will describe only the tropical rain forest. This biome is thought to be ecologically the oldest, and there is no question that it is the richest in number of species of plants and animals. There are more kinds of organisms in the tropical rain forest than in all other biomes combined. Although the diversity of kinds of organisms is very high, the density of any one species may be quite low. For example, in a given area of several square miles extent, one might find only one individual of a particular tree species. The same area may have more than a hundred different species of trees. The interrelationships of the animals and plants of the rain forest are, as you might expect, extraordinarily complex.

The most conspicuous plants in the rain forest are, of course, the trees.

The tallest of these may be 75 m high, and their crowns form a canopy which intercepts much of the light. Below these tallest trees may be smaller species, whose canopies often give the forest a layered appearance. Large woody vines called lianas are abundant, as are epiphytes, plants which grow attached to other plants. The latter range in size from shrubs to herbaceous plants. Some may be partial parasites, but most simply use the trees as a place to live without deriving any special nourishment from them.

Tropical rain forest trees grow in areas which are warm to hot the year round and where the rainfall, between 200 to 400 cm in amount, is evenly distributed throughout the year. They are generally evergreen, with large leaves. Contrary to what you might believe, their flowers usually are relatively small and white or greenish white. Because of the extensive canopies of the trees in the layered forest, little light, and no direct sunlight, reaches the forest floor. Therefore, there are few herbaceous plants growing upon the ground. Since the temperature is high and moisture abundant, the dead leaves and other organic material falling from the canopy is quickly decomposed, and the nutrients released are absorbed by the shallow root systems of the trees. Under these conditions, a characteristic type of soil called **laterite** is often formed. Several kinds of laterite can be recognized, but they all share the properties of being heavy red clays and largely infertile. If the trees are removed, lateritic soils either form hard, compacted soil or erode away very rapidly.

Because of the relative infertility of the floor of the tropical rain forest, the greatest assemblage of animals and herbaceous plants is found in the canopy of the trees. Stratification of light and humidity leads often to localization of a species to one layer of the leafy cover, and other species may occur in layers above or below. Birds, reptiles, and amphibians are abundant, as are the tremendously diverse invertebrates, especially insects. Small mammals, such as monkeys, are also very common. There are few very large predators, but the abundance of energy ensures that there will be higher-order carnivores of many kinds.

In the New World, the tropical rain forest is found in South America (the Amazon basin primarily), Central America, eastern Mexico, and the West Indies. In the Old World, rain forest occurs primarily in the Congo basin in Africa, as well as in southern Asia, Malaysia, the Philippine Islands, and on the northeast coast of Australia. Man is rapidly destroying the rain forest in all these areas by cutting down the trees. As mentioned above, this leads to widespread erosion and the formation of impenetrable soils, useless for agriculture. As the trees and soil go, so also go the vast numbers of animals and smaller plants.

In many parts of the tropics, the rainfall may be high (80 to 200 cm) but seasonal. That is, there is a period of drought during a part of the year. In such areas, the biome called **savanna** occurs. Savannas are areas of scattered trees and grassland. The trees usually are deciduous, with small leaves, spines or thorns, and thick insulating bark. Since there is not a continuous canopy, light is available at the ground level. Perennial herbs generally are abundant, and they usually have underground storage structures, such as bulbs, corms, or tubers. Annual plants and epiphytes are relatively rare. Natural and man-made fires are common in the savanna during the dry season.

Savannas are the home of grazing and browsing mammals, which often occur in great numbers. In the savannas of Africa, not only are the individuals numerous but the species are many and diverse. It is not uncommon to find mixed herds of many different kinds of antelopes, zebras, giraffes, wildebeest, etc. Buffalos and elephants also occur in this biome. Slight differences in the manner of feeding, the parts of the plants utilized, and the time of feeding make it possible for so many species to occur together. Savannas of the New World are not as diverse as those of the Old World. In both hemispheres, a characteristic assemblage of predators has evolved, usually consisting of dogs and their relatives, which hunt in packs, and the various large cats, usually more solitary in nature (except for the lion).

Savannas usually have a number of species of burrowing animals, which may go into a dormant phase (aestivation) during the dry season. Termites are especially abundant in the savanna biome. Often they build large mounds consisting of over 1,000 cu m of soil. They play an important role in breaking up the clayey soils, which often are lateritic in nature.

Using the term "savanna" in the broad sense, one can say that this biome generally occurs polewards of the tropical forest and is intermediate between such forests and true deserts. Thus savannas are found on all the continents. Notable examples are in northeastern South America, southern Africa, and Australia. A special sort of savanna, called *monsoon forest,* covers extensive areas of southeastern Asia and India.

Because of the characteristic patterns of atmospheric circulation, there are bands or zones of high barometric pressure at about 30°N and 30°S latitude. Here the air is predominantly descending and very dry. Precipitation consequently is very low. In such regions, the biome called **desert** is found. Some deserts, such as the Atacama Desert in Peru and Chile, have an average rainfall of less than 2 cm a year. Others have an annual rainfall of up to 10 cm.

The organisms which are able to survive in the desert biomes usually do so by being active only during that part of the day or year when the environmental conditions are least extreme. That is, the animals are usually nocturnal, spending the hot days beneath rocks or in burrows, protected from the extreme heat. Plants may be annuals, germinating quickly at the rainiest time of the year, growing and flowering in the space of a few weeks, and then maturing seeds. The remainder of the year is spent in the form of very resistant seed. Perennial plants are generally dormant during the driest season. Trees and shrubs generally have thorns and small, coarse-textured leaves; many shed their leaves for much of the year. Characteristic of deserts also are succulent plants, such as cacti, which resist drying out by reducing their surface area and by storing moisture.

A familiar biome to most of us is the **grassland,** because this is the biome most used for agricultural purposes. In addition to the conspicuous grasses of many different kinds, there are many other perennial herbs. The fauna of grasslands has a large component of burrowing animals, in addition to, usually, herds of grazing herbivores and a variety of their predators.

Grasslands are found on all continents, usually in interior regions, where the rainfall is insufficient for trees and shrubs but is enough to support the perennial herbs. By removing woody plants from forested areas, it is possible to produce grasslands artificially, and grazing of livestock or periodic plowing will usually prevent reforestation. In many areas of the world, man has created grasslands from woodlands by clearing and repeated burning. Indeed, some biologists theorize that all grasslands are man-made. In drier areas, overexploitation of grasslands may convert them swiftly to desert. In contradistinction to forest biomes, where organic material necessary for long-range productivity is stored in plant parts, such as woody stems, the organic reserve of grasslands is in the form of humus in the soil. If this humus is used up or allowed to erode away, the grassland may be rapidly destroyed.

Intergrading with the grassland biome in many parts of the Northern Hemisphere is the **temperate forest.** This biome is found in areas where the winters are moderately cold and the summers are warm and relatively wet. The trees, usually of a variety of species, are deciduous, losing their leaves with the onset of winter. Before leafing out occurs in the spring, a flora of perennial herbs often puts on a spectacular show of flowers, setting seed as the trees begin once again to function actively and to develop a leafy canopy which shades the forest floor. Grazing herbivores such as deer, omnivores such as pigs, racoons, and bears, and smaller carnivores are characteristic of the temperate forest biome.

Northward of the temperate forest is the biome called **taiga** (a Russian word), which is coniferous forest, swampy in summer and cold and snowy in winter. The conifers include such familiar trees as spruce, fir, hemlock, and pines. Thickets of angiosperm trees, such as alders, birches, and willows, also are found. The moose is a conspicuous herbivore; bears, wolves, and members of the weasel family—fishers, martens, weasels —are very common.

Between the taiga and the polar ice is the **tundra** biome. Trees are absent, and the vegetation consists of a wide variety of low-growing perennials which are able to take advantage of the brief growing period. The ground is permanently frozen—a layer of **permafrost** persisting a few feet below the surface even in summer. Many of the animals of the tundra spend only summer there. Permanent residents are few in number and are specialized in many ways for this most harsh of the biomes. Hares, lemmings, caribou, wolves, and musk oxen are characteristic animals.

*BIOMES ARE ASSOCIATIONS OF PLANTS AND ANIMALS OF CHARAC-
TERISTIC LIFE FORM WHICH OCCUPY THE MAJOR CLIMATIC AREAS
OF THE EARTH. THE MOST IMPORTANT TERRESTRIAL BIOMES ARE
TROPICAL FOREST, SAVANNA, DESERT, GRASSLAND, TEMPERATE
DECIDUOUS FOREST, TAIGA, AND TUNDRA.*

Biogeographic Realms

Geologists have given us a picture of how the surface of the earth has changed in the course of time. The distribution of lands and seas has changed. Large continents have split apart into smaller ones. Land areas have become covered with waters, and land connections between continents have been made and broken many times. As glaciers advanced and retreated, as mountains rose and were eroded away, plants and animals evolved and spread over the planet. Eventually the biogeographic realms we recognize today developed (Fig. 14-28).

On the heterogeneous and changing earth, newly opened habitats made possible the development of new species as organisms increased their range. For terrestrial life, north-south mountain ranges often have acted as routes of dispersal for organisms of higher latitudes. East-west mountain ranges, as we saw earlier, usually act as barriers to dispersal. For marine organisms, land masses are barriers. Less obvious features of the oceans, such as differences in temperature or salt content of the water, may also prevent dispersal. It is the combination of the changes in the distribution of routes of dispersal and barriers to movement of the evolving, migrating organisms which has set up the patterns we recognize today.

Figure 14-28 shows the major biogeographic realms as they have been described for terrestrial vertebrates. A somewhat similar pattern is shown by vascular plants. Marine organisms, invertebrates, and nonvascular

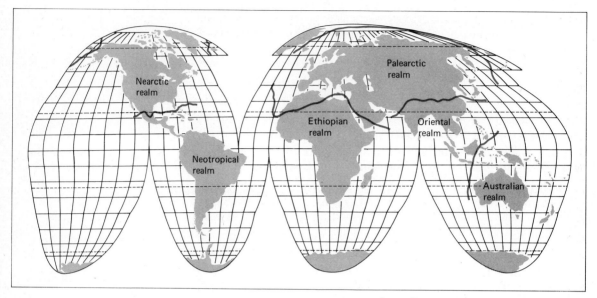

Fig. 14-28 The major biogeographic realms. *(Adapted from Relis B. Brown, GENERAL BIOLOGY. Copyright McGraw-Hill Book Company, 1970. Used by permission.)*

plants do not fit precisely into these realms for reasons that we cannot discuss here, such as different means of dispersal, etc. The distinctness of each biogeographic realm depends upon the degree and length of time of its isolation from other realms.

The Australian realm is the most distinctive since, as far as land verte-brates are concerned, it has been set apart the longest. The mammals are nearly all marsupials (pouched mammals), such as kangaroos, wom-bats, and the koala bear. Isolated for hundreds of thousands of years, the marsupials have evolved forms similar to virtually every type of placental mammal. The primitive egg-laying mammals, such as the duck-bill platypus and the spiny anteater *Echidna*, still persist in this realm, having been extinct elsewhere for millions of years.

The Oriental and Ethiopian realms have somewhat similar species—for example, the Indian elephant and the African elephant—but differ in a great number of forms. Among the apes, the Oriental realm has gibbons and the orangutan and the Ethiopian realm has chimpanzees and goril-las. The tiger is an important carnivore of the Oriental realm, and the lion plays this role in Africa (it was once important in the Oriental realm but has been hunted virtually to extinction). Especially characteristic of the Ethiopian realm are the grazing animals—zebras, giraffes, antelopes—mentioned above.

The Holarctic realm is the broadest of all, encompassing most of Europe, Asia, and North America. Certain animals—for example, moose, timber wolves, and hares—are found on all three continents. The North American area and the Eurasian area are sufficiently distinct, however, that they are often recognized as the Nearctic and Palearctic realms, respectively.

The Nearctic realm has had a history of intermittent connection with the Neotropical realm of Central and South America. Periodically, migrations have occurred north and south along the dispersal route of Central America. Although the Nearctic and Neotropical realms have contributed species to each other's fauna, the Neotropical realm remains a very distinctive one. Sloths, anteaters, armadillos, hummingbirds, and the New World monkeys are among the characteristic animals of this realm.

In each of the realms, characteristic and distinctive associations of organisms have evolved together. This coevolution has resulted in relatively stable communities, structured in the ways we have previously discussed. One of man's activities, in addition to restructuring or destroying biomes to suit himself, has been the moving of animals about the planet from one realm to another. Often these transplantations are unsuccessful, for the introduced organism finds no way of life, no ecological niche available to it.

In other instances, the alien species finds a perfectly suitable habitat, without the usual predators and parasites or other checks on population size. It then may undergo a population explosion in its new home. Unfortunately, most of the instances in which this has occurred have been ecologically and economically disastrous for man. One could enumerate a long list of pest species intentionally or accidentally introduced by man: the codling moth, the starling, the English sparrow in America; the prickly-pear cactus and the European rabbit in Australia; the mongoose and the African land snail in Hawaii. It is only with the greatest difficulty, if at all, that these pests can be controlled. The number of beneficial successful introductions is very small and includes such animals as the ring-necked pheasants and various kinds of trout introduced into North America.

BIOGEOGRAPHIC REALMS ARE CHARACTERISTIC TAXONOMIC GROUPINGS OF ORGANISMS COVERING LARGE AREAS OF THE EARTH. THE MOST IMPORTANT TERRESTRIAL BIOGEOGRAPHIC REALMS ARE THE PALEARCTIC, THE NEARCTIC, THE NEOTROPICAL, THE ETHIOPIAN, THE ORIENTAL, AND THE AUSTRALIAN.

SUMMARY

A community consists of all the organisms which live together in a particular place at a particular time. The organisms share the ability to live and reproduce under the local conditions. They also, however, have usually coevolved in relationship to one another as factors of the environment. An ecosystem is a community plus the physical environment with which the community exchanges energy and matter. Organisms in an ecosystem play the ecological roles of producers, primary consumers, secondary consumers, etc. The number of individuals, the total mass of living material, and the amount of energy contained therein generally all decrease from one trophic level to another as one goes up the food web. Energy is trapped by the green plants, moves up the trophic levels, and in the course of its use by organisms in the community, is eventually transformed into states no longer usable by biological systems. Chemical elements are not used up but, rather, cycle in ecosystems. Decomposer organisms, by breaking down complex molecules into simpler ones, play a crucial role in the cycling of matter.

Carbon cycles primarily from a gaseous reservoir in the atmosphere or water to organic molecules of living or fossil organisms. Oxidation of these compounds releases carbon dioxide back to the reservoir. Nitrogen cycles primarily from a gaseous reservoir in the atmosphere to complex organic molecules. In the course of breakdown of plant and animal tissues, nitrates are released, and intermediate breakdown products (humus) are essential for maintenance of soil structure and productivity. Phosphorus cycles from a sedimentary-rock reservoir to complex molecules of organisms. Because it is made available to land organisms only by geological processes, phosphorus may be lost for very long periods of time. Some substances become concentrated in organisms of higher trophic levels as a result of being very stable and not excreted by plants or animals. As such a substance is carried up the food web, virtually the same amount of it is contained within a smaller and smaller total biomass.

The habitat of an organism and also its way of life in that habitat make up the ecological niche of the organism. Ecological succession is the series of stages in the development of a relatively stable, long-lived climax community on land or in water. In the course of succession, the numbers and kinds of organisms increase and the habitat is greatly modified. It is generally assumed that in the course of evolution, communities have become more diversified and complex and that ecological niches have become more and more finely subdivided.

Organisms in a community act upon one another in a great many different ways. In most communities, some species benefit from the presence of another, which is neither harmed nor benefited in the relationship. Often two different kinds of organisms are involved in a relationship resulting in mutual benefit or in harm for one and benefit for the other. In either case, the relationship may become very specialized in the course of coevolution.

The orientation of the earth with respect to the sun, its rotation, and its revolution result in unequal heating and cooling, which establishes patterns of circulation of the atmosphere and the moisture it carries. This atmospheric circulation sets up the major climatic regions of the globe, which are modified by local features, such as mountain ranges, rivers, and oceans. Of the major habitat areas of the earth, the seas were presumably the first to be inhabited. From there, first plants and then animals occupied land and freshwater habitats. Terrestrial and freshwater habitats, in comparison with the oceans, appear to have many more species and a wealth of specialized ecological niches. Associations of plants and animals of characteristic life form which occupy the major climatic areas of the earth are called biomes. The most important terrestrial biomes are tropical forest, savanna, desert, grassland, temperate deciduous forest, taiga, and tundra. Biogeographic realms are characteristic taxonomic groupings of organisms covering large areas of the earth. The most important terrestrial biogeographic realms are the Palearctic, the Nearctic, the Neotropical, the Ethiopian, the Oriental, and the Australian.

SUPPLEMENTARY READINGS

Books

"The Biosphere," a *Sci. Amer.* book, Freeman, 1970.

Chambers, K. L. (ed.): "Biochemical Coevolution," Oregon State University Press, 1970.

"Continents Adrift," readings from *Sci. Amer.* with introductions by J. T. Wilson, Freeman, 1971.

Elton, C. S.: "The Ecology of Invasions by Animals and Plants," Methuen, 1958.

Farb, P.: "Face of North America," Harper, 1963.

Odum, H. T.: "Environment, Power, and Society," Interscience, 1971.

Whittaker, R. H.: "Communities and Eco-systems," Macmillan, 1970.

Wickler, W.: "Mimicry in Plants and Animals," McGraw-Hill, 1968.

Articles

Cole, L. C.: The Ecosphere, *Sci. Amer.,* **198,** no. 4, Offprint 144 (1958).

Ehrlich, P. R., and P. H. Raven: Butterflies and Plants, *Sci. Amer.* **216,** no. 6, Offprint 1076 (1967).

Hawking, F.: The Clock of the Malarial Parasite, *Sci. Amer.,* **222,** no. 6, Offprint 1181 (1970).

Stewart, R. W.: The Atmosphere and the Ocean, *Sci. Amer.,* **221,** no. 3, Offprint 881 (1969).

Woodwell, G. M.: Toxic Substances and Ecological Cycles, *Sci. Amer.,* **216,** no. 3, Offprint 1066 (1967).

CHAPTER 15

ORGANIZATION OF SOCIETIES

As we saw in Chap. 10, in those organisms which reproduce sexually, the gametes must be brought together in order for fertilization to occur. In many animals and all plants, fertilization is external. Males and females release their gametes into the environment, or the male gametes are released and make their way to the female reproductive structures. We have discussed how reproductive behavior of the sexes is coordinated. Obviously, with external or internal fertilization, there are advantages in having organisms grouped together. The chances of gametes getting together will be enhanced by aggregation of individuals. We will consider here how animals have solved the problems involved in the formation of groups.

Internal fertilization requires that the parents come together, at least long enough to copulate. As was discussed in Chap. 13, we would expect that selection in the course of evolution would tend to favor behavior patterns which would maximize reproductive success. Competition for food, space, and a mate would, in the long run, enhance aggressive behavior in general and, of course, predatory behavior in carnivores. Because of the aggression, finding a mate and copulating are often a risky business for animals, even those that are not carnivorous. For this reason, courtship and mating behaviors have evolved which function to reduce aggressive behavior to the point that fertilization can be accomplished.

As you are undoubtedly intuitively aware, however, there are advantages to group life other than that of copulation, which may, after all, take only a short time. These advantages have, in many animals, overweighed the advantages of a solitary, independent life. You can think of many different kinds of groups of animals: schools of fishes, flocks of birds, herds of mammals. In all these, and in other such groups, the aggressive, competitive behavior of the individuals is modified in such a way that group life is possible and the resources of the environment are efficiently exploited.

*THE ADVANTAGES OF GROUP LIFE MAY OUTWEIGH THE ADVANTAGES
OF A SOLITARY LIFE IN MANY ANIMALS. GROUP LIFE, AS WELL AS
REPRODUCTION, REQUIRES A MODIFICATION OF THE AGGRESSIVE,
COMPETITIVE BEHAVIOR ONE WOULD EXPECT TO BE ENHANCED BY
SELECTION.*

ADVANTAGES OF GROUP LIFE

It is possible to recognize two different kinds of groupings. Animals—
or, for that matter, plants—may occur in groups because of their individual
responses to particular factors of the environment. For example, butterflies attracted to a field of flowers or the attraction of flies to garbage become grouped because of their individual responses to a food source.
Such groups are called **aggregations.** Other groups are formed because
of the response of individuals to *one another.* These responses are
called **social responses,** and the group is known as a **social group.**
In the example of a troop of baboons to which we have repeatedly referred,
the individual baboons respond in similar ways to factors of the environment, such as food. But their social responses to one another form the
bonds that integrate them into a social group.

Aggregations and social groups have the advantages of reproductive
efficiency. Males and females are usually both present, and their behaviors can be correlated. There are other advantages, having to do with
protection of individuals and with efficient utilization of environmental
resources. It has been known for some time that there may be physiological benefits in grouping. Even in very simple invertebrates, such as
protozoans, it can be shown that groups of individuals are more resistant
to such things as poisons or irradiation than are single individuals. In
fact, this is true for vertebrates, such as fishes. These effects are not simply
physical ones; they often involve the secretion of substances into the
medium. Many aquatic organisms are known to "condition" the water in
which they live, that is, to alter it chemically and make it more suitable
for their life. A group of individuals may accomplish this more rapidly
for a given volume than an individual organism.

You can probably think of many other instances in which crowding may be
beneficial. Litters of kittens or puppies, as well as baby chicks, huddle
together. This reduces their heat loss and thereby increases metabolic
efficiency. Similarly, frogs and toads in a cage huddle together, reducing
moisture loss from their damp skins. Vertebrate individuals often have a
tendency to imitate each other. This may lead to what is called **social
facilitation.** Social facilitation tends to increase the activity of individuals
and thus increase their efficiency. It has been shown that animals learn-

ing a maze do so more rapidly as a group than as individuals. Indeed, in many situations, humans learn more rapidly in groups, where they can see and imitate one another.

Group action against predators, of course, is an important aspect of group life. You may have seen smaller birds flying about, encircling, and distracting a predator bird such as a falcon or owl. And you have heard of the defensive circle of musk-oxen with the young in the center. When a baboon troop feeds in the open, it assumes a characteristic structure (Fig. 15-1). The troop consists of females with young and juveniles in the center along with the most dominant males. Preceding and following them are less dominant males, together with adult females and older young. An approaching predator always encounters adult males. Should an actual attack occur, an impressive array of males with their enlarged canines is always interposed between the females and offspring and the predator. It is thought that schooling behavior in fishes may serve as a protective mechanism which makes it more difficult for a predator to select and catch a single victim in a rapidly changing array of potential prey, although schooling probably serves other functions such as facilitation of hunting and breeding.

It seems clear that groups of animals feed more efficiently, in most

Fig. 15-1 A baboon troop in Nairobi Park emerges from the grass with two adult males in the lead. (*Courtesy of DeVore and James.*)

instances, than single individuals. If one member of the group finds a particular source of food, its location is soon known to all. All-male groups of chimpanzees will leave the troop on scouting expeditions. Upon finding a source of food, such as a tree with ripe fruit, they set up a chorus of hollering that brings the rest of the troop.

The acceptance of a new food by a group is accelerated when a dominant male is the first to adopt it. Biologists studying the Japanese macaque, a social monkey similar to the baboon, noticed that the entire troop adopted a new kind of food within a few days of its "discovery" by a large male, whereas the same process took months when the discovery was made by a female. In humans, too, a new style is universally accepted if an important leader or his wife adopts it, whereas styles first adopted by minority groups or youth (such as long hair) may require up to a generation for cultural fixation. Groups which feed systematically and thoroughly across an area undoubtedly are more efficient than individual food gatherers. Animals stir up the substrate—soil, grass, etc.—thus raising insects and other prey for the group.

AMONG THE ADVANTAGES OF GROUP LIFE ARE SOCIAL FACILI-TATION, GROUP DEFENSES AGAINST ENVIRONMENTAL FACTORS INCLUDING PREDATORS, AND INCREASED EFFICIENCY IN UTILI-ZATION OF THE ENVIRONMENT.

**STRUCTURE AND
ORGANIZATION OF SOCIAL
GROUPS**
Parental Care

Formation of social groups usually begins with parental care in animals. The amount of parental care varies widely, however, from group to group. In some species, the eggs are abandoned by the parents after fertilization. The eggs may simply be left with whatever protective coverings they have, as in frogs and toads. Here they are deposited in the water in a jellylike coating in which the young develop. The eggs of many snakes and lizards are similarly abandoned. Many kinds of fishes, such as salmon, make a simple nest by scraping a depression in the sand or gravel into which the eggs are dropped. In salmon, the parents then die. In other fishes, one or perhaps both of the parents guard the nest until the eggs hatch. The young then disperse. The female python, after laying a large number of eggs, coils about them and incubates them with a temporarily increased body temperature. Alligators make a nest in which the eggs are sur-rounded by organic matter (humus), the decomposition of which warms the eggs. Birds known as mallee fowl make similar nests for their in-cubating eggs. By rearranging the organic matter in the nest, they are able to regulate the temperature quite closely.

There are many groups of fishes in which the female cares for the eggs and the young. This is also true of certain birds, such as chickens and turkeys. These birds are called **precocial** birds because the young are able to walk and feed themselves shortly after hatching. They remain with the mother, who broods them at night. By a series of vocal signals, she can call them to her, warn them of impending danger, etc. This sort of behavior is not uncommon in those species of mammals in which the male leaves the female after mating. Solitary carnivores, such as leopards and tigers, fall into this category. So also do bears. In these vertebrates, there is undoubtedly an important element of learning by the offspring of the skills of getting food. Eventually, after weaning and when they have learned to hunt, the young leave the mother to take up a life of their own.

Of course, in many animals, both parents remain with the eggs and the young. The cichlid fishes make simple nests, which both parents guard. After they hatch, the young form a small school which may remain with the parents for several weeks. In the mouthbreeder cichlids, the eggs are held in the mouth of the female or male parent until they hatch. The young remain with their parents, often retreating to the parental mouth when danger threatens (Fig. 15-2).

The birds known as **altricial** birds include those in which the young are not precocious. Unable to walk or feed themselves, they remain in a nest constructed by one or both parents, who feed them and brood them after having jointly incubated the eggs. This sort of bird family is clearly a

Fig. 15-2 An African mouthbreader fish (*Telapia*) with young fishes in and around its mouth. (*T. F. H.*)

social group, with social stimuli between parents and offspring holding the group together and structuring it. Songbirds and seabirds fall into this category, as do many others. Once the young are able to feed themselves and to fly, the young leave the parents, never to return to the family unit they grew up in.

Many altricial birds nest in large aggregations, or rookeries (Fig. 15–3). Such familiar seabirds as gulls, terns, gannets, and pelicans are examples. Very complex social integration occurs in these noisy, crowded nesting sites, where space is at a premium and each bird must find its own nest after feeding. Birds often have elaborate greetings for identifying their mates and for warning adjacent birds when they come too close.

The large breeding aggregations of seals are, at first sight, similar social groups. In fact, they are quite different. In these marine mammals, the males arrive first at the island or shore breeding places. There they establish defended areas (territories). When the females arrive, each male establishes a harem of as many females as he can attract. The larger, more dominant males have larger harems and spend much of their time retrieving females and defending their space. Once again, males recognize each other and the females. The latter, of course, are similarly familiar with each other and know their own offspring after they are born.

Fig. 15–3 A colony of gulls on Mohawk Island in Lake Erie. (*Courtesy of Grant Haist.*)

In terrestrial mammals, it appears to be a rare event for single-family groups to occur. When offspring remain with the parents after close parental care is no longer necessary, it is common for social groups larger than families to be established. For example, gibbons may have social groups consisting of the male and female and their offspring of various ages. On the other hand, mountain gorillas and baboons, together with most other primates, do not have family groups. One or more males and a group of females, together with infants and juveniles, make up one group. Males mate with any female in estrus. Elephants have similar social groups, as do many other mammals.

Chimpanzees are also sexually promiscuous, and occur in loosely structured groups. But often rather long term associations of females and offspring may occur, sometimes continuing even after the young have matured. Adult chimpanzees on occasion carry out organized hunts for young baboons, bush pigs, or other prey—meat is a great delicacy for them.

In some of the carnivores, complex hunting groups are formed. These are exemplified by the lion pride, the hyena clan, and the wolf pack. Once again there are one or more males, a larger group of females, and immature animals that are old enough to move with the group.

Ungulates have social groups known as herds or flocks. There is usually

Fig. 15–4 A pair of white-tail buck deer fighting. (*Courtesy of Leonard Lee Rue III.*)

one dominant male, at least during the breeding season (Fig. 15–4). The red deer herd structure, studied by Fraser Darling, is an example. The young remain with their mothers, and more than three generations of deer may make up a herd. After 3 years, the males are sexually mature. They leave the herd and form herds separate from those of the female. The female herd is tightly organized and guided by the older females. The rutting or breeding season begins in the fall. The male herds then break up, and each male attempts to round up a harem of females. Displaying and fighting other males, the male defends his harem area and mates with the females he has procured. Weaker or exhausted males may be displaced by younger or fresher males. In the red deer there may be three distinct social groups, each with its own characteristic structure.

In all these examples, the social groups prove to be structured and integrated by social stimuli eliciting social responses. Playing a prominent role in controlling aggression are social dominance relations and territory. Let us now discuss these phenomena in greater detail.

A COMMON SORT OF SOCIAL GROUP IN ANIMALS IS THE RESULT OF PARENTAL CARE. MALE AND FEMALE ANIMALS MAY ABANDON THEIR EGGS OR YOUNG, OR MORE COMPLEX GROUPS INVOLVING ONE OR BOTH PARENTS MAY BE FORMED. IN SOME SPECIES, LARGE BREEDING AGGREGATIONS ARE FOUND. IN OTHERS, SOCIAL GROUPS WITH COMPLEX RELATIONSHIPS AMONG PARENTS AND OFFSPRING OF DIFFERENT AGES OCCUR.

As we have already seen, individual animals commonly exhibit relationships of social dominance. When individuals of the same or different
social groups come together, there is usually aggressive behavior on
both sides. This commonly occurs among individuals of species which
do not form social groups. It also is found between individuals of a
social group. Aggressive behavior may or may not take the form of actual
physical fighting. Often it involves the "sizing-up" of both individuals.
There may be displays of hackles, canine teeth, horns, or antlers (Fig. 15–
5). The result of such an encounter is the establishment of a dominance
relationship between two individuals.

Dominance Relations

Dominance relations have been found in a wide variety of animals. Field
crickets' interactions, for example, have been studied by Richard Alexan-
der. After inspection of one another by the use of the antennae, one of the
males may retreat. If this does not occur, they go on to chirp at one
another. Eventually, they may reach the stage of actual fighting with the
mandibles until one male is defeated. Analogous behaviors have been
studied in every vertebrate group except amphibians. In addition to the
examples already cited, we might mention the dominance by large male
lizards of smaller males and females. In fish, dominance is seen in trout,
salmon, and dogfish sharks, as well as others.

As you perhaps know, dominance relations have been shown to be linear
in some birds and probably some mammals. The best example of the
linear dominance hierarchy is found in chickens. In any group of chickens, it is usually found that one rooster is dominant to and may peck all
other birds in the flock. He obtains first place at the feeding and watering
areas and has preference in mating. In a rather strictly linear descending
hierarchy are the subdominant birds. Each is able to peck less dominant
birds but may be pecked by those higher in the hierarchy. At the lower
end of the hierarchy is the poor hen who, though pecked by all, cannot
peck any other bird in the flock. In birds, a linear dominance hierarchy
is often referred to as a *peck order*.

In other kinds of social groups, as we have already mentioned, dominance
relations may be more complex than linear. Dominance relations also
change with the season in many organisms and with the situation in
others. In general, the stronger and larger individuals are dominant to
smaller and weaker ones. Males are usually dominant to females. A
dominant male loses dominance as he becomes weakened in age. Similarly, if an animal is weakened by injury or disease, his former dominance
level may be quickly and permanently lowered.

Once established, dominance is communicated in a variety of ways.
Actual rushing threats with open jaws are clear enough. There may also

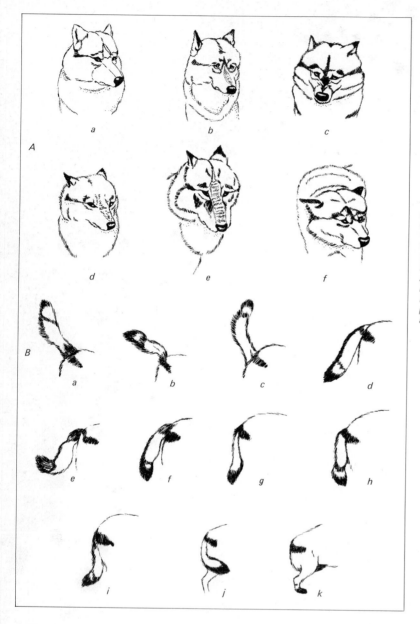

Fig. 15–5 Expression in wolves involves the head and tail (as well as other structures). (*Aa*) The leading wolf; (*Ab*) middle-ranking wolf with ears going back and gaze less steady; (*Ac*) rigid gaze, large pupils, and swelling of the brows and back of nose represent threat; (*Ad*) apprehensive humility; (*Ae*) readiness for flight indicated by the swollen forehead and flat back of nose; (*Af*) suspicion and inclination to resist. (*Ba*) Self confidence on meeting other volves; (*Bb*) confident threat; (*Bc*) threatening attitude with wag to side; (*Bd*) position without social tension; (*Be*) less-confident threat; (*Bf*) common position when wolf is eating or watching; (*Bg*) depressed mood; (*Bh*) between threat and defense; (*Bi*) active submission with tail wagging; (*Bj*) and (*Bk*) strong inhibition. (*Adapted from R. Schenkel, Behaviour, Vol. 1, pp. 81–130, © 1948 by E. J. Brill, as redrawn in ANIMALS AS SOCIAL BEINGS by Adolf Portmann, Copyright © 1961 by Hutchinson & Co. Ltd. Reprinted by permission of The Viking Press, Inc. and the Hutchinson & Co. Ltd.*)

be displays of teeth, tails, colored skin, or feathers whose significance is made clear in one encounter if it is not instinctual. In many vertebrates, all that is required is a fixed stare. The animal stared at either attempts to assert his dominance or, more usually, retreats.

In social groups of primates, and probably in other vertebrates, it is probably true that individuals thus associated have a very good picture of the physiological and psychological state of every other individual. This is so even in very large groups. Indeed, it has been reported that baboons living in the same area as chimpanzees understand quite well the dominance relations among the chimpanzees—in addition, of course, to their own.

Where dominance relations are important in social groups, various means of weakening aggressive tendencies have developed. For most animals, averting the eyes, turning the head aside, and slowly retreating seem to ameliorate aggression. In the wolf pack, assuming a submissive posture with tail between legs or even rolling over on the back, exposing the throat and belly, will often terminate aggression. Grooming in the primates also seems to serve this function, as well as others. If a less dominant female can get sufficiently close to a tense, aggressive dominant male to touch him safely, her grooming may soon reduce his aggression (see Fig. 1–5).

Once established, dominance relations usually effectively reduce the incidence of physical aggression. Every member of the group understands his position vis-à-vis all the others. By visual, vocal, and other signals, agression is confined to specific nonphysical interactions. Dominance relations enable normally aggressive, competitive individuals to live in groups and enjoy the various benefits of this kind of association.

Related to dominance is another aspect of behavior called **territoriality,** which was discussed in Chap. 12. In many animals, males, pairs, or social groups have a territory or space that is defended against others of the same species. Territories have been found in most of the major groups of animals. The size of the territory defended usually is roughly proportional to the size of the animal or, at least, to its powers of locomotion. Territories may be breeding areas or feeding areas.

Territoriality

The classic examples of territories are the nesting areas of songbirds. As the breeding season begins, male songbirds, such as robins, mockingbirds, and thrushes, distribute themselves over the landscape in territorial areas. Singing and visual displays at the margins of the territories have the dual function of establishing the limits of the territory and advertising the readiness of the male to breed. When a bird enters the territory of another male, he is clearly less dominant and is easily driven out by the owner. Territorial birds often have visual signs, such as the red breast of the robin or the yellow blotch on the chest of the meadowlark, which designate their maleness and readiness to defend.

Certain animals have clear-cut feeding territories. Dragonflies have an area which they patrol and feed in. Other dragonflies are chased out. Hummingbirds may drive others away from a favorite shrub.

Related to the concept of territory is that of **home range.** A home range is the area which an individual or social group uses in the course of its activities. Ordinarily, they do not leave the home range. In fact, even when pursued by predators, they may flee by zigzagging back and forth in their home range but not crossing its boundaries. The home range of large solitary predators may be very large. In these forms, the home range is probably the minimum area that can support the predator. All, none, or a portion of the home range may be defended against other animals of the same species. The defended area is the territory.

Finally, even when a territory or home range is not recognizable, this does not mean that individuals of a species freely associate with one another without restrictions. Most individual animals, including humans, have at least an **individual space** in which they usually do not tolerate others. This is sometimes expressed as an **interpersonal distance** which is maintained. Within a social group showing dominance and territoriality, the concept of individual space also applies. It should be clear that the dominant male baboon has a larger individual space than do the females and subdominant males.

Home range, territory, and individual space will vary as conditions change—they are not immutable. Even where the situation may appear to be clear-cut under certain circumstances, it may change dramatically under others. Territories shrink or expand, depending upon the number of individuals and the amount of food. Migratory birds may have territories during the breeding season but not when migrating. Territoriality and individual distance serve to spread animals more or less evenly over the environment and probably result in a more efficient utilization of its resources. They also confine aggression to specific areas and times and thus serve to regulate it. Serious physical contact is usually avoided as the boundaries of territories and individual space are learned.

SOCIAL GROUPINGS ARE STRUCTURED IN PART BY DOMINANCE RELATIONS, WHICH DETERMINE THE STATUS OF EACH INDIVIDUAL WITH RESPECT TO ALL OTHERS. ONCE ESTABLISHED, THESE RELATIONS SERVE TO REDUCE AGGRESSIVE BEHAVIOR IN THE GROUP. HOME RANGE, TERRITORY, AND INDIVIDUAL SPACE ARE MEANS OF DISTRIBUTING INDIVIDUALS IN SPACE AND THUS MAY LEAD TO MORE EFFICIENT UTILIZATION OF THE ENVIRONMENT.

In all the social groupings discussed above, there is division of labor among individuals. Males and females obviously play different roles, as do their offspring. This division of labor is not rigid, however, and roles overlap. Both males and females may defend the group in some species. In these social groups, also, rules may change with time and circumstances. Young animals mature and take their places as reproductive males and females. A dominant male loses his place with age or injury.

Division of labor, then, in vertebrate social groups is a changing phenomenon. Among invertebrates in which social groupings are found, the same kinds of relationships often occur. In certain groups of insects, however, division of labor and specialization of individuals have become rigid and elaborate. These insects are called **social insects.**

Many different degrees of social organization are found among the insects. These include simple aggregations, such as those of the caterpillars of butterflies or moths which live together in a webbed nest. Such aggregations as tent caterpillars show little complexity. They also include —for example, among the termites, ants, and bees—societies which are rivalled in complexity only by those of modern man. These insect societies have, as a major feature, a rigid division of labor. Different individuals are specialized for performing different tasks. Some, the workers, gather food, construct nests, and so on. Others may be soldiers, serving only in defense of the colony. Still others may perform only a reproductive function.

TERMITE SOCIETIES There is a fundamental difference in the way division of labor is achieved in these social insect colonies and in other social groupings, including man and the higher primates. In most social insects, individuals performing different tasks usually are morphologically differentiated from one another. Because they are easily recognized, they are said to belong to different **castes.** Members of a colony of termites, for example, normally include a primary reproductive caste, consisting of a queen and king, secondary reproductives, workers, and soldiers (Fig. 15–6).

The queen (Fig. 15–7) functions solely as a gigantic egg-laying machine. She normally occupies a special gallery in the termite nest, which has exits too small for her passage. There she is tended by workers, who feed her, clean her, and remove the eggs as she lays them. The secondary reproductives are individuals which take over the reproductive function if the primary reproductives should die. They are reproductive "insurance" for the colony.

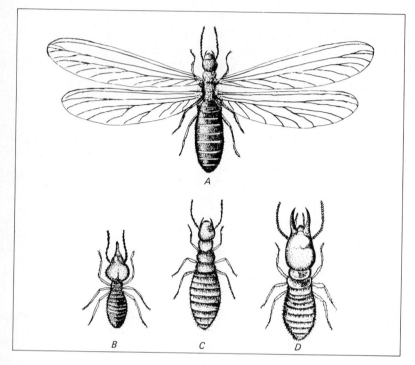

Fig. 15–6 Castes of termites as found in three genera: (*A*) sexual winged adult; (*B*) nasute soldier; (*C*) worker; (*D*) soldier.

The worker caste (or, more properly, castes) consists of immature individuals and adults that are sterile, the latter being found only in some termite species. These termites gather food, construct and maintain the nest, feed and groom the queen, move eggs and larvae.

Termite soldiers are sterile adults and consist of two general types. One type of soldier has an enlarged head with very powerful jaws with which to assault enemies. The other type is called the *nasute soldier*. This form has much reduced jaws, but the front of the head is developed into a sort of squirt gun, from which a repellant fluid can be sprayed onto enemies. In one species, each nasute soldier sacrifices himself in battle by self-destructing in the process of squirting a large glob of defensive secretion. Two distinctly different-sized subcastes, major and minor, often occur within both worker and soldier castes.

The control of the proportions of different castes in a termite colony raises interesting questions. Presumably, the eggs laid by the queen contain virtually identical sets of DNA instructions. Why, then, do some individuals become reproductives, others workers, and others soldiers. The most commonly accepted explanation is that development of individuals is modified by the chemical environment of the nest. Each caste is

Fig. 15-7 Museum model representing a queen termite in a gallery of the nest attended by workers. A male and several nasute soldiers are also shown. (*Buffalo Museum of Science.*)

assumed to release a volatile chemical stimulus (pheromone). In high enough concentration, this inhibits the development of young termites into the caste producing the hormone. Thus if there are enough soldiers in a colony, the level of "soldier" pheromone will be high and young termites will remain undifferentiated or will develop into workers or reproductives, if one of these castes is in short supply.

The role of such a pheromone, secreted by the queen and passed among the young termites as they exchange food with one another, has been demonstrated in one species. If young termites are denied access to the queen, secondary reproductives develop in the colony. It is not clear, however, if some sort of pheromonal balance in the colony explains the entire story of caste differentiation in termites. Nutritional differences may play a role in controlling paths of development, as may physical contact among individuals.

HONEYBEES   A second group of insects with very well defined castes and complex and integrated colonies are the honeybees. Chemical communication also plays a role in controlling the production of different castes in beehives. Queen bees secrete from glands in their jaws a pheromone called *queen substance.* As in termites, workers are always with the queen, grooming and feeding her. As in termites, also, workers constantly exchange small quantities of food with each other. This constant exchange of food and pheromones is critical to the social integration of colonies of social insects. It has been given the name of **trophallaxis.** In honeybees, a high level of queen substance in the hive inhibits workers from producing the special quarters in the hive in which queens develop. These **queen cells** are much larger than the cells in which worker larvae develop (Fig. 15-8).

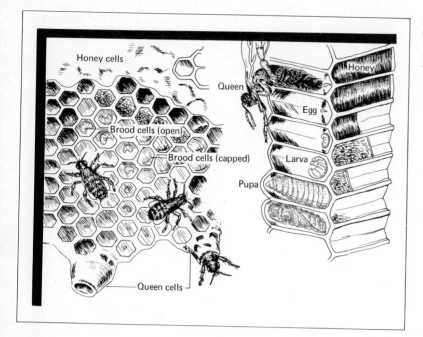

Fig. 15-8 Diagram of a honeybee comb as seen face on (*left*) and in section (*right*). (*Adapted from Kimball, BIOLOGY, 2/e, 1968, Addison-Wesley, Reading, Mass.*)

In honeybees, the difference between queens and workers is nutritional. Workers are sterile females which have not had a diet permitting sexual development. Larvae in the larger queen cells are fed a special diet by the workers, including the substance known as *royal jelly,* which they secrete from special glands. On this diet they become queen bees. A newly emerged queen kills all the other larvae in queen cells. Therefore, there is normally only one queen bee in a hive. If that queen is removed or if workers are denied contact with her, the inhibition of queen-cell construction, caused by queen substance, ceases. Some workers will then construct queen cells. New queens will be reared from the young female larvae (which developed from eggs laid by the former queen) that are transferred to the emergency queen cells constructed in response to the former queen's disappearance.

THE DANCE LANGUAGE Important as chemical communication is in the integration of colonies of social insects, perhaps the most interesting communication system used by these organisms is the dance language of the honeybees. The operation of this communication system has been worked out in great detail by the German biologist K. von Frisch. It is now known that when they return to the hive, workers that have been successful in searching for nectar perform a dance on the honeycombs.

The dance of the worker bee takes one of two basic forms. If the nectar source is nearby, a **round dance** is performed (Fig. 15–9A). Other bees clustering around the dancer can determine how abundant the nectar source is by the duration and speed of the dance. They learn of the richness of the nectar by trophallaxis and the nature of its source by the blossom smell clinging to the dancers. The excitement of the dance is contagious, and other workers soon fly off to find the source of nectar.

If, on the other hand, the foraging worker has found a relatively distant nectar source, a different kind of dance is performed. This dance, known as the **waggle dance,** takes the form of a flattened figure eight (Fig. 15–9B). The bee runs straight ahead a short distance and then turns alternately right and left before returning to the straight line. The information to be transmitted in this dance includes the nectar quality and blossom odor and also, very importantly, the distance of the source and its direction from the hive. Distance is communicated by the rapidity of the dance, the number of waggings of the body, and perhaps also by sounds made by the dancer. It is interesting that nonsocial moths have been found to vibrate after landing from a flight. The duration of the period of vibrations is proportional to the distance they have flown. The function of this is unknown, but it gives a clue about the evolutionary origin of one feature of bee communication.

Basically, the waggle dance is the round dance with a directional component added. The transmission of information in the waggle dance about direction of the nectar source is complex. Bees, as a result of the structure of their eyes, are able to detect the pattern of polarization of the sun's light, which changes as the sun moves. Therefore, they are able to determine the *position of the sun* as long as the entire sky is not obscured by cloud. If any blue sky is available, they can "compute" the position of the sun and then compare the direction of flight to the nectar source with the direction toward the sun.

If the returning foragers dance on a vertical comb of the hive, the direction of the sun is indicated by the orientation of the central portion of the figure eight. Basically, the plan of the dance is simple, but it is difficult to describe in words; refer to Fig. 15–10 for details. Should the nectar be in the direction of the sun, the central, straight-line portion of the dance is vertical, with the bee running up the comb before turning right or left. Should the nectar source be directly away from the sun, the bee dances in a downward direction. If the nectar is 90° to the left of the sun, the straight-line dance is horizontal, with the bee running to the left. A nectar source 52° to the right of the sun will be indicated by the straight line of the dance pointing up the comb and 52° to the right of vertical. The direction of the straight-line portion of the dance, therefore, always

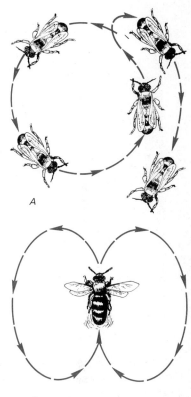

A

B

Fig. 15–9 Dance language of honeybees: the round dance (above) is used when a close food supply is found; the waggle dance is used to indicate distance and direction to food sources more than 90 m from the hive. *(Adapted from Kimball, BIOLOGY, 2/e, 1968, Addison-Wesley, Reading, Mass.)*

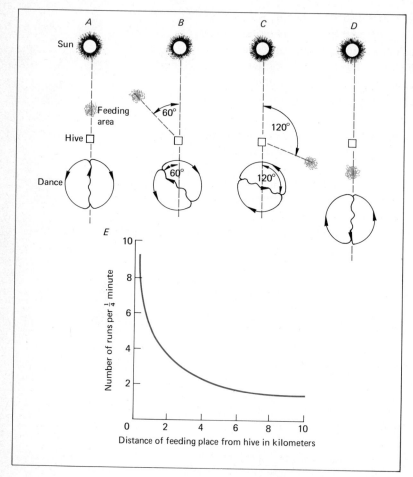

Fig. 15–10 Use of the waggle dance: (*A*), (*B*), (*C*), and (*D*), the position of a food source with respect to the sun is indicated by the angle of the straight portion of the dance. (*E*) The speed of the dance indicates the distance of the food source from the hive. (*Adapted from Paul B. Weisz, THE SCIENCE OF BIOLOGY. Copyright McGraw-Hill Book Company, 1971. Used by permission.*)

communicates the direction of the nectar source in comparison with the direction of the sun. As we shall see below, this form of communication in bees is called a "language" because the dance is delayed transmission of information to other bees, of a response to the environment, instead of the more usually immediate response, such as a cry of fright or a courtship song.

OTHER SOCIAL INSECTS We have barely introduced you to the fascinating world of social insects in this section. Much work has been done on the evolution of social behavior in bees, which show almost every grade of complexity in social organization. Social insects are among the most successful of all groups of organisms. In the tropics, termites are ubiquitous, quickly breaking down fallen trees and branches and destroying unprotected wooden structures. In warm, arid places, they often construct gigantic nests of great complexity, and specialized predators have evolved to feed upon them.

Ants are perhaps the most important predators in tropical forest areas, if we consider the ecosystem as a whole. Few biological phenomena are as impressive as raiding columns of ferocious army ants, often hundreds of yards long, marching along the forest floor and consuming every organism that cannot escape them. Leaf-cutter ants (which use the leaves as culture beds for the fungi they eat) are very important herbivores in tropical and subtropical America. They often make nests 10 or more yards across, and long columns of workers returning to the nest are a common sight (Fig. 15–11).

Fig. 15–11 A single leaf-cutter ant showing how the cut portion is held vertically. (*Courtesy of Treat Davidson from National Audubon Society.*)

Social Hymenoptera, especially wasps and bees, may pose a lethal threat to man on occasion. Females are usually equipped with potent stings, and some people are so allergic to their venom that one or two stings may kill them. Swarms of tropical hornets have been known to sting men to death, even those who did not suffer from an allergic reaction to the venom. Recently, the introduction of queens of an African strain of honeybees into Brazil has created a serious national problem. The new bees are much more aggressive than the local strains and are multiplying. The frequency of their attacks on humans is increasing, and deaths have been reported, especially among children.

Social insects represent an extreme in morphological and physiological division of labor and in the rigidity of their caste system. So highly integrated are their colonies that they have been analogized with organisms. Such an analogy should not be pursued in detail, but even from our brief description, you can perceive the similarities. Perhaps most notable is the homeostasis a colony of bees, ants, or termites is able to maintain as a result of the integration of the constituent individuals.

SOCIAL INSECTS HAVE RIGID AND ELABORATE DIVISION OF LABOR AND SPECIALIZATION OF INDIVIDUALS AS DIFFERENT CASTES. CONTROL OF THE NUMBER OF INDIVIDUALS IN EACH CASTE IS USUALLY ACCOMPLISHED BY CHEMICAL COMMUNICATION. IN THE COMPLEX COLONIES OF HONEYBEES, INFORMATION ABOUT FOOD SOURCES IS TRANSMITTED BY RETURNING BEES TO OTHER MEMBERS OF THE HIVE BY A KIND OF DANCE WHICH COMMUNICATES FOOD LOCATION WITH REFERENCE TO THE SUN.

COMMUNICATION SYSTEMS

Obviously the maintenance of social groups and such things as territories requires stimuli to which social responses are made. These constitute communication systems. In most animals, communication systems are multichannel. Nevertheless, there is usually a predominant sensory mode, or modes, through which communication is conducted. Therefore, for purposes of description, we will discuss the various sensory modes separately.

Chemical Communication

All animals excrete substances into the environment. These may be widely different sorts of compounds in different groups of animals. In many instances, the exact nature and concentration of these substances depend upon the specific individual's metabolism. Since every individual undoubtedly is slightly different from every other in metabolism, it follows that every individual will have a unique chemical aura by which it can be identified by other members of its group. There may be, of course, special scent glands in addition. You are familiar with the way in which male dogs mark stations they have passed by spraying them with urine and scent. Male cats have a similar behavior. Deer have special scent glands on their head and feet which they use to mark paths they have taken. Lemurs urinate on their feet and thus mark the trees in which they climb.

Many animals regularly produce external chemical messengers, analogous to hormones. These pheromones serve to integrate activities in groups in a variety of ways, as we have discussed in the preceding section. Ants, for example, lay down a trail substance which other ants can follow to a food source. Other pheromones produced by ants are an alarm substance and a grooming pheromone. Dead ants are recognized as such by their odor and are carried away to be disposed of.

We have already mentioned the pheromone which attracts male moths to the female. A similarly acting pheromone occurs in newts, which swim upstream in the chemical gradient created by the female. It is well known that female dogs in heat produce an odor which brings males to the vicinity very quickly. The males of giraffes and of duikers (small antelopes) taste the urine of the female to determine the presence of a substance that indicates she is in estrus.

Chemical substances are regularly exchanged in some animals. Termites maintain their cellulose-digesting intestinal protozoans in the course of such exchanges. Regurgitation (trophallaxis) is common in social insects, such as wasps, bees, and ants, which continually exchange "samples" from their digestive tracts. This serves to indicate to the group what food sources are being used, so that all may partake of them. It has been shown that there is reciprocal interchange between larval ants and the adults that feed them. In return for food, larvae provide secretions which the adults are eager to consume. This situation has been exploited by several parasites, which produce more of the substance than do the larval ants. Often they are fed at the expense of the larvae.

There is a growing body of evidence that pheromones, or at least chemical secretions, may be important in human populations. This is just beginning to be studied in man. Such interesting facts are emerging in other mammals. For example, in mice the smell of a male may integrate the reproductive cycles of all the females of the colony. However, the pregnancy of a female who has recently mated may be arrested by the odor of a strange male. In a reaction which occurs only if her olfactory lobes remain intact, the odor in some way inhibits the formation of gonadotrophins from the pituitary, and the corpus luteum fails to develop.

A special sort of chemical communication occurs between mother and offspring in many mammals. For example, in many ungulates, rodents, carnivores, and others, the mother eats the placenta and membranes upon the birth of the young. Even in those forms in which the placenta is not consumed, the offspring are thoroughly licked by the mother. The function of eating the placenta is not known. Licking the offspring, however, appears to result in her learning the characteristic chemical aura of her young. Females that fail to lick or are prevented from doing so usually fail to recognize their offspring and treat them as foreigners. Licking also provides the young with a thorough massage, which stimulates their respiration and blood circulation.

CHEMICAL COMMUNICATION CONTROLS OTHER SOCIAL GROUPS IN ANIMALS BESIDES SOCIAL INSECTS. MOST ANIMALS APPEAR TO HAVE A CHARACTERISTIC INDIVIDUAL CHEMICAL PRODUCTION, WHICH VARIES WITH THEIR STATE AND MAY AFFECT THE BEHAVIOR OF OTHERS.

Tactile Communication

Tactile communication is an important mode in many different groups. The licking of offspring by their mothers continues after birth in many groups and probably represents a combination of chemical and tactile communication important for social integration. We have seen that grooming is a common and important behavior in primates. Besides removing ectoparasites, it is a very significant integrative force. Grooming often serves to reduce aggressive behavior as well.

We have already discussed under reproductive behavior how some spiders vibrate the web of the female in such a way as to clearly indicate that a potential mate, rather than a prey organism, is present.

Electrical Communication

In some fishes, communication occurs through the production of and reception of electric fields. As was discussed in Chap. 8, electric fish have a perception of their environment in terms of the way objects distort their electric fields. Presumably they orient to one another, as well as to prey, using this sense. Needless to say, this is not a common mode of communication at the organism level.

Sound Communication

Communication via sound waves is a vital mode in both aquatic and terrestrial animals. The mode of sound production varies greatly from group to group. Basically, however, it depends upon the use of muscles to vibrate some structure of the body which imparts these vibrations to the air. In mammals, sound is produced in the larynx by air from the lungs flowing over the vocal cords. Tightening or loosening the vocal cords varies the pitch of the sound. Birds have a somewhat similar voice box, the syrinx. In other animals, sound is produced by vibrating a membrane, as in cicadas, frogs, and toads. Many animals produce sound by rubbing one surface against another. Crickets rub two wings together to make their chirp, and katydids and their relatives rub a row of pegs on the hind legs against the edge of a wing. And so on; there are many other ways in which sound can be produced.

Sound is used by animals in many different ways. We have seen how bats use sonar to echolocate their way about the environment and to find food. Whales and porpoises, as well as seals, use echolocation for the same purposes under water. It has recently been shown by R. S. Payne, however, that whales produce songs analogous to those of birds. The song lasts $9\frac{1}{2}$ min and then is repeated. The song ranges over 21 octaves, a greater range than that of a pipe organ. The use of this song is not clear as yet.

Bird songs apparently serve primarily to mark territories and indicate the male's readiness to breed. Similar patterns of behavior are known in carnivores and primates. When wandering groups of gibbons reach the limit of their territory, they produce a vigorous vocal display if they find another group there. In the New World howler monkeys, similar vocalizations indicate the presence of social groups in the forest canopy.

Vocal signals in birds and other organisms that are used in reproductive behavior or in dominance relations are often quite complex. Usually they are species-specific, of course. They share the property of being easily localized. The female songbird should have no difficulty in identifying and locating a male of her species. Breeding aggregations of cicadas, toads, and frogs are established and synchronized by vocal communications. Many animals also produce warning sounds, which announce the presence of a predator. It is not surprising that these have the property of being very difficult to localize. Such sounds are high-pitched, at a particular frequency which makes them difficult to localize by the common predators. It is not surprising either that many different kinds of animals produce very similar alarm calls. A variety of organisms may respond to the warning call of an alert member of one species.

Vocal communication in animals (except man) usually signifies the immediate state of an organism as a result of internal conditions or in response to a change in environment, such as the presence of a predator. The alarm cry of a startled animal is not made for the purpose of warning other organisms. In the course of selection and coevolution, it has come to have this function in the ecosystem. The special mode of vocal communication in man—language—goes far beyond this. It is not clear to what extent other animals have what can be considered a true language. Whales (and their close relatives) probably employ an elaborate vocal communication system, as work with porpoises has suggested. It appears quite probable that these whales are simply talking to each other. There is no reason why the marine habitat of whales should be incompatible with the development of a true language by these magnificent mammals.

Like sound communication, visual communication is a most important mode of interaction in both terrestrial and aquatic animals. We have already discussed visual communication in courtship and mating behavior. Other aspects of use of the visual sensory modality have also been considered. Cryptic coloration and shape are an attempt to prevent communication of the existence of a prey organism to a predator. Warning, or *aposematic,* coloration is intended to communicate that the individual so marked is toxic or dangerous in some way, even though this may not be true, as in mimicry. Visual signals are important in establishing dominance and in marking the boundaries of territories, as we have seen.

Visual Communication

One final example may be given of the use of this mode: its use for alarm signals. Many herd animals, such as antelope and deer, as well as many birds, utilize brightly colored areas to indicate potential danger. Thus, when antelope are frightened by a predator, they raise the tail and expose a strikingly lighter rump area. This serves as an alarm and also as a guide for following members of the herd. When alarmed, many birds fly into the air and at the same time display colored areas which may serve both to confuse a predator and to function as warning signals.

COMMUNICATION AMONG ORGANISMS INVOLVES ALL SENSORY MODES, BUT SOME SPECIES MAY BECOME HIGHLY SPECIALIZED FOR ONE OR A FEW. ORGANISMS LIVE IN SUCH DIFFERENT PERCEPTUAL WORLDS AS THOSE BASED UPON TOUCH, ELECTRIC FIELDS, SOUND, OR VISION.

Although the signals used by bees to indicate the location of and distance to a food source are sometimes called the "language" of the bees, no animal other than man is thought to have genuine language. The reason for referring to the bee dances as languages is that they represent, from the point of view of the observer, a delayed response. In most prehuman communication, no matter what the mode, the signal is a response to some environmental change and it evokes a response from other individuals which is immediate and relatively invariant.

Language

Man is able to remember events and to communicate information about them at some later time, often remote from the site of original occurrence. He does this by the use of symbols in addition to signals. As you remember, signals indicate the state of the *sending* individual. That they serve to evoke a change in the state of the receiving individual is significant only in an ecological sense and in terms of selection. Symbols with specific referents are used in an attempt on the part of the sender to influence the behavior of the receiver. Man is able to carry on abstract symbolization, that is, to deal with abstract symbols logically and apart from what we think of as the real world.

Cognition in man often involves the conception of a theoretical model of what goes on in the real world. This model can then be checked against the real world and modified. Actually, it is possible for man to construct a purely theoretical model and deal with it *as if it were* the real world. Often the language itself has, within its structure, assumptions about the real world.

The structure of languages often has acted to mold our view of nature into a form easily handled by the language. Ideas such as that an effect implies a cause or a creation a creator have, since Aristotle, been considered to be immutable laws of logic. It is interesting that Oriental religions usually have emphasized the artificiality of the subject-object dichotomy. Their philosophies aim to eliminate this division, the supposed result being similar in many ways to the goals of certain forms of psychotherapy in Western cultures. A linguistic need for a doer and the done, for objects and relationships among them, may have deeper and more significant effects than are presently realized. Our language requires us to put things into various relationships even when it is patent nonsense to do so. For instance, we are compelled to say "It is snowing," although the "it" is a meaningless word which soothes our sense of syntactic esthetics only. Similarly, we tend to think of natural selection as *something* that somehow changes a population, when it is merely shorthand for "differential reproduction of genotypes."

People of other cultures order natural phenomena in ways quite different from those we consider natural and proper. For instance, Eskimos have no generic term equivalent to "water," but they have a detailed and useful terminology describing the various kinds of frozen and liquid water. Gauchos have some 200 terms for horse colors, but they divide the vegetable world into four species: *pasta,* fodder; *paja,* bedding; *cardo,* woody materials; and *yuyos,* all other plants. As a language system develops, the effects of its structure seem to be invasive and widespread. All aspects of the culture eventually are involved, and a network develops that is difficult to escape. It has been suggested that the person most nearly free to describe nature impartially would be a linguist familiar with many widely different linguistic systems.

Some impression of the relation of language to behavior and to the description of nature can be gained by comparing even superficially the basic aspects of Indo-European languages, to which group English belongs, with a very different language. The language of the Hopi Indians has been studied in considerable detail by B. L. Whorf and offers revealing comparisons. It is difficult to describe the differences in English, for the languages are scarcely congruent. For example, our concept of plurality causes us to use cardinal numbers in referring to both real and imaginary pluralities. We count 10 objects and regard them as a group. (However, we say that there are "10 at a time," introducing the concept of time into group perception.) When we refer to 10 hours or to any cyclic sequence, actually only one item is experienced at a time; the others are remembered or predicted. We think of time in such a way as to "know" that there was a day yesterday and that there will be a day tomorrow. We can actually quantify "tomorrow" quite "precisely" in minutes, hours, days, months, and years. The Hopi Indian, on the other hand, would not think of using numbers for entities that do not form an objective group. He recognizes a group of 10 Indians. But if they stay for a visit, he reports that they "left *after* the tenth day," not that they "stayed 10 days."

It is interesting and important to realize that similar differences between the languages are manifest when physical quantity, phases of cycles, and other aspects of time such as duration are investigated. Our mass nouns, which we use to refer to unbounded homogeneous phenomena, imply, besides indefiniteness, lack of outline or size. When we particularize, we often must say "body of water," "dish of food," or "bag of oats." The relator "of" denotes or suggests *contents;* we must have a container for the portion of matter described. In Hopi, mass nouns also imply indefiniteness but not lack of outline and size. "Water" always means a specific mass or quantity of water. No container is implied. One could give examples almost without end. In Hopi, there is no basis for a formless item such as our "time"; our structuring of time with three verb tenses does not occur. Metaphors involving an imaginary space ("this discussion is *over* my head") are lacking.

It seems clear that such concepts as newtonian time, space, and matter are inherent in the language of the newtonian physicist. A scientist working in a language with very different structure conceivably might have been compelled to describe nature in, say, relativistic terms. It also seems clear that much of what we think of as "real," "commonsense," and "beyond doubt" in biology are recepts from our language and and culture. Biologists have much to learn from the study of the ways in which other cultures with different languages view nature.

In the broadest sense, however, language is just one aspect of communication in man. Different cultures have given different emphasis to the vocal use of language or its written surrogate. The fact remains, however, that human behavior evolved in a context of multichannel communication among the members of a group. We tend to think of human communication as predominantly vocal and primarily involving two people—a sender and a receiver. A perceptive observer, however, can distinguish chemical, tactile, and visual components even in simple communication interactions of people. Furthermore, all of us, consciously or unconsciously, express our state of being for all members of our social groupings to see. By the set of our shoulders, the wrinkling of our forehead, the pace of our walk, we broadcast how we feel.

Many observers, not all biologists or psychologists, have interpreted aspects of human behavior in terms of the known social behavior of other vertebrates. There is not, however, general agreement about how much social behavior in man, including aggression and behavioral sexual dimorphism, is learned in man and how much represents our phylogenetic heritage.

COMMUNICATION IN MAN IS DOMINATED BY LANGUAGE. THE STRUCTURE OF LANGUAGES IS THOUGHT TO AFFECT OUR PERCEPTION OF THE WORLD AND OUR MEANS OF DEALING WITH IT.

Social Integration as a Homeostatic Mechanism

We have seen that in the course of evolution there has been an increase in complexity. This increase in complexity has occurred at the organism level and at the population level. With increasing complexity must come increasing homeostasis or regulation. The development of hormonal and neuronal integrative systems is a good example at the organism level. In ecosystems, it is the web of complex interrelationships developed in coevolution that stabilize the community. As social groups become more complex, integrative systems of social behavior develop. This reaches a culmination in man, in whom language may be viewed as the most highly developed of all such integrative systems. Unfortunately, language also provides us with the opportunity to fool ourselves. For example, man may tell *himself* that he can control nature, and he may believe it. In the real world other things may be happening, however. We will discuss this aspect of man in the last chapter.

HOMEOSTASIS IS THE MAINTENANCE OF A STEADY STATE, OR EQUILIBRIUM. IN ADDITION TO THE MANY CELLULAR AND ORGANISMIC MEANS OF MAINTAINING HOMEOSTASIS, MANY POPULATIONAL MEANS HAVE ALSO EVOLVED. SOCIAL INTEGRATION AND CO-EVOLVED ECOLOGICAL RELATIONSHIPS ARE SUCH HOMEOSTATIC MECHANISMS.

SUMMARY

Most kinds of animals form groups or associations, either transient and often more or less incidental or long-lasting and often complexly structured. Among the advantages of group life are social facilitation, group defenses against environmental factors, including predators, and increased efficiency in utilization of the environment.

Social groups usually are based upon parental care. Male and female animals may abandon their eggs or young. Often, however, more complex groups involving one or both parents may be formed. In some species, large breeding aggregations are found. In others, social groups with complex relationships among parents and offspring of different ages occur.

Social groups are structured in part by dominance relations, which determine the status of each individual with respect to all others. Once established, these relations serve to reduce physical aggression in the group. Home range, territory, and individual space are related phenomena which provide a means of distributing individuals in space and thus may lead to efficient utilization of the environment.

Social insects have rigid and elaborate division of labor and specialization of individuals as different castes. Control of numbers of individuals in each caste is usually accomplished by chemical communication. In the complex colonies of honeybees, information about food resources is transmitted by returning bees to other members of the hive by a kind of dance, which communicates food location with reference to the position of the sun.

Chemical communication controls other social groups in animals besides social insects. Most animals appear to have a characteristic chemical individuality. This varies with their physiological and psychological state and may strongly affect other individuals. Communication in the broadest sense probably usually involves all the sensory modes to some extent. Some species may, however, become highly specialized for the use of one or a few senses. Organisms live in such different perceptual worlds as those based upon touch, electric fields, sound, and vision. Communication in man is dominated by language, the structure of which is thought to profoundly affect our perception of the world and our means of dealing with it.

Homeostasis is the maintenance of a steady state, or equilibrium. In addition to the many cellular and organismal means of homeostasis, many populational means have also evolved. Social integration is one such homeostatic mechanism.

Etkin, W.: "Social Behavior from Fish to Man," University of Chicago Press-Phoenix Books, 1967.
Van Lawick, H., and J. Van Lawick-Goodall: "Innocent Killers," Houghton-Mifflin, 1971.
Wilson, E. O.: "Insect Societies," Belknap Press, Harvard University Press, 1972.

Thorpe, W. H.: The Language of Birds, *Sci. Amer.,* **197,** no. 4, Offprint 145 (1956).
Todd, J. H.: The Chemical Languages of Fishes, *Sci. Amer.,* **244,** no. 5, Offprint 1222.
von Frisch, K.: Dialects in the Language of the Bees, *Sci. Amer.,* **206,** no. 2, Offprint 130 (1962).
Watts, C. R., and A. W. Stokes: The Social Order of Turkeys, *Sci. Amer.,* **224,** no. 6, Offprint 1224 (1971).
Wilson, E. O.: Pheromones, *Sci. Amer.,* **208,** no. 5, Offprint 157 (1963).

Articles

CHAPTER 16

HUMAN EVOLUTION AND ECOLOGY

Man, contrary to popular belief, does not dominate nature; he is part of nature. He is dependent on green plants for the food he eats and the oxygen he breathes. He is dependent on bacteria for the survival of those green plants, for the health of his bowels, and for the maintenance of the quality of the atmosphere. Indeed, he can survive only so long as some degree of stability is preserved in the planetary ecosystem. Man, after all, has been evolving in response to environmental changes originating during all the billions of years that life has existed on the earth. Throughout his own long evolutionary history, every adaptive change has been in the direction of a "better fit" for man into some role in ecosystems. It is therefore hardly surprising that today he has *not* somehow "risen above" nature.

MAN'S PHYSICAL EVOLUTIONARY HISTORY

If any other organism had a fossil history as complete as that of man, paleontologists would rank it among the best-known species historically. This is not because primate materials are common among fossils; quite to the contrary—they are rare. The effort that has gone into gathering and interpreting fossils which appear to throw light on human origins has been, however, enormous. Today it is possible to outline rather clearly the course of man's physical evolution from the time, several million years ago, when he was a small-brained, fully erect, but rather slight creature, to the present. True, there is still room for quibbling over details, and there probably always will be. But we can be certain that man did not appear suddenly and full-blown on the face of the earth.

Man shares a vast inheritance with all mammals. The conquest of land by vertebrates and the eventual appearance during the age of the dinosaur of our inconspicuous warm-blooded ancestors make a fascinating story, but one which we won't detail (Fig. 16-1). It was during the age of the dinosaurs that our ancestors acquired, together with warm-bloodedness, the many other characteristics that we share with mammals, from hair and

Fig. 16-1 Diagram to show the evolutionary relationships of the mammals and their reptile ancestors. (*Adapted from Kimball, BIOLOGY, 2/e, 1968, Addison-Wesley, Reading, Mass.*)

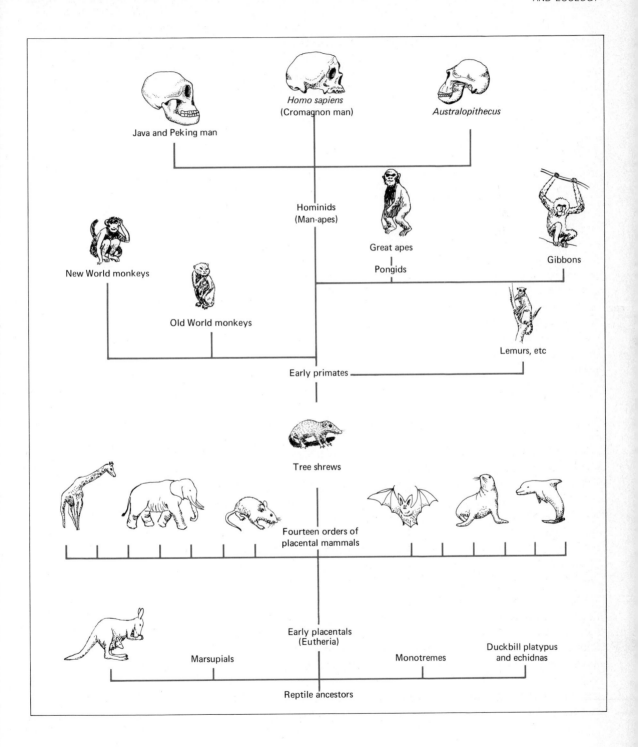

Java and Peking man

Homo sapiens
(Cromagnon man)

Australopithecus

Hominids
(Man-apes)

Great apes

Pongids

Gibbons

New World monkeys

Old World monkeys

Lemurs, etc

Early primates

Tree shrews

Fourteen orders of
placental mammals

Early placentals
(Eutheria)

Marsupials

Monotremes

Duckbill platypus
and echidnas

Reptile ancestors

placentas to speedier neurons and differentiated teeth. At some point early in mammalian history, quite likely the Paleocene (some 60 million years ago), came the first evolutionary step that was to lead eventually to the differentiation of man from the rest of the mammals. The evolution of culture, however, came very late in man's history.

THERE IS A CLEAR, IF SPORADIC, FOSSIL RECORD OF MAN'S EVOLUTION FROM SIMPLE, UNSPECIALIZED MAMMALS TO HIS PRESENT STATE. THIS EVOLUTION TOOK PLACE IN ABOUT 60 MILLION YEARS.

The Arboreal Period

The reason or reasons for the first step toward man can only be guessed. Perhaps the presence of efficient terrestrial predators, perhaps an abundance of fruit, perhaps something as yet unimagined was responsible. But for whatever reason, in one group of mammals individuals spending part of their day in the branches of trees and shrubs started to leave more offspring than their more terrestrially inclined relatives. Ascent into the trees meant the penetration of an entirely new environment, an environment in which the requirements for survival were strikingly different from those met by terrestrial animals. This early ascent into the trees on the part of our distant primate ancestors has left an indelible mark on man.

Many of the trends caused by natural selection in an arboreal creature need little explanation. Flexible grasping organs at the ends of arms and legs are useful devices for remaining in trees. When leaping from branch to branch, it obviously is necessary to be able to judge distances. Individuals with relatively good binocular vision were more likely to reproduce themselves than their less fortunate relatives, since such vision greatly enhances the ability to judge distances. Since variation in the degree of binocular vision undoubtedly had a genetic component, selection thus resulted in genotypes which developed into individuals with superior distance-judging ability. (In the rest of this discussion, "genotype" will be used as a shorthand for a kind of individual differing from others at least in part in the genetic information specifying the character under consideration.) Genotypes for the eyes rotated toward the front of the head tended to have binocular vision only if a large, long snout did not interfere (Fig. 16–2). Both binocular vision and grasping hands and feet lessened the need to possess a long snout for investigation and manipulation. Shortening of the snout and the resultant reduction of olfactory membranes would be a handicap for a ground-dwelling animal, and genotypes for such a reduction would probably be at a selective disadvantage. In the treetops, however, the loss of sense of smell was less serious, and presumably it was more than compensated for by improved vision.

Fig. 16-2 The tarsier (Tarsius) is a tree-dwelling primate with binocular vision and short snout. (*San Diego Zoo photo.*)

Binocular vision in itself would be of little use without a mechanism to evaluate sensory input and translate it into highly coordinated voluntary movements. Thus, in the arboreal primates, selection resulted in a trend toward high development of the cerebral cortex of the brain as a center of evaluation of sensory input and the formulation and initiation of responses to the environmental stimuli received. Because of the arboreal habitat, sight and touch came to override smell and hearing as sources of information.

Living in the trees presents some serious problems in the care of offspring. In primates this presumably gave a selective advantage to individuals that had smaller litters but gave them a high level of care. In many mammals, sexual activity is confined to a single season of the year, and the young are born at a time when a suitable food supply is available.

In the tropical forest environment of our distant ancestors, the supply of food (fruits and insects) presumably was relatively more constant than in temperate zones. Thus there was probably little selection in favor of a single period of sexual activity. On the other hand, there *was* selection which favored year-round sexuality. Such selection led to the continuous presence of males in the vicinity of females and increased the protection of the female with her small litter of helpless young.

Man is not the only descendant of the shrewlike animals that originally invaded the trees only to leave them again. Like man, baboons have returned to an almost wholly terrestrial life, although like most monkeys and apes they usually sleep in trees. Baboons are at home both in grassland and in forest. Gorillas are also primarily terrestrial, whereas chimpanzees are more at home in trees than are gorillas, although considerably less so than orangutans. Remaining primarily arboreal is an array of forms that includes such diverse primates as orangutans, gibbons, monkeys, marmosets, lemurs, tarsiers, and treeshrews.

DURING THE COURSE OF EVOLUTION, MAN'S ANCESTORS BECAME TREE DWELLERS. MANY IMPORTANT TRAITS—HANDS, BINOCULAR STEREOSCOPIC VISION, REDUCTION OF LITTER SIZE, AND OTHERS —RESULT FROM SELECTION IN AN ARBOREAL NICHE.

The Terrestrial Period

It is important that our early ancestors lived in trees, but it is also very important that they left them. Return to a ground-dwelling existence was a crucial stage in our evolution. It is difficult to see how a culture even vaguely resembling ours could have been developed by tree-dwelling organisms, if for no other reason than that there was an almost complete lack of raw materials for even a primitive technology in the arboreal habitat. The reasons why our ancestors left the trees are obscure— climatic changes may have reduced the forests, or smaller, more agile primates may have reduced the available food supply in the trees. Because of their increased size and intelligence, our ancestors leaving the trees were better able to cope with terrestrial living than were *their* ancestors which had first ascended them. Their size made them immune to the attacks of all but the largest and best organized predators and simultaneously made a wide size range of food articles available for their consumption.

The return to a terrestrial mode of existence was certainly gradual, and at one time our ancestors probably lived much as do modern chimps and baboons, foraging on the ground for much of the day but retreating to the trees to spend the night. It seems likely that the efficient two-legged posture so characteristic of man was achieved after our early ancestors left the shelter of the trees and began to forage in bands out on grassy savannas. Fossil evidence indicates that an abundance of food was available in the form of the other animals that roamed the open spaces, and selection probably favored any genotypes permitting, by whatever means, utilization of this food resource. An upright posture, providing reasonably rapid locomotion while at the same time freeing the hands to grip stones or clubs, apparently was at a selective premium. Intelligence and social organization would also have their reward in food.

It is important to remember, when one is casually discussing our family tree in this manner, that the evolutionary processes discussed are no different in principle from those accounting for bandless water snakes or banded snails. To say that our ancestors moved out of the trees to escape the competition of more agile foragers is merely a shorthand for the following: At one point in our evolutionary history, any recombinant individual that had the slightest behavioral tendency to descend from the trees and forage on the ground had a better chance of contributing to the gene pool of the following generations than other genotypes lacking this tendency. The frequency of the kind of genetic information producing this sort of behavior therefore increased in the populations concerned, and the behavioral norm was slowly shifted. A great many generations after the first pioneer individuals foraged briefly on the ground, the behavior of all individuals in the populations concerned became terrestrial.

AFTER A PERIOD OF EVOLUTION AS ARBOREAL MAMMALS, MAN'S ANCESTORS RETURNED TO A GROUND-DWELLING LIFE. MANY IMPORTANT TRAITS ARE THE RESULT OF SELECTION AND EVOLUTION IN THE TERRESTRIAL NICHE.

The Fossil Record

Let's briefly go over some of the salient features of the human fossil record. Figure 16–1 places man in perspective in the history of life. The separation of the evolutionary line leading to modern man from that leading to modern great apes (Pongidae) is not well documented in that record. It appears to have occurred sometime in the Miocene, perhaps 15 to 20 million years ago. This date is based on fragmentary material —mandibles and teeth—from India and Africa. These fossils, named *Ramapithecus* and *Kenyapithecus* (Fig. 16–3), are generally considered to be of a primitive human type and may well represent forms which are essentially direct ancestors of modern man.

For the start of the well-substantiated fossil record of modern man, we

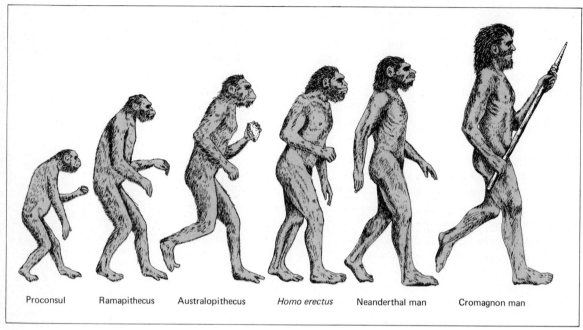

| Proconsul | Ramapithecus | Australopithecus | *Homo erectus* | Neanderthal man | Cromagnon man |

Fig. 16-3 An artist's conception of man's ancestors. (*Adapted with permission of The Macmillan Company from BIOLOGY: OBSERVATION AND CONCEPT by James F. Case and Vernon E. Stiers. Copyright © 1971 by The Macmillan Company.*)

must jump a gap of about 12 million years or more, to the border between the Pliocene and Miocene, some 5 million years ago. From this time period, we have a relative abundance of fossils of fully erect, bipedal, small-brained men (Fig. 16-3). Although they show some structural diversity, they may all reasonably be placed in the genus *Australopithecus*. Most of these fossils are from East and South Africa, but there is evidence that *Australopithecus* was widespread in the Old World.

Direct descendants of the australopithecines are the Java and Peking men and their relatives, *Homo erectus* (often called *Pithecanthropus*). They appear in the fossil record 1 million years ago or more. Material assigned to *Homo erectus* is now known from Europe, from North, East, and South Africa, and from China and Java. *Homo erectus* was fully erect, was intermediate in brain size between *Australopithecus* and modern man, and used fire and tools. The time of disappearance of *H. erectus* and the appearance of *H. sapiens* was in the late middle or early upper Pleistocene, some 200,000 to 400,000 years ago. The change was relatively small and continuous, and it did not necessarily occur in all populations or simultaneously in those populations where it did occur.

Fortunately for the taxonomist, the fossil record is not rich in this time period, which of course is what permits us to change names in the gap! Even so, there has been much confusion concerning the "emergence" of *Homo sapiens*. About 100,000 to 200,000 years ago, descendants of a geographically variable *H. erectus* were evolving into what we recognize as *H. sapiens*. These populations doubtless evolved at varying rates and suffered varying fates. Some must have died out; others met and fused. Some more *erectus*-like populations may have persisted in relative isolation until the last glaciation. The famous Neanderthal man seems to have been a geographic variant (or variants) of *H. sapiens*, one which probably disappeared from differing combinations of causes in different areas (interbreeding with "typical" *H. sapiens*, competing with the more modern *H. sapiens*, etc.).

Once our predecessors became upright, there remained only one major physical change necessary to convert them into modern *Homo sapiens*. This change was a great increase in brain size and skull volume. *Australopithecus* had a brain volume of some 450 to 600 cu cm, about that of a large ape. The cranial capacity of *Homo erectus* ran from some 775 to 1,225 cu cm, the latter well within the range of modern man. Estimates of the capacity of so-called *"Homo habilis"* (most probably best considered an *Australopithecus*) range from 643 to 723 cu cm, so that with this fossil placed in the series at the top of the *Australopithecus* range, *Homo erectus* can be seen as neatly bridging the brain-size gap between *Australopithecus* and *Homo sapiens*. The average capacity of modern man is about 1,450 cu cm. Changes in the shape and size of skulls in the hominid line are shown in Fig. 16–4. Because of the physical limitations on pelvic expansion in anthropoid females, most of the growth resulting in large brain size comes after birth. This great postnatal growth in skull capacity results in a very long period of helplessness in the infants of the larger-skulled forms, creating a mother-offspring relationship that has left a considerable mark upon our present-day culture.

Man, then, owes many of his most characteristic features to an ancestral sojourn in the trees followed by a period of terrestrial life. These are responsible for the well-developed association centers of his brain and the skillful manipulating devices on his forelimbs. They also gave him his family association, with its year-round sexuality and a strong mother-offspring relationship. Ultimately they gave him beer, airplanes, churches, and thermonuclear war—in short, they gave him culture.

THE FOSSIL RECORD INDICATES THAT MODERN MAN EVOLVED IN THE OLD WORLD FROM A SERIES OF PREHUMAN ANTHROPOIDS OF INCREASING BRAIN SIZE.

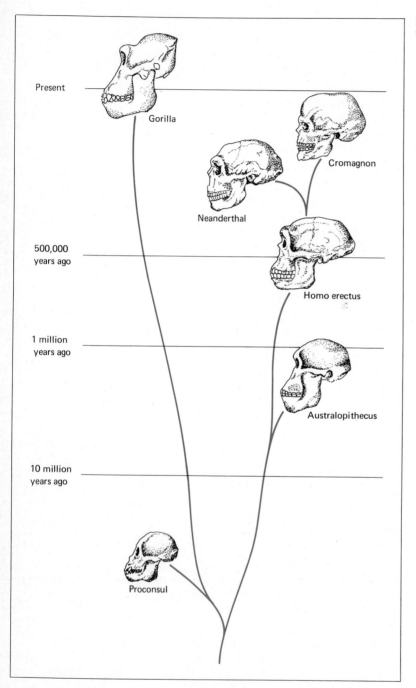

Fig. 16-4 Changes in the skull during the course of human evolution.

Present

Gorilla

Cromagnon

Neanderthal

500,000
years ago

Homo erectus

1 million
years ago

Australopithecus

10 million
years ago

Proconsul

At one time it was commonly thought that man's large brain made it possible for him to invent culture. It now seems that the reverse was true. The earliest presumed ancestors (or near-ancestors) of modern man, the australopithecines, were erect creatures with brains not differing appreciably in size from those of modern-day anthropoid apes. The evidence indicates that the australopithecines were animals of the plains and that they were primarily vegetarian. There is also evidence that their diet was supplemented somewhat with the meat of small animals. Little is known about the use of tools by these protohumans, but it seems unlikely that they could have left the shelter of the forests before they acquired reasonable security by employing rocks and clubs in their own defense. A variety of stone tools found in association with the australopithecine discovery called *Zinjanthropus* strongly indicates that australopithecines made and used tools. Such use would be *prima facie* evidence that australopithecines had at least a rudimentary culture.

MAN'S CULTURAL EVOLUTIONARY HISTORY

Development of Culture

There can be little doubt that an ape-brained anthropoid, quite possibly our own direct ancestor, was the possessor of a complex body of information that passed from generation to generation *nongenetically.* Such a nongenetic body of information is called **culture** and its change through time is **cultural evolution.** It also seems highly likely that these protohumans utilized a reasonably complex system of verbal communication. The making of stone implements is not as simple as the twentieth-century armchair observer might believe. While it is conceivable that young australopithecines learned to do this merely by careful observation and mimicry, it seems more likely that a certain amount of oral instruction went along with the demonstration. The possession of culture, and perhaps of speech, by these long extinct, very small-brained anthropoids clearly outlines the probable solution of one of the most vexing problems in human evolution: finding the "cause" of the roughly threefold increase in brain size between that of earliest fossil man and modern man.

As culture became important in prehuman society, genotypes with the mental characteristics permitting optimal utilization of this extragenetic information were more successful reproductively than other, less well-endowed genotypes. Genotypes were favored that produced brains with the highest ability to associate, integrate, and store incoming sensory data and to utilize these data in a manner that enhanced the survivability of the genotype. This selection pressure resulted in a trend toward great expansion of the cerebral cortex (Fig. 9–23) and an increase in the number and complexity of the neuronal systems necessary for thinking and speech.

It is not unreasonable to assume that much of this increased volume is the result of a premium being placed on storage capacity. One important part of man's brain, the *neopallium,* is relatively more free from commitment to special sensory and motor functions than the neopallium of the brain of other mammals. These "uncommitted" areas may be presumed to be concerned with association and memory. This presumption is supported by results obtained from electrical stimulation of the brain in conscious patients undergoing brain surgery. For example, stimulation of the temporal lobe may lead to the patient rehearing a complete symphony or reliving an event of the distant past. When a human being is subjected to a frontal lobotomy, his sensory and motor functions are relatively unimpaired but he may become "irresponsible."

A VERY IMPORTANT DEVELOPMENT IN MAN'S EVOLUTION WAS THE TRANSMISSION OF NONGENETIC INFORMATION FROM GENERATION TO GENERATION. THIS LED TO THE DEVELOPMENT OF CULTURE AND THE POSSIBILITY OF CULTURAL EVOLUTION.

There is a considerable body of literature on the reasoning power of chimpanzees and other nonhuman primates. On certain types of tests designed primarily to evaluate human reasoning power, some chimps score higher than many human adults. Indeed, as Dr. Harry Harlow succinctly puts it, if man is defined as the possessor of mental abilities that occur in other animals only in the most rudimentary forms, if at all, we "must of necessity disenfranchise many millions . . . from the society of *Homo sapiens.*" Chimpanzees may lack culture not because of any great absence of reasoning power but because of some other factor that inhibits the development of speech, the manufacture of standardized complex tools (chimps regularly use simple tools), or the reduction of intermale aggressiveness. There is growing evidence that chimpanzees may be able to learn to communicate by the use of sign language or by the use of plastic symbols for words and concepts. Nevertheless, the fact remains that a genuine culture developed only in *Homo sapiens.* The "invention" of rudimentary culture started a selective trend that led eventually to man's large brain; the large brain did not just mysteriously develop and then discover culture.

Cultural evolution and biological evolution cannot proceed independent of each other. Indeed, from the very beginning of culture, man's evolution has been characterized by the interactions of biological and cultural evolution. The existence of culture put a selective premium on certain types of brains; the evolution of the brain permitted an expansion and enrichment of the culture. Such interactions were certainly very important during the transition period from the australopithecines to *Homo sapiens,* but they are still very much with us. Before going further into such interactions, however, let us consider some characteristics of culture and some features of cultural evolution.

*BIOLOGICAL EVOLUTION AND CULTURAL EVOLUTION HAVE INTER-
ACTED THROUGHOUT THE COURSE OF MAN'S DEVELOPMENT.
CULTURE HAS LED TO CHANGES IN MAN'S PHYSICAL NATURE,
AND CHANGES IN MAN'S PHENOTYPE HAVE GREATLY INFLUENCED
HIS CULTURE.*

Language and Culture

One of the outstanding characteristics of human cultures is their tremendous diversity. Human beings speak some 2,800 different languages, describe their genetic relationships with each other with myriad complex kinship systems, believe in a great diversity of gods and spirits, and are organized into groups which practice every degree and kind of governmental control. Human beings fill their everyday lives with galaxies of taboos concerning everything from forms of greeting to the proper modes of dress. This cultural diversity is by no means superficial; indeed, people of different cultures often have basically different world views. This difference is frequently reflected in the language of a culture, as was discussed in Chap. 15. It has been claimed, especially by linguists E. Sapir and B. L. Whorf, that the language itself helps to create the world view.

Language differences are among the most important of all cultural isolating mechanisms. Communication of information about complex phenomena may be exceedingly difficult within a culture (as almost any teacher will freely testify), but between cultures with widely different languages the problems are immense. Contemplate the difficulties of explaining even a simple word such as "also" to an Eskimo if there are no dictionaries, no shared third language, nor even the certain knowledge that an equivalent concept exists in his mind. The intricate and highly developed language of the Eskimos does not have a structure congruent with that of our language.

Whatever the validity of the Sapir-Whorf hypothesis, there seems now to be little doubt that individuals of different cultures perceive things differently. That is, the same stimulus may be perceived differently by two different people simply because the individuals are members of different cultures. Considerable anecdotal evidence of the influence of culture on perception has been accumulated by anthropologists. For instance, there has been the repeated observation that black-and-white still photographs are not recognized as representing the object photographed by naïve people of many cultures. The conventions accepted by members of our culture, such as the representation of a large three-dimensional colorful object as a small two-dimensional black-gray-white rectangular field (as in a photograph), are not part of the perceptual repertoire of other cultures.

Recently, a careful study of the susceptibility to two-dimensional optical illusions of members of various cultures has very nearly removed all doubt that **perceptions** differ cross-culturally (rather than just the way people *describe* their perceptions). Anthropologists have been able to give convincing evidence that susceptibility to the illusions is related to the perceptual environment of the culture. For instance, people from "carpentered" worlds like ours, dominated by straight lines and right angles, were more susceptible to some illusions than were people from societies where huts are round and right angles rare (Fig. 16–5). This result is hardly surprising, as there is massive literature on the role of learning or experience in perception. A typical example is the demonstration that coins look larger to poor children than to rich children.

In spite of their great diversity, however, many similarities may be observed among cultures. Some form of religious belief is virtually universal. It has been suggested that these beliefs are based on observed differences between living and dead human beings, the assumption being that the absence of breathing and the lowering of the body temperature result from the desertion of the body by a spirit. This seemingly logical assumption, combined with, among other things, dreams and ignorance of the forces of the physical world, is thought by some to be the basis of all religions. The elaboration of these or other simple ideas into the complex pattern of religions that we have today was a long and complicated process. It seems eminently fair to say that, even with the flourishing of science in the last few centuries, man's creation of spirits, gods, and the related paraphernalia of religion has had the most far-reaching effects of any cultural phenomenon. In most societies of the past, and in the majority of societies today, organized systems of religion provide the principal means for the individual to orient intellectually to his physical and cultural surroundings. Among other things, such systems of orientation make it very difficult for the individual to appreciate the outlook of members of other societies.

Most people in every culture believe that their own way of doing things is, in some absolute sense, *right*. They are unaware (or, at most, only dimly aware) of their own biological and cultural history. They do not understand why they love their children, why the sun comes up in the morning, or why they must hate their group's enemies. They accept their biological heritage and the dicta of their culture without question. The acceptance of a particular set of cultural standards undoubtedly often

Müller–Lyer illusion

Sander parallelogram illusion

Fig. 16–5 Two optical "illusions" which have been used to study the effects of culture on perception. To observers from Western cultures, the thin line on the left in each illusion appears longer than it really is. Observers from non-Western cultures prove to be less susceptible to the illusions. Try measuring the thin lines after you decide which is longer in each pair.

added to the viability of the culture at one point in time and space. At another point, the same set of values may have proved suicidal.

Many of the important rules for living in *our* culture are believed by some to have been handed down from heaven a few thousand years ago. Most of them probably trace to the time when human beings gave up a nomadic hunting and food-gathering way of life and, with the invention of agriculture, began to settle down in rather large, organized groups. There were numerous advantages to living in such groups, among them cooperative defense, the ability to carry out projects requiring a great deal of manpower, and the opportunity for specialization into various trades and professions. For groups of any size, from family on up, to enjoy the fruits of cooperation, internal strife must be kept at a minimum. Thus, for instance, intergroup selection doubtless favored those groups that suppressed killing within the group. Such suppression in terms of biological evolution may have resulted in reduction in the size of man's canine teeth since these teeth may have been too dangerous a weapon for maintenance of dominance in an increasingly hairless social animal. In terms of cultural evolution, killing may have been discouraged by elders telling the young that killing a member of the ingroup would offend the spirits, and indeed that is essentially the way it is done today in parts of our own society. It should be noted carefully that in spite of a constant reiteration of "Thou shalt not kill," our society allows or even encourages certain kinds of killing. Thus society can kill its internal enemies (assorted "criminals"), and killing its external enemies is often encouraged. The social approval of killing outsiders at one time doubtless had considerable selective advantage for the society as a whole, but improved weapons have greatly increased the risks of such behavior.

CULTURES ARE EXTREMELY DIVERSE AND ARE SEPARATED BY, AMONG OTHER THINGS, LANGUAGE BARRIERS. PEOPLE ARE GENERALLY UNAWARE OF THEIR BIOLOGICAL AND CULTURAL HISTORY, AND MOST ASSUME THAT THEY WERE, MIRACULOUSLY, BORN INTO THE CULTURE WHICH HAS "THE TRUE WORD."

Selection and Culture

A major interaction of cultural and biological evolution has been in the change of selection pressures. The development of modern medical techniques, the elimination of many large predators, local increase of the food supply through improved agricultural methods, control and evaluation of the environment with furnaces, air conditioners, dams, radar weather-warning systems, and the like have permitted many otherwise nonviable genotypes to persist. The diabetic controls his disease with insulin and lives to reproduce. There is time to lead the congenitally blind man to the storm cellar when the tornado warning comes on the radio. Thus some differentials in reproduction have been ironed out by cultural factors in some societies. However, some recently introduced cultural factors have imposed other selective pressures. These, for instance, may favor genotypes that are relatively immune to insecticides in their food, air pollution, nervous tension, heart disease, and cancer. In the case of heart disease and cancer, differential reproduction is increased because, with increasing life-spans, reproduction is carried into the years when these diseases are prevalent.

The patterns of gene flow in human populations have also been tremendously changed by cultural developments. Systems of transportation have steadily improved, moving the entire human population more and more in a direction of panmixia (that is, random mating). However, this trend has been countered to some degree by immigration quotas and other cultural barriers to random mating. Other interactions of cultural evolution and biological evolution are obvious. Incest taboos tend to lower the amount of inbreeding. Social disapproval of interracial, interreligious, and interclass marriages tend to keep the population in many parts of the world divided into relatively small, partially interbreeding groups.

In recent years in Western countries, groups with the highest intellectual attainments are thought by some to have voluntarily limited their family size to a greater degree than other segments of the population. However, some evidence to the contrary also exists. Since intelligence has a genetic component, such a differential would be selection against intelligence. There are, however, many complications that make it impossible to predict the consequences of this trend, if it exists. Many hundreds of years would be required for such weak selection to detectably lower the average IQ of *Homo sapiens*.

AN IMPORTANT WAY IN WHICH CULTURE AFFECTS PHYSICAL EVOLUTION IS TO CHANGE SELECTION PRESSURES ON HUMAN POPULATIONS. CERTAIN SELECTIVE FORCES HAVE INCREASED AND OTHERS HAVE DECREASED IN INTENSITY. CULTURE HAS ALSO INTRODUCED ENTIRELY NEW SELECTIVE AGENTS.

Cultural Evolution and Biological Evolution

The influence of our primate background on our culture has been profound. The loss of a sharply defined estrous period in the female and the general helplessness of the primate infant, with the accompanying establishment of a family group stable throughout the year, have led to systems of interpersonal relationships that have been vastly elaborated in cultural evolution. On top of the relatively simple male-female and female-offspring relationships of prehuman family groups, cultural evolution has produced the monstrously complex set of phenomena usually included under such topics as love, sex, and kinship. That these phenomena have become deeply and basically interwoven into the entire fabric of our behavior has been amply demonstrated by anthropologists and psychologists. These phenomena enter into choices of political systems and political leaders, legal systems, the characteristics of the deities that men have devised, and even designs for automobiles.

Unfortunately, very little is known about the ways in which culture evolves. Some similarities with biological evolution are obvious, but the value of the analogies that follow is questioned by some. They are given here merely as food for thought. Many instances of **parallel** cultural evolution can be detected—that is, similar changes in homologous structures. All cultures have rites of passage marking certain key events of the life cycle: birth, puberty, marriage, death. Complex puberty rites, for instance, are widespread, ranging from severe tests of manhood involving torture and genital mutilation to ceremonies such as Christian confirmation and Jewish bar mitzvah. Ceremonial appeals to spirits in times of stress or danger are nearly ubiquitous. Consider Navajo dancers or ministers appealing to their gods for rain or the ritual aspects of war preparations involving dancing, martial music, war chiefs, and chaplains. There is no reason to believe that these features of culture do not trace continuously back to common evolutionary origins, although we do not understand these details of cultural evolution.

Some examples seem closer to the biological phenomenon of **convergence,** that is, the appearance of functionally similar phenomena in nonhomologous structures. An example of this might be the military dictatorships which sprang up in both Germany and Japan between World Wars I and II. The histories of the two cultures in which these phenomena appeared were widely divergent, and yet in many superficial aspects the dictatorships were similar. The stories of virgin births in different mythologies are probably better examples of convergence than parallelism. (The line is difficult to draw in any case.) Similarly, the idea that one may achieve a desired condition by eating a portion of a cadaver is widely and spottily distributed through human cultures. The bodies of victims put to death in Aztec religious ceremonies were devoured so that the eater could establish close contact with his gods. As Ralph Linton has said, "It was a religious concept not unlike that of the Christian communion except that the Aztecs were painfully literal about it."

An obvious cultural analog of natural selection can be found in the differential reproduction of entire cultures. The body of information making up some cultures has become more and more widespread (that is, has been possessed by more and more individuals) while others have decreased or become extinct. The cultures of some small groups have doubtless disappeared without a trace. However, cultural evolution is obviously *more* reticulate than biological evolution. Large cultures virtually never disappear without transferring some of their information to other cultures. Although American Indian cultures have been badly swamped and in some cases completely destroyed by the spread of Western Europeans, some of their elements have been transferred into Western European culture. The use of tobacco is a good example.

The ascendance of individuals with novel ideas may be a random phenomenon in human society. Thus one could consider the advent of Aristotle, Darwin, Hitler, Buddha, Tecumseh, Mao Tse-tung, etc., to be a sort of cultural analog of mutation. Men with their proclivities are doubtless present from time to time in all cultures, but as with gene mutations, the proper environment is necessary for them to gain prominence.

Another analogy can be drawn between a phenomenon known as *genetic homeostasis* and cultural integration. When strong selection is experimentally applied to a single character, only a certain amount of progress can be made before the effects of unbalancing a well-coordinated genotype counterbalance the selective pressure on the character. In other words, natural selection works to preserve a well-balanced genotype. It is possible that overdevelopment of some feature of a society may lead to the destruction of the integrated properties of the society. One might view the development of an extreme military dictatorship in Japan in this light. Another example is the promulgation of a fantastic taboo system by the Hawaiian priesthood, which, among other things, eventually led to the destruction of the entire religious system. Today this kind of imbalance can be seen in the overdevelopment of science and technology relative to other areas of our culture. This has led to a vast consumer demand for the products of technology, which has worsened the situation and resulted in considerable disruption, both cultural and physical, on the planet.

For many centuries our system of ethics (the tested rules under which a society operates) has been taught and maintained by church and state. It is either too much trouble or impossible to orient most people to an *evolutionary basis for ethics,* and so these institutions control human behavior through a combination of force, social pressure, and promised supernatural punishment. As scientific knowledge has increased over the past few centuries, governments and churches have slowly changed their ideas so that they do not usually conflict directly with the findings of science. In the past century or so, however, the progress in science has far outstripped the ability of our extremely conservative social systems to adjust to the changes. Advances in science, for instance, have produced an unprecedented surge in population size, which can be humanely halted only by intervening in the human birthrate. Unfortunately, the conservatism of our cultural structure will not permit us easily to counteract this deadly trend. The trend toward increase in population size, coupled with exploitation of the resources of the earth, has brought man to the point when the very life-support systems of the planet are threatened.

MANY ANALOGIES CAN BE DRAWN BETWEEN BIOLOGICAL EVOLUTION AND CULTURAL EVOLUTION. COMPARISON OF THE TWO INTERACTING PROCESSES MAY GIVE US A BETTER UNDERSTANDING OF OUR BIOLOGICAL AND CULTURAL FUTURE.

RACE One of the most important phenomena resulting from the interaction of the biological and cultural evolutions is that of "race." As noted in Chap. 13, different selection pressures have resulted in geographic variation in a great many human characteristics, including hemoglobin types, blood groups, body shape, skin color, and hair type. As a result, no two geographic samples of *Homo sapiens* are alike. The patterns of variation in these characters, however, show little similarity (Fig. 16–6). Man, therefore, can be divided into "races" only by arbitrarily choosing one or a few characters.

MAN SHOWS A GREAT DEAL OF GEOGRAPHIC VARIATION, BUT DISCORDANCE OF VARIATION MAKES CLASSIFICATION OF MEN INTO NATURAL RACES IMPOSSIBLE.

Arbitrary choosing of characters *has* occurred in cultural evolution, however, and has led to the belief that biological "races" of man exist. The character which has predominated has been one of the most obvious, skin color. Among the uneducated in the United States, any person known to have an ancestor with dark skin is arbitrarily classified as a "Negro." In Brazil, any person known to have a light-skinned ancestor is considered a member of the "white" race. Because of patterns of discrimination in various countries, peoples of such diverse origin as Mexicans of Spanish descent, blacks from Africa, and Arabs are sometimes lumped together as "colored." It is interesting to contemplate what patterns of prejudice might have arisen if human perceptual systems had focused more on stature or face form than on color, or if blood group had been the most striking feature of each individual.

A recurring question involves the possibility of genetically based geo-

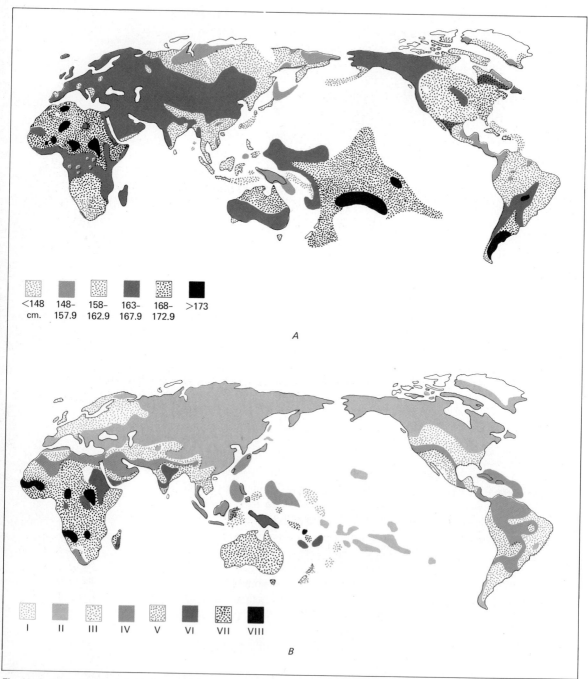

Fig. 16-6 Discordant variation in four characters in man: (*A*) height; (*B*) skin color (I-lightest, VIII-darkest); (*C*) shape of head (cephalic index); (*D*) shape of nose (nasal index). In *C* and *D*, the darker the area, the wider the head and nose are in proportion to their length. (*Adapted with permission of The Macmillan Company from THE CONCEPT OF RACE by Ashley Montagu, ed. Copyright © Ashley Montagu 1968.*)

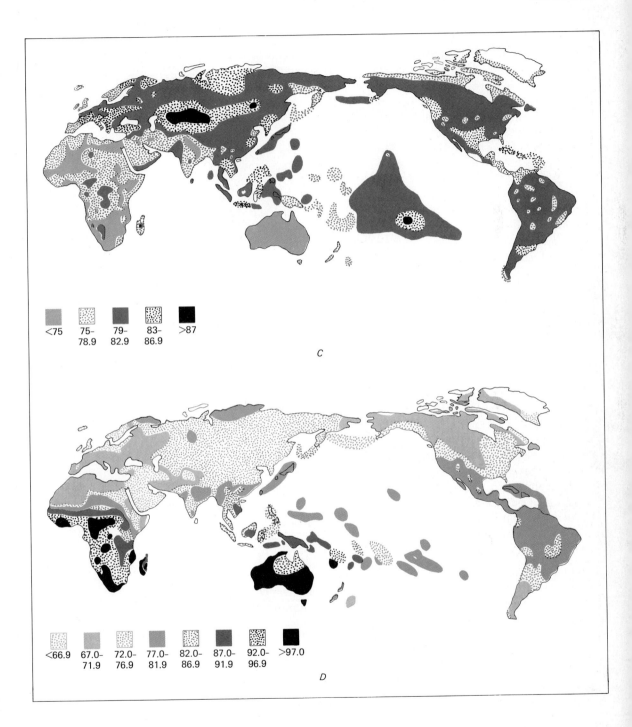

C

D

graphical variation in "mental" characteristics, especially in intelligence. This question cannot be answered with assurance. People cannot agree on what they mean by "intelligence," and no "intelligence test" has yet been designed which can fairly compare the innate ability of groups with different cultural backgrounds. Developers of IQ tests early in this century *thought* their tests measured genetic capacity, but more than half a century of psychological studies have shown they do nothing of the sort. In fact, standard IQ tests must be carefully adjusted so that the averages for men and women are the same, since women tend to do better on some elements of the test and men on others. Clearly, such a test would be unsatisfactory for answering that perennial question: Are men more intelligent than women? Lacking a test that will fairly compare men and women from the same culture, it is not surprising that no test has been devised to compare innate abilities of people of different cultures, people known to have dramatically different world views. Since people thought of as belonging to different races almost always have different cultures, no intelligence test will satisfactorily compare two races in any characteristic except performance on that test or similar tests. Such tests deal with an intellectual *phenotype*. Like all phenotypes, it is the result of a genotype interacting with an environment.

*RACES ARE ARBITRARILY DEFINED SOCIAL, NOT BIOLOGICAL, EN-
TITIES. NO INTELLIGENCE TEST HAS BEEN DEVISED WHICH CAN
FAIRLY COMPARE THE GENETIC ABILITY OF DIFFERENT RACES.*

A great deal of argument has occurred about possible differences in genetic intelligence among American blacks and whites. In some IQ tests, northern whites do better than northern blacks. Northern blacks do better than southern whites, and southern whites do better than southern blacks. These results indicate that a major determinant of performance is environmental rather than genetic. There is a great deal of other evidence that IQ is strongly influenced by environmental factors. For instance, a person's IQ score can be raised by education.

This is not to say that a person's IQ is not partly predetermined by his genetic endowment. There is clear evidence that it is. As a crude approximation, we might say that a range of intelligence, however defined, is established by heredity and that the IQ phenotype falls somewhere within that range at a point determined by environmental conditions.

With this background, the question of the comparative innate intelligence of whites and blacks can be put in perspective. It is of exactly the same *biological* interest as the question of comparative innate "intelligence" of people with blood type O and those with AB, of tall people and short people, of people with blue eyes and those with brown. It is a question of the interaction in various environments of the many genes which control intelligence (however defined) with those controlling whatever other character is considered. That is, it is an interesting but minor question.

INTELLIGENCE IS DETERMINED BY BOTH HEREDITY AND ENVIRON-MENT. THE COMPARATIVE GENETIC INTELLIGENCE OF GROUPS DIFFERING BY OTHER CHARACTERS IS AN INTERESTING BUT RELA-TIVELY UNIMPORTANT QUESTION BIOLOGICALLY.

This question has become important *sociologically* in recent years because of the claims by some that blacks are innately inferior in intelligence to whites and thus, for example, should be treated differently in school. Prominent people have called for investigations of the innate intelligence of blacks and whites, implying, of course, that the blacks will show a lower average. Competent biologists are badly disturbed by such ill-informed proposals, rooted as they often are in bigotry. As you can see from the above discussion, we have no way of comparing genetic intelligence of whites and blacks. What makes the whole discussion silly is that *if we could answer it, the question would be unimportant.*

In order to compare the "genetic IQ" (again remember, we can't even specify what it is we mean by that) of blacks and whites, we would first have to be able to determine the "genetic IQs" of individuals. Then we could measure a sample of blacks and whites and compare their averages. But, of course, once we knew the innate ability of *each individual,* we would have the information we desired. In our educational system, we would deal with people on the basis of their *individual* innate ability, hopefully working very hard on those with low genetically determined ranges to give them the maximum boost in the environmental component. *The color, blood type, or height of the individual would be of no significance.*

It is interesting to consider, though, just what may have happened to "genetic IQ" as a result of patterns of racism in American society over the last two centuries. Blacks have had little help in the environmental end: inadequate diets, poor or nonexistent schooling, etc. Those who managed to survive and reproduce may well have been those with the very highest genetic endowment. On the other hand, in a dominant white society, supported in part by the labor of blacks, even members of the Klu Klux Klan are sufficiently coddled by their environment to reproduce. It is not impossible that white racists ranting about low black genetic ability will be surprised if the answer they claim to seek is ever found.

*IN ORDER TO COMPARE THE "GENETIC IQ" OF BLACKS AND WHITES,
WE WOULD FIRST HAVE TO BE ABLE TO DEFINE AND DETERMINE
THE "GENETIC IQ" OF INDIVIDUALS. IF WE COULD DO THAT, WE
COULD TREAT INDIVIDUALS IN OUR EDUCATIONAL SYSTEM ACCORD-
ING TO THEIR INNATE ABILITY AND THE QUESTION OF THEIR COLOR
WOULD BE UNIMPORTANT.*

Cultural evolution has permitted man to become the dominant organism
on the earth. Man's most obvious effect on many ecosystems has been
their outright destruction. He does this in many ways: by building roads
and subdivisions, by plowing under prairies, by logging forests, by
building dams, and by herbiciding jungles, just to name a few. In these
cases, the effects on the original ecosystem are so severe that any remaining life will have to be integrated into whatever new ecosystem develops (and the new ecosystem may have cats, rats, and cockroaches as
its dominant animals).

**MAN'S IMPACT ON THE
ECOSPHERE**

Fig. 16-7 Heavy erosion of soil
occurs on burned-over lands. Underlying rock can be seen in many places
where the unprotected soil has been
washed away by rain. (*USDA SCS
photo.*)

A MAJOR DEVELOPMENT IN CULTURAL EVOLUTION HAS BEEN THE DEVELOPMENT OF SCIENCE AND TECHNOLOGY AND THE SEARCH FOR TECHNOLOGICAL SOLUTIONS TO MAN'S PROBLEMS. AS A RESULT, MAN HAS TOTALLY TRANSFORMED THE PLANET BY AGRICULTURE, MINING, INDUSTRY, AND THEIR BY-PRODUCTS.

In fact, very often the result of these kinds of activities is the replacement of a complex, stable ecosystem with a simple, unstable one. When a deciduous forest in the eastern United States is cut down and replaced by a crop of tobacco, this is precisely what happens. Instead of a situation in which long and interlinked food chains provide a mechanism for compensating for changes, food chains have been shortened and simplified. The number of insectivorous birds is reduced. A wide mix of plant species is replaced by a single one—making conditions for disease and pest outbreaks ideal. The farmer is immediately faced with the problem of attempting to restabilize the situation. He must, for instance, do something about the tobacco hornworms (sphinx moth larvae) which are threatening to devour his crop. He turns to pesticides, which as we will see below, create new problems.

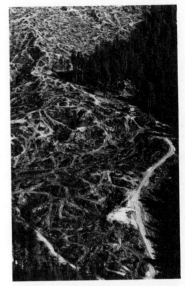

Fig. 16-8 Logging often may be very extensive, as seen in this redwood forest area in California. (*Courtesy of Donald F. Anthrop.*)

When forest cover is removed from an area and crops are not planted, other problems ensue (Fig. 16-7). Animals which depended on the trees for shelter and food disappear, as do shade-loving herbs and shrubs. Soil is exposed to the full force of sun, wind, and rain, and erosion takes place. This means that the moisture-retaining ability of the logged region is much reduced, and runoff from rainfall is much more rapid (Fig. 16-8). In turn, this leads to flooding in the logged drainage system. Flooding along the world's river systems, from the Yangtze in China to the Klamath in California, has been aggravated by logging in their watersheds.

Biocides and Ecosystems

Synthetic organic **biocides,** brought into use since World War II, have become one of the most potent weapons man has at his disposal for destroying his environment. We use the term "biocide" here because the terms "pesticide," "insecticide," and "herbicide" give a false impression of the selectivity of what are often broad-spectrum poisons. The more accustomed names are used below so as to show the *intended* purpose of the use of a particular biocide. Chlorinated hydrocarbon insecticides such as DDT, dieldrin, chlordane, and lindane are used all over the world. As was pointed out in Chap. 14, these long-lasting substances are concentrated by food chains, so that small concentrations in the general environment become massive concentrations in the bodies of high-order carnivores such as salmon and falcons. This means that the heaviest concentrations must be faced by those populations least able to do so, that is, populations of carnivores. Remember that the smaller a population is, the less likely it is to have the reserve genetic variability necessary for easy development of resistance. In addition, herbivores have been assaulted by plant pesticides for millions of years. Pests are, of course, mostly herbivores and as such tend to have mechanisms for combating poisons already developed. Predators are less likely to have been so equipped by evolution. Predatory birds, for example, suffer thinning of their egg shells when exposed to high levels of chlorinated hydrocarbons. The eggs crack and do not hatch (Fig. 16–9).

A very common result of patterns of pesticide usage which have been promoted by the manufacturers of the organic pesticides is as follows. At first, spectacular control of pests is achieved, since the chemical kills most of the pests—and virtually all the predators and parasites which normally control the pest. Then resistance of the pest population starts to develop, and in the absence of its natural enemies, the pest population grows rapidly. Larger and larger doses of poison are used, but the evolution of resistance keeps pace; the pesticide diminishes in its usefulness and may even become useless.

This is almost precisely what happened, for instance, when chlorinated hydrocarbons were used in attempts to control pests of cotton in the

Fig. 16–9 Two pelican eggs, thin-shelled as a result of chlorinated hydrocarbons, that did not hatch. (*Courtesy of Joseph R. Jehl, Jr.*)

coastal Cañete Valley of Peru. In the face of contrary advice from ecologically sophisticated entomologists, use of DDT, benzene hexachloride, and Toxaphene was introduced in 1949 and was at first very successful. But by the mid-1950s, resistance and kills of natural enemies had turned the program into a total disaster. In spite of tremendous pesticide inputs, cotton yields, which had originally climbed from 440 lb per acre in 1950 to 648 lb per acre in 1954, had dropped to 295 lb per acre in 1955–1956. At that time, a shift to a more ecologically rational control program, combining judicious use of pesticides in combination with various cultural practices and encouragement of natural enemies, got the situation under control.

One of the features of the disaster was the appearance of six brand-new pests, another common occurrence with standard pesticide practices. The enemies or competitors of some previously innocuous "bug" are killed off, and its populations surge until it reaches pest proportions. Mites, tiny eight-legged relatives of insects, were of little consequence as pests until the introduction of synthetic pesticides. These, especially DDT, decimated the predators and parasites of the mites, and *voilá*, instant pests. The pesticide industry now does a thriving business in miticides, pesticides designed to get rid of the Frankenstein monsters the industry itself had created!

Pesticide companies generally promote spraying by the calendar, that is, a certain number of times per season. This practice, as you can see, makes ideal conditions for the promotion of resistance. The pest population is subject to constant selection for resistance, and no opportunity is presented for natural selection to return the population to a less resistant level. In a nonpesticide environment, resistant insects are often at a selective disadvantage. Naturally, the appearance of resistance is generally countered by upping the dose, which is almost guaranteed to be disastrous in the long run. And, of course, both of these practices (calendar spraying and upping the dose) also enhance the prospects of promoting other organisms to pest status. It is not clear, however, how long these practices will continue to maximize the one factor they clearly maximize now: the profits of the pesticide industry.

Pesticides, then, have two major effects. First of all they destabilize agricultural ecosystems and, in the process, seem to make their own use seem even more desirable. After all, once predators and parasites have been all but exterminated, withdrawal of pesticide treatment permits a vast resurgence of the pest. But the second obvious effect is the more important. Pesticides cause high mortality in *nontarget* organisms, especially at the upper ends of food chains. Predatory birds now seem to be going into a worldwide decline, and pesticide effects have been recorded in populations of organisms as diverse as robins, earthworms, opossums, muskrats, turkeys, seagulls, crabs, salmon, snakes, and man.

Remember, whenever a population of organisms is driven to extinction by a pesticide, an ecosystem has become simplified. Unfortunately, we do not know exactly how much simplification the ecosystems of the world can stand and whether or not a stage will be reached at which an irreversible decline will begin. We should all remember the statement of the famous conservationist Aldo Leopold that the first rule of intelligent tinkering is to save all the parts. We also know all too little about the effects of pesticides on the flora and fauna of the soil—although we do know that the functioning of soil ecosystems is critical to the survival of most plants and many, if not all, animals. These problems are more subtle than those involved in the pest-control situation itself and in the simple deaths of nontarget organisms. They are also far more serious from the point of view of the survival of *Homo sapiens.*

Perhaps the most frightening (to ecologists) recent news on the pesticide front was the announcement in 1968 that DDT slowed down the photosynthetic rate of marine phytoplankton and that different species were differentially affected. Chlorinated hydrocarbons are highly fat-soluble, and it seems likely that in these single-celled plants the poison is attracted to the lipid layer of membrane structures and may concentrate in chloroplasts. In experiments, effects were found at DDT concentrations of a few parts per billion (ppb), quantities well below those commonly found offshore in areas treated with DDT. Water far at sea generally has DDT concentrations below 1 ppb, but this may well increase soon. DDT is already one of the most ubiquitous substances, being found in such unlikely places as the fat of Eskimos, Antarctic seals, and Antarctic penguins. Seals from the east coast of Scotland have been found with concentrations as high as 23 ppm in their blubber. Israelis have been found with as much as 19.2 ppm in their fat, and Americans average 11 ppm.

No one knows exactly what effect the DDT-phytoplankton interaction may have. Undoubtedly, some kinds of phytoplankton will be affected more than others, and this might result in disastrous algal blooms, such as those that have occurred in lakes, or the red tides of Florida waters but on a fantastically larger scale. Changes in the phytoplankton communities of the oceans could radically alter the character of marine food chains. One likely result would be a drastic decline in the amount of food which man could obtain from the sea.

PESTICIDES SIMPLIFY AND THUS DESTABILIZE AGRICULTURAL ECOSYSTEMS AND, BECAUSE THEY ARE OFTEN CONCENTRATED

IN HIGHER TROPHIC LEVELS, MAY HAVE WIDESPREAD EFFECTS ON NONTARGET ORGANISMS. THE DELETERIOUS EFFECTS OF MANY PESTICIDES ARE WORLDWIDE, AND THE EVENTUAL RESULTS ARE UNCLEAR.

Other Pollutants in Ecosystems

Synthetic organic insecticides are just one of the many classes of pollutants now entering ecosystems. They have been given a prominent place here because of the extreme seriousness of the problems they create and because they have been more thoroughly studied than most others. It was recently discovered, for instance, that polychlorinated biphenyls (PCBs), a class of nonpesticide chlorinated hydrocarbon compounds, have become serious pollutants, with effects similar to those of the related pesticides. Like DDT, PCBs now are found in the milk of nursing mothers, and they have been strongly implicated in the deaths of many kinds of carnivorous birds. PCBs are used in a variety of industrial processes and enter ecosystems by vaporization from containers, through factory smokestacks, through the dumping of industrial wastes into streams, and through the vaporization of tires by friction on the road.

In recent years, the rate of increase of herbicides has far outstripped the rate of growth of insecticide use. There are two widely used kinds of herbicides. One kind, which includes 2,4-D, 2,4,5-T and picloram, simulates a plant growth hormone, and its application causes the plants to grow abnormally and die. The other herbicides (simazine, monuron, etc.) interfere with the light reactions of photosynthesis. Plants attacked with these compounds starve to death. As animals don't carry on photosynthesis and since they do not respond to plant-growth hormones, the toxicity of these herbicides to animals is generally low, although there is growing evidence that some produce birth defects. Their direct effects on the soil flora are also little known, but some of them are degraded by bacteria.

But, of course, there are other than direct effects. For instance, the application of herbicides as jungle defoliants such as that which has occurred in Vietnam so that troop movements could be better observed (Fig. 16–10) inevitably leads to the extinction of many populations of insects, birds, arboreal reptiles, and arboreal mammals. These organisms are totally dependent on the forest canopy for their lives. Similarly, the loss of the foliage cover may have dramatic effects on the nutrient-poor tropical soils. In general, Temperate Zone forests do not have as distinct a canopy fauna as tropical jungles, nor does the vegetation contain the majority of the nutrient supply. Therefore, ecosystem effects of the use of herbicides can be expected to be less severe. Unbiased studies of her-

A

B

Fig. 16-10 (*A*) Aerial photograph of an unsprayed mangrove forest near the coast of Vietnam. Mangroves are important producers of fuel wood and charcoal, and they also provide food and nursery grounds for fish and shrimp; (*B*) the same general forest region three years after spraying with a defoliation agent. A few trees left alive are seen at the extreme top of the picture. Large dark areas are remaining deposits of humus. (*Herbicide Assessment Commission of the AAAS.*)

bicide effects are badly needed, especially on those, such as picloram, which do not break down readily.

It would be a dreary task, indeed, to recount the many other pollutants that have been shown to have serious effects on man and other organisms (Fig. 16-11). It cannot be assumed that any substance added to the environment intentionally or unintentionally will not have an effect in the ecosystem. The toxic effects of lead polluting the atmosphere as a result of the increase of automobiles and leaded gasoline are being reported from many parts of the world. Mercury poisoning of fish, birds, domesticated animals, and man is becoming increasingly common. Undoubtedly, many other instances of toxic effects of materials will be found. Substances often thought of as nontoxic, or even beneficial, may also have deleterious effects in ecosystems.

In Chap. 14, you were introduced to the problem of nitrate pollution stemming from the increased usage of inorganic nitrogen fertilizers. One of the results of this pollution is exemplified by the fate of Lake Erie (Fig. 16-12). Once-beautiful, it is now a stinking, septic mess. The lake waters are so foul that the U.S. Public Health Service recently urged ships not to take lake water aboard for cooking purposes within 5 miles of the U.S. shore. Probably no practical method will now purify that water. There are many industrial and municipal sources of Lake Erie's pollution, and sewage from Detroit, Cleveland, Toledo, Akron, and other cities adds to its load of nitrogen. But another important source of nitrogen pollution is runoff from the some 30,000 sq miles of farmland of the Lake Erie basin.

Fig. 16-11 Pollution may result in massive kills of fishes. (*Jeroboam, N.Y.*)

This runoff is rich in nitrogen; indeed, it is estimated to have a nitrogen content equivalent to that of the sewage of 20 million people. This tremendous quantity of nitrogen, plus a roughly equivalent amount from other sources, has immensely enriched the lake's waters. The growth of certain kinds of algae has been encouraged by this overfertilization and has been enhanced by overfertilization by phosphates also (some from fertilizers, some from detergents, some from industrial sources). Huge algal blooms resulted, which, when they died and decayed, depleted the lake waters of their oxygen and killed off fish and other animals with high oxygen requirements. A great deal of the phosphate and nitrate settled to the lake bottom with these masses of decaying algae. There is now a layer of muck 30 to 125 ft deep on the lake bottom, which is extremely rich in phosphorus and nitrogen compounds. These are bound by a chemical "skin" over the muck, made up of insoluble (Fe^{3+}) iron compounds. These compounds, unfortunately, change to the more soluble (Fe^{2+}) form in the absence of oxygen. Therefore, the depletion of oxygen as a result of overfertilization may further enrich the waters by

Fig. 16–12 Pollution from surrounding streams flows through the harbor of Ashtabula, Ohio, and on into Lake Erie. (*Environmental Protection Agency.*)

release of these trapped nutrients. As you learned in Chap. 14, this process of overfertilization is called **eutrophication.** In the case of Lake Erie, at least, eutrophication may itself lead to more rapid eutrophication. However, the process represents a worldwide problem. Although natural eutrophication occurs under appropriate conditions, the process is becoming much more common because of man's activities.

MOST OF THE MOLECULES INTRODUCED BY MAN INTO THE ENVIRONMENT TO CONTROL CERTAIN ORGANISMS HAVE WIDESPREAD DELETERIOUS EFFECTS ON OTHER ORGANISMS INCLUDING MAN. MISUSE OF OTHER SUBSTANCES, SUCH AS FERTILIZERS, MAY RESULT IN CHANGES IN ECOSYSTEMS, WHICH EVENTUALLY WILL BE IRREVERSIBLY DAMAGED.

Ecosystems and the Atmosphere

In Chap. 14, the biological origin and maintenance of the atmosphere were discussed. In any area, the atmosphere is an integral part of the ecosystem. In turn, the climate of an area is in part determined by the organisms living there. The rate at which the ground heats during the day, and thus the occurrence of updrafts, is in part a function of the plant cover. So are the pattern of horizontal airflow near the ground and the amount of water vapor in the air. Man, as we have already noted, changes the climate when he removes plant cover. And by changing the atmosphere, he may remove plant cover. Many air pollutants such as hydrofluoric acid, sulfur dioxide, ozone, and ethylene injure or kill plants. This, of course, leads to indirect effects on the animal populations dependent upon the plants.

But man's main potential for ecosystem modification through the atmosphere lies in those of his activities which are modifying the entire pattern of climate for the planet. The force that drives the weather, you will recall, is the unequal heating of the surface of the earth by solar energy. Man is doing many things to change both the overall heat balance and the pattern of heating. The normal slow patterns of change in climate are being accelerated, although the directions of change are not altogether clear. Man, for instance, has added a great deal of CO_2 to the atmosphere through his oxidation of fossil fuels. Carbon dioxide, like water vapor, absorbs some of the long-wavelength radiation being re-radiated by the earth's surface. The CO_2 then re-reradiates some of this energy back to the surface, contributing to the "greenhouse effect." The planetary result of increased CO_2 content in the atmosphere would be expected to have an overall warming effect, but through the changing and perhaps accelerating of atmospheric circulation, local cooling may also result.

Jet contrails may similarly add to the greenhouse effect, especially since there is growing evidence that they may form the nucleus for the development of clouds. They also, however, increase the amount of radiation reflected back into space. Whether their overall effect will be a warming or a cooling of the earth will vary with many other factors. There is no doubt, however, of the overall effect of the addition of dust and other particulate pollution to the atmosphere. Such particles, which now exist in a worldwide veil, absorb some incoming solar radiation and reradiate it to space. Particulate air pollution tends to cool the planetary surface. Large volcanic explosions of the past give us a foretaste of what we may expect if current trends in air pollution continue. For instance, the eruption of the volcano Tomboro, Sumbawa, in 1815 put an estimated 150 cu km of ash into the atmosphere. There was, as a result, virtually no summer in the most northern part of the United States in 1816, and the average July temperature in England was the lowest on record, 5° below the 250-year average.

Much of what happens to the climate of the earth will depend on the fates of the Antarctic and Greenland ice caps and the Arctic Ocean ice pack. The ice caps are a vast reservoir of cold—*there is enough ice in them to cover the entire planet with a 100-m-thick layer*. The extent of the caps and the pack is also a very important determinant of the planetary albedo, that is, the amount of radiation reflected by the earth. Should, for instance, the Antarctic ice cap spread outward (as some experts now think it may), about half of the world's population would die in the resulting tidal waves and a new ice age would be initiated because the area of the Antarctic ice "reflector" would be greatly increased. Or consider another possibility. If the Arctic Ocean ice pack should melt, as some experts predict, this could lead to northward shifts in the jet streams (rapidly flowing winds in the upper atmosphere). Such shifts could bring such severe drought to the plains of the United States that they would become deserts.

One thing is certain: *Any* major shifts in climate will lead to a reduction in agricultural output. Crops are grown where the climate suits them, and changes in climate mean, at the least, changes in crops. And there are few things that *Homo sapiens* is more conservative about than his agricultural practices.

POLLUTANTS OF THE ATMOSPHERE HAVE SHORT-RANGE EFFECTS ON MAN AND OTHER ORGANISMS. THEY ALSO HAVE LONG-RANGE EFFECTS ON THE CLIMATE OF THE PLANET, ANY CHANGE IN WHICH WILL LEAD TO REDUCED FOOD OUTPUT.

Direct Pollution Threats

Although the most serious threats to our existence posed by our activities come from the modification of ecosystems, the direct threats cannot be discounted. Not only do they produce immediate disability and death, but they also modify the selective regime to which man is subject. Air pollution causes increased death rates from respiratory diseases such as bronchitis (inflammation of the main air passages to the lungs) and emphysema (loss of elasticity and surface in the lungs). It has also been strongly implicated in cardiovascular disease and the production of certain cancers. Similar serious problems are caused by pollution of air and water with poisons such as fluorides (from many manufacturing processes), lead (a great deal of it from needlessly leaded gasolines), and pesticides. The industries producing these pollutants and the public agencies which most people presume are guarding them against their hazards show a monumental lack of concern and understanding for their possible long-term effects. Unfortunately, the economic aspects of our technology often take precedence over long-range ecological and biological effects. The public, of course, shares the responsibility for this situation.

A classic example of this is the dreary record of claims that DDT is harmless to humans. It is quite true that it does not show *acute* toxicity, but biologists have remained leery of it since it accumulates in fat and its mode of action is largely unknown. Recently, evidence has accumulated that DDT has serious effects on liver enzymes and sex hormones of mammals and birds. How serious this will prove for man remains to be seen, but the experiment is being run—virtually every human being on the face of the earth has been exposed to DDT.

Another interesting direct assault on our well-being has developed as a result of the use of inorganic nitrogen fertilizers. Because of the "loosening" of the nitrogen cycle in the soil, inorganic nitrate has become a major pollutant of drinking water. Nitrate itself is relatively harmless, but certain bacteria convert it to nitrite. These bacteria are especially likely to be present in the intestinal flora of infants. Nitrite absorbed into the bloodstream combines with hemoglobin to produce methemoglobin. Methemoglobin does not have the oxygen-carrying capacity of hemoglobin, and respiratory distress or, in some cases, suffocation occurs. Methemoglobinemia, as the disease is called, is an especially grave threat to babies in the Central Valley of California, and it is appearing in other areas as well.

Finally, a word must be said about radiation pollution. There is no such thing as a "safe" increase in the level of ionizing radiation. Such radiation induces mutations. There is no reason to believe at the moment that there is any need for a higher mutation rate in populations of *Homo sapiens*. Recombination is the main source of our genetic variability, and the size of our population is now so large that most mutations must occur many times each generation anyway. Therefore, each increment of radiation simply increases each individual's chance of dying of cancer and adds to the future crop of defective babies. Although the radiation problem will be partly solved if all atomic testing and other explosive uses of nuclear devices can be halted, it seems likely that there will be some increase associated with the upsurge in the use of nuclear power generators. These are, unhappily, being widely introduced with highly inadequate safety precautions.

THE ECOLOGICAL AND BIOLOGICAL EFFECTS OF TECHNOLOGY MUST BE CAREFULLY ASSESSED IN ORDER THAT MAN'S CULTURAL EVOLUTION DOES NOT LEAD TO HIS EXTINCTION.

CARRYING CAPACITY OF THE EARTH

A question which is of growing interest as the human population explosion continues is: Just how many people can the earth support? The question is phrased that way since it has long been abundantly clear that no substantial migration to other celestial bodies will be possible in the foreseeable future. Even under the most incredibly optimistic assumptions, holding the size of the human population constant by exporting the current annual surplus of people to other planets would cost at least $300 billion per day. That is twice the gross national product of the United States each week—and there's no place to send them. Ultimate limits on the earth seem to be set by thermal problems—the second law of thermodynamics again. All energy that man generates, even should he tap the almost limitless energy of fusion, is eventually degraded to heat, which must be disposed of (you may wish to refer to Box 4–1). The best guess is that if current trends of increasing energy consumption persist, a heat limit will be reached in less than 100 years. At that point, the release of energy by man will reach the magnitude of 1 percent of the incoming solar energy, and catastrophic atmospheric consequences will follow.

Hard Resources

The supply of "hard resources"—coal, oil, iron, copper, lead, and so forth—has a roughly similar temporal limit. We are supporting, after a fashion, 3.8 billion people today only because we are "living on capital." We are busily consuming nonrenewable resources at a fantastic rate. We are depleting stores of high-grade fossil fuels so rapidly that oil, aside from that in oil shales, may run out before the end of the twentieth century (coal reserves are considerably larger). In our needs for most other materials, substitutions and mining lower-grade ores should keep us going well into the next century, but the end will come in that department, too. If man survives that long, he will be forced to hold his population and technology at a level which can be maintained largely through the use of renewable resources—in particular, plant products and recycled metals.

BECAUSE MAN HAS NEARLY USED UP NONRENEWABLE FUEL RESOURCES, HE SOON WILL HAVE TO SWITCH TO OTHER SOURCES OF ENERGY. OTHER MINERAL RESOURCES WILL HAVE TO BE CONTINUALLY RECYCLED.

Food Supply

At present it seems likely that the limit on human population growth will be that classic resource limit, food supply. Of 3.8 billion human beings now alive, almost half at present have an inadequate diet. Somewhere between 10 and 20 million people die of hunger each year, most of them children, and many experts think this number will soon increase dramatically. There are grave inequities of food distribution both within and between countries. The most striking shortages are in the so-called "underdeveloped" countries (UDCs), although even there glut and famine may exist in adjacent areas because of transportation and economic problems. Most "developed" countries (DCs) are able to maintain a more than adequate average diet either by growing food, by importing it, or by harvesting it from the sea. Even in the DCs, however, hunger is widespread. It was recently revealed that in the United States, as many as 20 million people are "hungry," and many American children die of malnutrition.

Man is now hard-pressed to find the food for the 70 million more mouths that must be fed each year. A combination of pollution and overexploitation promises a steady reduction in the per capita yield from the sea after 1980, if not well before that (an absolute loss in yield was recorded in 1969, for the first time since 1950; there was a recovery in 1970 and another decline in 1971; whether this presages a final decline is not yet clear). The sea at the moment supplies only a few percent of man's total caloric input, but it provides fully 20 percent of his animal protein (animal protein includes eggs and dairy products) and 40 percent of his meat.

Increasing food production on the land is fraught with grave economic, political, anthropological, and above all, biological difficulties. These can be barely touched on in the space available here. One potential way of increasing food supplies is to bring more land into production. Unfortunately, virtually all the land that can be successfully and economically farmed is today in production. More land could be brought under the plow if massive irrigation projects were developed, but careful analyses show virtually all of these to be impractical, at least on the time scale that is critical.

Contrary to popular belief, vast areas of tropical jungles do not represent pools of potential farmland. Soils in most of these areas are lateritic and rapidly turn to hardpan when forests are cleared and permanent agricultural development is attempted. Unhappily, although we have considerable expertise at farming in the temperate regions, our knowledge of how to mass-produce food in the more diverse tropics is much more limited. And, as you can see from the discussions above, even Temperate Zone farming practices leave much to be desired.

The best hope for increasing food production on land lies in increasing yields from land already under production. A lot of publicity has been given to the introduction of high-yield "miracle" grain varieties in certain UDCs. These high-yield grains require careful cultivation and, especially, adequate fertilizer input if their potential is to be realized. And once yields are increased, the farmer must be rewarded economically. If higher production means lower prices, incentives to raise the new, more-difficult-to-grow grains will disappear.

In a very real sense, the problem of increasing agricultural production through the use of high-yield grains is a part of the problem of economic development of the UDCs. Fertilizer and pesticides must be produced or purchased, which means that capital or foreign exchange must be available. Roads must be constructed to carry these "inputs" to the fields and to carry produce to markets. Technicians must be trained and made available to counsel farmers. Irrigation systems must be constructed, and wells dug. And the level of aggregate demand (that is, of purchasing power) in the country must be increased. Merely distributing miracle seeds won't come close to doing the job. And, above all, the attitudes of agricultural peoples must be changed. They must change their ancient ways, must alter a pattern of life which very often is intimately tied with agricultural practices.

They must also accept new foods, something that is difficult even for people, such as Americans, who are accustomed to a vast diversity of food. Those in the world who are hungriest are also those who are most conservative in their eating habits; they recognize few things as food. As small a factor as whether or not grains of rice stick together or fall apart (depending on how starchy they are) may control acceptance or refusal. There are reports of rice-eating peoples starving to death rather than accepting wheat. And, of course, should the miracle of an agricultural revolution throughout the UDCs take place, the increase in pesticide pollution alone might be the ultimate disaster for the world ecosystem.

The newspapers are now full of panaceas for solving the world food problem: growing algae on sewage, growing bacteria on petroleum, processing leaves for protein, herding antelopes in Africa, and so forth. Some of these may make small, local contributions to the food problem, but none is likely to do much more than that. The overall food problem is best put into perspective by those two stunning statistics: almost half the world is hungry, and some 70 million more mouths to feed are added annually!

THERE IS NOT AT PRESENT SUFFICIENT FOOD BEING PRODUCED TO ADEQUATELY SUPPLY THE HUMAN POPULATION. SINCE THE PRESENT POPULATION CONTINUES TO INCREASE, FOOD PRODUCTION MUST BE INCREASED, AND THIS MUST BE DONE WITHOUT FURTHER DAMAGE TO THE LIFE-SUPPORT SYSTEMS OF THE PLANET.

In popular magazines, the contribution of biology to mankind's future is usually put in terms of heart transplants, kidney machines, or genetic engineering. There is not the slightest question that biology has the potential for making the life of the individual much more secure and, we hope, much more pleasant. However, with our present acceptance of technocracy and technological solutions for every problem, there is always the danger that "biological engineering" will *not* improve the lot of the individual. Piecemeal solutions to biological problems may have the same effects as piecemeal solutions to ecological problems. Obviously, as sophistication grows in biological-engineering techniques, it must be accompanied by a holistic philosophy which will always keep in view the long-range effects of modifying biological systems.

THE FUTURE

Given the time, biologists should be able to provide the means for permitting most people to live out a full 70- to 80-year life-span in comfort and, perhaps, by intervening in the genetic code, to substantially extend that life-span. With better understanding of the brain, antisocial behavior and mental disease could probably be much reduced. Educational methods could be vastly improved and used to inform people fully of their own individual and social natures. It is not inconceivable that under such circumstances most dangerous intergroup aggression, including organized warfare, could be eliminated. And finally, man's population size and activities could be brought into harmonious balance with his environment.

All this is possible, but improbable. It seems unlikely that the kind of organized activity necessary for the conduct of, say, research in molecular biology will persist for more than a few decades or so. And that isn't enough time. Current trends indicate that well before the year 2,000, worldwide famine, plague, thermonuclear war, or a combination of these will bring an end to civilization as we know it. The basic reason is the incredible growth of human population size.

There is, at the moment, considerable debate among experts as to when massive famines will be the dominant events on the planet. Some say as early as 1975, others as late as 1985 or 1990. What will happen as the UDCs get hungrier and the DCs struggle to maintain their position? Consider, for a moment, the plight of a UDC such as the Philippines, which is growing at a rate which will double its population in 20 years. In order to *stay even,* the Philippines must, in essence, duplicate in two decades all the amenities it now possesses for the support of human life. For every house and factory today, there must be two in 20 years. Road capacity must be doubled, jail capacity must be doubled, water and sewage-system capacities must be doubled, as must be the numbers of doctors, lawyers, judges, scientists, and so forth. Imports must be doubled, as must be farm production and exports. It would not be possible for the United States to double everything in 20 years. Our resources are rapidly being used up, and the taxes needed to support our present level of consumption are very high. And yet the United States is a modern technological society with great resources, a fine communications system, and an almost 100 percent literate public. The Philippines is none of these things—for her, such a doubling is an impossible dream. But even were the dream miraculously to be realized, Filipinos would not be satisfied. They don't want to stay even; they want to get ahead. They have "rising expectations."

As these expectations are dashed, the DCs are going to find it progressively more difficult to extract from the UDCs the raw materials the DCs need for their wasteful economies. We will not have food to give or sell them, and food will be their primary need. And they are all too aware that the United States, with some 6 percent of the world's population, consumes about one-third of its resources. They are aware that, since the DCs are the world's master consumers and polluters, the numerically smaller growth in DC populations has had more serious effects on the ability of the planet to sustain life than has their own profligate growth. The United States' miserly and poorly planned foreign-aid programs do not impress citizens of the hungry world nearly as much as our capacity to extract fish from protein-starved Peruvians and feed them to our cats and pigs. One can hardly expect the world political scene to brighten as the DC-UDC gap continues to grow. One "death-rate solution" to the population explosion would simply involve a massive increase in the death rate through starvation. It does not, however, seem likely that the rest of the world will starve gracefully while the DCs persist in affluence.

Another possible death-rate solution involves the potential for worldwide plague. The human population today is the largest, and thus the densest, in history. It is also the weakest; there are more hungry people today than there were people in 1875. The situation is made to order for a worldwide virus epidemic, especially since jet aircraft guarantee rapid transport for carriers of disease. Viruses often change their characteristics as they pass through populations, and there is no reason to believe that a lethal "supervirus" could not develop as a spontaneous mutant, wiping out a large portion of humanity. Even if the natural event does not occur, there is always the possibility of an escape of a viral or bacterial strain from one of the world's many biological warfare (BW) laboratories. Since it is theoretically possible now to construct a virus against which man has no natural resistance, it is small wonder that many prominent biologists have protested against continuing BW research. BW is essentially the poor man's H-bomb; its weapons can be developed by any nation with a few biologists and perhaps a quarter of a million dollars to spare. And since even virus labs manned by highly competent scientists have not turned out to be "escapeproof," there is considerable reason for apprehension.

Finally, after famine and plague, there is the third major possibility for a death-rate solution: thermonuclear warfare. There is no need to dwell on this here, except to point two things out. First, there is growing evidence that increasing population size in the face of finite resources adds to the chances of conflict. As the world population grows, so does the chance of thermonuclear war. Second, the biological and, especially, the ecological effects of thermonuclear war are not usually considered when its consequences are discussed. Picture, if you will, the effects on ecosystems of massive fire storms burning off vegetation, sterilizing the soil of watersheds, and lofting great amounts of radioactive debris into the atmosphere. Consider the runoff of silt-laden water carrying the contents of ruptured storage tanks of all descriptions into the sea. Think about the ruined offshore oil rigs leaking perpetually, the upsurge of rats and cockroaches, and the wild alterations of climate. Any survivors would face an unprecedentedly harsh, radioactive environment. Small, culturally and genetically depauperate groups probably could not long survive; a combination of inbreeding and genetic drift in the face of dramatic shifts in selection pressure might force them to extinction. *Homo sapiens* could pass from the scene.

ALTHOUGH WE NOW HAVE OR COULD SOON DEVELOP THE NECESSARY FACTS AND THEORY TO BRING MAN INTO HARMONIOUS BALANCE WITH THE EARTH, OUR RAPIDLY GROWING POPULATION SIZE MAKES IT IMPROBABLE THAT THIS WILL COME ABOUT. UNLESS OUR ATTITUDES TOWARD POPULATION CONTROL, THE ENVIRONMENT, AND OUR FELLOW MAN CHANGE AND WE ACT SWIFTLY, OUR SPECIES IS HEADING FOR CATASTROPHE.

But we cannot end this book on such a dismal note. Perhaps man's attitudes will change dramatically. Perhaps strong population–environmental programs and a new spirit of cooperation will help to ward off total catastrophe and ease man through the next few critical decades into a golden era beyond. In all parts of the world there is an increased interest in and a growing awareness of our ecological problems. A deep and serious concern with our social and economic situation and the effects of technology upon the quality of life has led to the evolution of a counterculture, determined to change the direction of man's cultural evolution. We can only hope that the members of this counterculture, often strident, violent, ineffective, and facing powerful repressive forces, can unite in common goals. They cannot succeed alone, however. All members of our society, all peoples of the world, must work together if the life of our species is to continue. The odds may be against it, but it is a goal worth striving for.

SUMMARY

Man evolved, in the course of some 60 million years, from a simple, unspecialized mammalian ancestor to his present state. There is a clear fossil record depicting the physical evolution of man, as well as aspects of his cultural evolution. Early in his evolution, man's ancestors became tree dwellers. Many important traits, such as grasping hands, binocular, stereoscopic vision, and small litter size, are the result of evolution in an arboreal niche. Eventually these arboreal forms returned to a ground-dwelling life. Selection and evolution in a terrestrial niche resulted in a variety of important characteristics of man today.

The fossil record indicates that modern man evolved in the Old World from a series of prehuman anthropoids with increasing brain size. A very important development in human evolution, occurring together with the increase of brain size, was the nongenetic transmission of information from generation to generation. This led to the development of culture and the possibility of cultural evolution.

Biological evolution and cultural evolution have interacted throughout the course of man's development. Culture has led to changes in man's physical nature, and changes in man's physical nature have greatly influenced his culture. An important way in which culture affects physical evolution is to change selection pressures in human populations. Certain selection forces have increased in intensity, and others have decreased. Culture has also introduced entirely new selective agents in man's environment. Many analogies can be drawn between biological evolution and cultural evolution. Comparison of the two interacting processes may result in a better understanding of our past and future evolution.

A major development in cultural evolution has been the rise of science and technology and the search for technological solutions to man's problems. As a result, man has totally transformed the planet by agriculture, mining, industry, and their by-products. Hoping to increase agricultural production, man has used pesticides and fertilizers heavily. Pesticides simplify and thus destabilize agricultural systems. Because they are often concentrated in organisms of the higher trophic levels, they may have widespread effects on nontarget organisms. Deleterious effects of many pesticides are worldwide, and the eventual results are not at all clear. Fertilizers in excess are also serious pollutants.

The ecological and biological effects of technology must be carefully assessed in order that man's cultural evolution does not prove his own undoing. At the present time, not enough food is being produced to adequately support the human population. Since the present population is rapidly increasing, food production must be greatly increased. This must be done without further damage to the life-support systems of the planet.

Although we now have, or could soon develop, the necessary facts and theory to bring man into harmonious balance with the earth, our rapidly growing population size makes it unlikely that this will happen. Unless our attitudes toward population control, the environment, and our fellow man change and we act swiftly, our species is heading for catastrophe.

SUPPLEMENTARY READINGS

Books

Ehrlich, P. R.: "The Population Bomb," 2d ed., Ballantine Books, New York, 1971.

Ehrlich, P. R., and A. H. Ehrlich: "Population, Resources, Environment," 2d ed., Freeman, 1972.

Ehrlich, P. R., and R. L. Harriman: "How to be a Survivor: A Plan to save Spaceship Earth," Ballantine Books, New York, 1971.

Ehrlich, P. R., J. P. Holdren, and R. W. Holm (eds.): "Man and the Ecosphere," readings from *Scientific American,* Freeman, 1941.

Holdren, J. P., and P. R. Ehrlich (eds.): "Global Ecology," Harcourt, Brace, Jovanovich, 1971.

Holdren, J. P., and P. Herrera: "Energy," Sierra Club Books, 1972.

Laughlin, W., and R. H. Osborne (eds.): "Human Variation and Origins," readings from *Scientific American,* Freeman, 1967.

Van Lawick-Goodall, Jane: "In The Shadow of Man," Houghton-Mifflin, 1971.

Articles

Bodmer, W. F., and L. L. Cavalli-Sforza: Intelligence and Race, *Sci. Amer.,* **223,** no. 4, Offprint 1199 (1970).

Howells, W. W.: Homo erectus, *Sci. Amer.,* **215,** no. 5, Offprint 630 (1966).

McVay, S.: The Last of the Great Whales, *Sci. Amer.,* **215,** no. 2, Offprint 1046 (1966).

INDEX